T0300117

Soil Invertebrates
Kaleidoscope of Adaptations

Nico M. van Straalen
Vrije Universiteit Amsterdam
Netherlands

CRC Press is an imprint of the
Taylor & Francis Group, an **informa** business
A SCIENCE PUBLISHERS BOOK

Cover images and chapter icons by Janine Mariën

First edition published 2023
by CRC Press
6000 Broken Sound Parkway NW, Suite 300, Boca Raton, FL 33487-2742

and by CRC Press
4 Park Square, Milton Park, Abingdon, Oxon, OX14 4RN

CRC Press is an imprint of Taylor & Francis Group, LLC

Library of Congress Cataloging-in-Publication Data (applied for)

ISBN: 978-1-4822-3123-6 (hbk)
ISBN: 978-1-032-48033-6 (pbk)
ISBN: 978-0-429-17283-0 (ebk)

DOI: 10.1201/9780429172830

Typeset in Times New Roman
by Radiant Productions

Preface

Soil is not just dirt, it is a precious resource, absolutely essential for the survival of mankind. However, we are losing fertile soils at an alarming rate; the assurance of food production for the global population is at risk. With these lessons from Jo Handelsman's alarming book in mind (Handelsman 2021), I intend to make my own contribution to soil protection. Soils harbour a great variety of microbes, fungi and miniscule animals. All functional processes in soil ecosystems are biological, catalysed by an army of "*little rotters*", that jointly regulate the health status and fertility of a soil (Guerra et al. 2021). This wonderful community has been close to me throughout my scientific career. This has now allowed me to synthesize the insights from soil ecological research accumulated over some 40 years.

I focus in this book on the animal component of the soil community, in particular the invertebrates. And I study soil invertebrates not from a utilitarian point of view (what do they do?), but from the perspective of a biologist, interested in their evolutionary history and their adaptations to the soil environment. Soil invertebrates make powerful model organisms for testing ideas of evolutionary biology. The evolution of soil invertebrates in the end determines also their ecological function.

Many subdisciplines of the life sciences rely heavily on a limited number of model organisms, for instance, animal physiology benefitted much from work on pigeons and dogs, genetics from peas and fruit flies, neurophysiology from zebra finches and mice, developmental biology from nematodes and zebrafish. Likewise, evolutionary biology has its favourite model systems, e.g., birds, lizards and ants. The animal category dearly missing in these lists is the one of soil-living invertebrates. This raises the question: is the progress of biological sciences biased because of a lack of insights specific to soil biology?

Fortunately, the relative underrepresentation of soil biota in general ecology, which was quite obvious before the turn of the century (Barot et al. 2007) has rapidly evened in the last two decades. The title of a 2000 paper published in Nature ("*Ecology goes underground*") (Copley 2000) is testimony to this. The call for "unifying principles in soil ecology" (Fierer et al. 2009) has worked out well. Contributions by eminent soil ecologists have brought soil to the forefront of ecological sciences. The concepts of functional redundancy, ecosystem services and above-belowground-coupling derive basically from insights gained from soil organisms.

Soil-specific issues have also penetrated the field of biodiversity science. In a recent analysis of the literature, Thakur et al. (2020) explored the possible validity of five biodiversity theories (island biogeography, niche theory, metacommunity theory, etc.) when tested against observations of soil biodiversity. Earthworms' place on earth is drawing wide attention in the context of soil protection (Phillips et al. 2019, Fierer 2019). One of the emerging insights is that soil biodiversity has a strong element of scale. Soil food-webs of different scales are nested inside each other (Pokarzhevskii et al. 2003). Things happening at the square millimetre can be very different from things happening at the scale of a hectare, let alone the continental scale. Realizing this element of scale will enrich general theories of global biodiversity patterns and the protection of soil biodiversity.

However, there are also aspects of soil systems that are at variance with "common ecological wisdom". Let us shortly visit some of these cases, to be discussed at length later in this book.

- *Biomass turnover related to body size.* A trend valid across a very wide range of organisms, from copepods to elephants, stipulates that biomass turnover of field populations decreases (according to an allometric relationship) with increasing body size of the organisms involved (Banse and Mosher 1980, Hendriks 2007). This implies that populations of small animals have a higher production to biomass turnover than large animals. In Chapter 3 we will see that this classical allometric relationship does not hold for soil invertebrates: biomass turnover depends only little on body size.
- *Genetic variation and dispersal.* It is often assumed that animals with a small home range and limited dispersal capacity will show populations that are highly locally differentiated (the law of *isolation by distance*, Hartl and Clark 1997). Since many soil invertebrates are assumed to live their lives locally and are not physiologically equipped to undertake long journeys, it may be expected that they will have populations differentiated on the scale of a centimetres to metres. This turns out not to be the case at all, as we will see in Chapter 3. Many soil invertebrate populations are hardly differentiated by distance.
- *Parthenogenesis in food-enriched environments.* The wide distribution of parthenogenetic reproduction is a puzzle waiting to be explained satisfactorily. One of the few rules of thumb, inspired by water fleas and aphids, is that parthenogenesis is associated with environments ensuring a high food supply, while sexual reproduction is favoured in unstable, food-limited environments. In Chapter 4 we will see that this trend does not hold for soil invertebrates. In many lineages, the soil environment, with its resource-limiting conditions, is the locus for parthenogenesis.
- *Iteroparity.* In life-history theory it is often assumed that enhanced reproductive output should trade-off with growth, survival, or future reproductive output (Stearns 1992). Iteroparity is considered a bet-hedging strategy which is selected for in unpredictable environments. In soil invertebrates, however, iteroparity is associated with the stable environment of the soil. It is associated with indeterminate growth rather than with resource allocation.
- *Female choice and sexual selection.* The theory of sexual selection, driven by choice for "good genes", explains the abundant and sometimes excessive decoration and display in sexual partners. This theory is mostly inspired by

research on birds, where this phenomenon is obvious. In Chapter 5 of this book we will see that soil invertebrates, with their mostly chemical communication between the sexes, offer an equally bewildering array of selective, cryptic, and manipulative strategies to ensure reproductive success. Hermaphrodites especially, represent some quite bizarre examples of sexual conflict never seen in other animals.

- *Universal temperature dependence.* Several authors have emphasized the "universal" temperature dependence of metabolism, expressed as the Boltzmann factor (e.g., Gillooly et al. 2001). However, comparison of species within soil invertebrate lineages clearly demonstrates that this factor is not constant at all; it is subject to evolution and shows a clear pattern across the soil profile. Multiple examples, discussed in Chapter 6, show this to be the case.
- *Phylogeny and physiological adaptation.* In classical comparative physiology it is often assumed that soil invertebrates, e.g., isopods, can be arranged in a linear sequence, from marine, aquatic, drought-sensitive species to fully terrestrial, drought-tolerant species (Warburg 1968, Little 1983). However, as we will see in Chapter 2, phylogenetic reconstructions have revealed multiple terrestrialization events within soil invertebrate groups. The interpretation of trends in adaptive physiology should synchronize with the terrestrialization pathway of the group.
- *Importance of parasites and pathogens.* The soil is a medium full of microbial activity, including many pathogenic, toxic and parasitic life-forms. The regulatory capacity emanating from the microbial community might be much greater in soil than in any other medium. In Chapter 6 the microbial world of soil invertebrates is explored. Resistance to pathogens in some lineages can be considered a typical adaptation to the soil environment.
- *Anthropogenic toxicity.* The soil is a habitat with a great buffering capacity and therefore acts as the "ultimate sink" for many anthropogenic pollutants. Soil invertebrates are exposed to higher concentrations of such pollutants than in water or air. Selection for dealing with stress associated with toxicants is particularly strong, as illustrated in Chapter 6 by several cases of resistance evolution.
- *Major transitions in evolution.* According to the classical Darwinian principle, adaptation to changing conditions is driven by natural selection on standing genetic variation and proceeds in small steps. However, in certain lineages, soil invertebrates show large phenotypic changes, seemingly without intermediate steps. The acquisition and loss of symbionts is such a change, illustrated by diet choice in termites. Horizontal gene transfer is another factor facilitating evolutionary change (Chapter 6).

This list of issues illustrates that the study of soil biology might enrich general biology with trends and theories that do not easily come up by focussing on marine, aquatic and terrestrial habitats only. This represents the main reason why the author felt that this book had to be written. Soil invertebrates, by showing their host of adaptations to the soil environment illustrate several evolutionary and ecological principles with relevance to biology as a whole.

Acknowledgements. I dedicate this book to Prof. Dr. Els N.G. Joosse, who initiated my interest in soil invertebrates and supervised my first steps in the wonderful world of soil ecology. Drs. Martin Holmstrup and Joris Koene made useful comments on parts of the text. I owe greatly to my former colleagues in the department of Ecology and Ecotoxicology at VU University Amsterdam, where I spent most of my career. To the international community of soil biologists I owe many useful insights. Finally, I feel a warm gratitude to all my students and PhD graduates in soil ecology who joined me, each for a while, to enjoy the kaleidoscope of adaptations shown by soil invertebrates.

The sonnets adorning each chapter title page were inspired by John Satchell's "Pedosphere Harmony" (Satchell 1977).

Nomen est Omen - about the poet of service. Jasper Aertsz is a descendant of a venerable family of gravediggers. He himself still regularly uses the spade. This literally pushes him with his nose into the earth's layers, where so many soil animals can be found - vertebrates and invertebrates. The contact with this world prompts Aertsz to express in a poetic way what is going on and what moves there. Over the years, Aertsz became a strong advocate of cremation.

Contents

Chapter 1

The Selective Environment of the Soil

King of Soil

Ashes to ashes ... Stardust, yet earthbound
we are, beggar, billionaire, king or knave.
But will your bones be Last Day-proof and -safe,
then there's much ado, nine feet underground!

So, all kinds of invertebrates come to rave
-pods, -grades -todes, slugs in great amount.
Odd jobs together: you will be astound!
they grate, cleave, saw, drill. No delay as they slave.

Deconstructing you, years and years around
omnipresent and persistent, they outbrave
every other species on Earth to be found.

And worms are the best in how they behave.
Amongst them Annelid has to be crowned.
He, the King of Soil. Guardian of your grave.

Jasper Aertsz

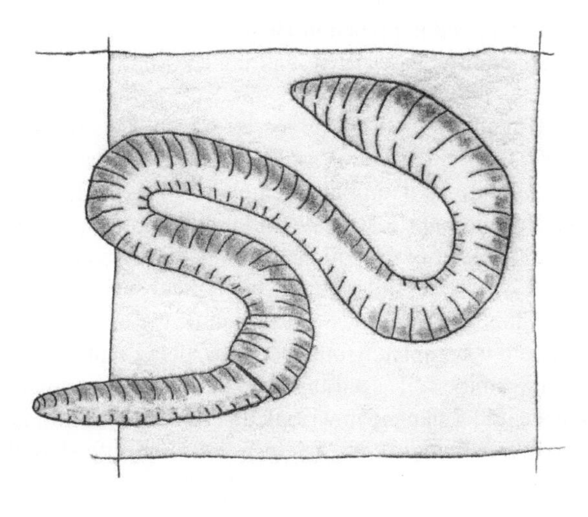

Chapter 1

The Selective Environment of the Soil

"Up we go! Up we go!", till at last, pop! His snout came out into the sunlight, and he found himself rolling in the warm grass of a great meadow

Kenneth Grahame, The Wind in the Willows

Soil is a transition zone. It presents very strong gradients of abiotic and concomitant biotic factors. To live in soil is to live in a twilight zone: you can either go down and seek stability, constancy, rest (and boredom), or you can go up and face environmental variation, heat, cold, challenges and activity. And it all happens on a few decimetres, centimetres, millimetres sometimes. The sharp cline of living conditions with depth calls for a kaleidoscope of adaptations, and this is precisely what this whole book is about. In this first chapter we will review the environmental factors that select for adaptations in the organisms that inhabit the twilight zone.

1.1 The Ghilarov-Kennedy-Gisin-Bouché hypothesis

Many scientists have recognized the transition aspect of soils. The great Russian soil biologist, Mercurii Sergeivich Ghilarov (1912–1985), considered by many the founding father of soil biology, formulated a theory of insect evolution in which he pictured the soil as a transitional medium between the aquatic environment and the xeric aerial environment. From his book published in Russian in 1949 (Ghilarov 1949), which he summarized in French in 1958 (Ghilarov 1958). I copy here a figure which illustrates his point with a clarity that unnecessitates all further explanation (Figure 1.1).

A resumé of Ghilarov's theory is given by his student, Andrei Pokarzhevskii (Pokarzhevskii 2002). The French soil zoologist Guy Vannier also gave a summary (Vannier 1987). In Ghilarov's tradition, he emphasized the soil as a "*porosphere*", a medium characterized by pores that allow all kinds of gradients between water and air. The porosphere character of the soil matrix calls for organisms that can deal both with free water and with near-saturated air, the typical selective gradient in which soil animals live. Although Ghilarov's conception of the evolutionary relationships in the animal kingdom is completely outdated now, being replaced by modern DNA phylogenies, his view of the soil as a transition zone and as a template for adaptation could still inspire an author like me in writing this book. Ghilarov's idea matches the modern concept of "*terrestrialization*" (Selden 2005): all life started in the marine environment and at different points in time, invertebrates made their way to the land.

Another, even older, phrasing of the same phenomenon is due to the American entomologist Clarence Hamilton Kennedy (1879–1952), who specialized in

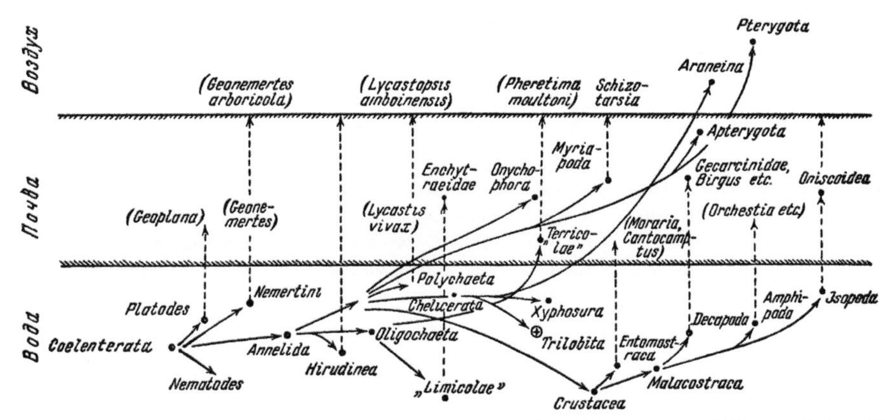

Figure 1.1. Summary of M.S. Ghilarov's theory on the evolution of insects. The middle zone is the soil (*почва*), which is placed in between the water (*вода*) and the air (*воздух*). Figure copied from Ghilarov's book (1949), courtesy of Dr. Andrei Pokarzhevskii.

dragonflies and conducted numerous surveys in the Western US, at a time when urbanization there had hardly begun. One of his better-known synthetic papers, published in 1928 in the journal Ecology, describes insect evolution as being driven by the "*energy intensity*" of the environment. In his own words:

"A hot sun-baked environment has a greater energy intensity than a cold lightless cave. The tropics have a greater energy intensity than the boreal regions. The hot and light middle of the day has a greater energy intensity than the cool dark night." (Kennedy 1928)

Kennedy categorized the various groups of insects according to three energy intensity gradients: geographic, seasonal and diurnal. His argument was that nocturnal and hidden insects are more often found in the primitive orders, while sun and heat-loving insects are more prominent in the latest and highest groups evolved (Table 1.1). The trend would hold across orders, but also, in some orders (Lepidoptera, Diptera and Hymenoptera), across families. Kennedy notes that crepuscular, nocturnal or hidden insects, such as found among springtails, bristletails and crickets, in other words the soil-dwelling groups of the hexapods, are to be considered as ancestral to the "higher" insect groups, such as bees, wasps, flies and butterflies. This trend would be correlated with enhanced adaptations to the "energy intensity" that is higher above-ground than in the soil.

Obviously, like in the case of Ghilarov, the classification of insects that Kennedy refers to is completely outdated, but the core of his theory, like Ghilarov's, is still valid as a starting point for the argument in this book. As the soil represents a gradient of energy intensity *par excellence*, the Ghilarov-Kennedy hypothesis would predict that we find not only drought tolerance but also metabolism, activity and reproductive strategies to be correlated with depth in the soil profile. Is this true? We shall see in the coming chapters.

The evolutionary significance of the soil profile, as a template along which species adapt and differentiate, was also recognized by the Swiss Collembola specialist, Hermann Gisin (1917–1967). In his mainly taxonomic work on European

Table 1.1. Classification of insects of different evolutionary stages by the "energy intensity" of the environment in which they are most commonly found (diurnal distribution, seasonal distribution, geographic distribution), according to Kennedy (1928).

Crepescular, nocturnal or hidden insects	Shade-loving or wood insects	Sun and heat loving insects
Insects of the following large groups avoid light and heat, and are slow, or active for short periods only. These are the primitive orders or are the lowest families of higher orders.	Insects of the following large groups are diurnal but prefer shady cool habitats. They are active for longer periods than the primitive and in age are intermediate.	Insects of the following large groups are sunshine and heat loving, adults being very active for long hours. These are the latest and highest groups evolved and are the conspicuous insects of open, sunny and hot fields in mid-summer.
Thysanura Collembola Ephemerida Plecoptera low Orthoptera (Grylloblattidae, Blattidae) Dermaptera Isoptera Zoraptera Embiidina Corrodentia Thysanoptera Trichoptera low Lepidoptera (heterocerous families) low Diptera (nematocerous families)	Zygoptera mid-Orthoptera (Gryllidae, Mantidae, Phasmidae, Locustidae) Homoptera Heteroptera Neuroptera Mecoptera Coleoptera low and mid Hymenoptera (Tenthredinoidea, Ichneumenoidea, Proctotrypoidea, Chalcidoidea, Cynipoidea, Evanoidea) mid-Diptera (acalyptrate families)	Anisoptera high Orthoptera (Acrididae) high Hymenoptera (Sphecoidea, Vespoidea, ants) high Lepidoptera Rophalocera high Diptera (calyptrate families, also some acalyptrate and brachycerous families)

Collembola he also had an eye on the ecology of the species (Gisin 1943). Gisin divided the Collembola in three classes of "*Lebensforme*" (life-forms): *atmobios*, *hemiedaphon* and *euedaphon*. The atmobios, also called *epigeon*, comprises species with colour patterns, a large furca and long antennae. There are more of these species in the upper layers of soil and the vegetation. The euedaphon, however, is unpigmented, has a reduced furca and no eyes; these species are typical for below-ground habitats (eu-edaphon = what is truly in the soil). The hemiedaphon includes species with intermediate properties (hemi-edaphon = what is halfway in the soil). It is remarkable that Gisin's classification does not correlate with the taxonomic status of the species, since we find unrelated families in the same ecological group (Figure 1.2).

The life-form system of Gisin has been used by many soil ecologists, as a template to classify ecological functions and life-history strategies of soil invertebrates. For example, Faber (1991) proposed that soil fauna stratification could be used as a tool to evaluate depth-dependent ecological functions, Siepel (1994) developed an alternative system, to classify life-history tactics of Collembola, and Potapov et al. (2016) connected the life forms to trophic niches, indicated by stable isotope ratios.

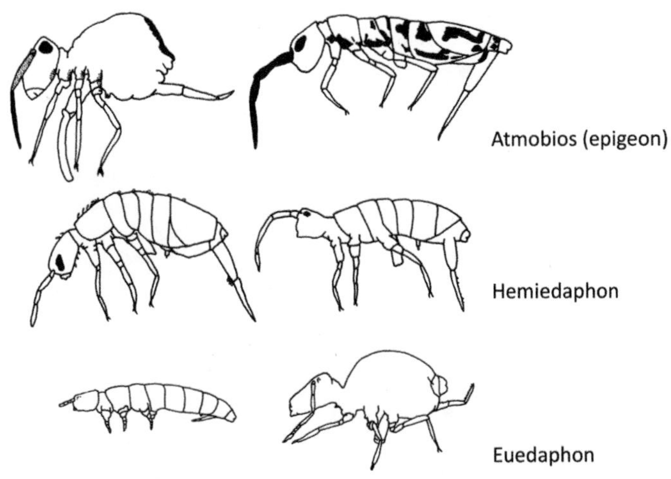

Figure 1.2. Illustrating the life-form classification of Collembola (springtails) developed by Hermann Gisin (1943). The length of the antennae, the presence of eyes, the colour pattern and the length of the furca were used to classify species as living on top of the soil (epigeon), at intermediate positions in the soil profile (hemiedaphon) and in the soil proper (euedaphon). Figures redrawn from Gisin (1960), with modifications.

Finally I refer to the work of the French soil biologist Marcel Bouché, who elaborated a system for earthworms (Bouché 1977). Like Gisin he classified earthworms according to pigment distribution, body size, musculature, light avoidance and feeding strategies, into three categories, entitled "*épigés*" (feeding and living in the litter and at the surface), "*endogés*" (feeding belowground and living in the soil) and "*anéciques*" (feeding on surface materials but living mostly in deep burrows). Bouché placed 15 common species of earthworm in a triangle defined by these axes (Figure 1.3).

Obviously, in earthworms, like in Collembola, the morphology is correlated closely with the position of the species in the soil profile, and this again is correlated with the ecological function of the species. Bouché, like Gisin and his Collembola, also noted that the taxonomic classification of earthworms does not synchronize with their ecological classification. In several taxa one may find representatives of all three ecological groupings. This even holds, in some cases, for a single species.

The concepts developed by Ghilarov, Kennedy, Gisin and Bouché illustrate that among soil zoologists the idea of the soil profile acting as a template for adaptation has prevailed already for many years. I summarize this as the Ghilarov-Kennedy-Gisin-Bouché hypothesis (abbreviated *GKGB hypothesis*): the soil is a transition medium allowing adaptations across the soil profile where the evolutionary trend is towards more advanced life forms living at the surface. This book is basically an illustration, a confirmation and a qualification of the hypotheses formulated in different wordings by the four scientists mentioned.

The soil invertebrate classification systems developed in the past were mainly derived from morphology, behaviour and feeding preferences and they have been discussed mostly in an ecological context, that is, in relation to soil functions such

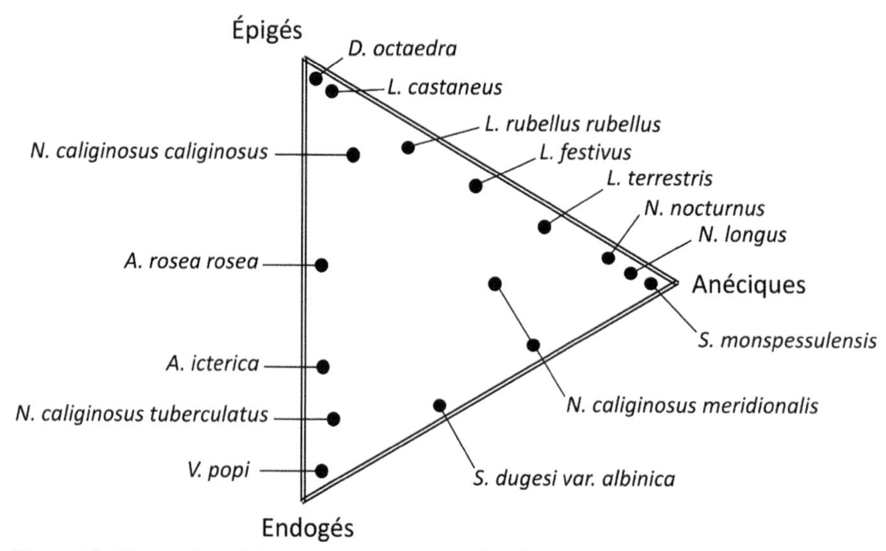

Figure 1.3. Diagram in which some common species of earthworm are positioned with respect to three ecological dimensions: soil-living (*endogés*), surface feeding and living (*épigés*) and surface feeding but living in deep burrows (*anéciques*). Redrawn from Bouché (1977), with modifications.

as litter decomposition. In this book I will leave the functional-ecological aspects of soil invertebrates aside; the reader is referred to the many excellent books on soil ecology, e.g., Curry (1994), Lavelle and Spain (2005), Bardgett (2005) and Orgiazzi et al. (2016). The present book departs from the observation that soil invertebrates are more than "*little rotters*". They are animals with a biology (physiology, biochemistry, genetics) that allows them to evolve, i.e., to develop adaptations to the conditions in the soil profile. This book explores the kaleidoscope of soil invertebrate adaptations as an exercise in evolutionary analysis.

1.2 Heterogeneity of soil temperature in space and time

Das Klima der bodennahen Luftschicht, a textbook of microclimatology by the German meteorologist Rudolf Geiger, might as well be awarded for bearing one of the most beautiful titles of any scientific book. Ever since that classical text we know that atmospheric conditions near ground level may differ substantially from the standard conditions measured at 1.50 m above ground (Geiger 1961, Geiger et al. 2009). A specialized branch of earth sciences, biometeorology, aims to measure and explain the temperature regimes as experienced by biological entities in the environment.

Soil and surface-active invertebrates, due to their small body size, have little heat capacity and so will quickly assume the temperature of their immediate environment. This can be seen from the equation that defines the temperature change of an object (ΔT) as a function of the heat flux into it (ΔQ):

$$\Delta T = \frac{1}{c}\Delta Q$$

where C is the *heat capacity* of the object (in Joule per degree Celsius). Heat capacity is more or less proportional to mass, so the equation shows that for objects of small mass, e.g., microarthropods, the increase in temperature at a given heat flux can be quite large. Whereas dromedary camels, just by being big, can postpone excessive heating of their body until nightfall, this option is unattainable for almost all soil invertebrates.

Under constant conditions, as the organism warms up, it will radiate heat according to its temperature, until incoming and outgoing radiation cancel, but in addition to radiation there are many other heat fluxes. The temperature equilibrium of a living organism in the environment is obtained by balancing positive fluxes with negative fluxes. The positive fluxes are:

- direct solar radiation
- radiation from the atmosphere and the clouds
- radiation from objects in the environment, e.g., shrubs and rocks
- metabolic heat production

and the negative fluxes are:

- reflection
- emission of radiation (based on body temperature)
- loss of heat due to conduction (mainly to the surface)
- heat loss due to convection (air movement, especially wind)
- evaporative heat loss.

The erection of such budgets for specific objects in the environment is the task of microclimatologists. Figure 1.4 shows the various components in a theoretical example. Some fluxes can be both positive and negative, e.g., heat loss due to conduction (10) is positive when outgoing, i.e., when the soil is colder than the animal, but negative when the soil is warmer than the animal and heat flows inwards. Radiation coming from the sky (4) is due to the atmosphere having a certain temperature. We know from physics that every object will radiate energy in terms of its temperature. For an ideal object (a black body) the law of Boltzmann holds:

$$R = \sigma T^4$$

where R is radiation energy emitted per time unit per surface area, T is Kelvin temperature and σ is the Stefan-Boltzmann constant ($W\ m^{-2}\ K^{-4}$). For the atmosphere a virtual temperature, called *radiation temperature*, is defined, which is the temperature that, according to Boltzmann's law, corresponds with the observed radiation flux from the sky. On a cold night and under a clear sky, the high, cold layers of the atmosphere contribute to radiation and so the radiation temperature of the sky is much lower (e.g., –40°C) than the temperature of an animal living at the surface. This implies that the animal loses heat to the atmosphere, the well-known cooling effect of clear skies.

It is obvious from Figure 1.4 that the heat budget of an invertebrate at the surface in the presence of solar radiation will be mostly positive, and the animal will warm

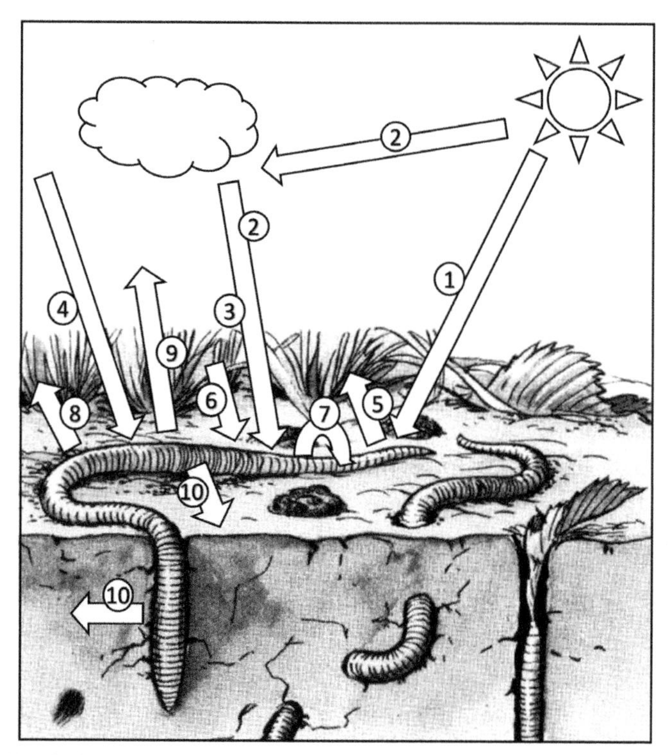

Figure 1.4. Heat budget for an earthworm at the surface of a soil. 1 direct solar radiation, 2 radiation reflected by the clouds, 3 radiation emitted by the clouds, 4 radiation from the sky, 5 reflection, 6 radiation from nearby objects in the environment (vegetation, rocks), 7 metabolic heat production, 8 heat loss due to evapotranspiration and convection, 9 emitted radiation due to body temperature, 10 heat loss due to conduction. Worm and soil drawing from Kutschera and Elliott (2010), reproduced with permission from Ulrich Kutschera.

up quickly (and lose water). That is why earthworms usually crawl out their burrows only at night (and are called "night crawlers" by lawn owners).

When considering the heat budget at the scale of an individual small animal, that is, at the level of cm to mm, a tremendous variability is revealed. By way of example, a picture is shown of temperature measurements made by the Dutch biometeorologist Ph. Stoutjesdijk in a field at the southern border of a pine forest in the Netherlands, on a sunny day in wintertime (Figure 1.5).

The ambient air temperature was 12°C, but locally temperatures at ground level were measured varying from –2°C (at the foot of a juniper shrub), to +62°C (in a piece of dry leaf litter exposed perpendicular to the sunbeams). This extreme range of variation is what small invertebrates have to deal with. At non-exposed spots aboveground temperatures were about equal to ambient, which is due to the fact that air masses in the shade receive mostly indirect solar radiation emitted by the sky and objects in the environment. However, the exposed side of the tree received the full solar radiation of the day and temperatures were considerably higher than ambient, up to 45°C. The sun-exposed parts of the soil likewise show large differences, which relate to the thermal properties of the materials. For instance, layers of undecomposed

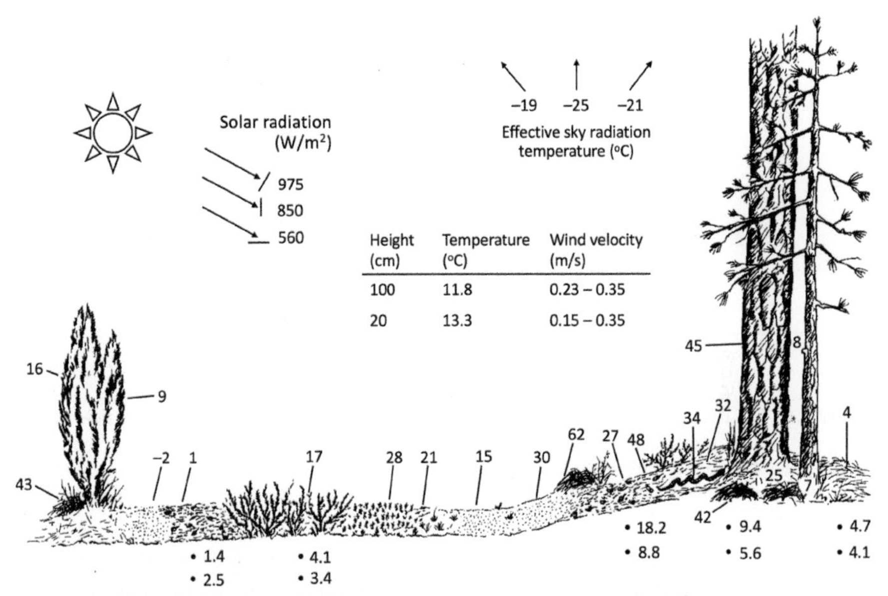

Figure 1.5. Micrometeorological measurements in a heathland bordering a pine forest, on a sunny day in March. Reproduced, with modifications, from Stoutjesdijk (1977), with permission from Springer Nature.

organic matter are often dark and do not conduct heat easily, and therefore may become considerably warmer than a piece of bare sand that reflects most of the radiation.

It is obvious from such measurements that the amplitude of temperature variation is much larger at the surface than in the soil. The contrast between surface temperature and soil temperature can be dramatic in desert environments. It is not uncommon, and measurements in deserts show this, that the surface temperature varies between +60 and –10°C (Hill et al. 2008). Many desert animals escape the scorching surface temperatures during the day and also the low temperatures during the night by digging burrows.

The absolute necessity of life in a burrow is illustrated nicely by the spider *Seothyra henscheli* (Araneae, Eresidae), which is common in the Namib desert dunes in Namibia. Long-term studies by Yael Lubin and Joh Henschel at Walvis Bay, Namibia and Ben Gurion University, Israel, have revealed its intricate behavioural adaptations to the extreme thermal conditions in this environment (Lubin and Henschel 1990, Birkhofer et al. 2012). To escape the high temperatures at the surface, the spider (shown right in Figure 1.6) digs a vertical hole, around 15 cm deep. A horizontal web, covered with sand, is displayed at the surface leaving a pattern of two semicircles (suggesting an antilope's footprint). The spider catches prey from an ambush below the border of the web. The catch is just quick enough for the spider not to be immobilized by a heat shock. Micromeasurements show that the surface temperatures may reach 73°C at the hottest moment of the day, however, below 6 cm depth in the burrow, temperatures are in a spider-safe zone between 30 and 40 degrees (Figure 1.6). This example again illustrates the extremely steep temperature gradients, over a few cm of the profile, that face life in soil.

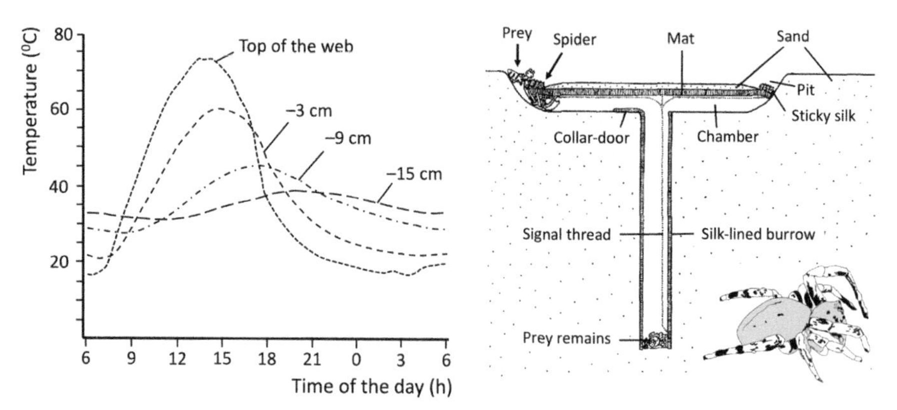

Figure 1.6. The desert spider *Seothyra henscheli* (Eresidae), shown right, digs a vertical hole in sand dunes and catches prey (mainly desert ants) using a horizontal web covered with sand. On the left the diurnal change of temperature at the top of the web and at three different depths in the burrow is shown. Reproduced, with modifications, from Lubin and Henschel (1990), with permission of Springer Nature.

Not only (horizontal and vertical) spatial variation, also temporal variation of temperature is a significant factor for soil-living animals. In temperate environments soil temperatures on the average follow the seasonal cycle of the air temperature. However, not only the mean, but also the amplitude shows a significant seasonal variation. In general, the temperature amplitude (daily maximum minus daily minimum) is greatest in spring and smallest in autumn. This fact is of great importance to small invertebrates.

The argument is illustrated in Figure 1.7. Temperature recordings are shown, measured in the litter layer of a pine forest, at 5 cm depth during 2 weeks in spring (Van Straalen 1985a). These are compared with the threshold temperature for egg development in the collembolan *Orchesella cincta* as estimated from laboratory experiments at different constant incubation temperatures (Van Straalen and Joosse 1985). Judged from the mean temperature in the field, there would be no egg development possible until April 12, because only on that day the average temperature passes the threshold. However, due to daily cycling, egg development is already possible during limited times of the day from April 5 onwards. To estimate the vernal activity of ectothermic invertebrates, it is essential not to rely on daily averages, but to take the hourly changes into account.

The principle that fluctuating temperatures may speed-up development compared to the corresponding constant mean has puzzled many ecologists in the past. It is often interpreted as an effect of fluctuations as such; however, this is a wrong conception. Often the sole reason is non-linearity in the temperature response, for instance due to a threshold or an exponential increase with temperature. Even if organisms just respond to the instantaneous temperature, fluctuating temperatures will always yield a faster development if the temperature response is non-linear. We will see in Chapter 6 that temperature responses of soil invertebrates are often non-linear indeed. Actually, the arithmetic mean is a wrong summary statistic in these cases. This principle was already noted and discussed with commendable clarity by

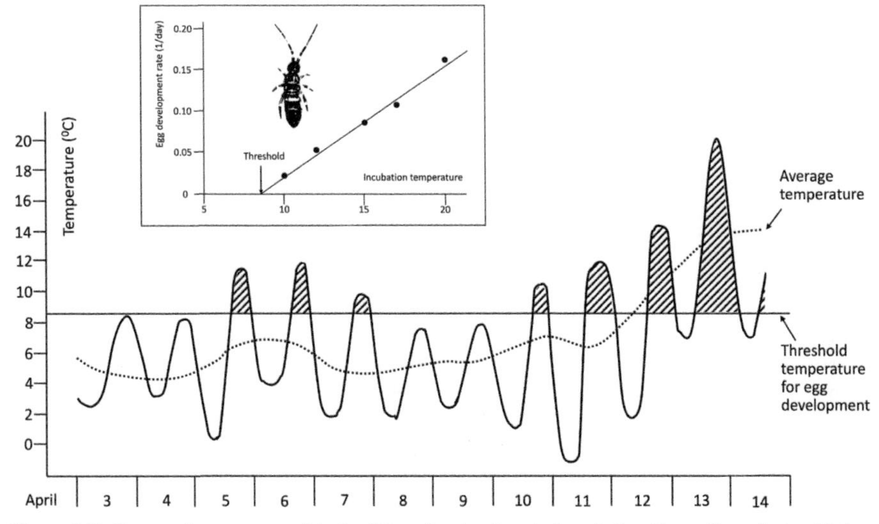

Figure 1.7. Temperatures measured in the litter of a pine forest, 5 cm below the surface, during 12 days in spring (Van Straalen 1985a). The data are compared to the threshold temperature for egg development estimated from laboratory incubations for the collembolan *Orchesella cincta*, shown in the inset (Van Straalen and Joosse 1985). Although the daily average temperature only crosses the threshold on April 12, development is already possible before, in the hatched time windows.

Kaufmann (1932). The reader is also referred to Hagstrum and Hagstrum (1970) and Laudien (1973).

Finally, the last aspect of temperature discussed here is the seasonal cycle. As an example, I show a study of seasonal variation of soil temperature measured to explain the community composition and depth distribution of oribatid mites (Jakšová et al. 2019). Temperatures were measured at different depths in five different deciduous woodlands growing on scree slopes below mountain cliffs, in the Western Carpathians in Slovakia (Figure 1.8).

It is obvious from these measurements that the surface temperatures are both lower (in winter time) and higher (in summertime) than the temperatures deeper in the soil profile. The surface temperature shows a clear periodic cycle due to the seasons, and on top of that there are fluctuations with shorter periods (3–4 weeks). The seasonal cycle can also be recognized in the soil temperatures (although with a much smaller amplitude), but the 3-week fluctuations are completely absent. Like in the case of the desert environment discussed above, these data again confirm the pronounced environmental variability at the surface and the buffered conditions deeper in the soil.

1.3 Soil pore water and acidity

Soil moisture content is one of the most determining factors for the life of soil invertebrates. Like temperature, it shows a depth gradient, from saturated conditions in groundwater to limiting humidity conditions at the surface. Humidity is a strong selective factor in the soil transition zone, with great relevance to the evolutionary hypothesis of this book. In all large-scale inventories of soil invertebrates, annual

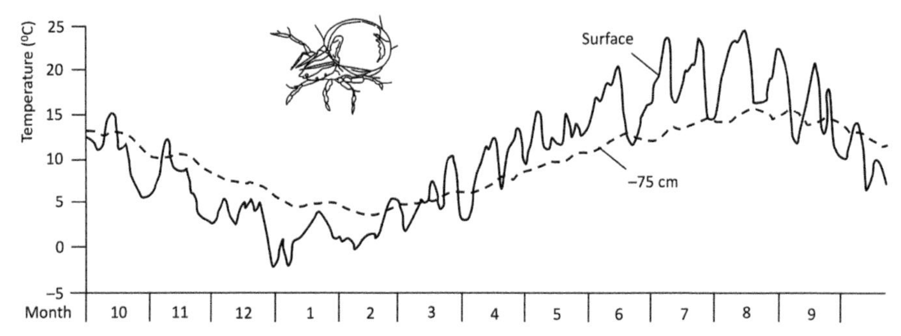

Figure 1.8. Changes of temperature measured at the surface of an organic soil in deciduous forest in the Western Carpathians, Slovakia, and at 75 cm depth. The annual cycle shows both higher (in summer) and lower (in winter) temperatures at the surface, compared to the –75 cm profile. The mite shown is *Ceratoppia bipilis*, a common oribatid in this system. Redrawn with modifications from Jakšová et al. (2019).

precipitation comes out as one of the main factors explaining abundance and species composition (Phillips et al. 2019), however, it must be realized that such correlations are only indirect. Soil invertebrates do not "feel" the rainfall or the climate, they feel the water potential of the soil and the relative humidity of the air pores.

Interestingly, just what is a wet soil and what is a dry soil is not that easy to tell. There are many different ways to express the amount of moisture in a soil: relative to the total fresh weight, relative to the dry weight, relative to the volume of the soil, or relative to the water holding capacity. Of these four measures, the first is used most often, but it tells us relatively little about the ecological effects of soil humidity. The better measure is to relate the total amount of water to the capacity of the soil to hold that water, which depends on soil type and a host of compositional properties of the soil.

Figure 1.9 illustrates the principle of water retention in a soil. When a soil is wetted until water drips out freely, we say that the soil is *water-saturated*: all pores are filled with water. When the soil is slowly dried, the dripping will stop. There is no loss of gravitational water anymore, although the soil is still wet. We say that soil is at *field capacity*. Any water left in the soil is called *capillary water*. When the soil is dried even more, there will come a point at which plants cannot extract water anymore and will wilt permanently. This is the *permanent wilting point*. There is still some water in the soil, but it is bound so tightly that it has become unavailable; it is called *hygroscopic water*. Upon further drying in an oven even that water may be removed and we get the dry weight of the soil.

Water holding capacity (*WHC*), is defined as the volume of water in a soil at field capacity, relative to the effective total volume of the soil. It is usually expressed as a volumetric ratio (L water per L soil), but sometimes also as a gravimetric ratio (kg water per kg soil). It is a distinctive property, often determined in soil ecology studies. It can be estimated experimentally by saturating a soil column with water (like in the left panel of Figure 1.9), or it can be theoretically derived from compositional properties.

The water holding capacity depends on soil properties such as particle size distribution, pore volume and organic matter content. A clayey organic soil has a

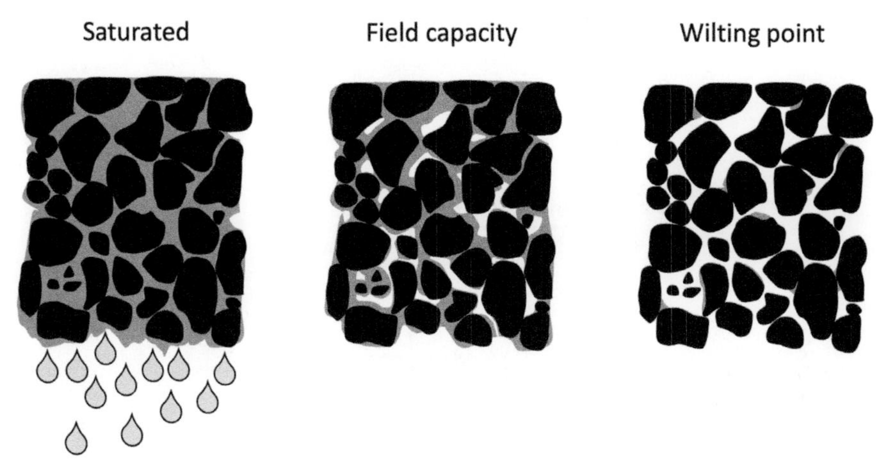

Figure 1.9. Illustration of the concept of water potential. A water-saturated soil (left) has all soil pores filled and loses gravitational water. At field capacity (centre) there is no spontaneous water loss, the soil is moist and said to be at 100% of its water holding capacity (WHC). The water retained is called capillary water. At the permanent wilting point (right) little water is left in the soil and this is unavailable to plants. This water is called hygroscopic water.

much higher WHC than a sandy soil. Expressing the moisture content of a soil as a percentage of the water holding capacity removes some of the differences between soil types, allowing comparisons across soil types. For instance, a soil at 60% of WHC is moist but not too wet and usually suitable for culturing earthworms and growing plants and this holds for a great range of soil types.

The permanent wilting point too depends greatly on the type of soil. It is usually somewhere around 30% of WHC, of course depending very much on the plant species. To express these measures in a more precise manner, agronomists use the term "*root-zone plant-available water holding capacity*" (RZPA-WHC), which limits the WHC to the root zone and to the water that is available to a specific plant or crop.

The water retention capacity of a soil can also be expressed as the pressure (actually under-pressure or suction) that is needed to extract water from the soil. For a completely saturated soil, that pressure is zero. For a soil at permanent wilting point the pressure is very large (larger than the pressure exerted by plants to suck up water). Because the relationship between suction pressure and water content is nonlinear (suction needs to increase disproportionally at lower moisture contents), it has become common to define the logarithm of the suction power as the *pF-value* of the soil. This is not a standard unit because, for historical and practical reasons, pF is expressed in cm of water pressure; for example, at 100 cm suction pressure, pF is $\log 100 = 2$. The SI unit of suction pressure is Pascal (Newton per square meter). When expressed in Pascal (usually MPa), pF is equivalent to the *water potential*, designated by the symbol Ψ (psi).

The relationship between pF and water content is shown in Figure 1.10. Curves are drawn for two types of soil, one sandy, the other clayey. The graph shows that the two soils have their permanent wilting point and field capacity at the same pF value

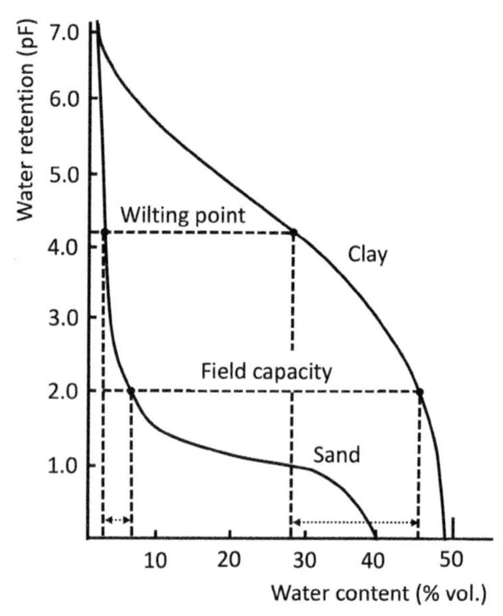

Figure 1.10. Theoretical soil moisture retention curves for two soils, relating suction pressure pF (^{10}log of cm water) to volumetric water content. The permanent wilting point at pF 4.2 and the field capacity at pF 2.0 are the same for the two soils, but the corresponding volumetric water contents are very different.

(4.2 and 2.0, respectively), despite a large difference in the water content at which these points are reached. The range of water contents allowing adequate plant growth is much smaller for sandy soils than for clay soils.

While pF characterizes a soil, invertebrates living in the soil pores do not experience pF directly, but feel the "drying power" of soil as the humidity of the air. *Relative air humidity* is another commonly measured but physically complicated property. We know that the maximum vapour pressure of water decreases with temperature in a non-linear manner according the general gas equation, so the same amount of water in air makes a dry air at high temperature and a humid air at low temperature. Therefore, air humidity is usually expressed as the amount of water relative to the saturation value at the temperature under consideration (%RH).

As long as there is sufficient free water in the soil (around field capacity and above) soil invertebrates living in the water phase (nematodes, rotifers, tardigrades, etc.) will not notice changes in soil moisture content. Invertebrates living in aerial soil pores will remain unaffected as long as the air in the soil pores is water-saturated. However, if the moisture content decreases from field capacity towards the wilting point, more and more of the water becomes tightly bound to soil particles, while the air sachets, especially the macropores, will become undersaturated with water vapour. This poses an evaporation stress on the invertebrates. We will see in Chapter 6 how invertebrates have evolved different degrees of drought tolerance to cope with these conditions.

With soil water come a range of dissolved substances, simple ones like atomic ions and complicated ones such as phenolic acids. These substances are more relevant for plants than for animals, since plants take up all their nutrients from the soil, while animals take them in mostly with the food. However, some dissolved substances affect animals directly, including protons (pH), calcium, carbon dioxide and carbonate (alkalinity), chloride, potassium and sodium (osmolarity), and ammonium and nitrate (mineral nitrogen). We consider here only pH and leave the other components for Chapter 6.

After temperature and soil moisture, pH is the next most important factor in determining abundance and diversity of soil invertebrates, although it is obviously of more importance to soft-bodied invertebrates in direct contact with soil water (earthworms, enchytraeids, nematodes), than to well-sclerotized animals (crustaceans, mites, spiders, insects). There are large differences between invertebrate groups in pH preference and tolerance. Many arthropods prefer a circumneutral, or slightly acidic, pH but among Collembola and oribatid mites both *acidophilic* species (preferring pH 3) and *alkalophilic* species (preferring pH 8) are present (Soejono Sastrodihardjo and Van Straalen 1993, Van Straalen and Verhoef 1997).

Acidity of natural soils (forests, heathlands) is greatly determined by the humus form (see Section 1.5). Agricultural soils usually are more alkaline than natural soils, as their pH is influenced by anthropogenic management, e.g., liming, fertilization, cultivation and growing crops. Land taken out of agriculture often shows a gradual decrease of pH over the years and a similar trend is seen in natural succession series.

In depth profiles, one usually observes an increase of pH with depth (Figure 1.11). This is often associated with gradients of other dissolved ions, in this case with nitrate and ammonium. These gradients are due to leaf litter accumulation at the surface and the production of acidic compounds during decomposition (respiration of organic carbon), *fermentation* (anaerobic respiration) and *nitrification* (conversion of ammonia to nitrate).

Soil pH, although basically a property of the pore water, interacts with the solid phase by all kinds of exchange reactions. Addition of H^+ to a soil, by natural or anthropogenic acidification, does not lead directly to a decrease of soil pH because the pore water is buffered by the solid phase. Broadly speaking, a soil has three different buffer ranges. At neutral pH any added protons will associate with calcium carbonate of the solid phase, which dissociates and dissolves (Figure 1.12). Because H^+ is "consumed" by the dissolution reaction, the pH does not change until all available $CaCO_3$ is dissolved (A to B). Thereafter, pH will decrease until around pH 5 where the next buffer system enters, consisting of K^+, Mg^{2+} and Ca^{2+} associated with the solid phase (C to D). The system of cations electrostatically bound to negative surface charges of clay minerals is called the *exchange complex* of the soil. When this system is also depleted, pH will decrease again to around 3 or lower (D to E), where clay minerals will start to dissolve, leading to free Fe^{3+} and Al^{3+} in the pore water (the iron-aluminium buffer range). As free Al^{3+} is very toxic to plant roots, this is also the range where serious effects on plants are beginning to be noted.

From the titration curve in Figure 1.12 we may conclude that acidification is not the same as pH decrease. A soil may absorb a large amount of acid and still maintain its pH. This is acknowledged in the concept of *acid neutralizing capacity* (ANC, Van Breemen et al. 1983). The ANC is defined as the molar number of protons that can be added to a soil without change of pH, up to a certain reference point. It can

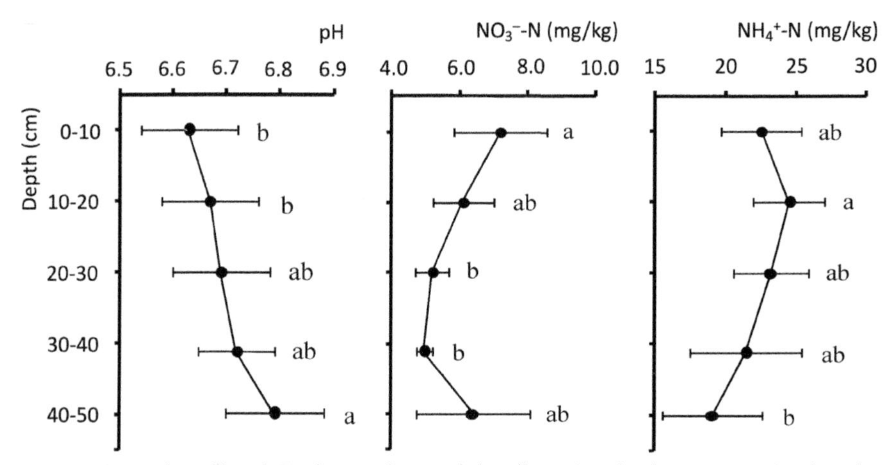

Figure 1.11. Depth profiles of pH, nitrate and ammonia in soil samples taken in a young poplar plantation in east-central China. Means bearing the same letter (a,b,...) do not differ significantly from each other. Reproduced, with modifications, from Feng et al. (2019), with permission from Weinfeng Wang.

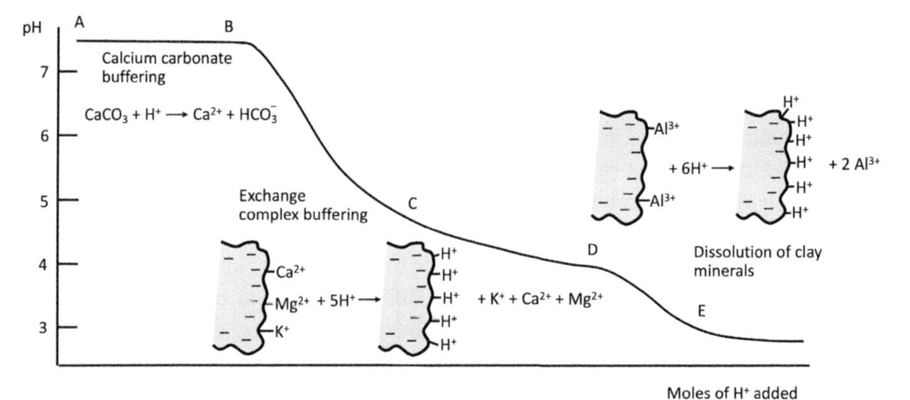

Figure 1.12. Titration curve for a calcareous soil, showing pH changes with increasing addition of acid. Three successive buffer systems stabilize the pH. Inspired by Van Breemen et al. (1983).

be estimated from the concentrations of calcium carbonate and the various exchange complexes.

Despite pH of the bulk soil being buffered, there may still be significant spatial variation and dynamics at the microscale (Bringmark 1989). Bulk pH is measured after shaking a sample of a few grams of soil with water, or a solution of KCl or $CaCl_2$. This assumes that the capillary water in contact with the solid phase is sufficiently buffered and insensitive to dilution. The validity of this assumption is difficult to check. Bulk pH gives an impression of the average soil acidity on the scale of centimetres to metres, but what happens on the scale of an individual invertebrate remains unclear.

Luckily, since a decade or so, pH microelectrodes have been developed that allow pH measurements on a very small scale (mm to µm). This technology has revealed a tremendous variability of local pH gradients and dynamic changes. It turns out that acidic and alkaline microenvironments are present at the surface of minerals, where bacteria are attached in a biofilm (Kim and Or 2019). Some of the reactions conducted by bacteria, e.g., ammonium oxidation (nitrification) produce significant amounts of acid (HNO_2 and HNO_3) that may alter the local pH without much change of the bulk pH.

Another interesting development is the use of "*planar optodes*" (Blossfeld and Gansert 2007). These consist of a foil with embedded fluorophores that when excited by light of a specific wavelength, emit a fluorescence pulse that is dependent on the local proton concentration. By mounting the foil in a slab of soil (*rhizotron*) and scanning the profile though a transparent plate a two-dimensional image can be created of pH dynamics on the sale of millimetres. Up to now these new micro-techniques work best for water-saturated soils and have not yet been applied to soils through which invertebrates can crawl.

Soil acidity is a complex property which involves a whole microworld of chemistry. How exactly a soil invertebrate perceives the soil pH is not very clear still at the moment, despite pH being one of the major factors explaining invertebrate abundance differences across soil types.

1.4 Bulk density, pores and burrows

Among the factors that limit the distribution of soil invertebrates, the next one to be considered is pore size. All invertebrates that cannot dig burrows, tunnels or holes rely on the natural air-filled pores and pockets in a soil. Even for burrowing species such as earthworms, the soil needs to have a certain consistency that allows digging. Available space is a prime factor for soil invertebrates; however, it is seldomly considered in soil surveys because it is not easy to measure on a large scale.

Space in soil depends on *porosity*. The most obvious indicator of porosity is *bulk density* (mass per volume, expressed in g/cm^3 or kg/L). This is a classical measure of soil structure, often determined simply by measuring both volume and mass of a sample. It is related to porosity by the following equation:

$$Pt = 1 - \frac{\rho_b}{\rho_d} = \frac{V_p}{V_T}$$

where Pt = porosity (a number between zero and 1, also expressed as a percentage when multiplied by 100), ρ_b is *bulk density* (g/cm^3), ρ_d is *particle density* (g/cm^3), V_p is pore volume and V_T is total soil volume. However, this formula is difficult to use, since particle density (the weight of all particles in a certain volume) is difficult to measure and can only be calculated precisely for media of homogenous composition. Soil does not consist of identical particles, instead it has a *particle size distribution* covering very small clay particles of less than a micrometre, and very large aggregates, up to gravel.

The particle size distribution is usually divided into four different categories: coarse sand, fine sand, silt and clay (Table 1.2). The mass distribution over these categories defines the *texture* of a soil. For instance, a soil with 60% sand, 30% silt and 10% clay (by weight) is considered a sandy loam. The texture of a soil varies with depth, due to erosion, leaching, organic matter infiltration and bioturbation. Profile

Table 1.2. Classes of particle sizes usually distinguished in soil structure analysis. From Lavelle and Spain (2005).

Class	Range of particle diameters (μm)	Approximate number of particles per g	Approximate surface area (m^2/g)
Coarse sand	200 – 2,000	5.4×10^2	0.21
Sand	20 – 200	5.4×10^5	2.1
Silt	2 – 20	5.4×10^8	21.0
Clay	0.2 – 2	7.2×10^{11}	230.0

and texture together define *soil type*, of which there are too many to recapitulate here; the reader is referred to a textbook of soil science.

Bulk density and porosity are very central properties of a soil since they determine water holding capacity and infiltration capacity. Water and nutrients bind to the internal surface of the soil and this is much larger for clayey soils than for sandy soils (Table 1.2). Soil compaction, caused by anthropogenic infrastructure, machinery and trampling, destroys porosity and is one of the most serious threats in the protection of soil health. Compacted soils also do not provide a home to soil invertebrates, which even worsens the situation.

Non-digging soil invertebrates live their life in *macropores*, i.e., pores larger than 0.5 mm diameter. These are either made by burrowing animals or formed by fissures in the soil, voids between larger aggregates or plant roots. They often take the form of vertical channels, through which water is quickly drained to the subsoil. While micropores take up most of the rainwater when the soil is dry, macropores become the preferred pathway for soils close to saturation (Jarvis 2020). Macropores drain freely with gravity, in contrast to micropores which withhold the water by capillary action. An impression of the soil as a complex of water-filled and air-filled macropores and micropores in between mineral aggregates is given in Figure 1.13.

Although macropores are "macro", they are still small relative to the dimensions of most soil animals. Pore size may well be considered a selective factor in the evolution of soil invertebrates. For all non-digging species, life in soil requires a small body size, or at least precludes body size from increasing. There may be advantages of living a secluded life in soil, but the limitations on body size come with several secondary consequences. Are there any trends in body size along evolutionary lineages of soil invertebrates? We will discuss this question in Chapter 3.

A great deal of the macropores are made by burrowing invertebrates. Earthworms are well-known for this. The galleries of earthworms measure 2–12 mm in diameter and they can introduce a porosity of 4% in the upper 50 cm of soil (Kooistra and Brussaard 1995, Jarvis 2020). In addition to worms, a great variety of other invertebrates make a habit of digging: holes, burrows, tunnels, galleries, pits, nests, hills, mounds, etc. In Table 1.3 a number of soil-digging invertebrates are listed. Ground-nesting and living in burrows is often supported by specific morphologies such as a flattened, cylindrical or vermiform body and specialized scoops, tooths or lobes on the legs, depending on the mode of digging (Villani et al. 1999).

The tremendous variety of burrowing structures illustrates the great diversity of purposes for which they are made. Many serve to provide shelter, either from

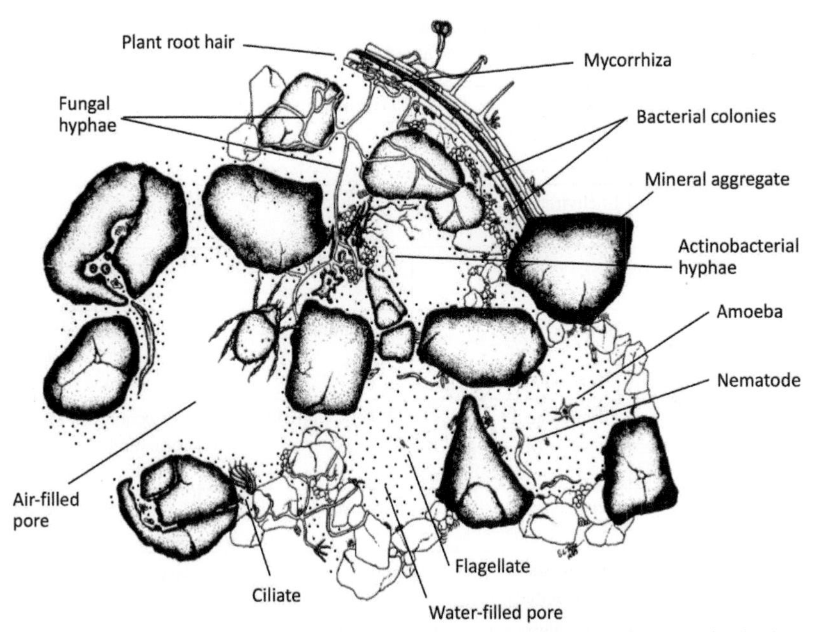

Figure 1.13. Microscopic view of the soil ecosystem, from Paul and Clark (1989); drawing by S. Rose and T. Elliot, courtesy of The Soil and Water Conservation Society.

predation or from heat, cold or floods. Often a considerable part of life is spent underground, in other cases the real activity is aboveground and the burrow is just a place to hide during part of the day. In spiders, some scorpions and ant lions, the burrow is constructed as a trap to catch surface-active prey. Ants and termites build a colony in self-constructed mounds and hills, and often drag considerable amounts of leaf material inside the nest, to feed upon or to culture fungi.

To illustrate the intricacy of these constructions, Figure 1.14 shows a drawing of the burrow system of the mole cricket, *Scapteriscus acletus*, an American species of the orthopteran family Gryllotalpidae. The burrow of these animals involves superficial tunnels but also chambers the shape of a horn, opening at the surface, which the male uses to call from. The shape of the burrow increases the male's acoustic output. Different species sit at different positions in the burrow, by which the stridulation is modulated differently, allowing species-specific calls (Nickerson et al. 1979).

The activity of burrowing, digging and tunnelling by soil invertebrates is designated as *bioturbation*. An important ecological effect of these activities is that soil particles and organic matter are moved from one place to another, from the surface to deeper layers, mixed in with the soil profile, chemically altered, inoculated with microbes, taken apart, fragmented, stacked in heaps, etc.

Ever since Darwin the bioturbation of earthworms has been valued as a significant contribution to soil fertility. In his book "*The Formation of Vegetable Mould through the Action of Worms*" Darwin investigated a yard near Leith Hill Place in Surry, England, where "*A lady, on whose accuracy I can implicitly rely, offered to collect during a year all the castings thrown up on two separate square yards.*"

Table 1.3. List of invertebrate groups that are known for their soil-digging and bioturbation behaviours. From various sources.

Common name	Scientific name	Type of construction	Function
Earthworms	Lumbricidae, Megascolecidae	Burrow	Shelter, feeding
Potworms	Enchytraeidae	Burrow	Shelter, feeding
Cockroaches	Dictyoptera	Gallery in dead wood	Protection, feeding, reproduction
Termites	Isoptera	Nest, mound, hill	Protection, colony formation, feeding, reproduction
Ants	Formicidae	Nest, mound, hill	Protection, colony formation, feeding, reproduction
Earwigs	Dermaptera	Shallow burrow (nest)	Larval growth, parental care
Dung beetles	Geotrupidae, Scarabaeinae, Aphodiinae	Tunnel	Reproduction, larval growth
Carrion beetles	Silphidae	Burrow underneath carrion	Reproduction, larval growth
Rove beetles	*Bledius* species (Staphylinidae)	Tunnel	Shelter
Darkling beetles	Tenebrionidae	Burrow	Shelter
Passalid beetles	Passalidae	Galleries in dead wood	Wood feeding, shelter
Mole crickets	Gryllotalpidae	Sound chamber, tunnel	Calling, nymphal growth
Ant lion larvae	Myrmeleontidae	Pit, trap	Predation, nymphal growth
Cicada nymphs	Cicadoidea	Burrow, mud towers	Feeding, nymphal growth
Tarantulas, trapdoor spiders, desert spiders	Theraphosidae, Ctenizidae, Eresidae	Pit, trap, tunnel, burrow	Predation, shelter
Scorpions	Scorpiones	Burrow	Shelter, predation
Isopods	Trachelipodidae	Burrow	Shelter, feeding
Intertidal amphipods	Corophiidae, Talitridae	Tunnel	Shelter, feeding, group-living
Sand crabs, mole crabs	Albuneidae, Blepharipodidae, Hippidae	Burrow	Shelter, feeding

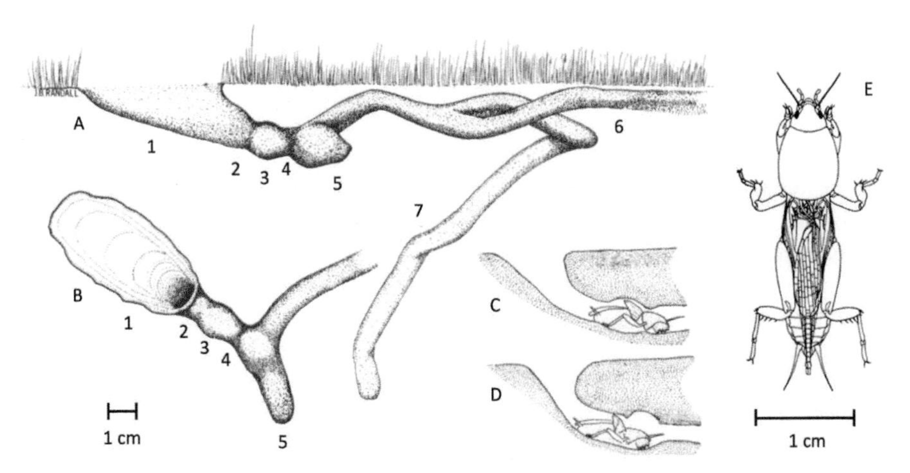

Figure 1.14. Schematic drawing of the tunnelling system of the mole cricket, *Scapteriscus acletus*, with the remarkable calling chambers, used by males to amplify their stridulation calls. A lateral view, B top view, C position of *S. acletus* and *S. vicinus* when singing, D singing position of *Gryllotalpa vineae*, E habitus of *Scapteriscus vicinus*. 1. horn, 2.1st constriction, 3. bulb (calling chamber), 4. 2nd constriction, 5. turn-around chamber, 6. superficial tunnel, 7. deep tunnel. Burrow drawing by J.B. Randall, from Nickerson et al. (1979) reproduced by permission of Oxford University Press; drawing of *S. vicinus* from Capinera and Leppla (2001), by permission of the University of Florida.

(Darwin 1881). From the data he estimated that earthworms ejected annually 16.1 tons per acre of castings, which corresponds to about 4 kg per m^2.

The digging behaviour of soil invertebrates is considered a case of ecological engineering. Ecological engineers take a key position in the ecosystem since by alteration of soil structure they contribute to ecological functions of the soil, such as litter decomposition, nutrient cycling, distribution of microbial inoculates, grazing of fungi, aeration, water infiltration, drainage and fertility. We will however, not dwell on the functional aspects of soil invertebrates in this book; the reader is referred to the many excellent review papers and soil ecology textbooks (Paul and Clark 1989, Jones et al. 1994, Brussaard et al. 1997, Lavelle et al. 1997, Brussaard 1998, Lavelle and Spain 2005, Bardgett 2005, Blouin et al. 2013, Coleman et al. 2017).

1.5 Humus forms

Organic matter content is one of the most important soil properties, although it is not a property of the soil alone; it is the result of interactions between soil organisms and "true" soil properties such as mineral composition and pore size. The type of vegetation is the prime factor determining the nature of organic matter developing from shed leaves and other plant remains reaching the soil surface. Conversely, the type of vegetation in itself is determined by soil properties and climate. Soil organic matter is also derived from root exudates, microbial synthesis and excreta from soil macro-invertebrates.

Soil organic matter protects the soil against erosion and run-off, it ensures aggregation of soil particles, has a role in nutrient cycling and improves infiltration and water retention. It also contributes to the ion exchange complex of a soil (*cf.* Figure 1.12) and it has the potential to bind potentially toxic compounds and store them in non-bioavailable form. In summary, soil organic matter and the physicochemical factors associated with it, are one of the most determining factors for the ecological functions of a soil, including the life of soil invertebrates.

"*Humus*", strictly speaking, is the fraction of soil organic matter that is transformed by microbial and animal activity. In a wider sense, "humus" is the entirety of organic horizons of a soil including the depth distribution of organic matter. "Humus" is equivalent to the "vegetable mould" of Darwin in his book on earthworms cited above. Humus can be approached from a botanical perspective (starting with the type of leaves that fall on the ground), from a chemical perspective (fractionation into a multitude of chemical compounds) and from a physical perspective (its contribution to soil structure by interaction with mineral particles). Historical overviews of the concept of humus are given by Kononova (1961) and Feller (1997).

The simplest quantitative overall measure for soil organic matter (*SOM*) content is amount of mass lost by burning, called *loss on ignition* (*l.o.i.*). Alternatively, SOM is estimated from the carbon content of a soil determined by chromate oxidation in strong acid and assuming that SOM contains 58% carbon. Although l.o.i. may overestimate SOM in case of soils with a high concentration of free carbonates (by as much as 25%), it is still a widely used and reproducible measure (Wander 2019). When comparing soils of different types, l.o.i. shows a strong positive correlation with clay content and a strong negative correlation with bulk density. This is due to the association of SOM with mineral particles and the larger voids in soils with SOM-mediated aggregates.

Table 1.4 gives an overview of the various chemical and biochemical approaches that are applied in soil science to characterize soil organic matter. The residence time of the various components varies by orders of magnitude (from days to centuries). Usually, a distinction is made between labile (quickly decomposed) and recalcitrant (persistent) matter, with an intermediate category in between. Within each category a variety of biochemical procedures is applied to define certain fractions. They all characterize different (but overlapping) aspects of soil organic matter, illustrating the enormous chemical complexity.

One particular fraction often studied consists of *humic acid*. This is a group of chemical substances that can be extracted as sodium, potassium or ammonium salts by alkali solutions and forming amorphous precipitates upon acidification to pH 1. Organic compounds that remain in solution at this low pH are designated as *fulvic acids*, while the organic matter that is not solubilized by alkali extraction is called *humin* (Kononova 1961). Humic acids and fulvic acids are not single compounds, they consist of many different ones, all with a high abundance of aromatic hydroxy (phenolic) and carboxylic groups. These functional groups will bind soluble nutrients such as calcium, iron and trace metals; therefore, humic and fulvic acids, with their pH-dependent mobility, contribute to the vertical transport of these nutrients in acid soils.

The concentrations of water-soluble phenolic compounds depend greatly on the aboveground vegetation, for instance, the amount of monomeric phenolics is much higher under pine and spruce than under beech and birch, and intermediate under oak (Kuiters and Denneman 1987, Whitehead et al. 1982).

Table 1.4. Soil organic matter pools and fractions. Modified from Wander (2019).

Pool	Half life	Description	Procedural defined fractions
Labile or active SOM	Days to a few years	Material of recent plant origin Living components of SOM High nutrient status Physically unprotected Involved in biological activity	Chloroform-labile microbial biomass Microwave-irradiation-labile microbial biomass Amino acids Phospholipids Mineralizable C and N, estimated by incubation Substrate-induced mineralizable carbon SOM extractable by hot water or dilute salts Polysaccharides
Slow or intermediate SOM	Few years to decades	Physically protected SOM Separated from the other two fractions by physical location Partially decomposed residues and decay products	Amino-compounds Glycolproteins Particulate organic matter in aggregates Humic materials hydrolyzable by acid/ base extraction Mobile humic materials
Recalcitrant, passive, stable and inert SOM	Decades to centuries	Compounds with chemically-defined recalcitrance, minerally associated	Aliphatic macromolecules (lipids, cutans, algaenans, suberans) Polycyclic aromatic compounds Charcoal Sporopollenins Lignins Strongly condensed SOM Non-hydrolyzable SOM

These chemical methods of soil science are not often applied in studies of soil invertebrates. To what extent certain chemically defined components of SOM create a selective environment to soil invertebrates, e.g., in terms of nutritional value or negative impacts on growth, is not known, except for compounds with a clear toxic action. Soil ecologists study the interaction between invertebrates and SOM on a more holistic level, involving concepts like decomposition, nutrient cycling and ecological engineering (Lavelle et al. 1997, Brussaard 1998, Wolters 2000), not looking into the interactions of single compounds.

Another way to characterize the organic matter of a soil is by the *mull-moder-mor* gradient. This concept can be traced back to the Danish forester Peter Erasmus Müller who first presented it in a lecture before the Royal Danish Agricultural Society in 1879 (Petersen 1991). His book "*Studies on the natural humus forms and their effects on vegetation and soil*", originally published in Danish in 1887, but later translated into German and French, became a classic of soil science. Müller developed the notion that different soils develop under different types of forests. He observed that in beech forests, soils tended to have a shallow surface cover and a deep, dark, layer of well-mixed organic matter. In heathlands, however, soils had a thick surface layer of very slowly decomposing litter, contrasting sharply with the light-coloured mineral layer below it, from which minerals leach to the groundwater or accumulate lower in the profile. Müller called these two types of humus "*mull*" and "*mor*". Later the term "*moder*" was introduced as an intermediate humus form.

The mull-mor gradient is one of the best predictors of the gross composition of both the microbial community and the soil invertebrate community (Ponge 2003). Due to the slower rate of decomposition in mor-type humus and the accumulation of phenolics, pH is lower than in mull humus. As a consequence, earthworms do not thrive well in mor humus, as there are few acid-tolerant species, while enchytraeid worms, which are more acidophilic in general, replace the earthworms. The earthworm/enchytraeid ratio is very strongly correlated with the humus type (Ponge 2003). In addition, because bacteria in general are less tolerant of acidity than fungi, the bacteria/fungi ratio likewise shows a strong correlation with the mull-mor gradient.

Despite the strong correlation of soil invertebrates and microorganisms with humus type, it seems that humus forms exert their selective effect mainly through the pH. Soil pH is the second-best predictor (after soil humidity) of bacterial and archaeal biomass worldwide (Fierer 2017). These trends are summarized in Table 1.5.

Table 1.5. Biological features associated with the three main humus forms in soil. From Ponge (2003) and Kononova (1961).

	Mull	**Moder**	**Mor**
Type of vegetation	Grasslands, deciduous woodland with rich herb layer	Deciduous and coniferous woodland with little undergrowth	Heathlands, coniferous forests, sphagnum bogs, alpine meadows
Aboveground biodiversity	High	Medium	Low
Productivity	High	Medium	Low
Soil type	Brown soils	Grey-brown podzolic soils, löss	Podzols
Phenolic content of litter	Low	Intermediate	High
Humification	Rapid	Slow	Very slow
Acidity	Neutral	Neutral	Acid
Buffer range	Carbonate	Silicate	Iron/aluminium
C/N ratio of organic matter	10–15	15–25	25–40
Annelid worm community dominated by	Earthworms	Enchytraeids	Few macroinvertebrates
Microbial community dominated by	Bacteria	Fungi	Acidophilic fungi

1.6 The rhizosphere

The soil zone that is directly affected by plant roots deserves separate treatment here since the conditions and selective factors are very different from the bulk soil. Natural soils are always crowded with plant roots, which compete with each other for space, water and nutrients. A variable but significant portion of any vegetation is belowground (50 to 90% of total plant biomass), with functions in mechanical support, vegetative reproduction, water absorption and nutrient uptake. Belowground storage organs of crops such as potato, yam and cassava constitute important food sources for the human world population. It is no wonder that this belowground biomass is also exploited by soil-living invertebrates.

A very diverse community of invertebrates has specialized in root-feeding. This includes first of all nematodes (roundworms), which are notorious for the damage they may cause to valuable crops. Some nematodes not only pierce the root cells but also form egg-containing cysts inside the plant tissue, e.g., potato cyst nematodes of the genus *Globodera*. Plant parasitism is not the ancestral feeding habit of nematodes, however, which most likely is bacterivory. Within the phylum Nematoda plant-root feeding has evolved at least three times, in the groups Tylenchida, Dorylaimida and Triplonchida (Blaxter et al. 1998, Haegeman et al. 2011). Root-feeding nematodes can be recognized for the presence of a stylet in their buccal cavity, a structure that they use to pierce plant cell walls (Figure 1.15). Because of their great economic importance, research on nematodes has developed into a specialized agricultural discipline, *nematology*. But also in natural environments, nematodes are a conspicuous and ecologically important component of the rhizosphere (Yeates 1981, Bongers and Ferris 1999, Bonkowski et al. 2009).

Many root herbivores are found within the insects, especially the orders Hemiptera, Coleoptera, Diptera and Lepidoptera. Within the Hymenoptera there are gall wasps that cause galls on roots. Also some Collembola and prostigmatid mites will feed on knots and bulbs. Root feeders are lacking in earthworms and isopods. Most likely the mouthparts of the various clades allow one group more likely than another to adopt a root-feeding habit. This is especially applicable to Hemiptera (bugs) that all have sucking and piercing mouthparts and include many root-feeders. Figure 1.16 gives an impression of the biological diversity of the root herbivore community.

Living in the soil matrix like root herbivores do, requires behavioural and metabolic adaptations since they cannot disperse rapidly from adverse conditions like their aboveground relatives can (Barnett and Johnson 2013). However, many root herbivores are linked to aboveground communities because they are juvenile stages of species that have their adult life aboveground. This holds true for the root-feeding cicadas, beetles, flies, midges and moths. In other cases the belowground animals represent an asexual generation of a species that has its sexual generation in the above-ground vegetation (e.g., the gall wasp *Biorhiza pallida*). Some root herbivores are specialized on a single plant species or a limited group (e.g., cabbage fly on

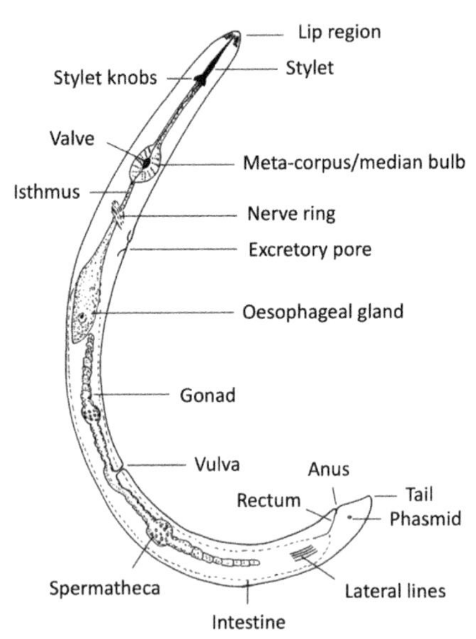

Figure 1.15. Typical morphology of a (female) plant-parasitic nematode. The sharp stylet with basal knobs is characteristic for these animals. Reproduced, with modifications, from Stirling et al. (2002).

Brassicaceae), others have a more generalized feeding habit (e.g., the collembolan *Protaphorura* is actually a detritivore but will nibble from knots and bulbs when these are available).

With their adaptations to life underground, root herbivores are specialized animals, although their morphology is often simplified (Figure 1.16), compared to the above-ground stages. Their specializations relate to identification of plant roots by chemical cues, piercing plant walls, detoxifying possible toxins released by plants and feeding on a one-sided diet. The rhizosphere herbivores do not fit in the GKGB hypothesis presented in the first paragraphs of this Chapter. A look at Figure 1.16 confirms that in many cases, they are derived from groups that have their ancestry aboveground. The community of root-feeding invertebrates consists of many lineages of secondary evolutionary specialization; it is not a community that represents the soil-living ancestors of these lineages.

Root herbivores are estimated to consume 6 to 30% of belowground plant production (Andersen 1987). Although this may seem significant, root herbivory in natural ecosystems is rarely a factor that destroys significant parts of the vegetation (Blossey and Hunt-Joshi 2003). Many plants are well equipped in fending off root herbivores by the production of feeding deterrents or toxic compounds.

Herbivores are not the only organisms in the rhizosphere. Organic carbon compounds excreted by plant roots (exudates) attract a lot of heterotrophic bacteria and fungi. These again attract bacterivores (protists, tardigrades, nematodes) and fungivores (many mites and collembolans). Predators, parasites and parasitoids feast on these bacterivores and fungivores. Even earthworms benefit from the higher abundance of biomass in the root zone. The selective factors in the rhizosphere are therefore essentially different from those in the bulk soil. Competition, predation,

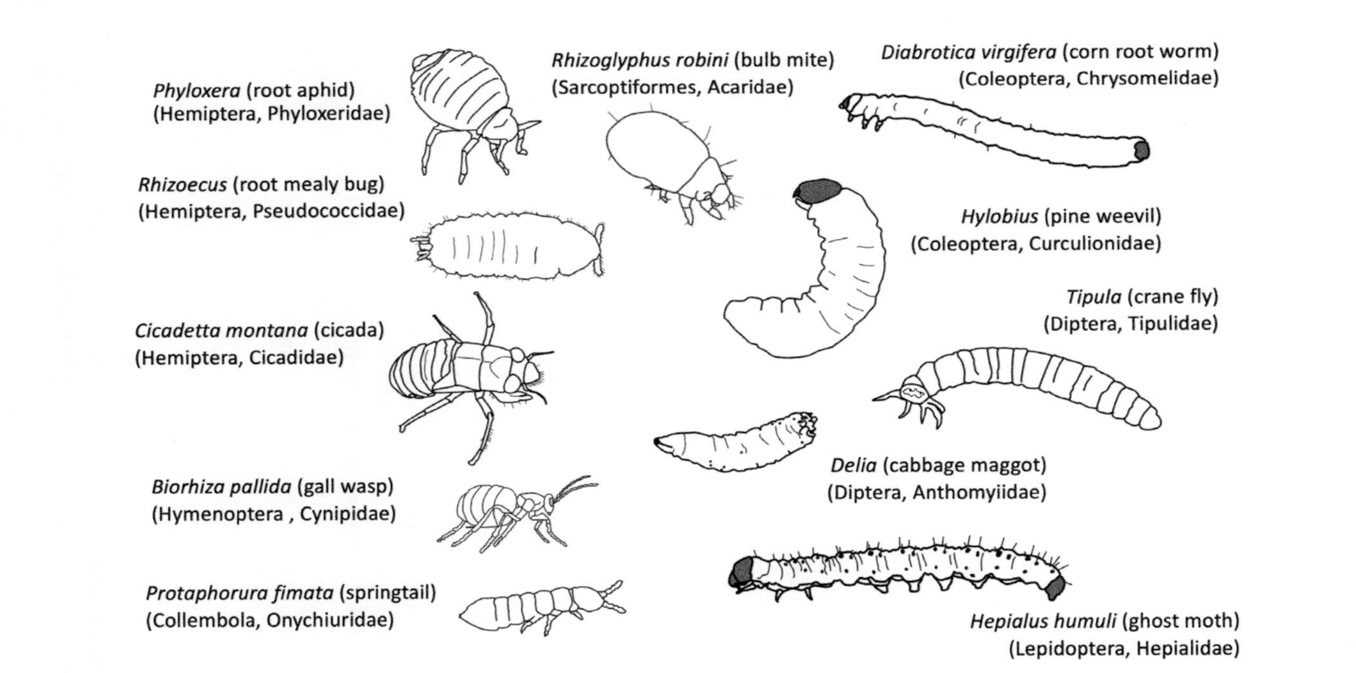

Figure 1.16. Bestiary of root herbivores (not drawn to the same scale), illustrating the diversity. Specialized root-feeding evolved in many different arthropod groups independently.

Table 1.6. The five groups of fungi that engage in mutualistic symbioses with plant roots. The classification is based on the morphology of contact structures. From Audet (2012) and https://mycorrhizas.info.

Category of mycorhiza	Morphology	Fungal group	Plant hosts
Arbuscular mycorrhizae	Arbuscules and vesicles in plant cells	Glomeromycota	Vascular plants
Ectomycorrhizae	Hyphal sheet on lateral roots, labyrinthine hyphae penetrating root cells	Basidiomycota	Trees
Orchid mycorrhizae	Coils of hyphae (pelotons) penetrating root cells	Basidiomycota	Orchids
Ericoid mycorrhizae	Coiled hyphae in association with hair roots	Ascomycota	Ericaceae
Subepidermal mycorrhizae	Fungal hyphae in cavities under epidermal cells	Unclear group	*Thysanotus* (perennial herb of family Asparagaceae)

chemical antagonism, parasitism and mutualism dominate the community more than abiotic gradients.

The study of interactions within this community, and especially the consequences for aboveground plant growth is a lively field of study, that arose since around the beginning of this century, when ecologists discovered the many causal links between belowground and aboveground ecological processes (De Deyn et al. 2003, Wardle et al. 2004, Van der Heijden et al. 2008, Philippot et al. 2013).

A prominent biotic component of the rhizosphere community is formed by fungi that interact with plant roots in some form of symbiosis. These symbiotic interactions can be *commensal* (the fungus profits from plant exudates without causing damage), *parasitic* (the fungus profits from plant tissues and causes damage), or *mutualistic* (both fungus and plant profit from the symbiosis). The mutualistic symbiosis is called *mycorrhiza*. Phosphate and ammonia are the basic commodities provided by fungi, while sugars (hexoses) are provided by the plant (Parniske 2008). The plant profits from the extensive contact area of fungal mycelium with the soil, as well as the action of specific phosphate chelating compounds produced by fungi. The fungus profits from the near unlimited provision of carbohydrates from plant photosynthesis. It is estimated that no less than 80% of plant species profit from interactions with fungi and the mycorrhizal association is estimated to date back to the early Devonian, 400 million years ago (Strullu-Derrien et al. 2018). It might have played a major role in the terrestrialization of plant life.

There are five different evolutionary lineages within the kingdom of Fungi in which a significant part of the species is mycorrhizal (Table 1.6). One of them, Glomeromycota, is most abundant and this group was raised to the phylum level in 2001 (Schüßler et al. 2001). These fungi induce mycorrhizae that are characterized by specific structures in the plant roots, called *arbuscules*, which serve a role in exchange processes between the fungus and the plant, hence the fungi are called *arbuscular mycorrhizal fungi*, abbreviated AMF.

The study of mycorrhizae has developed into a model for the evolution of symbiotic interactions in general (Kiers and Van der Heijden 2006). An important

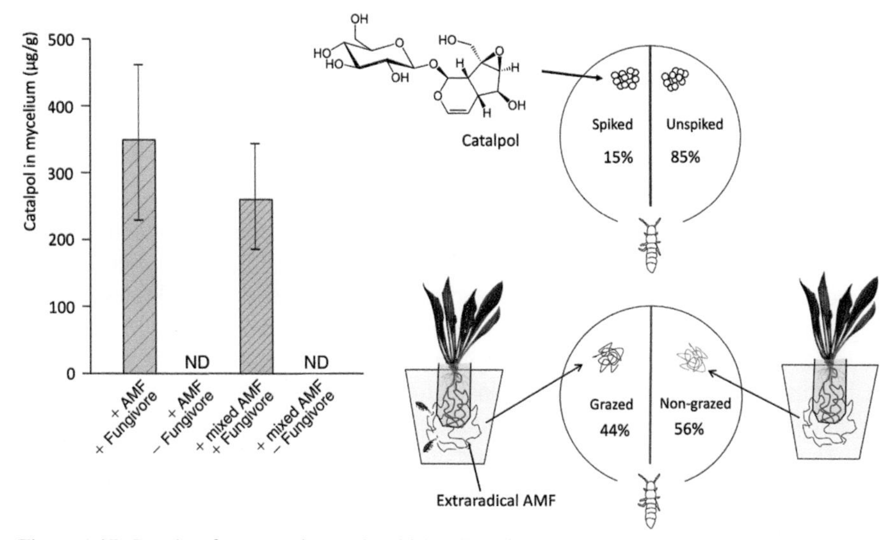

Figure 1.17. Results of pot experiments in which springtails (*Folsomia candida*) were allowed to graze on extraradical mycelium of *arbuscular mycorrhizal* fungi (*Rhizophagus irregularis*) in symbiosis with roots of *Plantago lanceolata*. Plant roots were excluded from the bulk soil by a gauze. Left: the iridoid glucoside catalpol (shown in structural formula) is measured in the mycelium only when grazed by springtails. Right: springtails prefer unspiked food and ungrazed mycelium when offered a choice. AMF = *arbuscular mycorrhizal* fungus, mixed AMF = *R. regularis* plus *Glomus custos*. Data from Duhamel et al. (2013) and unpublished work by Roel Pel.

conclusion from this work is that mutualisms are stabilized not only by reciprocal rewards, but also by the ability, in both partners, to recognize and exclude cheaters. The work of Toby Kiers at Vrije Universiteit Amsterdam (Kiers et al. 2003, Kiers and Denison 2008, Kiers et al. 2011) has demonstrated that such mechanisms indeed exist in plant-AMF symbioses and other mutualistic interactions.

Interestingly, the mycorrhizal symbiosis is also influenced by soil invertebrates. Many soil invertebrates feed on fungal mycelium (nematodes, Collembola, oribatid mites, isopods). Fungivory generally diminishes the positive effect of mycorrhizae on plant growth (Warnock et al. 1982, Johnson et al. 2005), a tendency that is counteracted by the production of feeding deterrents, synthesized by the plant and transferred to the fungus for protection against herbivores. It has even been suggested (Duhamel et al. 2013) that plants might select the most profitable AMF strains in this manner, by regulating the supply of feeding deterrents to them. Figure 1.17 summarizes experimental data (Duhamel et al. 2013) that provide evidence for this. Feeding deterrents synthesized by the plant, such as the iridoid glucoside *catalpol*, are transferred to mycorrhizae when these are grazed by fungivores. This provides a mechanism by which plants can “reward” cooperative fungi and “punish” less-cooperative fungi, and thus stabilize the mutualism.

The rhizosphere is a separate ecological unit, different from the bulk soil. It is dominated by plant-microbe interactions much more than the rest of the soil. Biotic interactions of invertebrates with plant roots, and with microbial communities associated with plant roots, play a major role. Soil invertebrates live in a complex rhizosphere community consisting of different evolutionary lineages, each with their own evolutionary adaptations, separately acquired. Many belowground herbivores descend from terrestrial lineages or represent juvenile stages of terrestrial species. These points have to be taken into account when we explore the evolution of soil invertebrates in Chapter 2.

1.7 Microbial communities

The major nutritional resource for soil invertebrates is dead organic matter, including the heterotrophic bacteria and fungi associated with it. Organic matter is degraded by microbes, facilitated by fragmentation and predigestion by soil invertebrates. But the degradation also gives rise to new organic matter, metabolic products and leachates, of which some polymerize to high-molecular weight compounds, the ingredients of humus. In addition, the soil contains autotrophic bacteria and Archaea that add organic matter from carbon dioxide and nitrogen gas.

The richness of this complex and unseen microbial community remained cloudy for a long time. Seven "*grand questions*" were formulated in 1927 by the Ukranian-American microbiologist and Nobel prize winner Selman Waksman, but for each of these questions significant gaps of knowledge still exist today (Nannipieri 2020). Since about the year 2000, new DNA sequencing technology applied to soil communities as a whole ("*metagenomics*") has blown-up the estimates of biodiversity by orders of magnitude, especially among bacteria (Hug et al. 2016). Estimates of global biodiversity now indicate that ¾ of all species on earth are bacteria (Larsen et al. 2017). Needless to say, soil invertebrates live in a microbial world.

The microbiome of a soil is not a single entity, there are many different microbiomes depending on soil properties and climatic factors. Large-scale inventories have shown that the most decisive factor for total microbial biomass is soil humidity (Fierer 2017, Delgado-Baquerizo et al. 2018). Bacterial biodiversity is highest in the temperate zones of the world; it decreases towards the poles and also towards the equator. For fungi this trend is reversed: the temperate zones have the lowest biodiversity (Bahram et al. 2018). The negative correlation between bacteria and fungi is mainly due to their differential responses to soil pH.

Regarding specific taxonomic groups, the largest number of bacterial DNA sequences in soil is due to the phylum Acidobacteria (Figure 1.18). This came as a surprise to many microbiologists. The first acidobacterium was discovered in 1991 and up to the year 2003 only three lineages were known (Quaiser et al. 2003). Microbiologists never expected it would turn out to be the most abundant bacterial group in soil ecosystems. They are slow-growing bacteria, difficult to study by traditional microbiological methods, and very much underrepresented in cultures.

Soil pH has a major influence on the composition of bacterial communities: the contribution of Actinobacteria and Bacteriodetes decreases with decreasing pH, while Acidobacteria increase and Proteobacteria reach their highest abundance at circumneutral pH (Lauber et al. 2009). Apart from these main trends, the composition of a microbial community varies significantly from one place to another: there is biogeography in microbiology. This was another surprise to microbiologists, who for good reasons had always assumed that microbes are everywhere.

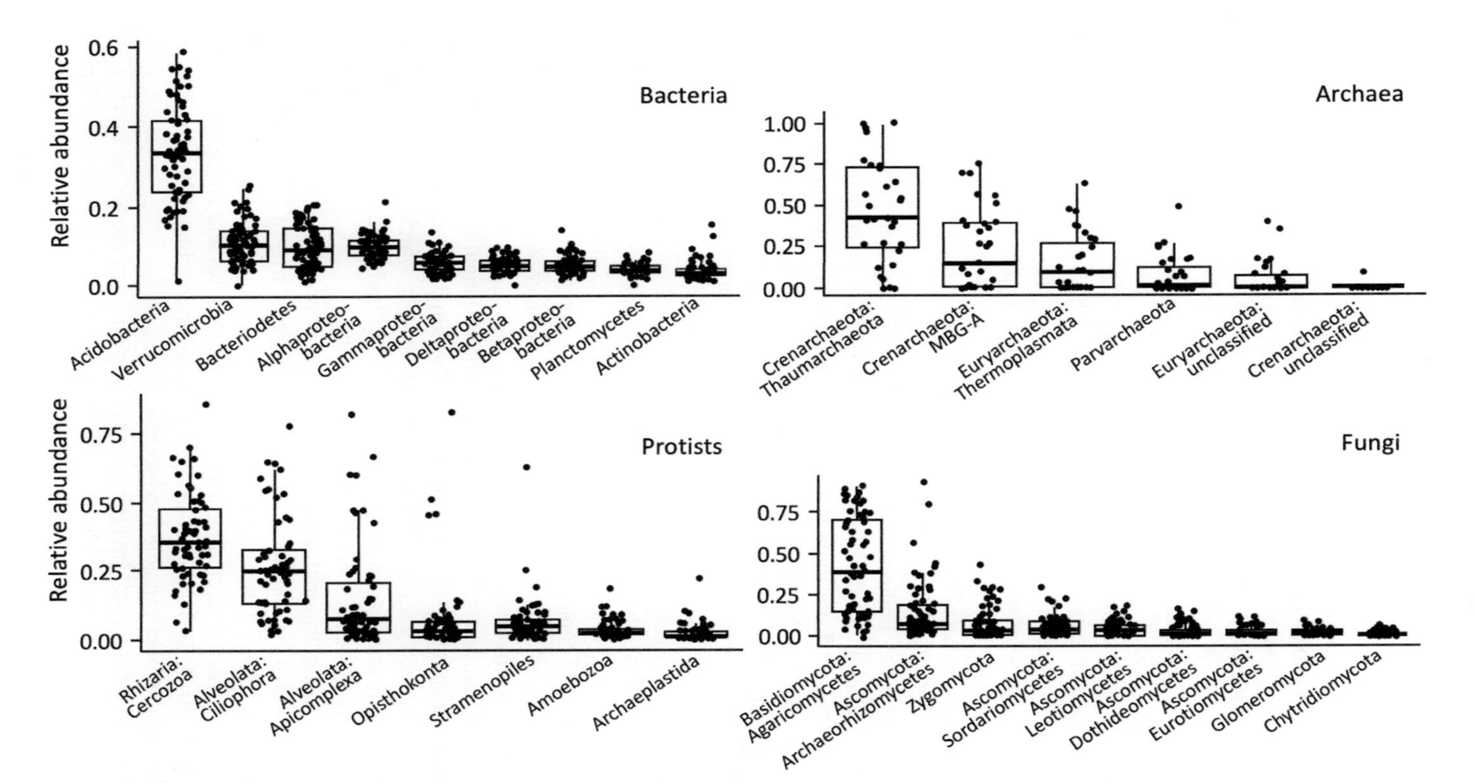

Figure 1.18. Survey of microbial diversity in soil ecosystems throughout the world, classified by main groupings. Abundance is expressed as the relative contribution of DNA marker sequence reads to the total number per regnum (Bacteria, Archaea, Protista and Fungi). The bacterial and archaeal markers were 16S rRNA genes, for protists it was the 18S rRNA gene and fungi were classified based on ITS1 sequences. Data for individual samples are shown; the variation is indicated by box plots. Redrawn with modifications from Fierer (2017), based on data from Crowther et al. (2014).

Archaea in soil are dominated by the phylum Crenarchaeota (Figure 1.18). This group was initially known only from extreme environments, especially in sulfaroles, but it is now shown to be very common in soils as well. The phylum Euryarchaeota includes the ecologically important lineage of methanogens, the only organisms that can generate methane from carbon dioxide or acetate. It was assumed for a long time that microbial methanogenesis could only occur in anoxic sediments, but it is now known that well-oxygenated wetlands emit significant amounts of methane (Angle et al. 2017). Obviously, Archaea as a whole is an understudied group and their possible contribution to soil functions is only beginning to be explored.

Among the microbial protists, most DNA sequences are from Cercozoa (heterotrophic flagellated and amoeboid forms) and ciliates (Ciliophora). They represent a large part of the bacterivorous functions in soil, in addition to animals such as nematodes. Interestingly, soil ecosystems also harbour quite a number of Apicomplexa, a phylum of unicellular organisms to which the human parasites *Plasmodium* and *Toxoplasma* belong. Whether soil apicomplexans are all parasites, e.g., on earthworms or other animals, is unclear at the moment. Apicomplexa are another enigmatic member of the microbial community surrounding soil invertebrates (Del Campo et al. 2019).

Among fungi, the highest abundance is due to basidiomycetes; in many cases these constitute more than half of all the fungal DNA sequences. We know basidiomycetes mostly from their aboveground fruiting bodies (mushrooms), but DNA analysis indicates that their mycelia also dominate the belowground fungal biomass. Ascomycota is the second most abundant group. Glomeromycota (to which the arbuscular mycorrhizal fungi belong) are not among the most common representatives, despite their ecological relevance (Figure 1.18).

How do soil invertebrates deal with this enormous diversity of microbial life around them? The first observation is that microbial biomass, specifically fungal biomass, makes a good food source for an animal. Table 1.7 summarizes some data on the main nutritional components of fungal biomass, compared to plant and animal biomass. Fungi contain relatively high amounts of nitrogen and protein compared to plants, while ascomycetes contain even slightly more nitrogen and phosphorus compared to basidiomycetes. The lipid concentration of fungal biomass is variable; it depends very much on the species, but in general it is higher than in plants, though lower than in animals. Fungi are able to synthesize sterols, which many insects are unable to do. Fungi are also known for their relatively high content of potassium, magnesium and trace metals such as zinc and copper.

By and large, the nutritional composition of fungi is more like animal tissue than like plant tissue, which explains to a certain extent the popularity of fungivory among soil invertebrates. Specialized fungus-invertebrate relationships, up to obligate mutualisms, have evolved, including leaf-cutter ants, termites and ambrosia beetles (Stokland et al. 2012, Birkemoe et al. 2018). These invertebrates have perfected fungal culture as the most valuable food source.

The challenges facing soil invertebrates feeding on fungi are not so much with the nutritional composition but with the many feeding deterrents and toxins in fungal biomass. The attractiveness of fungi is counteracted by a host of chemical defence mechanisms (*mycotoxins*). The diversity of these substances, sometimes even

Table 1.7. Ranges (lowest and highest values reported) of the nutritional composition of fungal biomass. compared to plants and invertebrate animals. From Christias et al. (1975), Wassef (1977), Swift et al. (1979), Lavy and Verhoef (1996), Lavy et al. (2001), Zhang and Elser (2017) and Ma et al. (2018). Values are in percentages of dry weight.

	Ascomycota (mycelium)	**Basidiomycota (mycelium)**	**Invertebrates (whole body)**	**Plants (leaves)**
Carbon	42.4 – 50.3%	39.7 – 44.2%	37.6% – 41.8%	43.0 – 50.3%
Nitrogen	1.0 – 8.0%	1.1 – 5.0%	7.7 – 10.5%	0.56 – 1.31%
Phosphorus	0.10 – 1.55%	0.10 – 1.00%	1.1 – 6.9%	0.03 – 0.32%
Protein	22.7 – 39.0%	19 – 35%	21 – 70%	2 – 9%
Lipid	1.2 – 45.0%	0.9 – 49.1%	7.1 – 23.7%	2 – 8%

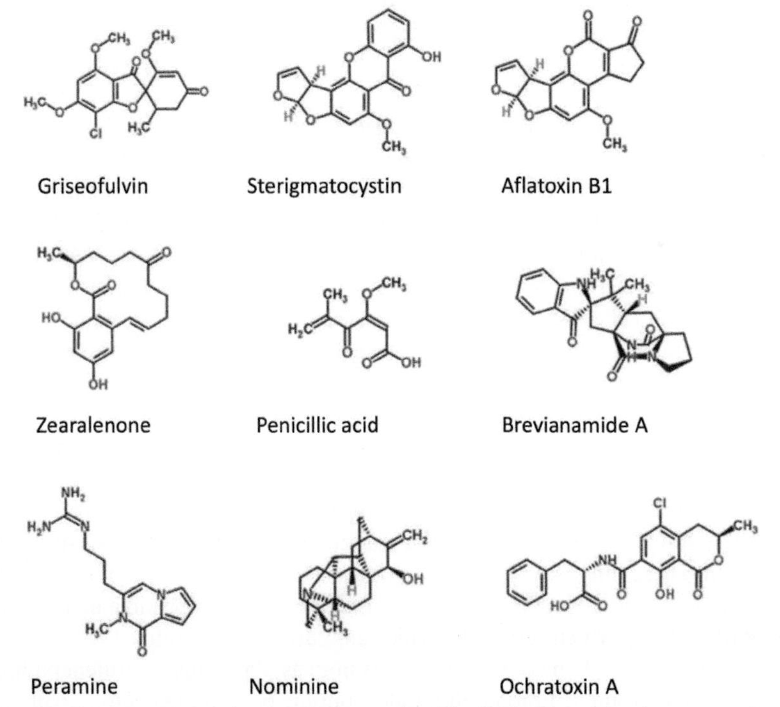

Figure 1.19. Examples of secondary metabolites produced by fungi with anti-fungivore properties. From Rohlfs (2015), with modifications.

produced by the same fungal colony, is tremendous. Figure 1.19 shows a number of the better-known compounds. The aflatoxins produced by *Aspergillus* species are a case in point. The LC_{50} value of aflatoxin B1 to honeybees was estimated as 6.8 μg/g in the diet (Niu et al. 2011), suggesting it could be quite toxic to at least some invertebrates in soil. However, the hairy fungus beetle *Typhaea stercorea* (Mycetophagidae) is able to complete its entire life cycle while feeding on *Aspergillus flavus* (Rohlfs 2015). There are extremely large differences between species in their susceptibility to mycotoxins, which are due to the activity and inducibility of metabolic pathways that degrade the compounds. We will review these adaptations in Chapter 6.

Secondary metabolites of fungi are highly inducible, that is, their production is enhanced by tissue damage, e.g., due to fungivory. In fungi, the biochemical pathway that transduces a wounding signal to a stress response is *oxylipin signalling*. Oxylipins are derived from polyunsaturated fatty acids by oxidation. In fungi there are two categories, non-volatile oxylipins and volatile oxylipins. The non-volatile group mainly influences morphological differentiation, e.g., the formation of conidiophores. The volatile group, mainly C8-compounds, regulates the stress response, including activation of genes encoding stress defence proteins and the biosynthesis of secondary metabolites (Holighaus and Rohlfs 2019).

Oxylipins are widespread in the tree of life. For instance, *jasmonic acid*, a plant hormone, is an oxylipin. The fact that fungi use the oxylipin signalling pathway in their stress response may explain why grazing on mycorrhizae by fungivores can induce the synthesis of feeding repellents by the host plant (*cf.* Figure 1.17). The totality of volatiles emanating from soil is designated as the '*volatilome*"; it is a reflection of the, mostly fungal-driven, metabolic processes in soil and as such could be of diagnostic value (Honeker et al. 2021).

The volatile oxylipins that disperse into air form a typical fungal "bouquet" that is specific to the species but also reflects the physiological state of the fungal mycelium. The volatiles can be perceived by invertebrates and serve as "*infochemicals*", i.e., chemicals that inform the fungivore about the phenotype of the fungus (Holighaus and Rohlfs 2019). Choice experiments have shown that Collembola are attracted by these odours and can discriminate between different species of fungus in this way (Bengtsson et al. 1988, Bengtsson et al. 1991, Hedlund et al. 1995). Moreover, when fungal mycelium is stressed by toxicants in the medium, e.g., heavy metals, this changes the fungal odour and consequently the palatability to fungivores (Bengtsson et al. 1985a).

In addition to fungi producing potent toxins, we also know several fungi that are right away pathogenic, i.e., they are able to penetrate the cuticle of an invertebrate, grow inside the body, kill the host and leave the cadaver to disperse. This is especially known for so-called entomopathogenic fungi, which attack insects, spiders and harvestmen. These fungi have received great attention in studies of biological control, since they are potentially able to suppress insect pest populations in a natural way. A number of well-known entomopathogenic fungi are listed in Table 1.8. Five groups within the fungal kingdom contain entomopathogens. Particularly virulent pathogens are found in the group of Entomophthorales, but the best studied entomopathogenic fungi are Ascomycota. To this group belong genera like *Metarhizium* and *Beauveria*, which are used in organic farming to control pest insects. Entomopathogenic "fungi" are also found outside the Fungi proper, within the Oomycetes (water moulds). In total there are no less than 1604 entomopathogenic fungi known to date (Table 1.8). How soil invertebrates respond to parasites and pathogens (fungi, bacteria and viruses) will be discussed in Chapter 6.

As a final example of the belowground chemical warfare we mention the case of toxins produced by *Bacillus*, *Vibrio* and *Clostridium* species. Botulin toxin A produced by *Clostridium botulinum* is the most potent toxin that we know. *Bacillus anthracis* produces a hardly less dangerous toxin called *anthrax*. *Vibrio cholerae*

Table 1.8. List of fungal groups and genera known for their entomopathogenic representatives. From Scholte et al. (2004), Araújo and Hughes (2016) and Litwin et al. (2020).

Phylum	Estimated number of entomopathogenic species	Examples of genera
Oomycetes	12	*Leptolegnia, Pythium, Lagenidium, Crypticola*
Chytridiomycota	65	*Coelomomyces*
Microsporidia	339	*Pleistophora, Cystospogenes, Nosema*
Entomophthoromycota	474	*Furia, Conidiobolus, Entomophaga, Erynia*
Basidiomycota	238	*Uredinella, Septobasidium*
Ascomycota	476	*Metarhizium, Beauveria, Isaria, Cordyceps, Lecanicillium, Paecilomyces*
Total	**1604**	

produces a range of toxins causing serious gastroenteritis in humans. While these bacteria can be reckoned as true members of soil microbial communities, the question may be asked whether soil invertebrates are affected by them. Despite the fact that we know a lot about the pathogenicity of these bacteria to vertebrates, their ecological role in the soil ecosystem is unclear.

The bacterium *Bacillus thuringiensis* is known for its insecticidal proteins called *Bt toxins*. This group includes the so-called *Cry-proteins* ("crystalline proteins"), of which there are no less than 174 different ones (Van Frankenhuyzen 2009). The proteins, when ingested by an animal, are activated by digestion in the gut and bind to receptors in the gut epithelium, after which they cause perforations in the cell membrane and finally destruction of the gut (Soberón et al. 2009). Thanks to their high activity against certain insects, Cry proteins are often used as biological insecticides. Another well-known application is to transform plants with Cry-encoding genes which will provide resistance against pest insects.

Although the Cry proteins are highly toxic to several terrestrial insects, there is no evidence that in natural soils they have significant effects on soil invertebrates such as earthworms and nematodes (Zeilinger et al. 2010, Höss et al. 2013). They are also hardly toxic to Collembola (Yang et al. 2015). This is likely due to the absence of specific receptors in these animals but this remains unknown to date. Effects of Bt toxins on root herbivores, of which many descend from aboveground terrestrial groups (*cf.* Figure 1.16), cannot be excluded. The interaction between bacterial toxins and the soil fauna is still underexplored.

In summary, the microbial communities in soil offer an important food source for soil invertebrates, but they come with sophisticated chemical warfare and virulent pathogens in some cases. In addition to the selective factors associated with humidity and temperature and on top of the competition and predation pressures in the rhizosphere, soil invertebrates have to deal with the chemical environment created by microbes. How they have succeeded in adapting to this threefold selective pressure will be discussed in consequent chapters.

CHAPTER 2

Evolution of Terrestrialized Invertebrate Lineages

Tardigrades

Are there anaerobes in Venus' haze,
are there bacteria on the moon,
are there on Mars microbes in a lagoon.
Are there tardigrades in space?

Are they from every danger immune,
are they evolving fast in a high-brow race,
are they all damn suckers in any case.
Are they close by or as far as planet Dune?

What if they are programmed to erase,
what if they are the size of a baboon,
what if they are to moult here and re-phase?

Then we must brave the coming of The Green Goon.
Then we must abide a terrible coup de grace.
Tart and great, they sweep us out very soon...

Oh God!

Jasper Aertsz

CHAPTER 2
Evolution of Terrestrialized Invertebrate Lineages

"That's all right, bless you!" responded the Rat cheerily. "What's a little wet to a Water Rat? I'm more in the water than out of it most days. Don't you think any more about it."

Kenneth Grahame, The Wind in the Willows

In agreement with the basic premise of this book, we shall start our exploration of soil invertebrates in the water. Knowing that animal life started in the sea, invertebrates have conquered the land from an aquatic (marine or freshwater) habitat, a process called *terrestrialization*. But unlike terrestrial vertebrates (amphibians and tetrapods), soil invertebrates do not all descend from the same terrestrialization event. How many of such events have there been in evolutionary history and when did they take place? Did they all go together, or one after the other?

Terrestrializations were accompanied by significant changes in anatomy and physiology (Little 1983). To deal with the new conditions, adaptations were required in osmoregulation, respiration, transpiration, temperature regulation, reproduction, development, food acquisition, digestion, metabolism, predatory escape and pathogen defence. Not all invertebrates succeeded in overcoming all these requirements, as we will see in the Chapters to come. Some groups only solved the problem of osmoregulation and were able to live in freshwater, groundwater and cave habitats, but never made the transition to a fully terrestrial life which requires at least a certain degree of drought resistance and air-breathing. In permanently humid ecosystems such as tropical rain forests, animals showing "incomplete terrestrialization" can nevertheless make out a conspicuous part of the aboveground fauna (e.g., leeches, snails, amphipods).

Due to the complex adaptations necessary for complete terrestrialization, it is not surprising that in many cases the marine-freshwater-terrestrial transition was irreversible and happened only once in a lineage. However, extensive marine regressions have occurred in some groups, as we will see below. Common to all terrestrializations is that it allowed invertebrates to undergo enormous adaptive radiations, leading to new body plans and new evolutionary lineages.

In this Chapter we will not analyse the various adaptations themselves, but document the terrestrializations as events in evolutionary history and use them to test Ghilarov's hypothesis, as pictured in Figure 1.1. Ghilarov's hypothesis predicts that:

- Above-ground invertebrates show traits that evolved in their ancestors as adaptations to soil-living.

- Soil-living invertebrates show traits that they share with their freshwater or marine ancestors.
- Soil invertebrates are younger than their marine ancestors but older than their above-ground evolutionary descendants.
- Soil invertebrates are clustered in the phylogeny of their group in between marine or freshwater groups and above-ground terrestrial groups.

In this Chapter we will explore the invertebrate terrestrializations using phylogenetic analysis. The scenario that we are looking for is the following. In case of a terrestrialization, there will be a monophyletic lineage of mostly terrestrial species that clusters within a larger group of freshwater or marine species. If species have secondarily returned to aquatic life, we should find aquatic habitat use at the tips of a tree, clustered inside a basically terrestrial group. By plotting habitat use on a phylogeny of the group such patterns may be revealed. In cases where a molecular clock is available, the terrestrialization may be dated in time.

2.1 Animals of the pore water: nematodes and tardigrades

We start our analysis by considering the fauna of pore water and groundwater. These animals are true soil invertebrates but they have not always evolved specific adaptations to life outside the water phase, instead they live in soil pores and subsoil microhabitats as if these were aquatic or marine habitats. Among the pore water invertebrates, we reckon nematodes, tardigrades, and several groups of groundwater crustaceans. In this section we discuss nematodes and tardigrades, along with their evolutionary sister groups, nematomorphs and onychophorans, *resp.*

The phylum Nematoda is a very large monophyletic lineage of worm-like, mostly small animals with a chitin cuticle, comprising no less than 27,000 described species (Figure 2.1). Their development passes through discrete stages and so in order to grow they need to moult, like insects. It is one of the main groups classified under the supergroup Ecdysozoa ("moulting animals"), to which also arthropods and tardigrades belong.

Nematodes are ubiquitous in any soil and sediment and represent one of the most abundant animal groups on earth in terms of numbers. They are so numerous that the U.S. "father of nematology", Nathan Cobb, in 1915 remarked that if you swept away the universe with all matter except the nematodes, you would still dimly see our globe, and all its mountains and valleys, as a film of nematodes (Huettel and Golden 1991).

Classical taxonomy of the phylum, based upon morphological traits, has always been very difficult because the discriminatory characters are microscopical and often rather subtle. In addition, there is a good deal of convergence, especially in the mouth parts, so morphology often puts you on the wrong track. The works of Blaxter et al. (1998) and Holterman et al. (2006) have revealed the true phylogeny of the phylum, based on 18S rRNA gene sequences. Nevertheless, several inconsistencies and unresolved divergences remained, which could not be resolved by rRNA sequences alone. More recent phylogenomic and transcriptomic analyses have improved upon

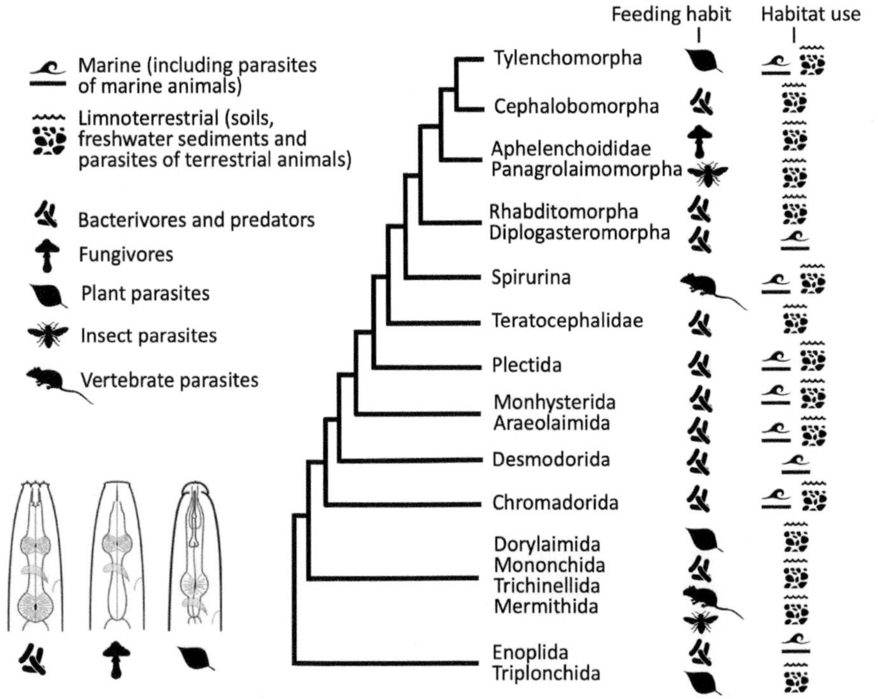

Figure 2.1. Simplified phylogeny of the phylum Nematoda on the basis of 18S rRNA gene sequences, indicating the twelve main clades and the evolution of feeding habits, as well as habitat use. On the left three types of nematode mouthparts are shown in relation to feeding groups, according to Zaborski. The phylogeny was modelled after Bert et al. (2011), Holterman et al. (2019) and Schratzberger et al. (2019). Nematode mouthparts figure credit Edmond R. Zaborski, University of Illinois.

this (Ahmed et al. 2022). The backbone of the nematode tree appears now to be stable across studies. A simplified phylogeny is given in Figure 2.1 (Bert et al. 2011).

The nematode backbone tree shows twelve evolutionary lineages, each a monophyletic group with several families. Mapping the feeding habits of nematodes (judged from the mouthparts) upon these lineages reveals that switches of feeding habits have occurred more than once. Most likely bacterivory is the ancestral feeding habit, while plant-parasitic and animal-parasitic forms have evolved several times. In a similar fashion, habitat use is irregularly distributed over the tree. There is not a single clade of terrestrial species clustering inside an aquatic or marine clade. In fact, Holterman et al. (2019) have estimated that no less than 30 independent transitions between marine and limnoterrestrial habitats have occurred during nematode evolution (and the other way around). Nematodes have been qualified as "*habitat commuters*" (Holterman et al. 2008, 2019, Schratzberger et al. 2019).

The absence of a single terrestrialization event in the phylum Nematoda and the frequent ecological switches to and fro, point at an inherent flexibility of the body plan. The water-permeability of the cuticle and the relatively simple secretory-excretory system may have contributed to that. The multiple origin of parasitic lineages is another indication of ecological flexibility. Alternatively, one might argue that the colonization of the terrestrial environment did not require major biological changes, as the animals, also in soil and moss cushions, continued to live in a water film. Anyway, the situation in nematodes is very different from flatworms and annelid worms, as we will see below.

That all "terrestrial" nematodes still live in the pore water does not exclude that some species are markedly drought-resistant during part of their life. For example, the model species *Caenorhabditis elegans* knows a resistant larval stage called "*dauer*", which is a stage-3 larva with arrested development and specific adaptations for dispersal. The *dauer* larva is induced by food shortage and is long-lived. The gene expression programme characteristic for the dauer can also (by mutation) be induced in the adult, which then shows a greatly extended lifetime (McElwee et al. 2004). The dauer larva has a sealed buccal opening, shrunken muscles and arrested germline development. It also shows "*nictation*": it crawls out of the soil, up along grass stems and starts swaying with its body from the tip of the vegetation, often in swarms. This behaviour enhances the likelihood of being blown somewhere else by the wind.

Also, eggs and larvae of parasitic nematodes that disperse outside their host can be quite tolerant to atmospheric, out-of-soil, conditions. For instance, parasites of the genus *Strongyloides* have a normal sexual life-cycle in the soil, but their filariform larvae can emerge from the soil and penetrate the skin of mammals. Eggs of the pinworm *Enterobius* are easily dispersed under nails and on the hands of children. However, these drought-resistant specializations, allowing life outside the soil, are not clustered in the phylogeny. That is why we may conclude that nematodes, assuming they originated in the marine environment, have conquered freshwater and terrestrial habitats many times, but without much morphological change.

The phylum Nematomorpha, or *horse-hair worms*, is a sister group of Nematoda. The two phyla form a monophyletic lineage in the superphylum Ecdysozoa. Nematomorphs are long and thin worm-like animals, all parasites of terrestrial arthropods such as ground beetles, crickets, grasshoppers and millipedes. They live curled up in the gut or the body cavity, their total length being much larger than the body length of their host. They mature inside the host but reproduction is obligatory in freshwater. To achieve this, the worms impose behavioural changes that cause the arthropod to seek out freshwater habitats. Once in the water, the worms are triggered to leave the host through the anus. The worm larvae hatching from aquatic eggs infect insect larvae with an aerial adult stage, e.g., midges. This host, called *paratenic,* effectively transports the parasite from the aquatic environment back to the terrestrial environment where it may find a new arthropod host.

The genera *Gordius* and *Gordonius* are common nematomorph representatives. They are named after the tangle of worms mating in the water, looking like Alexander the Great's gordian knot. Since reproduction takes place in freshwater, nematomorphs should be considered aquatic animals that live part of their life as a parasite in a terrestrial host. In very wet environments, e.g., tropical cloud forests, adult nematomorphs are also seen to reproduce on the ground (Chiu et al. 2020). We will revisit these worms and other parasites in Section 6.9.

Tardigrades ("water bears") represent one of the most enticing soil invertebrates, once you have seen their droll crawling behaviour under the microscope. With 1190 different species described in 2013 and an estimated total diversity of 2654 species it is a relatively small phylum, although numerous (Nelson 2002, Bartels et al. 2016).

The animals are characterized by four pairs of fleshy legs, which end in claws. The fine morphology of the claws, and the buccopharyngeal apparatus have been used for taxonomic delineation. Several tardigrades have Malpighian tubules, a trait which links them to hexapod arthropods such as proturans and insects. It seems unlikely that this is true homology although the similarity is very large (Møbjerg and Dahl 1996).

A simplified phylogeny of the phylum, based on DNA sequences, is provided in Figure 2.2 (Guil and Giribet 2012, Guil et al. 2019). It shows that marine tardigrades cluster in one of the clades (Arthrotardigrada and Echiniscoidea, taken together as Heterotardigrada). As this is also the most ancestral group, relative to outgroups, this confirms that tardigrades originated in the marine environment and have moved onto the land after the split between Heterotardigrada and Eutardigrada (indicated by a star in Figure 2.2). All Eutardigrada are limnetic (freshwater-living), or soil species. There are only few species (62 are mentioned by Garey et al. 2008), that are exclusively found in freshwater; most of the species are truly terrestrial. However also in the terrestrial environment, tardigrades live in wet microhabitats such as moss cushions and lichens. In the soil, they live in water films around soil particles.

Apart from this major terrestrialization of Eutardigarda, there are also land colonizations in Heterotardigrada. Several species of the genus *Echiniscoides* are intertidal, living in barnacles and shoreline sea weeds (Kristensen and Hallas 1980, Faurby et al. 2012). Other echiniscoids live in moss, in coexistence with eutardigrades (Rebecchi et al. 2008). The terrestrialization in the Echiniscoidea likely occurred separate from the major event in the eutardigrades (Figure 2.2).

From the viewpoint of the tardigrade the differences between aquatic and terrestrial habitats are minimal. The splits between soil and freshwater species are on the species level and so can be seen as evolutionary specializations, not as main splits

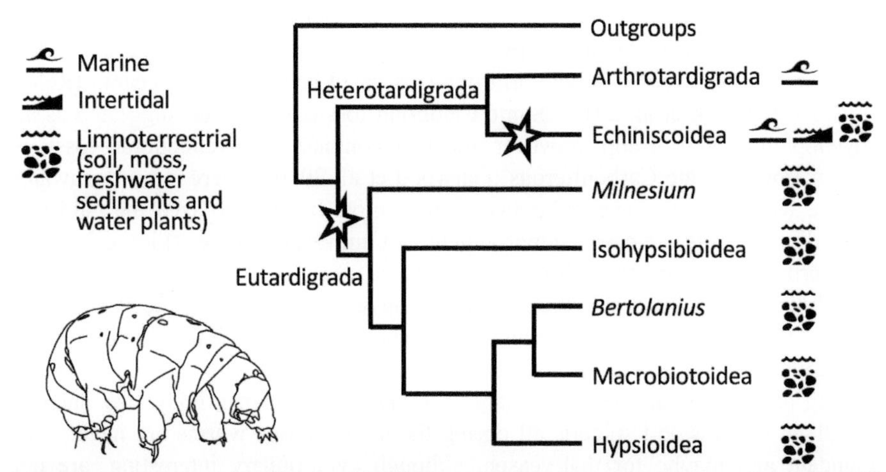

Figure 2.2. Phylogeny of Tardigrada, with only the main splits shown. The phylogeny is based on 18S rRNA, 28S rRNA and COI sequences. The drawing, after a SEM image on Wikimedia Commons, shows *Milnesium tardigradum* (body size 0.2 – 0.7 mm). Phylogeny and habitat information was compiled from Guil and Giribet (2012), Guil et al. (2019), Marley et al. (2011), Czechowski et al. (2012), Nichols et al. (2006), Garey et al. (2008) and Rebecchi et al. (2008). Two putative (undated) terrestrializations are indicated by asterisks.

in the phylogeny (Figure 2.2). That tardigrades colonized the land only few times, maybe only twice, and hardly ever returned to the marine environment, distinguishes them from nematodes, which is striking, as the habitats of these two groups are very comparable.

In some species of terrestrial tardigrade, the legs are shortened and the claws reduced, possibly as an adaptation to the interstitial environment (Nelson et al. 2018). However, there are no unique characters distinguishing soil tardigrades. Like in nematodes the colonization of soils opened up a great diversity of ecological niches which allowed diversifying speciation, without profound changes in the body plan. This contrasts greatly with the situation in arthropods, as we will see below.

Like nematodes, tardigrades know a stage of developmental arrest, with extreme drought resistance. When slowly drying out, the animal may lose all its free water and becomes resistant not only against desiccation but also against frost and even vacuum. In this state of *anhydrobiosis* it can survive for many years. Tardigrades are considered the toughest animals on earth. They are even used by NASA as a model for biological survival in space (Erdmann and Kaczmarek 2017). Anhydrobiosis is also known in nematodes, rotifers and Collembola. We will explore the genetic and physiological background of this phenomenon further in Chapter 6.

Related to the phylum Tardigrada are the enigmatic Onychophora or *velvet worms*. They share several features with tardigrades, especially the lobopods and the claws, although their appendages are more numerous than those of water bears. Velvet worms are also much larger (a few cm, compared to the microscopic tardigrades). The animals have some typical arthropod traits, such a (thin) chitinous exoskeleton, absence of a coelom and a similar organization of the ventral nervous system. In the past, velvet worms were considered an evolutionary transition between annelids and arthropods, but that view is outdated now. Molecular-phylogenetic analyses place them firmly with tardigrades and arthropods in the Panarthropoda superphylum, with onychophorans more related to arthropods than tardigrades (Edgecombe 2010).

The phylum Onychophora most likely represents a very old lineage of the Ecdysozoa. The divergence with arthropods is dated back in the Ediacaran (Rehm et al. 2011, Erwin et al. 2011). Some Cambrian fossils have been suggested as an onychophoran stem group; however, the first convincing onychophoran fossil is known from the late Carboniferous (Garwood et al. 2016). There is no freshwater lineage within extant onychophorans, so they most likely colonized the land just once, directly from the marine environment, not unlike tardigrades (Figure 2.2) and some arthropod groups to be discussed below.

Unlike their sister group, the tardigrades, onychophorans live at the soil surface, in litter, under stones, in termite tunnels and other crevices. With their flexible body they are able to squeeze themselves through small openings. They are all predators, feeding on invertebrates relatively large compared to their own body size, such as woodlice, crickets and spiders, all during the night. Velvet worms are never very abundant and maybe for that reason, although evolutionary interesting, are not popular animals for ecological studies. However, they do represent a classical showcase of biogeography, as we will see in Section 3.3.

2.2 Groundwater fauna: so many crustaceans

Many animals of the pore water also live in aquifers below the soil, crevices eroded in limestone (karst), subterranean streams and riverine gravel layers. In addition to nematodes and tardigrades, several macroinvertebrates, specifically crustaceans, are home to these environments. The community consists of highly specialized animals, adapted to the spatial restrictions of the underground habitat and the oxygen-poor and oligotrophic environment Gibert (2001).

Animals that are completely restricted to life in groundwater are called *stygobionts*. They all share a set of typical characters: elongated (*vermiform*) bodies, well developed sensory organs, absence of eyes (*anophthalmia*), lack of body pigment, delayed maturity, increased longevity, low rate of reproduction, parthenogenesis, and low metabolic rate. Some animals have these adaptations to a lesser extent; they can live in groundwater but also occur in surface waters and are called *stygophilic*.

The groundwater fauna shows a great similarity to the fauna of caves (*troglobionts*), in fact there are gradual transitions between cave habitats, karstic habitats, gravel habitats, and sediments. The larger the pore size, the more macrofauna species are seen in the groundwater fauna.

Many macroscopic animals have representatives colonizing belowground water, including fish and amphibians. Among the invertebrate groundwater fauna are tiny annelid worms (Martínez-Ansemil et al. 2012) and snails, but crustaceans are the most diverse; they are represented by various taxa: Cladocera, Ostracoda, Copepoda, Syncarida, Isopoda, Amphipoda and Remipedia. Figure 2.3 illustrates some of these creatures. Interestingly, the groundwater crustaceans often look quite different from the typical appearance of their group. On first sight it is even difficult to recognize *Parastenocaris* as a copepod and *Caecidotea* as an isopod. The similarity in body form is obviously the result of convergent evolution.

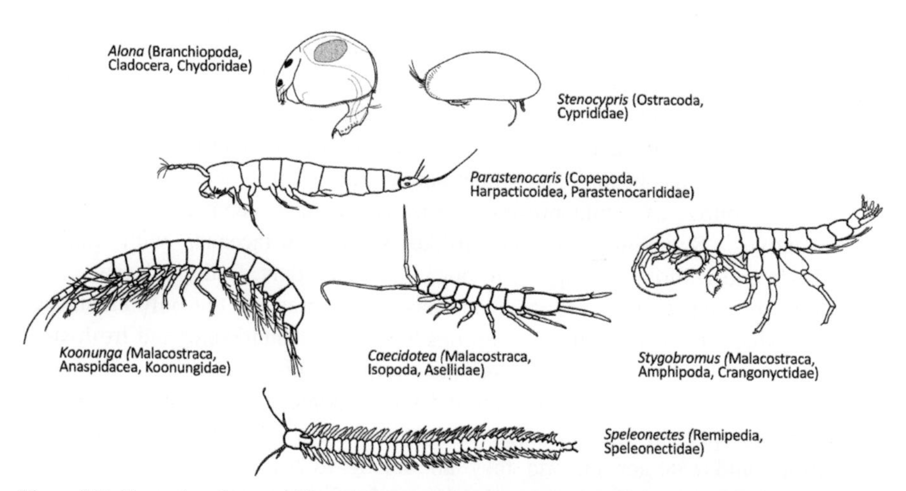

Figure 2.3. Examples of stygophilic and stygobiontic representatives within the Crustacea. Redrawn with modifications from Dumont and Negrea (1996), R.J. Smith (Lake Biwa Museum Japan), Cottarelli et al. (2012), O.A. Sayce (Wikipedia), Dante Fenolio (Science Library), New York Natural Heritage Program and Joris van der Ham (Wikipedia) (not drawn to the same scale).

The transition to subterranean life has evolved many times, independently, within each of the crustacean groups mentioned. Groundwater crustaceans have evolved both out of freshwater ancestors (such as in the cladocerans) and directly out of marine ancestors (such as in the amphipods). Even within the same group, the subterranean adaptations might have occurred independently more than once. For example, Notenboom (1991), in a review of groundwater amphipod evolution, argues that these animals evolved when caves colonized by marine amphipods became isolated from the sea and by geological changes gradually lost their marine character, while the amphipods adapted to become stygobiontic. The author argues that this has taken place at least two times within the Amphipoda, and in addition, *marine regressions* have occurred, by which some stygobiontic amphipods became marine again.

Among Cladocera 94 species have been reported to live in the groundwater, but many more may be discovered (Dumont 1995). Stygobiontic species are found in the family Chydoridae, especially the genus *Alona*. The groundwater chydorids, which are often found in the underflow of rivers, are assumed to have evolved out of benthic freshwater species. This is apparent from apomorphic characters such as the loss of eyes and pigment, however, other morphological structures do not indicate a great degree of specialization but rather represent conserved primitive conditions (Dumont 1995).

Ostracoda comprise approximately 2000 species, but how many of these are restricted to groundwater is not known. Stygobiontic and stygophilic species of ostracods can be found in many families; three families that have several representatives in the groundwater are Cyprididae, Entocytheridae and Candonidae. Figure 2.3 shows a stygophilic ostracod of the genus *Stenocypris*, with its typical two-valved shell, giving it a mussel-like appearance.

The subclass Copepoda is a very large group of crustaceans, with both marine, freshwater, groundwater and parasitic representatives. Among the seven orders the subclasses Calanoidea, Harpacticoidea and Cyclopoidea are best known. The calanoid family Diaptomidae and the harpacticoid families Amereidae, Canthocamptidae and Parastenocaridae have several representatives in the groundwater. The species *Parastenocaris germanica* is one of the few groundwater invertebrates on which experimental work has been done (Notenboom et al. 1992).

Some groups of malacostracan crustaceans also contain groundwater representatives. Syncarida are shrimp-like small crustaceans, with reduced appendages and without a carapace. In one of the suborders, Bathynellacea, all the species are restricted to groundwater and caves; it represents the only crustacean group for which this is true, all other groups have both groundwater and freshwater species. Figure 2.3 shows a species from the group Anaspidacea.

The order Amphipoda contains about 10,000 marine, freshwater, groundwater and terrestrial species. The subterranean species, about 740, are distributed over 36 families and 138 genera, and they all belong to the suborder Gammaridea (Holsinger 1993). Groundwater amphipods may be found in all parts of the world, however, the Mediterranean region in Europe and the East and South parts of North America and Mexico are clearly superior in taxonomic diversity; these two regions alone give home to some 573 subterranean amphipod species, or 77% of the groundwater amphipod fauna. This regional differentiation of biodiversity is explained from the presence of karst formations, absence of glaciation and exposure to marine sources of colonization.

The order Isopoda, best known for the terrestrial woodlice, actually has a wide distribution in marine, freshwater and soil habitats. It is also represented in the groundwater. The families Asellidae, Sphaeromidae and Parasellidae include species that live in groundwater, wells and caves. The genus *Caecidotea* (Figure 2.3) belongs to the family Asellidae, which has also the well-known freshwater species *Asellus aquaticus*.

Finally, we point out the curious Remipedia, a group only discovered in 1981. They are known from landlocked marine caves in the Caribbean region, the Canary Islands and Western Australia. They are typical troglobionts, with many apomorphic characters. The animals don't look like modern crustaceans at all; they bear a superficial resemblance to annelids. However, they also have some typically crustacean traits such as biramous legs and two antennas (Figure 2.3). The uniform segmentation without distinction between thorax and abdomen is reminiscent of Burgess Shale fossils from the Cambrian.

How are all these groundwater crustaceans positioned in Ghilarov's scheme (Figure 1.1)? Figure 2.4 shows a recent phylogeny of Crustacea (Giribet and Edgecombe 2019), developed using phylogenomic analysis. The phylogeny shows that two crustacean lineages have evolved groups that we consider soil (terrestrial) invertebrates: Hexapoda and Malacostraca. However, while Hexapoda is completely terrestrial (save a few marine and several freshwater regressions discussed below) only some groups of Malacostraca became terrestrial (lineages of isopods, amphipods and decapods); these will be discussed later in this Chapter. It is also obvious that groundwater crustaceans are not ancestral to terrestrial lineages. There is only one example where a terrestrial animal seems to have evolved from a groundwater ancestor: the blind amphipod *Niphargus talikadzei* lives in soils of alder bushes in Russia, but looks so much like a groundwater amphipod that a phreatic origin is likely (Friend and Richardson 1986).

The crustacean phylogeny also implies that Crustacea (usually considered a subphylum or a class of Arthropoda) is paraphyletic. Insects derive from crustaceans and do not form an arthropod lineage separate from Crustacea. There is now very good evidence for this reconstruction (Cook et al. 2005, Carapelli et al. 2007, Timmermans et al. 2008, Regier et al. 2010, Rehm et al. 2011, Schwentner et al. 2017, Lozano-Fernandez et al. 2019, Giribet and Edgecombe 2019). From a cladistic point of view, the term *Pancrustacea* (which includes hexapods) should be used rather than Crustacea.

The terrestrialization of hexapods is dated to the beginning of the Ordovician, around 480 million years ago (Rota-Stabelli et al. 2013, Giribet and Edgecombe 2019), while the diversification of Malacostraca is much later, in the Carboniferous (Figure 2.4). These datings depend on the calibration of the molecular clock used and differ somewhat between studies, e.g., the split between hexapods and crustaceans is also dated back to the Cambrian, 520 million years ago (Schwentner et al. 2017).

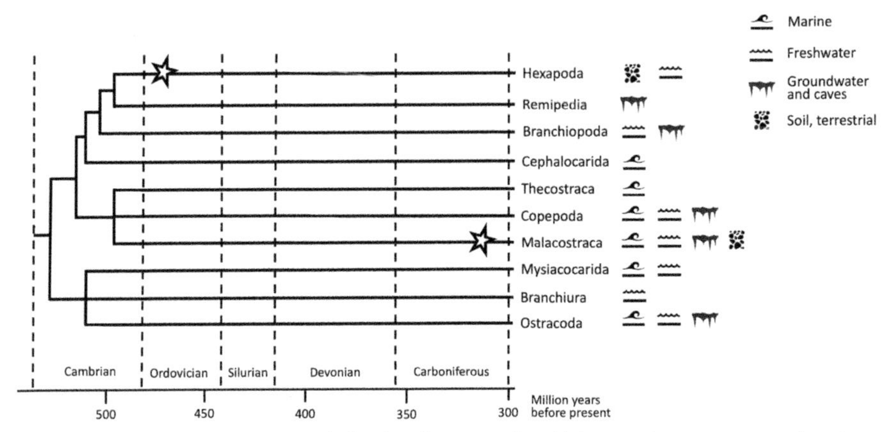

Figure 2.4. Phylogeny of Crustacea indicating the groups in which groundwater representatives have evolved. Tree derived from Giribet and Edgecombe (2019); habitat information from various sources including Notenboom (1991), Holsinger (1993) and Dumont (1995). The stars indicate terrestrializations (not suggesting that all animals in the lineage became terrestrial).

A special case is the sister relationship between Hexapoda and the stygobiontic Remipedia, which is now confirmed by several phylogenomic studies, although not all (Schwentner et al. 2017, Lozano-Fernandez et al. 2016, 2019). This relationship suggests that hexapods evolved from groundwater or cave-living ancestors, related to present-day remipedes. A possible scenario is that during the Cambrian or the Ordovician, geological processes caused coastal seas to be enclosed by land and isolated. These systems gradually developed into caves or brackish karst systems, while the crustaceans split into a lineage that specialized to cave-living and another lineage that managed to colonize the land, when the caves dried up (Glenner et al. 2006).

Overviewing the biodiversity of groundwater crustaceans, a number of conclusions may be drawn.

- The great majority of groundwater species evolved from freshwater, benthic habitats. This seems to be true for all groundwater cladocerans, ostracods and copepods.
- Among malacostracan crustaceans, many amphipods and a few isopods have adopted a life in the groundwater. These lineages most likely evolved from marine populations that became isolated in subterranean habitats like caves.
- Within the amphipods, regression to the marine environment took place in several cases.
- Groundwater-adapted animals do not form an intermediate stage between freshwater and terrestrial lineages, they must rather be considered specialized aquatic animals.
- The presence of a troglobiontic group at the origin of the hexapods suggests that cave life, for this group, preceded the colonization of soils.

2.3 From crustaceans to insects: Collembola and other apterygotes

The evolutionary relationships between wingless and pterygote hexapods have been obscure for a long time. In the past, Protura, Diplura, Collembola, and Thysanura were just considered separate orders of the insects, but this classification was dismissed a long time ago. We now know that the old grouping of Thysanura is not evolutionary correct; thysanurans were split into Archaeognatha ("jumping bristletails", also called Microcoryphia) and Zygentoma (silverfish). However, the interrelationships among the apterygote groups and the insects are still not certain. Many different schemes have been proposed on the basis morphological characters (Massoud 1976, Koch 1997). Mitochondrial genome sequences suggest that Hexapoda is paraphyletic, and that the hexapod body plan originated twice: once in the insects and once in Collembola (Nardi et al. 2003). However, nuclear markers indicate that Hexapoda is monophyletic and Collembola are related to either Protura or Diplura (Luan et al. 2005, Timmermans et al. 2008).

Here we adopt the scheme developed by Misof et al. (2014), who used sequences of 1478 genes across several insect groups as well as apterygotes and crustaceans. In this phylogeny Collembola and Protura are sisters ("Ellipura"), to the exclusion of Diplura. In other phylogenies Collembola are sister to Diplura (e.g., Giribet et al. 2004). In still other cases, Diplura, Collembola and Protura are a monophyletic lineage, sister to the rest of the hexapods. This lineage is called *Entognatha*, which refers to the mouthparts kept inside the head in these animals. The other apterygote hexapods (Archaeognatha and Zygentoma) are then joined with the Pterygota as *Ectognatha* (having the mouthparts externally). These relationships are not yet stable. There seems to be a discrepancy between molecular analyses (Carapelli et al. 2000), which often support the Ellipura hypothesis and morphological analyses (Dallai and Burroni 1982, Bitsch and Bitsch 2000), which provide only weak support for this hypothesis (Carapelli et al. 2006).

Among these apterygote groups, Collembola, or springtails, are by far the best investigated. There is a substantial body of ecological literature about them, covering population dynamics and community structure in a variety of habitats. Collembola are also among the soil animals with a good record of experimental work, which owes to the fact that they can be cultured with relative ease in the laboratory and withstand experimental manipulation. The genome of the model species *Folsomia candida* shows many crustacean signatures, which confirms the transitional position of Collembola in between crustaceans and insects (Faddeeva-Vakhrusheva et al. 2017).

Collembola are also known from the fossil record. The first known hexapod, *Rhyniella praecursor*, is a collembolan from the early Devonian of the Rhynie Chert in Scotland (dated 407–396 million years before present), described in 1926. In the same deposition, slightly younger, remains have been found that were assigned to an ectognathic insect, *Rhyniognatha*. This suggest that not only Collembola but also pterygote insects were diversifying already in the early Devonian, maybe even in the Silurian (Engel and Grimaldi 2004). This is perfectly possible within the time window of hexapod terrestrialization, which was estimated from DNA data as

480 Ma BP (Figure 2.5). Hopkin (1997) points out that there are also springtail specimens preserved in amber from the Cretaceous, present in the National History Museum of London.

Rhyniella looks very much like a modern collembolan. It was assigned to various families, but after thorough re-analysis, Greenslade and Whalley (1986) concluded that it should be classified with the Isotomidae. The striking resemblance to present-day Collembola suggests that the group has changed little in outward morphology over millions of years (Hopkin 1997).

Collembola derive their scientific name from a unique organ they possess, the *ventral tube,* or *collophore*, which was assumed by early researchers to be an organ by which the animals adhere to smooth surfaces (κoλλα = glue). We know now that the ventral tube has nothing to do with locomotion but is instead an organ involved in water and ionic regulation (see Chapter 6). The common name, springtails, is actually more appropriate since it refers to the ability to jump, using another unique organ, the *furca*. This structure is well developed in the majority of surface-active springtails and it allows the animal to jump into the air, and so to escape from disturbance, notably predators. The furca is not moved directly by muscles, but it flips from an anterior to a posterior position following a change in the articulation of the hinge formed by abdominal sclerites, which in turn is brought about by integument muscle action and haemolymph pressure change (Bretfeld 1963). In rest, the furca is held in place by a small abdominal structure, the *retinaculum* (Figure 2.6).

The three abdominal appendages, ventral tube, retinaculum and furca are serial homologs of the legs. They can be subject to homeotic transformations by modifying the expression of *Hox* genes such as *Ultrabithorax* and *Abdominal A* (Konopova and Akam 2014). We will explore the developmental evolution of springtails in more detail in Chapter 4 (Figure 4.20).

Another unique feature of Collembola is the *post-antennal organ*, a roundish or elongate structure consisting of lobes set around a depression in the cuticle, which is positioned in between the base of the antennae and the eyes; it may take various forms in different species and is absent in some families, e.g., Entomobryidae. The

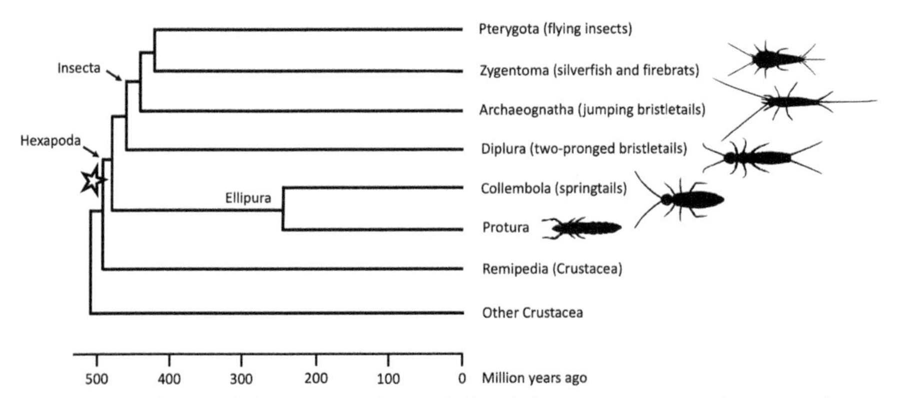

Figure 2.5. Phylogeny of the apterygote hexapods in relation to crustaceans and pterygote insects, based on data in Misof et al. (2014) and Katz (2020). In this phylogeny Collembola and Protura form a monophyletic group, designated as Ellipura, separate from Diplura. The terrestrialization of hexapods, indicated by an asterisk, is dated around 480 million years ago, i.e., in the beginning of the Ordovician.

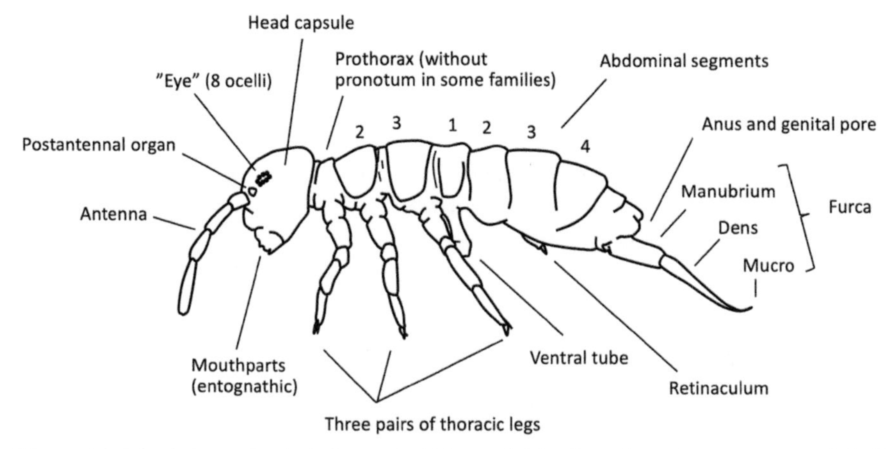

Figure 2.6. Morphology of a typical springtail (*Isotoma*). The furca, here shown swept-out, is held forward under the body when in rest. Redrawn, with modifications, from Gisin (1960).

post-antennal organ is assumed to have a chemoreceptive function. The general external morphology of springtails is pictured in Figure 2.6.

Collembola are classified in 18 families (Hopkin 1997, Potapov et al. 2020). These again are grouped into two main suborders, *Arthropleona* and *Symphypleona*. The first group represents the species which have an elongated body shape, the second comprises species with a globular body and fused abdominal segments. Within Arthropleona two lineages are often distinguished: Entomobryomorpha and Poduromorpha (Figure 2.7). Phylogenetic analysis suggests that Symphypleona represents the ancestral group, while Arthropleona are nested inside it. This is somewhat unexpected; the fused abdominal segments of Symphypleona suggest a derived condition, relative to outgroups such as Remipedia. This issue has to wait resolution by an ancestral state reconstruction involving more species.

The taxonomy of Collembola is based on a variety of cuticular structures, including the mouthparts, the claws, and the furca. These structures are best observed after clearing the specimens in lactic acid and preparing a slide. Equally important are spines, *bothriotrichs* (long hairs standing in a cup-like cuticular structure) and *chaetae* (bristles). In some groups the taxonomy is exclusively based on *chaetotaxy*, i.e., the position and patterns of bristle hairs on the body.

Hopkin (1997) estimated the total number of collembolan species at 6,500, however, the Belgian taxonomist Frans Janssens, on the website *collembola.org*, mentions a total number of 9,300 species described up to 2022. Several areas of the world with a high degree of endemism have not yet been explored extensively, and given the many cryptic species discovered through molecular analysis, the total may even exceed 50,000 (Cicconardi et al. 2013).

Collembola are small animals, their adult body length varies from 0.5 mm to 3 mm, although a few species can reach a size of nearly 1 cm (the "giant" collembolan *Tetrodontophora bielanensis*). Some species have conspicuous body colouring, especially members of the families Neanuridae, Entomobryidae and Sminthuridae, others are evenly pigmented or completely white. Collembola occur in almost

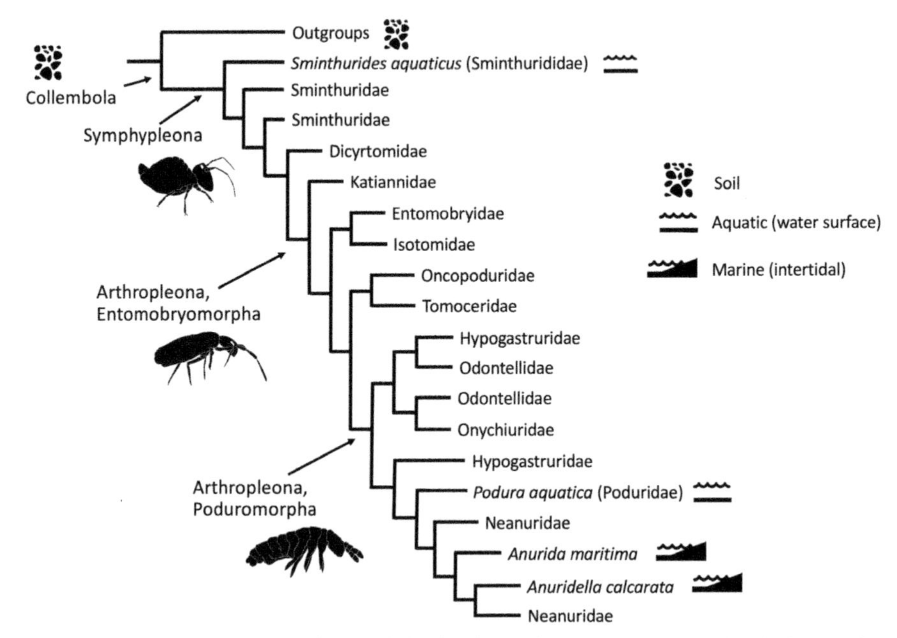

Figure 2.7. Schematic phylogeny of Collembola, showing marine and aquatic regressions in four species. The tree is based on 28S rRNA gene sequences, derived from D'Haese (2002). For simplicity only the aquatic and marine species are shown separately, while all other species have been collapsed into families. Note that several families are not monophyletic, as presently defined. Also shown is the main division into suborders and superfamilies. This division is confirmed by a recent genome-wide analysis of model species (Zhang et al. 2019). All Collembola are terrestrial (soil-living), except the four species indicated. Collembola silhouettes courtesy of Ninon Robin (Robin et al. 2019).

all habitats, everywhere where there is dead vegetation and sufficient humidity. Although most of the species are true soil dwellers, some representatives feed in the vegetation (e.g., the "lucerne flea", *Sminthurus viridis*), and Collembola are also common among the canopy fauna of tropical forests.

There is some debate about the degree of food specialization in Collembola. On the basis of a gut content analysis Vegter (1983) concluded that diet choice is more a matter of humidity-driven microhabitat choice than of direct food choice; this is supported by the relatively small interspecies differences in digestive enzyme activities (Urbásek and Rusek 1994). On the other hand, there are several reports showing preferences for certain species of fungus (Bengtsson et al. 1991, Hedlund et al. 1995, Kaneko et al. 1995). Some Collembola may feed on carrion or are even predatory. *Folsomia candida*, often viewed as a typical fungivore, feeds preferentially on nematodes when offered a choice (Lee and Widden 1996). Other Collembola may feed on plant roots, especially in the absence of dead vegetation (Joosse and Koelman 1979, *cf.* Figure 1.16). It may be best to view Collembola in general as opportunistic feeders.

Although Collembola represent the first really terrestrial lineage of the hexapods, there are some exceptions. For example, the species *Anurida maritima* and *Anuridella calcarata* are salt-adapted intertidal species, living in crevices of mud banks and exploring the open space of tidal flats in search for food (Witteveen et al. 1987, 1988). Other exceptions are *Podura aquatica* and *Sminthurides aquaticus*, species with a similar lifestyle, but roaming the surface of fresh waters. All "aquatic" collembolans lay their eggs on land, not in the water. It is therefore most logical to consider them as marine and freshwater regressions, rather than reflecting the ancestral conditions of Collembola. There is good evidence from phylogenetic analysis to support this interpretation (D'Haese 2002, Figure 2.7). The marine and freshwater species are nested in a fully terrestrial family structure and the transitions to non-soil habitats have occurred at least four times.

The sister group of Collembola (at least according to most phylogenies) is the peculiar group of *Protura*. These apterygotes lack antennae; they use their first pair of legs, held in front of the body, as such (Figure 2.6). Interestingly, Malpighian tubules (which are lacking in Collembola) are present in rudimentary form in Protura. During early life, segments are added to the abdomen, making a final complement of eleven segments plus the telson. The animals are very small (0.7 – 2.5 mm), whitish and found in moist litter and under bark, mostly in forest ecosystems. They seem to favour temperate and boreal conditions more than warm and tropical climates. Little is known about the feeding habits of proturans, but it may be assumed that they feed on fungi and decaying vegetation. Only in 1901 Protura was recognized as a separate order and up to 1906 only one species was described (Szeptycki 1997). At the moment there are 748 species known, but the global biodiversity is estimated to be above 900 (Galli et al. 2020).

Diplura derive their name from the two long cerci on the abdomen, which are sometimes as long as the body itself (Figure 2.5). The abdomen has 11 segments in all stages, in contrast to Protura (which increase the number of abdominal segments during post-embryonic growth) and Collembola (which have only six abdominal segments). Diplurans have a variety of feeding habits, rotting plant material, fungi and carrion; even live animals belong to their menu. The feeding habits of diplurans are not clearly divided over the families, each family displays the complete spectrum (Carpenter 1988). The family Campodeidae includes the species *Campodea staphylinus*, which is the most common dipluran in woodland and grassland soil in temperate climates. Another well-known family is Japygidae, which have relatively short cerci, sometimes in the shape of pincers, resembling earwigs. These diplurans are common in the tropics and in Mediterranean climates. The total number of species is estimated at around 800.

In Archaeognatha and Zygentoma, the two long cerci on the abdomen are supplemented by a very long outgrowth of the *epiproct* (the dorsal plate part of the telson), which forms a third, median thread. Although the three abdominal threads are an obvious similarity between the two orders, there is an important distinction between them in the mouthparts. In Zygentoma there is a dicondylic articulation of the mandible, which thus bites transversally, like in true insects. In Archaeognatha, the mandible has a single articulation with the head capsule and the action of the mandibles is similar to the apterygote (primitive) insects. In addition, the compound eyes of Zygentoma are small and inconspicuous, while those of Archaeognatha are relatively large. Larink (1997) provides a review of the external and internal morphology of the two groups.

Zygentoma include common species such as silverfish (*Lepisma saccharina*) and firebrats (*Thermobia domestica*), which are often found inside houses, feeding on paper, glue and old photographs. They can digest lignocellulose, like termites (*cf.* Section 6.10). Much less is known about the feeding habits of wild species of Zygentoma, but it is assumed that they are generally omnivorous. Archaeognatha, which include the family Machilidae or bristletails, have similar feeding habits, they are non-specialized feeders, including some predators. The number of species of Zygentoma and Archaeognatha together is estimated as 900, most of which can be considered as true members of the soil fauna. The taxonomy of Archaeognatha is far from complete; detailed analysis of allozyme loci revealed cryptic species which can hardly be discriminated from each other using external morphology (Fanciulli et al. 1997).

2.4 Soil and surface-active insects

The evolutionary movement of hexapods onto the land resulted in a completely non-marine lineage, that became the most speciose group of all animals, the insects. A significant number of insects may be considered soil invertebrates, as they spend their whole life or a part of their life in the soil. If there are life-stages in the soil distinct from terrestrial stages, it is always the eggs or the larvae that are in the soil. This may be seen as evidence supporting the GKGB hypothesis introduced in Chapter 1. As reproduction is a conservative phase of the life-cycle, the fact that so many terrestrial invertebrates return to the soil for their reproduction suggests that their evolutionary ancestors were soil-living.

An overview of all insect groups that have some relationship with the soil is impossible to give. We provide here a simple classification by functional groups ("guilds"), just to give the reader an impression of the diversity. That diversity is illustrated in Figure 2.8.

- *Soil-living detritivores*. Several insects consume dead organic matter of plant origin; they are called detritivores or saprotrophs. Among flies and midges, saprotrophic larvae are common. The maggots of soldier flies (Diptera: Stratiomyidae), hoverflies (Syrphidae), and chironomids (Chironomidae) are sometimes an abundant component of the belowground insect fauna, especially in wet soils and swamps. Other detritivore insects include darkling beetles (Coleoptera: Tenebrionidae), a very large family of soil, litter and waste-associated beetles, around 20,000 species. Also, Psocoptera ("book lice") and drain flies (Diptera: Psychodidae) are detrital feeders in the soil.
- *Saproxylic insects*. A lot of insect groups are associated with dead wood, including termites (Isoptera) and several families of beetles (Coleoptera: Elateridae, Passalidae, Cerambycidae, Anobiidae, Lucanidae, Scarabaeidae, Buprestidae, Curculionidae). The boring and tunnelling activities of these animals contribute to the penetration by fungi and the decomposition of wood. Often saproxylic insects work along with fungi on which they feed, sometimes in mutualistic symbiosis (Birkemoe et al. 2018). Wood-boring families also exist among Hymenoptera (Siricidae, Xiphydriidae), Lepidoptera (Sesiidae, Cossidae) and flies (Diptera: Tipulidae, Syrphidae, Chironomidae). Within the Hymenoptera,

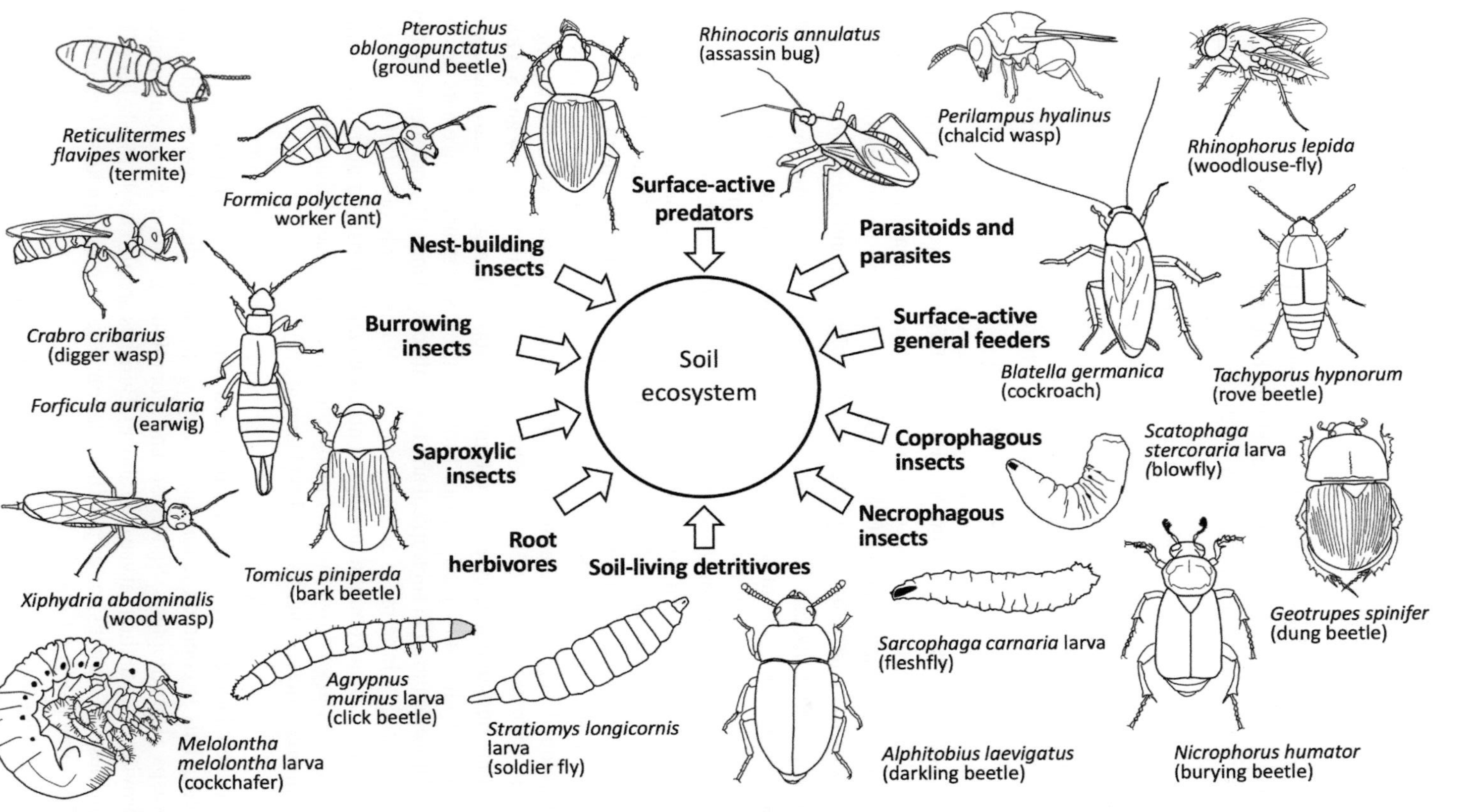

Figure 2.8. Illustrating the diversity of insects with a functional significance to the soil ecosystem, categorized by guild. For every guild two representatives are shown.

there is an evolutionary transition from herbivore sawflies to families associated with dead wood. The saproxylic hymenopterans do not represent a primary soil-living group but instead a secondary adaptation, derived from leaf-eating wasps (Stokland et al. 2012). The same is likely true for other saproxylic insects.

- *Necrophagous insects.* Blowflies (Diptera: Calliphoridae, Muscidae and Sarcophagidae) are known for their ability to locate carrion or cadavers from a distance and deposit their eggs in rotting flesh. Larval development is adapted to the short life of the substrate. Also, some groups of beetles are necrophagous (Coleoptera: Silphidae, Dermestidae, Staphylinidae). As necrophagy is usually a localized phenomenon, these organisms are not often studied in the context of soil ecology, but they do represent an important functionality of the soil: cleaning up animal carcasses.
- *Coprophagous insects.* Several groups of beetles are specialized in feeding on dung. This habit is especially common in ecosystems that are populated by large herbivores such as savannas. Dung-foraging beetles dig tunnels and deposit pellets of dung on the bottom as feed for the larvae. They belong to the family Geotrupidae and two subfamilies of the Scarabaeidae, Scarabaeinae en Aphodiinae (Nichols et al. 2008). Among flies also there are groups of which the larvae live in herbivore dung (Diptera: Scatophagidae and Sphaeroceridae).
- *Nest-building insects.* Ants are known for their engineering function in the soil ecosystem. An ant nest represents a profound alteration of the local soil conditions. In addition, a colony of ants consumes a significant amount of organic material collected from the surroundings (leaves, live prey, corpses, plant and animal secretions). All ants belong to one family (Hymenoptera: Formicidae), which is split into around 21 subfamilies. By 2007, 12,000 described species were noted, but the worldwide total is estimated as 22,000 (Ward 2007). Because of their complex social behaviour and peculiar genetics, the study of ants has developed into a separate discipline of ecology, too extensive to review here, although some aspects of their evolution will return in Chapters 4 and 5. Other famous nest-building insects are termites (Isoptera), a group of around 2,400 species, with a unique body plan. Termites affect soil structure like ants do, but are more common in subtropical and tropical ecosystems. The parallel evolution of eusociality in termites and ants presents another fertile object of study. Nest building is also common among bumblebees and wasps; the nests of *Bombus terrestris* can go down to 1 m. Finally, there are some insects that create smaller, temporary, nests, such as earwigs (Dermaptera).
- *Burrowing insects.* Several insects dig burrows in order to create a safe place for their larvae. The larvae are provided with killed or immobilized prey for feed. This is common among the many sand wasps, digger wasps and spider wasps (Hymenoptera: Sphecidae and Pompilidae). Other burrowing insects provide dung to their larvae (see above). Burrows are also made for shelter (e.g., the nests of earwigs) or as a calling station, e.g., in mole crickets (Orthoptera: Gryllotalpidae, *cf.* Figure 1.14).
- *Root herbivores.* These have been discussed in Chapter 1 (Figure 1.16); they are usually larvae of species that have their adult stages aboveground.

- *Surface active predators*. Ground beetles (Coleoptera: Carabidae and Cicindelidae) are active at the surface and many of them are voracious predators. They will forage on populations of microarthropods (springtails, mites, insect larvae). Carabidae is a species-rich and well-investigated group of beetles, 40,000 species worldwide. They are often studied using pitfall traps and a good deal is known about their population dynamics. We will illustrate the use of carabids in soil ecology studies in Chapter 3. Other surface-active predators can be found among rove beetles (Coleoptera: Staphylinidae), another large family (63,650 species as off 2018). Also, predatory bugs (Hemiptera: Reduviidae) are known for their voracious behaviour. While carabids and staphylinids run towards a prey to catch it with their protruding mandibles, reduviids hide in an ambush and pierce prey with their rostrum (hence "assassin bugs"). For more on predatory behaviour, see Section 5.10.
- *Surface-active general feeders*. In both carabids and staphylinids we find not only specialized predators but also many polyphagous predators, scavengers and detritivores, eating a great variety of food items, including dead organic matter, plant leaves, fungi and insect eggs. In addition, omnivorous feeding is common among cockroaches (Dictyoptera) and crickets (Orthoptera: Gryllidae) Also, several scorpion flies (Mecoptera: Panorpidae) are considered scavengers that can be found in leaf litter and moss. Snow scorpion flies, family Boreidae, are seen on the surface during winter time.
- *Parasitoids and parasites*. Several hymenopteran wasps lay their eggs in larvae or eggs of other insects. They are tiny creatures (from 3 mm to less than 1 mm) and are therefore able to make their way through the litter to reach their hosts. The community of soil and litter-living Hymenoptera, abbreviated SLH, is very rich in species; it involves no less than 15 families (Ceraphronidae, Diapriidae, Chalcididae, etc.). They are assumed to contribute to the regulation of fungivorous, detritivore and root herbivore populations and are therefore often considered in biological control programs. Also, some families of flies exist of which the larvae are parasites of isopods and insects (Diptera: Rhinophoridae, Tachinidae).

Most of the insect orders underwent their evolutionary radiation long after the first terrestrialization of the hexapods. They all remained terrestrial; there are hardly any marine regressions known for hexapods. A few species of Collembola (Figure 2.7) and some flies in the family Dolichopodidae have become intertidal and there are also insect parasites of marine mammals, but the great majority of hexapods have stayed true to the terrestrial environment. However, in several orders insects have adopted a life in freshwater, often as larva (damselflies, dragonflies, caddisflies, stoneflies, midges), but also in some cases as adults (water bugs, water beetles). The question may be raised: are these groups to be considered descendants of ancestral lineages that moved from the sea directly into freshwater and stayed there, or do they represent primarily terrestrial species that secondarily adopted a life in the water? An answer may come from a consideration of the phylogeny of insects (Figure 2.9).

We present a simplified phylogenetic tree, derived from the phylogenomic study by Misof et al. (2014), which was also used in Figure 2.5. Only the main lineages

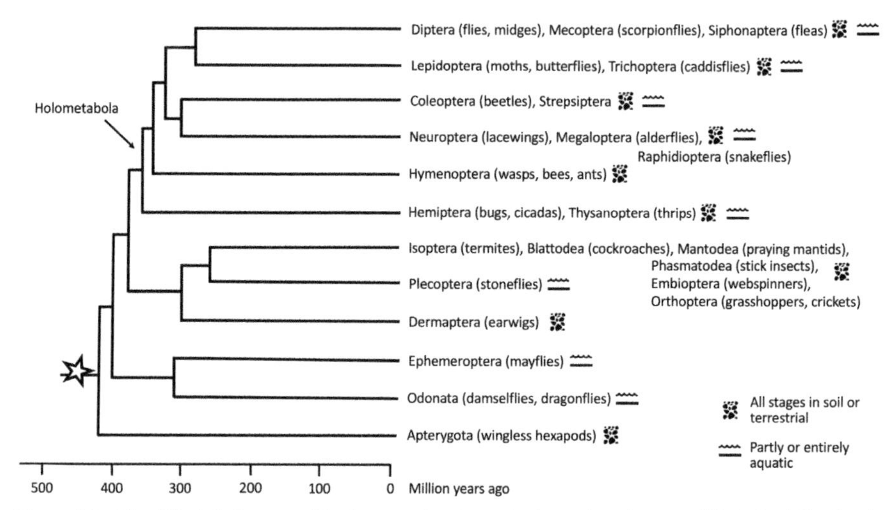

Figure 2.9. Simplified phylogeny of the insects, showing only the main orders, modified after Misof et al. (2014), a phylogenomic study using 1478 nuclear gene sequences and 144 taxa. The habitat uses added suggest that while the insects are assumed to represent an originally terrestrial lineage, regressions to the aquatic environment have occurred many times.

and the better-known orders are shown. The positions of the various insect lineages have often switched relative to each other in different evolutionary analyses. For example, the monophyly of earwigs, stoneflies, termites and the various orthopteran groups, is a relatively new insight. Also, the relationship of thrips (Thysanoptera) to bugs (Hemiptera) is now well established. The reconstruction of lineages within the Holometabola, shown in Figure 2.9, is in line with previous insights. Holometabola are confirmed to be monophyletic, and the origin of the clade is dated in the early Carboniferous (360 Ma BP). However, the diversification of the various holometabolan lineages (Hymenoptera, Lepidoptera and Diptera) occurred only much later, during the rise of flowering plants early in the Cretaceous (140 Ma BP).

At the basis of the insect tree we find a monophyletic lineage of groups with freshwater nymphal stages: damselflies, dragonflies and mayflies. Do these groups represent an ancient lineage that descended directly from an aquatic crustacean group related to Remipedia? In that case the terrestrialization of apterygotes (Collembola and the like) would have occurred separately from the insects. There are phylogenetic analyses that actually support such a reconstruction (Nardi et al. 2003). It would imply that also the sex-legged body plan has evolved twice, because no crustacean has six legs. In agreement with this, the node support for a sister relationship between Odonata + Ephemeroptera and the rest of the insects is not particularly strong (Misof et al. 2014).

However, most modern phylogenomic analyses position the apterygotes clearly in between insects and crustaceans, leaving Insecta monophyletic. As apterygotes are all terrestrial, the aquatic nymphal stages of the basal insect orders must be explained as a regression, i.e., a return to water by an originally terrestrial lineage. This is indeed the most common opinion among entomologists. A review of evidence in favour of a terrestrial origin of insects is given by Pritchard et al. (1993). One could also argue that the difference between aquatic and terrestrial life may not have been so great during the time at which the first insect lineages evolved, the Carboniferous, which was characterized by extensive swamp forests.

Mapping the habitat use on the insect phylogenetic tree shows that freshwater insects can actually be found in almost every order: in addition to Odonata and Ephemeroptera the larval stages of midges, caddisflies, alderflies and stoneflies are aquatic, as well as the larval and adult stages of water bugs and water beetles. Since the freshwater stages in the holometabolan orders are all nested inside completely terrestrial lineages, the most parsimonious explanation is that they represent secondary aquatic adaptations, like in the Collembola.

2.5 Terrestrializations among isopods, amphipods and decapods

Another set of terrestrializations coming out of the crustaceans took place in the Malacostraca, as we have seen before (Figure 2.4). We may distinguish three independent events, among three different orders of Malacostraca. In all cases the terrestrializations occurred directly from marine ancestors and they are much more recent than in hexapods. Also, none of these three orders is completely terrestrial; the maximum is seen in isopods, where the number of terrestrial species is more than one-third of the total.

Isopods have been mentioned above as members of the groundwater fauna, however, the order as a whole is best known for woodlice found in litter and under tree bark. These animals belong to the most common soil invertebrates, well known to the general public. Several representatives, such as *Armadillidium vulgare* and *Porcellio scaber* are important experimental models for ecological research. The name isopod derives from the fact that all legs are more or less similar. Unlike in other crustaceans they are not differentiated in walking legs, chelicerate legs and swimming legs. Isopods differ from amphipods in that the body is mostly dorso-ventrally flattened, compared to the laterally compressed amphipods.

The body of isopods (Figure 2.10) is divided into a small head, a large *pereion* (equivalent to the thorax) and a shorter abdomen (*pleon*), which often articulates with the pereion like a "tail". The head bears a pair of compounds eyes (although some families have only simple eyes, ocelli), and two pairs of antennae. The first of these, the *antennulae*, are often very small and difficult to see, the second pair of antennae are large, with many joints. The mouthparts consist of a pair of mandibles, two pairs of maxillae and one pair of *maxillipedes*. The maxillipedes have a plate-like structure which forms the "floor" of the mouth cavity and is visible in ventral view (Figure 2.10). There is no carapace covering the thoracic segments like in crabs and lobsters, and so all segments (*pereionites* and *pleonites*) are visible from above.

The seven segments of the pereion each carry a walking leg (*pereiopod*). In species that breed their eggs and larvae in a brood pouch under the body, the coxae of the pereiopods carry a medial plate-like extension called *oostegite*. It is actually a modified *endite* (see Figure 4.18). The oostegites of subsequent perionites form the bottom of the brood pouch. This system is also common in amphipods.

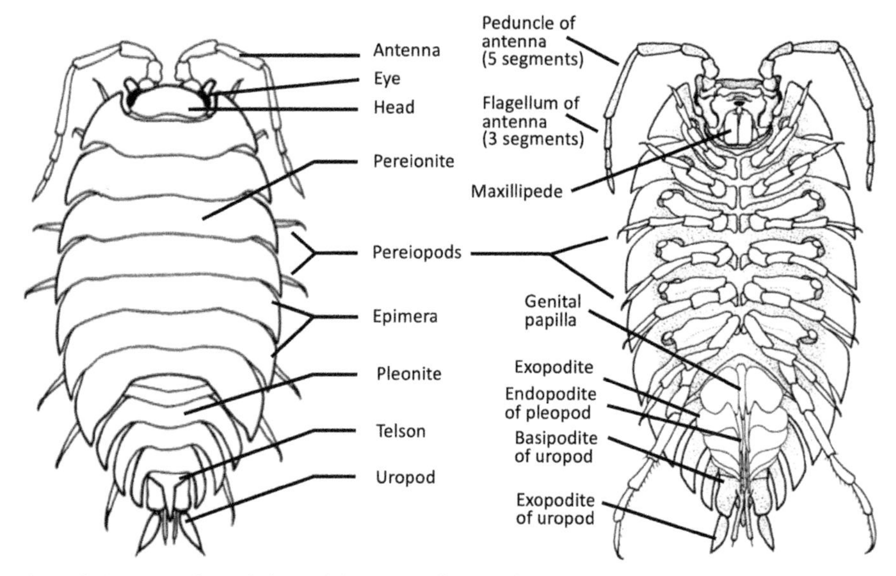

Figure 2.10. External morphology of *Oniscus asellus*, a typical terrestrial isopod (woodlouse), in dorsal view (left) and ventral view (right). Modified after Sutton (1972), reproduced by permission from Elsevier.

The appendages of the pleon are flattened and modified for respiratory purposes; these legs are often called *pleopods*, although in terrestrial isopods they do not have a function in swimming. The pleopods still show the classical biramous organization of crustacean legs, a basal basipodite with a median endopodite and an outward exopodite (Figure 2.10, *cf.* Figure 4.18). In many species some of the pleopods are sexually dimorphic: in males the endopodites of the first and second pleopods are elongated to a stylet supporting the transfer of sperm during copulation. At the end of the body we find a pair of biramous *uropods*, which extend behind the body on both sides of the telson.

The order Isopoda contains more than 10,300 species worldwide (Broly et al. 2013a) and the number of terrestrial species is estimated as 3,710, classified in 527 genera and 37 families (Sfenthourakis and Hornung 2018). Species identification is based on the general body outline, the structure of antennae, eyes, maxillae, pleopods and uropods, and some aspects of body pigmentation. There are more than 10 families, among which Ligiidae, Trichoniscidae, Oniscidae, Philoscidae, Porcellionidae, Trachelipodidae and Armadillidiidae are the best known.

The family Ligiidae contains the large *Ligia oceanica*, a species living in rock crevices on the sea shore, feeding on shoreline detritus. The family Trichoniscidae is a species-rich group of mostly small isopods, all characterized by the conically tapering distal part of the antenna (the *flagellum*); the parthenogenetic species *Trichoniscus pusillus* is one of the most common representatives. The family Oniscidae is characterized by a flagellum with three limbs; *Oniscus asellus* (Figure 2.10) is one of the most common species in temperate climates. Among the Philosciidae is *Philoscia muscorum*, a fast running forest floor isopod with a median stripe on the body, along with many tropical species, among which representatives from the genus *Burmoniscus*. *Porcellio scaber* and other species of the genera *Porcellio* and

Porcellionides belong to the family Porcellionidae; this family has a flagellum with two limbs. The pill woodlouse, *Armadillidium vulgare*, which is able to roll itself in a ball like a glomerid millipede, is classified in the family Armadillidiidae.

Terrestrial isopods are commonly joined together in one suborder, Oniscidea, which is the only terrestrial suborder of Isopoda. Within the Oniscidea sequences of species can be identified, showing a progression of adaptation to terrestrial conditions, ranging from hardly more than aquatic to truly terrestrial, and even adapted to desert climates (Warburg 1968). In fact, woodlice represent a classical example of comparative physiology, illustrating how drought resistance, nitrogen metabolism and water relations change during an evolutionary transition from marine to terrestrial habitats (see Section 6.4 of Chapter 6).

Oniscidea has long been assumed monophyletic. The absence of terrestrial isopods outside Oniscidea and the many morphological *synapomorphies* (shared derived traits) strongly suggested that all woodlice descend from a single terrestrialization event, directly from the marine environment. The intertidal genus *Ligia* (sea slater), commonly found around rocks and stones at the shoreline, is often seen as a transitory species (Wilson 2009, Broly et al. 2013a). The invasion of the land is projected in the late Carboniferous, close to the Permian (299 Ma BP), although the earliest fossil woodlice are much more recent, stemming from the early Cretaceous (Broly et al. 2013a).

Nevertheless, not all scientists agree on monophyly of Oniscidea. The French zoologist and biospeleologist Albert Vandel (1894–1980) had suggested already in 1965 that woodlice represent not a single but at least three independent colonisations of the land. Later this scenario received more and more support from molecular data, although the relations are different from what Vandel proposed.

In a study of two protein-encoding genes plus two rRNA genes (Dimitriou et al. 2019) and in another study of complete mitochondrial genomes (Zhang et al. 2022), strong support was obtained for a sister relationship between the marine genus *Idotea* and the intertidal/terrestrial genus *Ligia* (Figure 2.11). Only if *Ligia* is excluded, Oniscidea can be "saved" as a monophyletic lineage. The analysis confirms Vandel's claim that there was more than one terrestrialization in isopods. In fact, there were three, one leading to the intertidal genus *Ligia*, a second in the group of hygrophilic woodlice living in waterlogged soils (e.g., the genus *Ligidium*) and a third giving rise to common woodlice of the genera *Oniscus* and *Porcellio* (Figure 2.11). Another striking property of this phylogeny is the basal position of two freshwater lineages (Asellota and Phreatoicidea). Although it is usually assumed that the terrestrialization of isopods came directly from the marine environment, an intermediate freshwater phase cannot be excluded.

Because woodlice are such important components of the decomposer fauna, their feeding ecology has been the subject of much research (Sutton 1972, Warburg 1987). All woodlice are considered detritivores, but many species have particular preferences for leaves infested with fungi, for certain species of leaves, for a certain stage of decomposition and for a high nitrogen content (low C/N ratio). Isopods are also known for their *coprophagy* (eating of faeces). It has been suggested that this may be an adaptation to the poor quality of their food and that they gain additional nutrients from the faecal pellets when these are degraded externally by

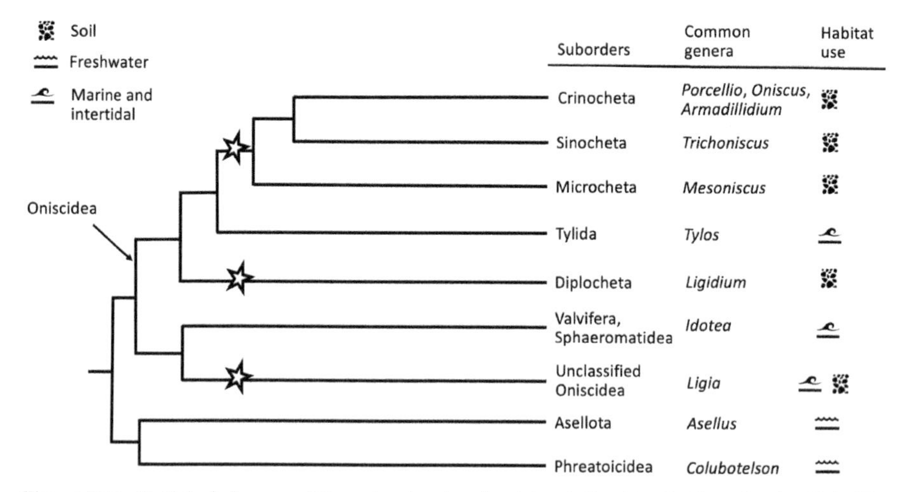

Figure 2.11. Partial phylogeny of isopods, showing the interrelationships in the suborder Oniscidea (terrestrial isopods). Tree topology is derived from Dimitriou et al. (2019), developed using 16S rRNA, 28S rRNA and two protein-coding genes. As *Ligia* is classified within the Oniscidea, this group is not monophyletic. The stars indicate three independent invasions of the land.

microorganisms (Hassall and Rushton 1985). It is also a common observation that woodlice will show cannibalistic behaviour if the conditions become less favourable and that growth is stimulated by adding extra protein to the food. This indicates that woodlice may be under continuous limitation of nitrogen. Due to their voracious feeding on dead leaves, with low assimilation efficiency, woodlice contribute significantly to the fragmentation and decomposition of organic material in soil.

The second malacostracan order with terrestrial representatives is Amphipoda. Since this order clusters away from Isopoda in the phylogeny of Malacostraca (Wilson 2009), the terrestrialization of amphipods was certainly independent from isopods. Most of the terrestrial amphipods (often denoted as “landhoppers”) are restricted to estuaries and very moist habitats. The families Talitridae, Hyalidae and Corophiidae live in coastal flood plains and beaches, feeding on shoreline detritus, algae and remains of dead animals. Most likely these lineages represent two or more independent colonization events, since Corophiidae are classified separately from Talitridae (Lowry and Myers 2013). However, only Talitridae (including the genera *Talitrus* and *Orchestia*) are usually considered “truly terrestrial” in the sense that they are obligatory land-living (Friend and Richardson 1986, Spicer et al. 1987).

Intertidal amphipods feed mostly during the night and spend the day hiding in crevices or in burrows. The species *Corophium volutator* lives in vertical tunnels dug in tidal mud flats and uses its very large antennae to collect organic material from the surface (Figure 2.12). Due to its high densities in some places *Corophium* is an important bioturbator as it significantly increases the concentration of suspended solids in the overlying water (Ciarelli et al. 1999).

In tropical rain forests and other moist habitats such as cloud forests, especially in the Indo-pacific region, amphipods occur as truly terrestrial animals, populating the litter in the same way as isopods do in temperate forests. There are at least 120 species of such “truly terrestrial” amphipods and, due to their relatively large size,

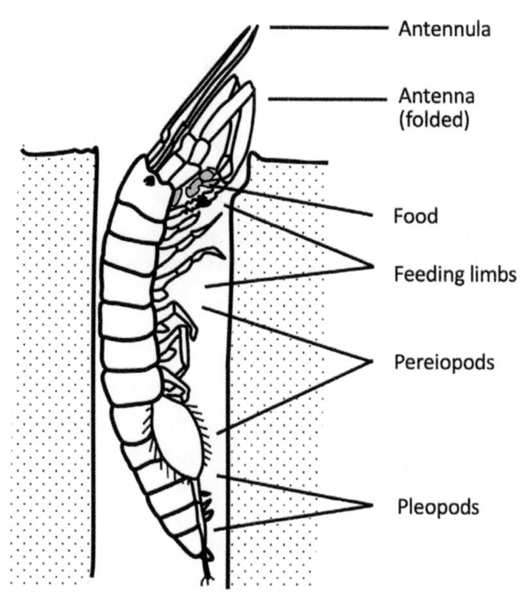

Figure 2.12. A representative of the terrestrial amphipods, *Corophium volutator*, a common species on tidal mud flats, at the mouth of its burrow. The animal collects food from the surface using its very large antennae, the size of its body.

they have a significant effect on litter decomposition (Friend and Richardson 1986). Several species of landhopper occur in Australia and New Zealand and some, e.g., *Orchestia hurleyi*, are genuinely adapted to grassland habitats and may reach a density of 3000 individuals per square meter in such environments. It is often argued that talitrid amphipods were "preadapted" to terrestriality due to their life in intertidal mud plains, in which they developed the leg adaptations for locomotion in an aerial rather than freshwater environment. Amphipods used to same evolutionary pathway to terrestrial life as isopods did (via the intertidal habitat of beaches and mud plains).

A trait specific to amphipods is that the first thoracic segment is fused with the head. The body is laterally compressed and often takes the form of a sickle. The two antennae are both developed, although in some species the antennula is reduced, like in isopods. There are seven pereion segments, like in isopods, each bearing a leg, which are differentiated into feeding limbs, walking legs, jumping legs and gnathopods. One of the pereiopods is sometimes modified as a large claw for the male to cling to the female during copulation. Terrestrial amphipods have a jumping mode of locomotion, for which they use the last pereiopods, which often have the second segment (basis) significantly enlarged (Figure 2.12). The pleopods, of which there are no more than three, are better developed as in isopods, although terrestrial amphipods are not good swimmers.

The last terrestrialization arising from the Crustacea to be discussed here is found within the order Decapoda. This is another large group of crustaceans, all with five pairs of walking legs (hence the name ten-legged). The number of species is estimated to amount to 15,000, almost all of them marine, including the well-known crabs, lobsters and shrimps. However, within the crabs several species have ventured

Figure 2.13. *Geograpsus lividus*, a relatively small land crab (2,5 – 3,5 cm) from the coasts of Florida to Brazil, the Gulf of Mexico, and Galapagos to Chile.

onto the land. They are especially known from rocky shores of the Caribbean, South American coasts and California.

Land crabs are commonly assigned to a single family, Grapsidae, however, this may not constitute a monophyletic lineage (Schubart et al. 1998). It is more likely that the shore crabs that ventured onto the land did so on specific shores, where they became tied to local conditions by reproducing on land. For example, some species of grapsid crab breed in water-filled *Bromelia* leaves, where they nurse their young. Consequently, there is a large degree of endemism in these land crabs. On the island of Jamaica alone, there are no less than nine endemic land crab species (Schubart et al. 1998), each with exceptional adaptations.

In their external morphology land crabs are hardly different from coastal and oceanic crabs. The hind legs lack the flattened parts typical for swimming carbs, and the front legs are sometimes adapted for digging. In some species of land crab, the eyes are stalked longer. However, in general the body plan remained basically the same. This is illustrated in Figure 2.13, which shows a common species of Meso-American land crab, *Geograpsus lividus*. It seems that the adaptations that allowed crabs to colonize the land were mainly behavioural, rather than morphological or physiological. Their reliance on gill respiration made it impossible to penetrate the land beyond moist terrestrial habitats.

Overviewing the various terrestrializations within the Pancrustacea, we are left with a diverse picture. While the invasion of the land by hexapods was associated with major changes in the body plan (six legs) and the evolution of many novel body parts (tracheal system, Malpighian tubules), land crabs remained crabs, terrestrial amphipods remained amphipods and woodlice changed only little compared to marine isopods. It is appropriate to remember though that the hexapod terrestrialization was much earlier in evolutionary history than the terrestrializations in amphipods, isopods and decapods (Figure 2.4). Overall, there is no unique evolutionary pathway to terrestriality. We will explore the physiological and genomic adaptations to terrestrial life in hexapods and isopods further in Chapter 6.

2.6 Soil invertebrates with many legs

Up to now we have the discussed the arthropods' share of soil invertebrates only with reference to Pancrustacea (hexapods and crustaceans), however, the other two main groups of arthropods, chelicerates and myriapods, are also abundantly represented in the soil ecosystem. According to several phylogenetic analyses (Edgecombe 2010, Regier et al. 2010, Rota-Stabelli et al. 2011, Misof et al. 2014) *Myriapoda*

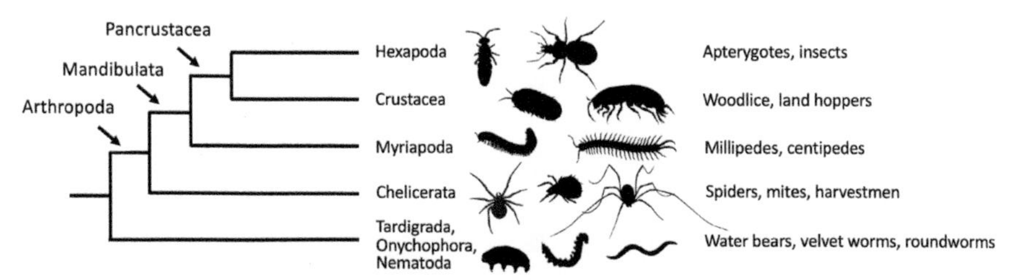

Figure 2.14. Relationships between the main evolutionary lineages of arthropods, according to the Mandibulata hypothesis, as supported by Regier et al. (2010), Rota-Stabelli et al. (2011), Misof et al. (2014) and others. The tree shows the topology of relationships, not scaled to the evolutionary distances between the groups.

are related to Pancrustacea and are united with them in the Mandibulata supergroup. This group includes all arthropods with mandibles, to the exclusion of arthropods with cheliceres. The Mandibulata hypothesis implies that Chelicerata are a separate, earlier split in the arthropod phylum (Figure 2.14).

In this grand scheme of arthropod evolution, it is assumed that terrestrializations from the marine environment have taken place independently in hexapods, myriapods and chelicerates. This has to be so since the majority of crustaceans is marine, while also chelicerates contain a basal marine group (see below). All three terrestrializations are projected in the late Cambrian and early Ordovician, between 500 and 480 Ma BP, most likely first the chelicerates, then the myriapods, then the hexapods. That myriapods terrestrialized separate from insects implies that certain shared characters (tracheae, Malpighian tubules, postantennal organs) evolved twice, independently in the stem lineages of Hexapoda and Myriapoda (Sombke and Edgecombe 2014).

The subphylum Myriapoda ("many-legged" arthropods) is defined by the multitude of legs. It includes some 16,000 described species (Qu et al. 2020), but the diversity may be much larger, especially due to the many tropical millipedes not yet described. The marine stem group of extant myriapods appears to be the Euthycarcinoidea, a group of fossils with a superficial resemblance to sea scorpions, present from the Cambrian to the Triassic. The affiliation of this group to millipedes was revealed by close inspection of *Heterocrania rhyniensis*, an euthycarcinoid from the Early Devonian Rhynie Chert of Scotland (Edgecombe et al. 2020).

There are four extant classes which all have important groups of soil-living representatives. Figure 2.15 shows the evolutionary relationships among these groups (Edgecombe and Giribet 2007, Shear and Edgecombe 2010, Fernández et al. 2017, Edgecombe et al. 2020). All myriapods outside euthycarcinoids are terrestrial, so the evolution of myriapods likely involved just a single terrestrialization. According to this reconstruction, centipedes would be closer to the ancestral body plan, while millipedes, with their many legs, would be more derived. The topology of the myriapod tree of life is now stable, since Fernández et al. (2017) showed that, provided the right outgroups are chosen, molecular data coincide with morphological data.

All myriapods have many legs, but millipedes (*Diplopoda*) have more than centipedes (*Chilopoda*). The scientific name refers to the fact that there are two pairs of legs for each segment; the body rings are therefore called *diplosegments*.

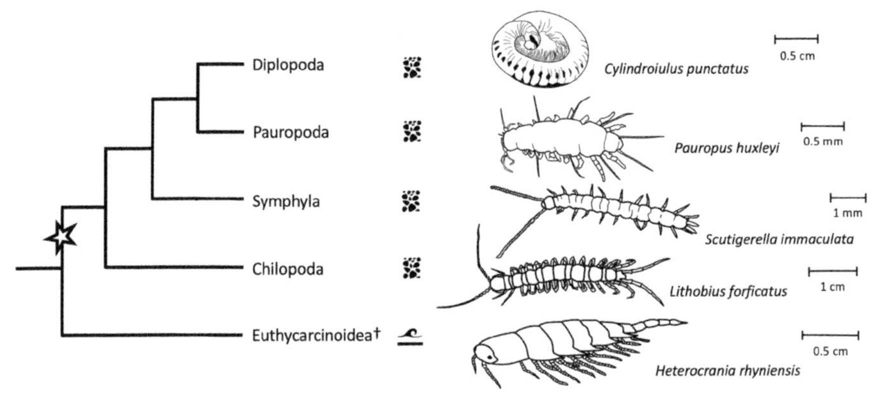

Figure 2.15. Evolutionary relationships within the myriapods, showing the positions of pauropods and symphylans relative to centipedes and millipedes, as well as the assumed (extinct) marine stem group Euthycarcinoidea. A representative species of each group is shown for illustration. Tree topology is based on Edgecombe and Giribet (2007), Shear and Edgecombe (2010), Fernández et al. (2017) and Edgecombe et al. (2020).

There is a tendency both in millipedes and in centipedes for the number of legs to increase; this is known as the "*elongation theory*", which holds that the evolutionary trend is from fewer segments to many, not the other way around (Hopkin and Read 1992). In millipedes, excessive length growth has evolved at least two times, in the orders Siphonophorida and Polyzoniida. The present record is held by an Australian groundwater species aptly named *Eumillipes persephone*, with 1306 legs (Marek et al. 2021). In centipedes, excessive length growth is found in the order Geophilomorpha in association with soil-living (Figure 5.24).

The formation of segments in centipedes is controlled by a segmentation oscillator, a genetic network generating waves of regulatory molecules. Towards the end of development this clock is suppressed and the last segments are added individually (Brena and Akam 2013). One may speculate that the excessive number of segments in some myriapod lineages is due to loss of inhibitory control over the segmentation clock, a mechanism suggested for snakes (Mansfield 2013).

The body of millipedes consists of a small, heavily calcified head, bearing one pair of antennae and primitive eyes (actually fields of ocelli), followed by an equally thick and tough segment, the *collum*. The front part of the body is also important for reproduction; the gonoducts open ventrally in the third segment or in the coxae of the second pair of legs and this region is often modified to build structures that may support the transfer of sperm during copulation. In addition, the legs of the seventh segment may be modified to form *gonopods*; these structures, which also support copulation, may be very complex and often do not resemble legs at all. An extensive account of the external and internal morphology of millipedes is given by Hopkin and Read (1992).

Diplopoda are the most successful myriapod group: there are 12,000 known species worldwide, classified into 14 different orders. Enghoff (1984) presents a cladistic system of the Diplopoda, in which the orders are grouped into superorders and infraclasses in a way that reflects the putative evolutionary relationships among

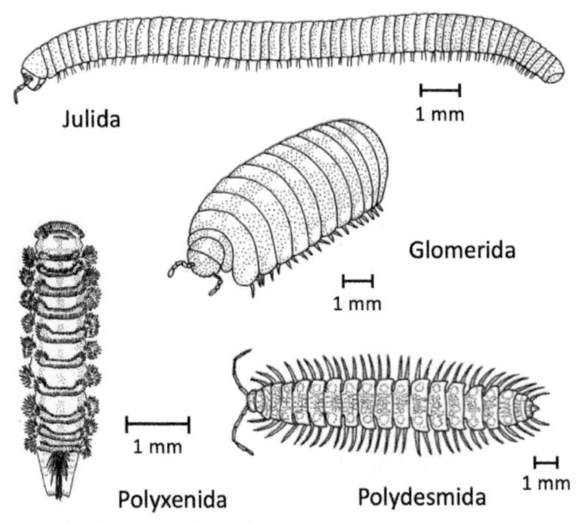

Figure 2.16. Four different body forms of millipedes. Reproduced from Hopkin and Read (1992), with permission of Oxford University Press, through PLSclear.

the species. This classification comes close to a division in four different body forms. The "bristly" millipedes (order *Polyxenida*) are considered to be the most primitive. They lack calcification of the exoskeleton which is characteristic of all other millipedes. The dorsal side of the body carries a conspicuous set of bristles (Figure 2.16). The pill millipedes (typified by the order *Glomerida*) are known for their ability to roll into a ball. A common species in temperate regions is *Glomeris marginata*, which has been a model for many population and decomposition studies. A third body-form, the flatback millipedes, includes the order *Polydesmida*. The body is dorso-ventrally flattened, and the legs are relatively long compared to the body (Figure 2.16). The diplosegmental arrangement is very clearly seen in these species. This order is the largest of all millipede orders (2,700 species). Finally, the cylindrical millipedes have the body form of a snake, with short legs and a heavily calcified exoskeleton. The order *Julida* is a typical example of this body form (Figure 2.16).

Diplopoda span a range of body sizes not seen in any other soil invertebrate group. There are tiny diplopods, measuring a few mm, for example many polyxenids, but also very large species, up to 30 cm in length, which can be found in the orders Spirostreptida and Spirobolida. Diplopods thus cover the full range of mesofauna to macrofauna. Some of the tropical species have spectacular colour patterns, for example the "*Bombay train*", a yellow and orange coloured spirobolid occurring in South-East Asian rain forests.

Almost all millipedes feed on dead plant material, and the mandibles are of the biting and grinding type. However, exceptions to this general rule do occur. Hopkin and Read (1992) mention several examples of "atypical" millipedes with mouthparts specialized to feed on plant roots or fungi and even to be used in filter feeding. The cylindrical species are able to force their bodies into solid substrates such as wood and soil, by the action of the many legs. Several species of millipedes are able to dig into soil and feed on dead organic matter much like earthworms. Because of their relatively large size and often high densities (50 to 250 individuals per m^2, see Chapter 3), millipedes are considered an important factor in processing of organic material, especially in undisturbed habitats such as woodlands and ungrazed chalk grasslands. Because of their high calcium content, millipedes are an important food source for ground-feeding birds.

Millipedes occur in a variety of habitats. The most likely ancestral habitat is the surface and litter layer of wet forests and the species living there are called "*stratobionts*". From these life-forms five evolutionary tendencies can be discerned (Kime and Golovatch 2000, Golovatch and Kime 2009):

- increase in body size with either shortening or elongation of antennae and legs
- living in caves, with loss of eyes and decolouration of the integument (*troglobionts*)
- living in soil, with a smaller or elongated body (*geobionts*)
- living in dead wood (*xylobionts*)
- living in the vegetation (*epiphytobionts*)

This system is comparable to the classification of life-forms in Collembola and earthworms (*cf.* Chapter 1).

The second major group of Myriapoda is *Chilopoda*. The scientific name (derived from the Greek χιλιοι = thousand) suggests a larger number of legs than the English name centipedes (from the latin: centum = hundred), however, in both names the number of legs is exaggerated because most centipedes actually have 15, 21 or 23 pairs of legs and only in some geophilomorphs it reaches a maximum of 177 pairs. Interestingly, the number of leg-bearing segments is always uneven (Edgecombe and Giribet 2007).

The most conspicuous feature of centipedes is the poisonous jaw, actually the first leg, which is used in immobilizing prey after capture. All centipedes are carnivorous and most of the surface-living species can run very fast. They are usually among the first animals to scare students of cryptobiota when lifting the bark of a dead tree trunk. Some tropical centipedes, which may reach a size of more than 20 cm, are indeed dangerous because of the amount of poison they can inject in a human body. In contrast to millipedes, centipedes do not accumulate calcium in their exoskeleton, and the tergites (dorsal plates of each segment) are sclerotized in the same way as in insects. All centipedes have a dorso-ventrally flattened body shape.

The class Chilopoda comprises around 3.300 species worldwide, all can be considered members of the soil fauna. Identification is based on ridges and pore fields on the body surface, the number of ocelli, the coxosternum (the underside of the head), and the gonopods. Centipedes show a great diversity in form, size and behaviour, which justifies their grouping in five orders (*cf.* Figure 5.24). These orders differ from each other in the number of legs, the length of the legs, the eyes, the spiracles and the form of the tergites. In Lithobiomorpha and Scutigeromorpha the number of segments and legs increases during development (*anamorphic development*), while in Scolopendromorpha and Geophilomorpha the number of segments is fixed once hatched (*epimorphic development*). In many groups there is an alternation of small and larger tergites, however each segment bears a pair of legs.

The last pair of legs extends considerably behind the body, similar to the antennae in front of the body.

The prey of centipedes varies with their size and microhabitat. Mites and springtails are favoured by lithobiid species, whereas earthworms, enchytraeids and dipteran larvae are the main prey for geophilomorphs (Poser 1988). The geophilomorph *Strigimia acuminata* has been reported to forage in groups, and this is assumed to allow capturing larger worms than when they would forage solitarily. Large scolopendrids may even attack mice. Due to their fierce hunting behaviour, centipedes have a profound impact on their prey populations.

Like in millipedes and many other soil invertebrates, centipedes show an interesting transition series of typically surface-active species (e.g., Lithobiomorpha), to true soil living species (many geophilomorphs). The phylogeny of Chilopoda (Edgecombe and Giribet 2007) suggests that the surface-active lifestyle, with fewer segments, represents the ancestral state, while soil-living, with an elongated body and an increased number of segments, is derived therefrom. This trend is also apparent within one order, for example the ecological differences between two lithobiid species, *Lithobius forficatus* and *L. variegatus*, can be seen as adaptations to a surface-active mode of life, versus a more secluded life in the soil itself (Lewis 1965).

In addition to the better-known millipedes and centipedes, Myriapoda includes two other classes, *Pauropoda* and *Symphyla*. Pauropods are related to millipedes in that some of their segments are fused to form diplosegments, which accordingly bear two pairs of legs. The number of these segments is smaller than in most diplopods, only 9 to 11. A peculiar trait seen only in pauropods is the branching of the antenna (*cf.* Figure 2.15). Pauropods are tiny (up to 1.5 mm in length) and may develop populations of around 1000 individuals per m^2 in deciduous woodlands and grassland soils (Lagerlöf and Scheller 1989). They are true soil invertebrates, eyeless, breathing through the skin and feeding on a great variety of organic resources in soil, including decaying plant material, fungi and carrion. Scheller (2008) proposed pauropods to be classified into two orders, Hexamerocerata, with one family, and Tetramerocerata with eight families. Around 780 species are known.

The other group of "lower myriapods", symphylans, resembles centipedes more than millipedes (Figure 2.15). Like pauropods, symphylans are small in size: 0.5 to 8 mm, however, their body organization has some features which indicate a higher state of development, such as tracheae in the first three segments. They may reach densities of around 300 individuals per m^2 in grasslands and savanna soils. Also, like centipedes, symphylans are mostly predatory but being confined to soil and litter they are not active at the surface. The class Symphyla includes two families, the larger-sized Scutigerellidae and the smaller Scolopendrillidae, which together contain approximately 120 species.

2.7 Spiders, mites and the like

The last terrestrialization of the arthropods to be discussed here, but actually the first one in evolutionary time, is due to chelicerates (*cf.* Figure 2.14). This is a highly diverse group with both very small and fairly large animals, with segmented legs and

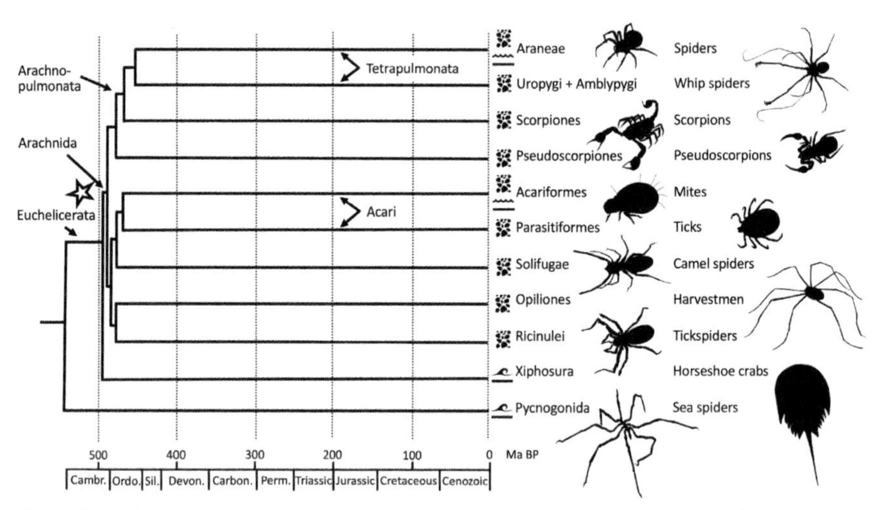

Figure 2.17. Evolutionary relationships between the chelicerate main groups. Tree topology and dating derived from Lozano-Fernandez et al. (2020).

a chitinous exoskeleton, thus easily recognized as arthropods, but also quite different from the other main groups, myriapods, crustaceans and hexapods. Chelicerates owe their name to the chelicerae, two pincers or fangs at the front of the body, which often carry poison glands and are used to immobilize prey. Chelicerates include well-known organisms such as spiders, mites, ticks, harvestmen and scorpions, plus a few smaller groups. The evolutionary relationships among these groups have been obscure for a long time, especially with respect to the positions of horseshoe crabs and scorpions. Some authors have maintained that horseshoe crabs (Xiphosura) cluster inside the Arachnida and so represent a marine regression within the terrestrial chelicerates (Ballesteros and Sharma 2019). However, we follow the majority point of view with Xiphosura as sister to Arachnida (Figure 2.17). The topology of our tree is taken from a recent study by Lozano-Fernandez et al. (2020), which involved 233 slow-evolving genes of 75 species of chelicerate and 14 outgroup arthropods. In addition, the splits in the phylogeny were dated with a molecular clock analysis calibrated on 27 fossil markers.

It turns out that many groups colloquially designated as spiders are not spiders in the strict sense (whip spiders, camel spiders, tick spiders and sea spiders). Only *Araneae* are considered true spiders. Araneae are part of the larger group *Tetrapulmonata*, which includes true spiders, whip spiders (*Uropygi*) and whip scorpions (*Amblypygi*); all have two pairs of book lungs, although some spiders have lost one pair. Based on the morphological similarity of their book lungs (Scholtz and Kamenz 2006), Tetrapulmonata are joined with scorpions in the group *Arachnopulmonata*. The remaining chelicerate groups are called *Apulmonata*. Pseudoscorpions are apulmonate and not monophyletic with scorpions, despite their outward resemblance. Further in the tree, ticks and mites are usually considered sister groups joined in the monophyletic lineage of Acari. Harvestmen appear to be related to tickspiders, not to mites, as the shape of their body would suggest. Tickspiders and harvestmen are at the basis of the Arachnida, which includes all terrestrial chelicerates, to the exclusion only of horseshoe crabs (Xiphosura) and sea spiders (Pycnogonida).

In this phylogeny, Arachnida is a perfectly monophyletic lineage nested in a marine stem group and so it is very likely that all terrestrial chelicerates descend from a single terrestrialization. This event is projected in the late Cambrian (Figure 2.17). It follows that aquatic groups among mites (water mites) and spiders (water spiders) are secondary adaptations, very much like the freshwater insects and aquatic Collembola.

The colonization of the land by chelicerates, like in myriapods and hexapods, opened up a lot of new opportunities. A rapid evolutionary radiation took place in the Ordovician and all major lineages present today date back to that period. At the same time, hardly anything happened with the stem groups that remained in the sea (horseshoe crabs and sea spiders). Horseshoe crabs are often presented in textbooks as a prime example of evolutionary stability, evolving through anagenesis, a mode of speciation in which subsequent species are connected by a single line of descent, with only one species being present at a time (Strickberger 2000). This contrasts with cladogenesis, the formation of clades, which is the typical pattern we see in the Arachnida (Figure 2.17). Horseshoe crabs are related to sea scorpions (*Eurypterida*), a now extinct group that dominated the seas in the Silurian and lived up to the Permian mass extinction of 252 Ma BP. Eurypterida and Xiphosura are commonly joined in the clade *Merostomata*.

A most peculiar group of chelicerates is present at the base of the tree, the sea spiders or *Pycnogonida*. These animals, sometimes jokingly referred to as "*no bodies*", have a free-swimming larva called *protonymphon* (Brenneis et al. 2017). This larval stage resembles a *nauplius*, the one-eyed free-swimming larva of crustaceans. This similarity has been used to argue for a relationship between chelicerates and crustaceans, however, this is not supported by molecular phylogenies (*cf.* Figure 2.14). One might speculate that a nauplius stage was ancestral to all arthropods, and was lost in the terrestrial lineages, because they all switched to direct development (see Section 4.8).

The body plan of a chelicerate is illustrated here by the anatomy of a spider (Figure 2.18). The body is divided into two "somas", *prosoma* and *opisthosoma*. These terms are preferred over cephalothorax and abdomen, because the division into tagmata is not homologous across crustaceans and chelicerates. The prosoma carries the *chelicerae*, a pair of *pedipalps* and four pairs of legs. While the eight legs serve the purpose of walking, the pedipalps usually have a sensory function, although they are sometimes involved in prey catching, food handling or carrying an egg cocoon. Internally, muscles dominate the prosoma, as well as a system of blind sacs from the intestinal tract that run up to and partly into the legs. These diverticula seem to have a circulatory function, distributing digestive products directly to the muscles, a function which is normally performed by the blood.

The opisthosoma has no legs, but caudally it carries six (sometimes two, four or eight) spinnerets, the "tits" of the silk glands. Just in front, a sieve plate (*cribellum*) adds very fine silk to the threads coming from the spinnerets. The spiders that have such a plate are called *cribellate*. More towards the front, the ventral side of the opisthosoma has the genital opening, which in females is surrounded by a plate with

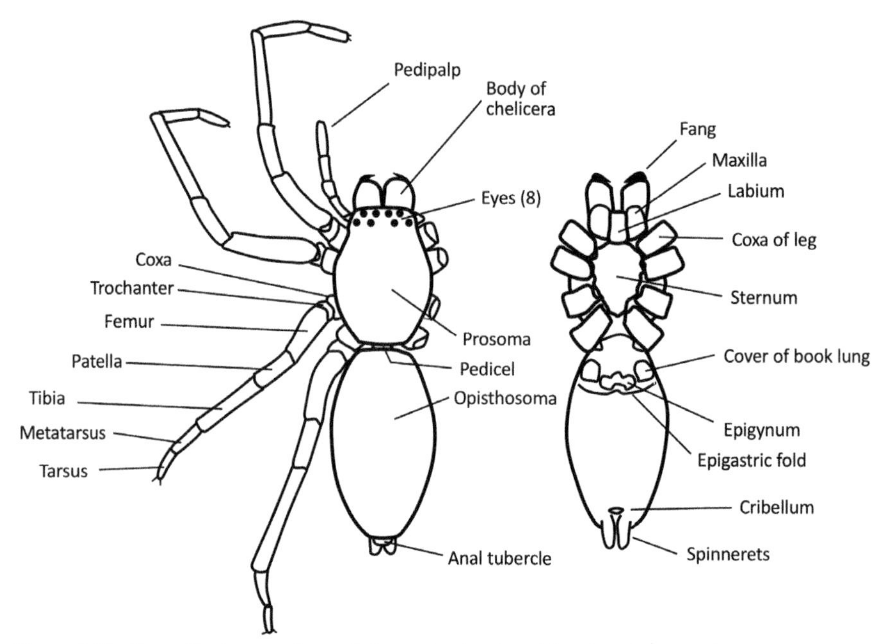

Figure 2.18. Generalized external morphology of a female spider, dorsal view (left) and ventral view (right).

a species-specific structure, the *epigynum*. The shape of the epigynum matches the genital palp of the male.

Archnopulmonate chelicerates (spiders and scorpions) have one, two or four pairs of *book lungs*. These consist of infoldings of the integument, stacked like the leaves of a book. Apulmonate chelicerates such as pseudoscorpions and harvestmen do not have book lungs but a series of *tracheal tubules* opening in spiracles in the opisthosoma. In horseshoe crabs the folds extend outward and are visible as a package of "*book gills*" under the opisthosoma. All these respiratory structures plus the spinnerets can be considered serial homologs of the legs like the abdominal appendages of Collembola. However, despite earlier suggestions (Carroll et al. 2005) they are not homologous to the tracheal tubes of insects, as these develop from the epipodite of the leg, rather than from the leg itself (Sharma 2017). It seems that chelicerates and hexapods have found different solutions for respiratory efficiency on land. The homology relationships of arthropod appendages will be discussed further in Section 4.9.

Internally, the opisthosoma is occupied by the *midgut gland*, the gonads and the silk glands. An *anal tubercle* is present at the very end of the opisthosoma, but as many chelicerates digest their prey externally, they do not produce a lot of faeces and the actual anus is difficult to discern.

Below we visit in a little more detail the three most common chelicerate groups in the soil ecosystem: harvestmen, mites and spiders. *Harvestmen* ("daddy long-legs"), officially designated as Opiliones, are characterized by a fusion of prosoma and opisthosoma into a single body, the front part of which carries the extremely long legs. In some species the pedipalps are also elongated, giving the impression that the animal walks on ten legs.

All harvestmen have well-developed chelicerae, but they differ from the chelicerae of spiders. In spiders the chelicera consists of a basal segment and a fang (Figure 2.18), but in Opiliones the chelicera has three segments. Molecular studies suggest that a chelicera with three of four segments is the ancestral condition, while loss of the proximal segment has occurred in several chelicerate lineages, including the spiders (Giribet and Sharma 2015).

Opiliones are a large order of Chelicerata, with an estimated number of 6,500 species in 40 families. They have limited dispersal ability and therefore show a great degree of endemism. Their geographic distribution coincides well with their phylogenetic history, which is exceptional for organisms that have been on earth for such a long time. Recently the species *Phalangium opilio* has emerged as a developmental model due to its tractability under laboratory conditions (Giribet and Sharma 2015). Research on this species has provided important insights into the evolution of chelicerate structures, such as the chelicera and the book lungs (*cf.* Section 4.9).

All Opiliones are predators of small arthropods and worms; they are also scavengers, feeding on dead insects. Harvestmen are often seen to carry mites on their legs. Some of these mites are ectoparasites, but most of them are phoretic; they just use the long legs of daddy to get somewhere else. The role of *phoresis* in soil invertebrate dispersal will be discussed further in Section 3.9.

Mites and ticks (Acari) are the most abundant arthropod group in soil, challenged only by Collembola. Their evolutionary success is apparent from the fact that they are well represented not only in soil but also in above-ground vegetation, including the *phyllosphere*. To date, 40,000 species have been named, classified into 540 families, however, the true number could be 500,000, or even more (Van Dam et al. 2018). Ticks are usually separated from mites, but the difference is small. Both suborders, Acariformes and Parasitiformes, include free-living species as well as ectoparasites, although all true ticks (Ixodoidea) are classified under Parasitiformes. Ticks receive special attention outside soil ecology because many present a threat to human health (including soil ecologists frequenting the field!), not so much because of their blood-feeding but because of transmission of pathogenic bacteria, e.g., *Borellia burgdorferi*, causing Lyme's disease and viruses such as TBE virus, causing tick-borne encephalitis (Bogovic and Strle 2015).

The classical chelicerate division of the body into prosoma and opisthosoma is not applicable to mites, instead, the main articulation of the body is between the *gnathosoma*, which carries the chelicerae and pedipalps and is homologous to the frontal part of the prosoma, and the *idiosoma*, the rest of the body, which comprises the four walking legs and the opisthosoma. This body division also holds for ticks, which insert the gnathosoma into the host.

The evolutionary position of the different parasitic and free-living lineages of mites, and their relationship to other Arachnida, have raised many a debate among acaralogists. The consensus view from morphological cladistics is that Acari, as well as Parasitiformes and Acariformes, are monophyletic (Dunlop and Alberti 2007). Several attempts have been made to establish mite phylogenies based on DNA data, but

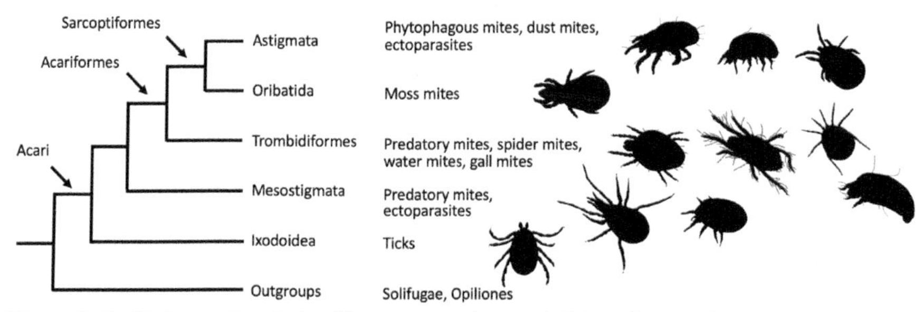

Figure 2.19. Phylogenetic relationships among mites and ticks. The tree is one of the possible phylogenies discussed in Van Dam et al. (2018). Note that Parasitiformes (Ixodoidea plus Mesostigmata) is not monophyletic in this tree. Gall mites are classified here under Trombidiformes, but their position is uncertain. For outgroups see Figure 2.17.

this has not yet provided a stable topology, even in the age of genomics (Dabert et al. 2010, Klimov et al. 2018, Van Dam et al. 2018). One of the problems comes from differences between lineages in the rate of evolution (*heterotachy*). The diversity of habitat and food specializations within the Acari (parasitic, herbivore, detritivore and predatory lineages all mixed together) causes a diversity of selection pressures and evolutionary rates. In a molecular phylogeny, fast-evolving groups tend to get clustered together, a phenomenon known as "*long branch attraction*" (Lopez et al. 2002); this distorts the relationships with other lineages. For the moment we reproduce here a provisional phylogenomics-based tree discussed by Van Dam et al. (2018) (Figure 2.19).

According to this phylogeny, the Acari split into five different lineages, two of which (Ixodoidea and Mesostigmata) fall under the Parasitiformes, while the other three (Trombidiformes, Oribatida and Astigmata) are grouped as Acariformes. However, Parasitiformes is paraphyletic in this analysis and so not an appropriate category. The tree is obtained under one particular bioinformatics model, while other models give different outcomes. For instance, Astigmata is sometimes clustered as a pedomorphic lineage within Oribatida (Maraun et al. 2004), however, in other analyses Oribatida and Astigmata are resolved as monophyletic lineages in sistergroup relationship (Domes et al. 2007a, Van Dam et al. 2018). More extensive taxon sampling is needed to resolve this instability.

The ancient origin of mites suggests that they took part in the very first terrestrial food-webs that developed on land. Their small body size might have helped them to penetrate the interstitial media of the terrestrial environment (Schaefer et al. 2010). All mites became terrestrial, except some families that have secondarily adopted a freshwater lifestyle. An example of this are the *Hydrachnidia* (water mites), which fall into the group *Trombidiformes*. This is a diverse lineage including also many predatory mites, as well as gall mites (*Eryophioidea*), another specialized group, which has only two pairs of legs and an elongated, worm-like body. However, it is also possible that gall mites constitute a separate lineage outside Trombidiformes, together with some other vermiform mites (Klimov et al. 2018).

Mites in soil are dominated by the suborder *Oribatida* (formerly *Cryptostigmata*), a group of some 9,500 species, many with very large parthenogenetic populations and reaching densities well over 20,000 per m^2. Also, several species of oribatid

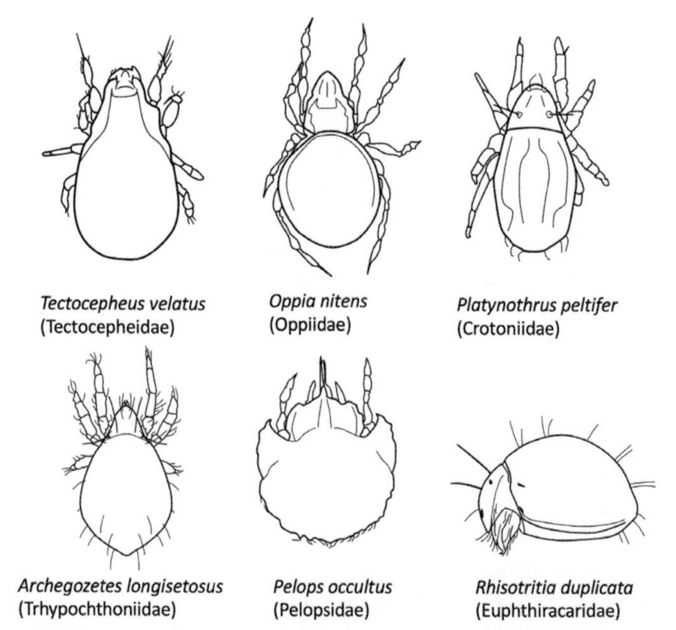

Figure 2.20. Examples of oribatid mite species used in experimental biological research.

live among moss and lichens on tree bark (hence the common name "*moss mites*"). Oribatid communities are known for their high local species richness, a phenomenon which seems to conflict with the hypothesis that similar species cannot coexist in ecosystems with relatively little physical complexity. This has been described as the "*enigma of soil animal species diversity*" (Anderson 1975). However, trophic level indicators such as $^{15}N/^{14}N$ ratios indicate that a wide niche differentiation exists among oribatids. In fact, oribatids alone span the complete range of feeding habits: carnivory, scavenging, fungivory, detritivory, as well as herbivory (Schneider et al. 2004).

Oribatid mites also illustrate several basic principles of evolutionary biology, such as the evolution of parthenogenesis, as we will see in Chapter 4. Because oribatids are closely associated with the organic matter and the microbes upon which they feed, their community composition is a fair reflection of the soil's condition, a fact that is exploited in countless bioindication studies (Van Straalen 1998). In Figure 2.20 some of the better investigated laboratory mites are shown.

Oribatida include the toughest animals on earth, the box mites or phthiracarids (Schmelzle and Blüthgen 2019). These mites are able to double-fold their body, retract their appendages in a groove, and turn themselves into a smooth ovoid ball. In this condition, they are able to resist pressures up to 560,000 times their body weight, obviously a strategy to escape predation. By the same mechanism, but not as an adaptation, they escape from the forceps of a soil ecologist trying to sort soil-extracted samples! Oribatids are also known for their peculiar mineral composition. The species *Platynothrus peltifer* has body concentrations of manganese and zinc higher than any other animal known (Janssen and Hogervorst 1993).

Another abundant group of mites is the largely predatory Mesostigmata (*cf.* Figure 2.19). This group includes Gamasina (also called Parasitidae), Uropodina and Phytoseiidae. While the latter mites are mainly associated with the phyllosphere, gamasids and Uropodina are common in soils. Uropodina include (in addition to predators) many tiny, slow-moving fungivorous scale mites and phoretic ectoparasites of insects, e.g., the varroa mite of bees. Gamasids are voracious feeders on oribatids, springtails, insect larvae and small worms (Koehler 1999). Their chelicerae are modified to pierce any moderately soft-skinned prey. Predatory mites are ideal natural enemies of plant pests, since they are extremely flexible in their choice of prey, yet can be selected to prefer specific pest species (Lesna and Sabelis 1999, Lesna et al. 2000).

Finally, many soil-living mites are found in the group Astigmata. This class consists of fungal and detritus feeders, as well as some predators. Several species are pests of stored products (e.g., the cheese mite *Tyrophagus putrescentiae*). Others are associated with specific habitats, e.g., feather mites, living as ectoparasites on birds and in bird nests, and house dust mite, *Dermatophagoides pteronyssinus*.

Switching to the last chelicerate group, we meet the impressive diversity of spiders. The order Araneae (*cf.* Figure 2.17) contains more than 41,000 described species, while 170,000 are expected in total (Coddington and Levi 1991). This illustrates the success of the spider body plan. There are more than 30 different families, which are distinguished from each other by the number and position of the eyes, the structure of the chelicerae, the shape of the spinnerets and many other characters. The epigynum (Figure 2.18, Figure 5.3) is a very important character for identifying female spiders to species. In the males, the tarsal members of the palps are modified to form a specialized structure which can receive and store sperm and which is used to pump sperm into the female genital opening during copulation (Figure 5.3). The shape of the sperm pump is a very important diagnostic character for identifying male spiders.

Among the enormous diversity of spiders we mention six families that are often encountered in field studies using core samples or pitfalls. Gnaphosidae and Clubionidae are nocturnal hunters on the ground and spend the day sitting in a little sack under a stone or under bark. They are usually dull coloured, brown, grey or black, however, some species that hunt during the day are brightly coloured. Agelenidae weave a funnel-like web close to the ground and wait for prey falling in. Thomisidae or crab-spiders are recognizable by their body outline; the animals hold their legs like a crab and often walk sideways. They do not spin a web but catch their prey from an ambush. Lycosidae or wolf spiders are known by their brown and black colours, their basking and rapid running across flat and open terrain. Wolf spiders are fierce hunters. The females of these spiders are often seen carrying an egg cocoon fixed to their abdomen. Linyphiidae constitute a very large family, the second largest family of spiders, unfortunately difficult to identify, especially the juveniles. Spiders of this family weave a web the shape of a sheet or hammock; many of these webs can be seen on a dewy morning in the open field or on the forest floor. The spider rests under the web while prey that falls in from above is dragged through the web.

All spiders are predatory and given their abundance it is no wonder that they have a considerable controlling influence on prey populations, which may consist of Collembola, mites and other small arthropods. Spiders may also be cannibalistic, especially as a strategy to survive periods of low prey abundance. The regulating capacity of spiders on their prey is obvious from a classical study conducted by Clarke and Grant (1968) who observed an increase in the density of springtails in the litter layer of a woodland after removal of spiders. Spiders are often considered as "beneficial" arthropods in agricultural systems, where they may contribute to the suppression of pest insects (Marc et al. 1999).

Spiders are known for the webs made from the product of their six silk glands, but they use the silken threads not only for preparing a web, but also for encapsulating prey, for making egg cocoons, for moving about (using "drag-lines") and for dispersal. The latter activity involves the so-called *ballooning* behaviour, seen in many erigonids and linyphiids, by which small spiders can sail through the air while hanging on a thread. We will discuss the dispersal of soil invertebrates in more detail in Chapter 3.

2.8 Flatworms, earthworms and potworms

The phylum Platyhelminthes includes many parasitic worm-like animals, such as trematodes and cestodes, but also free-living groups, historically classified as Turbellaria, although this taxon is not monophyletic. Among the various free-living flatworms, the order Tricladida includes common representatives from the benthic environment of freshwater habitats, such as *Polycelis* and *Planaria*. A separate lineage of tricladids, *Terricola*, deserves a position among soil invertebrates. An introduction to the systematics and biogeography of these peculiar animals, that are mostly known from humid tropical ecosystems, is given by Winsor (1998), Winsor et al. (1998) and Sluys (2019).

Terrestrial flatworms gained attention when in the 1970s a species from New Zealand (*Arthurdendyus triangulatus*) invaded the British Isles. It turned out to be a fierce predator of earthworms. Another species, introduced from Australia, had a similar effect (Jones et al. 2001a,b). These introductions most likely resulted from transport of plants and associated soil. Flatworms also established themselves in mainland Scotland, Orkney and the Faroer islands and in some places depleted earthworm populations, especially in soils of anthropogenically influenced habitats (Boag et al. 1994, Christensen and Mather 1995). Concern arose that elimination of earthworms by flatworms might have a detrimental effect on soil structure, however, a large-scale comparative analysis suggested that most earthworm species are able to adapt to the predation pressure exerted by flatworms and that increased water-logging observed in flatworm-infested fields might also be caused by lack of ploughing (Santoro and Jones 2001, Jones et al. 2001a).

Recent reconstructions of the evolutionary relationships among free-living flatworms have opened up interesting insights on terrestrialization scenarios in this group (Álvarez-Presas et al. 2008, Riutort et al. 2012, Benítez-Álvarez et al. 2020). Figure 2.21 shows a simplified phylogeny based upon these studies. At the basis of the Tricladida is a mostly marine group, *Maricola*. Next, we find a group of cave-associated freshwater flatworms with only few species and a fragmented circumtropical distribution, *Cavernicola*. The remainder of the tree is populated by

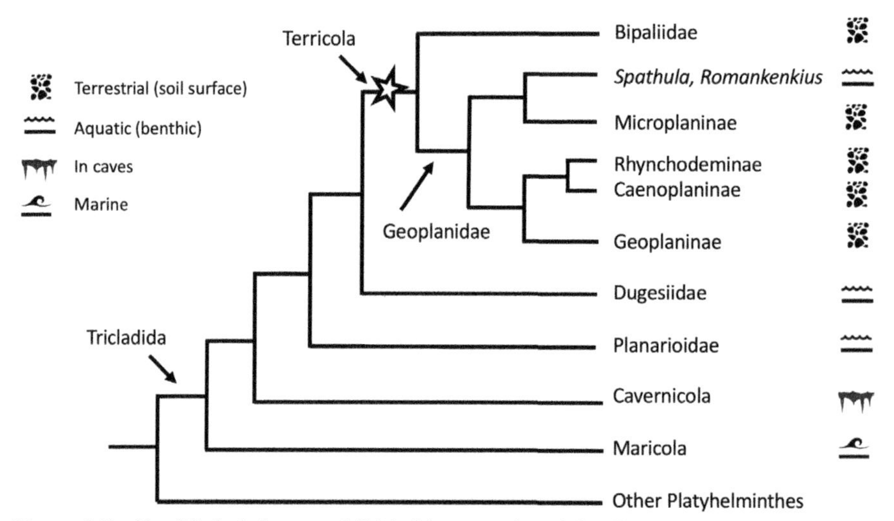

Figure 2.21. Simplified phylogeny of Tricladida, an order of free-living flatworms, showing the interrelationships of the main lineages, with special attention for terrestrial representatives (Terricola). The tree is composed out of molecular phylogenies (16S rRNA, 28S rRNA and COI genes) published by Álvarez-Presas et al. (2008), Riutort et al. (2012) and Benítez-Álvarez et al. (2020), condensed to the family level, except for the subfamilies of Geoplanidae that are specified to visualize the position of two aquatic genera (*Spathula* and *Romankenkius*) within a cluster of terrestrial species.

freshwater species, except for the *Terricola,* which are nested inside the freshwater groups. Within the terrestrial Terricola there are again two freshwater genera.

Ancestral state reconstruction based on this tree suggests that a single colonization took place from the marine environment towards a freshwater habitat. From there, one lineage colonized caves and groundwater (present Cavernicola), but many remained aquatic (Planarioidea and Dugesiidae) (Benítez-Álvarez et al. 2020). Out of the latter a new lineage colonized the terrestrial environment (present Terricola), while in that group a few genera returned to the freshwater environment (Álvarez-Presas et al. 2008). This scenario is very comparable to the terrestrialization of hexapods discussed earlier in this Chapter. The terrestrialization of flatworms may be one of the oldest among soil invertebrates (Sluys 2019), but a calibrated time-tree showing this is lacking at the moment.

Terricola have a slug-like appearance, are often very colourful and live in moist but not completely wet habitats, mainly in the tropics. They are all predatory. The mouth is located in the ventral surface, in the middle of the body (Figure 2.22). When feeding, the *pharynx* is everted into the prey, which is then digested externally. Digestive products are assimilated in the gut which consists of three limbs, one extending median to the front of the body, the other two extending posterior (hence the name Tricladida). The ecology of Terricola is still badly known, the reader is referred to Winsor et al. (1998) for more information. Terrestrial flatworms are extremely difficult to sample, because they are sticky and may easily fragment into pieces. They may also induce themselves to autolysis ("*melting*") when neglected. Winsor (1998) describes methods for collecting, handling, fixation and storage of these little-known animals.

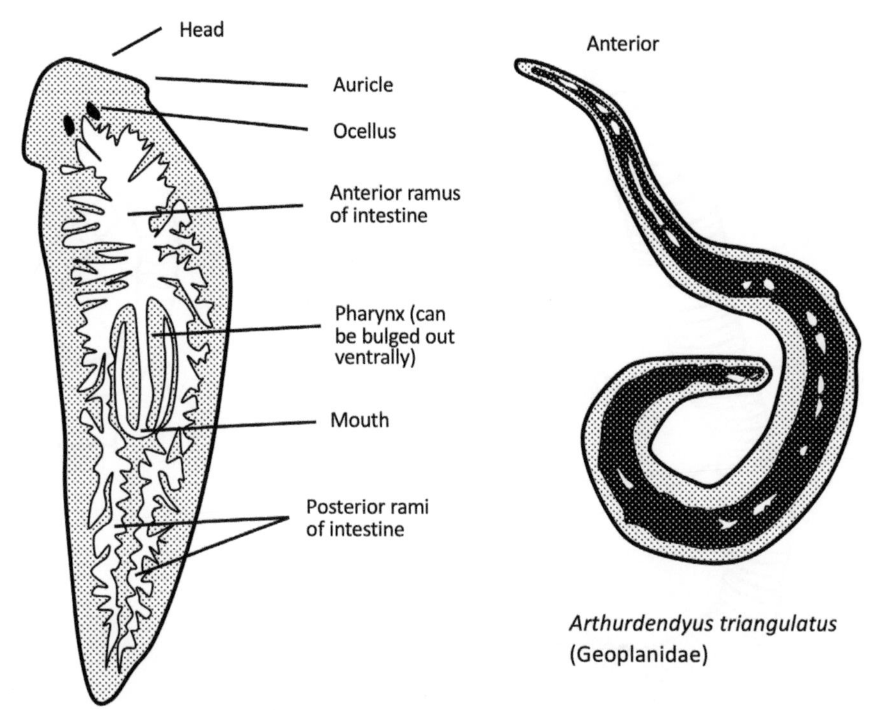

Figure 2.22. Left: general body plan of a tricladid flatworm (Platyhelminthes, Tricladida), showing the position of the mouth halfway the animal, absence of anus and the three gut diverticula (one to the front, two backwards). Right: habitus of the New Zealand flatworm (real size 5 – 20 cm).

Although flatworms are called worms, the true worms, as any zoologist will agree, are in the phylum Annelida. It is the annelid worm that deserves the name "*King of the Soil*". To celebrate the importance of earthworms, British soil ecologist and multitalented thinker John Satchell (1923–2003) called on the audience of the VIth International Soil Zoology Colloquium held in Uppsala, 1976, to consider the "*pedosphere harmony*", with dipteran larvae playing flute, springtails playing saxophone, but earthworms, sounding above all, playing the trombones of the grave (Satchell 1977).

Despite the tremendous diversity of soil-living arthropods discussed above, earthworms are considered the greatest engineers of soil ecosystems, controlling soil porosity, vertical mixing of organic material, redistribution of microbial communities, and organic matter decomposition. These functions of earthworms, investigated ever since Darwin (1881), are highlighted in several books (Edwards and Lofty 1972, Satchell 1983, Edwards 2004, Lavelle and Spain 2005, Karaca 2011) and will not be repeated here. In this book we will deal with the evolutionary questions: how did earthworms conquer the land, what terrestrialization scenario applies to them and how did they come to be soil organisms *par excellence*?

The external morphology of earthworms is only little diversified. Species identification uses the placement of the clitellum, bristles and gonopores, but this offers only few diagnostic characters. Reliable identification must use internal

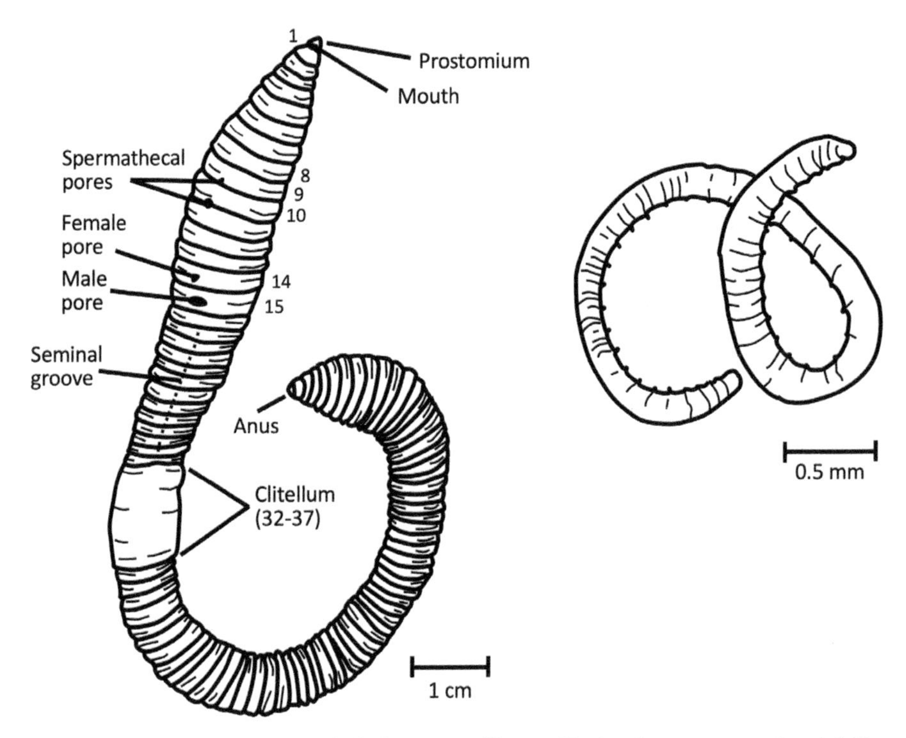

Figure 2.23. External morphology of clitellate worms illustrated by *Lumbricus terrestris* (Lumbricidae, left) and *Cognettia sphagnetorum* (Enchytraeidae, right).

characters such the morphology of the *spermatheca* and accessory glands. Juveniles are notoriously difficult to identify. A considerable degree of cryptic diversity has therefore remained concealed in many earthworm taxa until the advent of molecular approaches (Huang et al. 2007, King et al. 2008, Klarica et al. 2012). Even common earthworms such as *Lumbricus rubellus*, *Allolobophora chlorotica* and *Allolobophora caliginosa* appear to consist of two, even three lineages, that differ from each other enough to call them separate species (King et al. 2008, Fernández et al. 2012). The situation is even worse in the Enchytraeidae, where hybridizations between species add another layer of complexity (Martinsson and Erséus 2014, 2021).

Figure 2.23 shows some of the external characteristics of earthworms and potworms. The most conspicuous feature in adult worms is of course the *clitellum*, which is a glandular portion of the body associated with cocoon production. In lumbricids it is usually swollen but in other groups it is hardly differentiated from the rest of the body. In megascolecid worms it appears as a constriction, rather than a swelling. The position of the clitellum and the number of segments it covers, differ considerably between species. In lumbricids the clitellum is in the front of the body, behind the genital pores, starting at segment 22 or further on (Edwards and Lofty 1972).

As both earthworms and potworms are *hermaphroditic* (although with several exceptions, to be discussed in Chapter 4), the external morphology shows both female and male gonopores, which are in different segments or grooves behind each other.

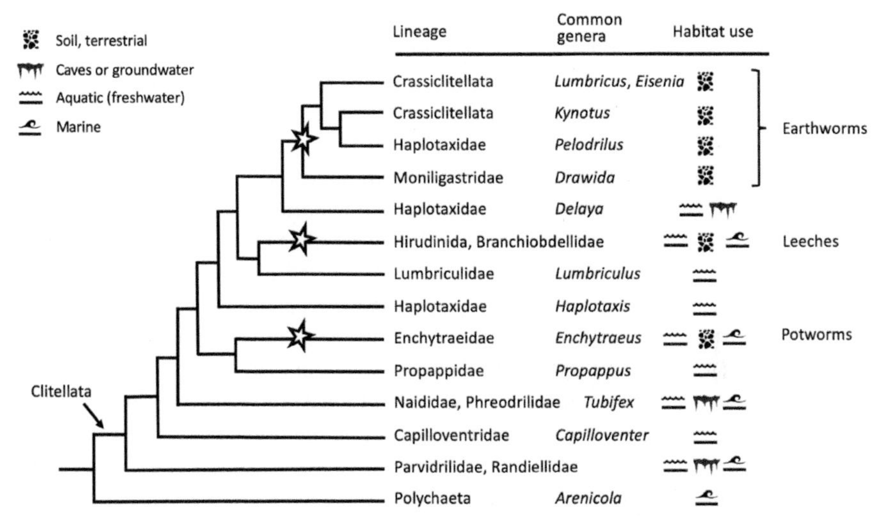

Figure 2.24. Phylogeny of Clitellata, derived from transcriptome analysis (Erséus et al. 2020) for 63 clitellate species plus 10 outgroup taxa (polychaetes). Only the main lineages are shown (aquatic species were condensed to families). The habitat use indicated was adapted with slight changes, from ancestral state reconstructions in Erséus et al. (2020). The terrestrialization events in Crassiclitellata, Enchytraeidae and Hirudinida are indicated by asterisks (not implying that the whole lineage is terrestrial). The family Haplotaxidae is not yet resolved (polyphyletic in this phylogeny). The suborder Crassiclitellata, although usually considered monophyletic, is paraphyletic in this phylogeny, unless *Pelodrilus* is included.

During hermaphroditic mating, two worms align with the front ends in opposite directions, and the ventral sides pressed to each other such that the spermatheca of one worm is close to the clitellum of the other. A seminal groove extends along the body from the male aperture to the clitellum. After fertilization a cocoon is formed around the eggs in the shape of a half-ring, which is then forced sliding to the front of the animal and shed in the soil. There are many interesting evolutionary questions surrounding the hermaphroditism of earthworms and potworms that we will discuss in Chapter 4.

Earthworms are annelid worms belonging to the class *Clitellata* (worms with a clitellum, see Figure 2.23). This monophyletic group includes all freshwater and soil species, to the exclusion of almost all marine annelids, which belong to the class *Polychaeta*. The taxon Oligochaeta is not used anymore because this excludes leeches (Hirudinea), which according to modern phylogenetic insights cluster within the Oligochaeta (Erséus 2005). The larger (monophyletic) group, including leeches, was therefore renamed Clitellata.

Within the Clitellata there are two groups, both monophyletic lineages, that we recognize as soil-living worms, Crassiclitellata and Enchytraeidae; the other groups are all freshwater and marine (Figure 2.24). Crassiclitellata includes only terrestrial species. Enchytraeids (also designated as potworms) are not directly related to Crassiclitellata but have their own origin in the tree; this family includes marine, freshwater as well as terrestrial species. Among leeches there are also several terrestrial species, but as these move mostly in the vegetation, they are usually not included in the soil invertebrates.

How do the terrestrial clitellate worms relate to the marine and freshwater groups? Several authors have addressed this question by means of molecular phylogenetics, mostly using mitochondrial genes such as 16S rRNA and cytochrome c oxidase subunit I (Jamieson et al. 2002, Erséus and Källersjö 2004, Kaygorodova and Sherbakov 2006, Marotta et al. 2008, Struck et al. 2011, Christoffersen 2012, James and Davidson 2012, Kvist and Siddall 2013, Domínguez et al. 2015). We focus here on a recent analysis published in 2020, which used genome-wide transcriptomics for 64 clitellate species plus 10 outgroup taxa (Erséus et al. 2020). A simplified version of the phylogeny in this paper is reproduced in Figure 2.24.

The figure shows that the lineage of Crassiclitellata, which in itself is wholly terrestrial, is nested inside freshwater lineages of clitellate worms. It is therefore highly likely that all present earthworms find their origin in a single terrestrialization event by a freshwater, not marine, ancestor. That ancestor itself must have had its origin in a marine lineage, related to a worm group in the Polychaeta, most likely the tiny polychaetes in marine sediments. The phylogeny of Crassiclitellata is a superb illustration of the GKGB hypothesis discussed in Chapter 1.

The origin of the Clitellata is projected in the Devonian, between 371 and 403 Ma BP, while the terrestrialization of Crassiclitellata is dated between 141 and 204 Ma BP, in the Jurassic (Erséus et al. 2020). So, both the marine-freshwater transition and the freshwater-terrestrial transition happened in annelids much later than in arthropods. For more than 250 million years, from the Ordovician to the Jurassic, terrestrial arthropods never met any earthworm.

Crassiclitellata includes 18 families of earthworm (James and Davidson 2012), which show a distinctive biogeography and an appreciable degree of endemism. The correlation between earthworm communities and geography was already noted by the German zoologist Johann Wilhelm Michaelsen (1860–1937) whose ideas about earthworms were seen as evidence for the then radical theory on plate tectonics by his colleague at the University of Hamburg Alfred Wegener (1880–1930) (James 1998). Ecologists have also noted that latitude is the principal factor discriminating earthworm communities worldwide (Lavelle 1983). The main split is between the families of the Northern Hemisphere (Lumbricidae, Homogastridae and Criodillidae) and the Southern Hemisphere (Megascolecidae, Microchaetidae, Rhinodrilidae, Almidae, Glossoscolecidae and Eudrilidae) (Anderson et al. 2017). The date of this split falls in the Mesozoic (161–185 Ma BP) and correlates with the break-up of Pangea 175 Ma BP.

It is also obvious from Figure 2.24 that enchytraeids have an origin different from crassiclitellates and they are not a sister group of the earthworms. Instead, enchytraeids descend from another freshwater group. A molecular phylogeny developed by Christensen and Glenner (2010) suggests that enchytraeids colonized the land more than once. One group (species of the genera *Enchytraeus* and *Lumbricillus*) specialized on feeding on sea weeds washed ashore in intertidal flats and beaches, another group (*Mesenchytraeus*, *Cognettia*, *Fridericia*, *Buchholzia* and *Achaeta*) turned to feed on organic litter of inland forests. In addition, maybe because they have retained more aquatic characters than earthworms, we also see several freshwater and marine regressions in this group. So, the terrestrializations among

Enchytraeidae are more difficult to track than in earthworms (Erséus et al. 2010), not unlike the situation in isopods compared to hexapods.

2.9 Snails and slugs, finally

While among molluscs Bivalvia and Cephalopoda never made it to the land, Gastropoda have done so abundantly. The number of terrestrialization events in gastropods amounts to ten or more, and gastropods should be considered the most successful group of all soil invertebrates, except nematodes. However, like in most of the groups reviewed above, but unlike nematodes, secondary marine regressions are particularly rare in gastropods (Vermeij and Dudley 2000). Here we will review only the most obvious and best investigated gastropod terrestrializations.

A crucial innovation among land snails was the use of the mantle cavity (also called pallial cavity) as a respiratory chamber. All molluscs have a *mantle*, a fold of the body wall that runs from the dorsal side of the animal (where it makes the shell) to lateral, forming a cavity around the visceral mass. This cavity may be closed on the ventral side, leaving only a siphon to connect it to the outside. In the mantle cavity the gills are suspended, while also the anal, genital and renal pores open on its brim. When used as a lung, the dorsal surface is vascularized to expedite the uptake of oxygen. In fully terrestrial, air-breathing, snails, the gills have disappeared; these animals rely on their lung completely. Some gastropod lineages show interesting transitional stages. In Section 4.10 we will discuss the respiratory adaptations of land snails in more detail.

The lung of land snails did however, not evolve on land, it evolved in the water, as witnessed by so many freshwater lung snails ancestral to land snails. Most aquatic snails will also respire through the skin, e.g., pond snails of the genus *Lymnaea*. They preferably open their breathing hole of the lung (*pneumostome*) at the water surface, but if forced underwater will survive by skin breathing. In humid environments, land snails will also breath partly through their skin.

The external morphology of land snails is shown in Figure 2.25. The best-known group is the monophyletic lineage Stylommatophora, which includes many common members of the soil invertebrate community, snails and slugs of the genera *Helix*, *Limax*, *Cepaea* and *Arion*. They are all scavengers and litter feeders, not spurning green leaves of low vegetation. However, terrestrial snails are also found outside the Stylommatophora, in the lineages Systellommatophora and Ellobioidea (Figure 2.26, Romero et al. 2016a). These three groups fall into the monophyletic clade Panpulmonata, which includes the (paraphyletic) Pulmonata plus some lineages previously classified as Opisthobranchia, e.g., Pyramidelloidea (Kocot et al. 2013; Figure 2.26). The classical subclass subdivision of Gastropoda into Pulmonata, Opisthobranchia and Prosobranchia has become obsolete since none of these groups turned out to be monophyletic when DNA analysis made its way to the snails.

The superfamily Ellobioidae illustrates some other interesting aspects of snail terrestrialization (Romero et al. 2016b). Ellobiidae ("hollow-shelled snails") are a characteristic component of mangrove forests in tropical regions worldwide but also occur in temperate forests. The group includes several genera that are truly terrestrial, e.g., *Pythia* (in Papua New Guinea rainforests) and *Carychium* (in forest floors of

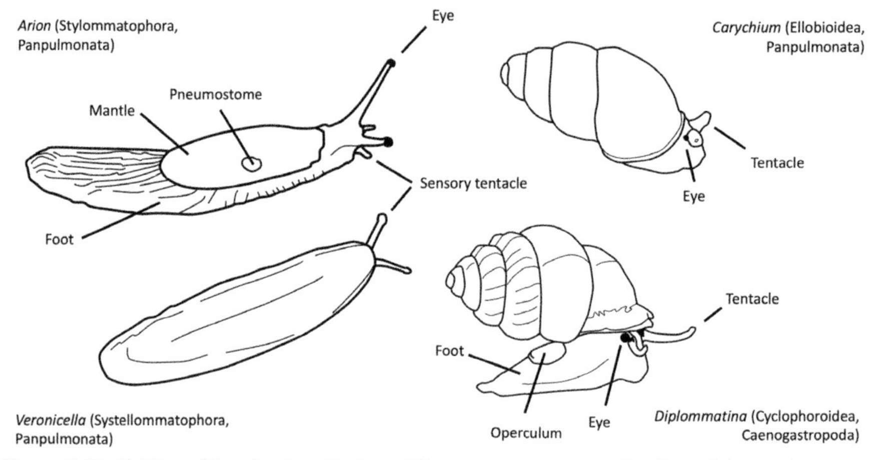

Figure 2.25. Habitus of four land snails from different groups, representing four of the maybe ten or twelve independent terrestrializations in gastropods.

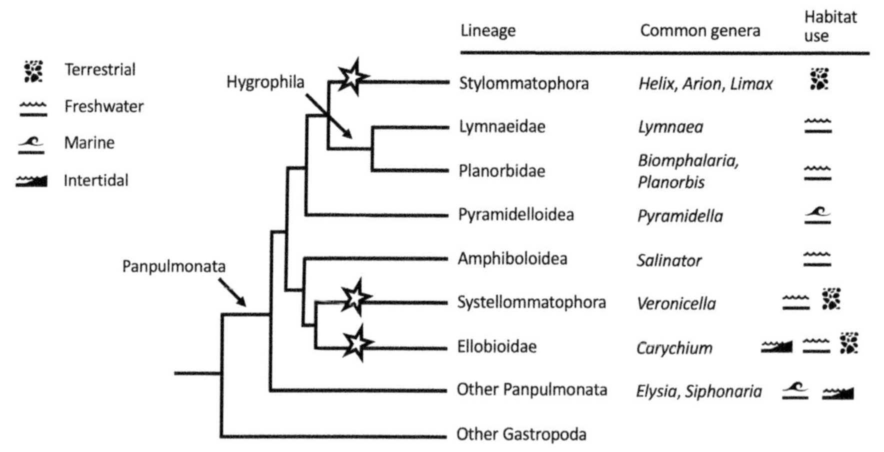

Figure 2.26. Simplified phylogeny of Panpulmonata, one of the major monophyletic clades of Gastropoda. For sake of clarity several of the aquatic lineages were condensed. The phylogeny shows three major terrestrialization events, indicated by stars. Tree topology is derived from Romero et al. (2016a).

the Holarctic). Phylogenetic analysis of Ellobioidea shows that these genera alone represent two independent terrestrializations (Romero et al. 2016b).

A simplified rendering of Panpulmonata phylogeny, derived from Romero et al. (2016a) is shown in Figure 2.26. It is obvious from this graph that the three major terrestrial lineages (Stylommatophora, Systellommatophora and Ellobioidae) are closely associated with freshwater groups. The superfamily *Hygrophila*, which includes many common freshwater snails of the temperate region, is a direct sister group of the land snails. The suggestion from this topology is that land snails of the Panpulmonata evolved out of lineages that had already colonized freshwater environments. The fact that they were already equipped with a lung while in the water may have contributed to the success of their terrestrial colonization. Terrestrialization

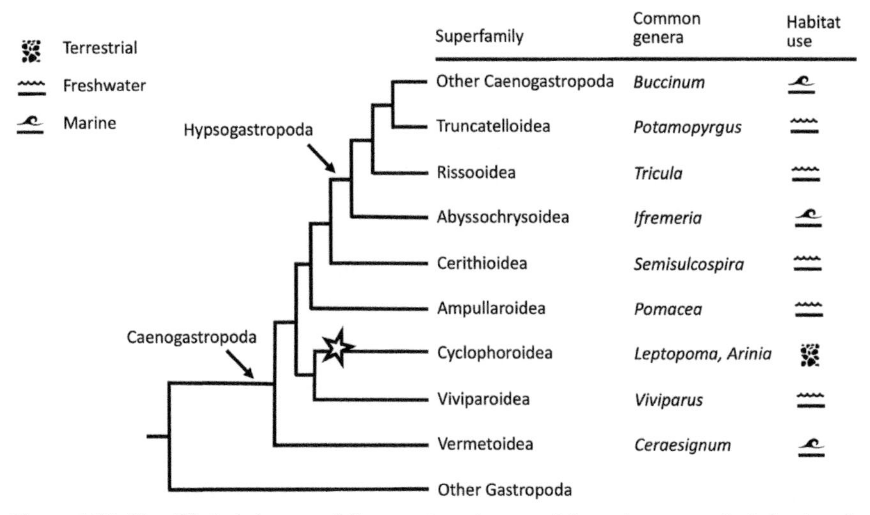

Figure 2.27. Simplified phylogeny of Caenogastropoda, one of the main gastropod clades, based on mitochondrial gene sequences. Tree topology is greatly condensed for marine lineages to highlight the position of terrestrial Cyclophoroidea among the many freshwater lineages. Adapted from Osca et al. (2015).

of Ellobioidae might have passed through an intertidal stage, given their abundance in tropical mangrove forests and temperate salt marshes.

Also outside the Panpulmonata several terrestrializations have occurred. The best known is the group Cyclophoroidea within the large clade of Caenogastropoda. This clade includes many lineages that previously fell under Prosobranchia, a now abandoned category (Osca et al. 2015). Like panpulmonate snails, Cyclophoroidea use the mantle cavity as a lung and have lost their gills. However, they still have the *operculum*, a shield to close off the shell when disturbed, very much like their marine ancestors. The large clade Caenogastropoda is mostly marine, but it includes several freshwater groups. In the caenogastropod phylogeny, the terrestrial Cyclophoroidea are completely surrounded by freshwater lineages (Figure 2.27), suggesting, like in the Panpulmonata, that terrestrial snails evolved out of freshwater ancestors, not directly from marine ones.

Cyclophoroidea is a diverse and species-rich terrestrial snail group, maybe the richest in species of any land snail group. They are often found on limestone rocks, but are also quite diverse in tropical rain forests where they are seen in the litter and on tree bark (Schilthuizen and Rutjes 2001). Most likely their biodiversity is grossly underestimated at the moment. The group has a wide geographic distribution in the Southern Hemisphere and is assumed to have arisen around the Permian-Triassic boundary, 250 Ma BP (Gervascio et al. 2017), making it the oldest terrestrialization among gastropods, anyway older than the land invasions by Panpulmonata, which are dated in the Mesozoic to Cenozoic (Romero et al. 2016a,b).

As a finishing touch on gastropod evolution, some remarks on the peculiar mechanism of *torsion* should be made. Torsion is the process, by which, early in the *veliger* stage, the visceral mass is twisted relative to the foot and the head. The turning is brought about by high cell proliferation at the right side of the animal, under the control of asymmetrically expressed developmental regulator TGF-β (Kurita and Wada 2011, Link et al. 2019). The turning is over 180 degrees, which brings the anus to the front of the animal, above the head. The direction of turning is linked to the coiling of the shell: while the shell usually coils dextral, the torsion is sinistral. The asymmetry finds its origin in a spatial shift of the spindle during the third (spiral) cleavage of the egg, triggered by a maternal factor added to the yolk (Abe and Kuroda 2019).

The issue of torsion has occupied zoologists for a long time. What is its function? The most popular hypothesis of all times is Garstang's theory which holds that the torsion allows the soft-bodied animal to retract itself inside the shell, and so is a predation escape mechanism. However, this suggestion does not find support in experiments (Pennington and Chia 1985). Lever (1979) forwarded a more attractive hypothesis, which holds that it is not the torsion itself, but the coiled shell that was selected. A coiled conical shell would provide a hydrodynamic advantage to snails living on exposed rocky shores, at the times when gastropods evolved in the Cambrian. The torsion was subsequently fixed in the developmental program and, not providing a great disadvantage, was maintained up to the present day, even when, once on the land, snails never saw the whirls of waves anymore.

2.10 Terrestrialization scenarios

The lineage-by-lineage review in the preceding sections generally supports the GKGB hypothesis formulated in Chapter 1 (Van Straalen 2021). Many soil invertebrates have their evolutionary ancestors in a freshwater environment. So, soil can indeed be considered as an evolutionary transition zone between water and air (Figure 1.1). However, the situation is more complex than envisaged by Ghilarov, as there are different scenarios for the pathway. First of all, root herbivores may be seen as an exception, as all of them except Collembola are secondarily adapted to soil conditions; their ancestors are aboveground, not in the water (Figure 1.16).

Another qualification of Ghilarov's hypothesis is that there are major and minor terrestrializations. The major events have occurred only once in a lineage, happened a long time ago (early in the Paleozoic) and came with new body plans and significant adaptive radiation on land. Examples are the terrestrializations of hexapods, myriapods and chelicerates. The "minor" terrestrializations happened many times in the same clade, are more recent (in the Mesozoic or Cenozoic) and did not change the body plan much. Examples are the terrestrializations of potworms, land snails and amphipods. The major terrestrializations are also more successful in the sense that the changes in body plan were more profound and gave rise to entirely new classes or new orders of animals, while in the minor events a worm stayed a worm and an isopod remained isopod.

Vermeij and Dudley (2000) discussed the question of rarity of marine regressions. They noted that in many cases, once a lineage has become terrestrial, there is hardly a way back. Insects are a case in point. This view is only partly supported by the present analysis. Nematodes, enchytraeids and groundwater amphipods are examples of groups where marine regressions have been quite frequent. In addition, marine regressions seem to be more common among minor terrestrializations than after major events. The fact that some metazoan clades show marine regression and others not suggests that it is not the increased competition pressure in the marine environment that is preventing them from going back (as suggested by Vermeij and Dudley 2000), but some aspect of the physiology or the body plan that became canalized in the developmental programme that prevents recurrence to a marine lifestyle.

A third way to look upon the GKGB hypothesis is to classify the terrestrializations with respect to pathway. We may delineate six modes:

1. *Commuting*. Nematodes have colonized the land so often and returned to the marine environment so many times that they may be considered "commuters". As a consequence, it is hardly possible to distinguish terrestrial and marine clades in the phylogenetic tree (Figure 2.1). This is related to a great physiological flexibility aided by a permeable integument (see Section 2.1).
2. *Direct*. Under this scenario marine invertebrates going on land skipped any intertidal or freshwater phase; terrestrial lineages evolved directly out of marine ancestors. The tardigrade phylogenetic tree (Figure 2.2) suggests this type of evolutionary pathway for Eutardigrada; the trees of onychophorans, myriapods and chelicerates (Figure 2.15 and Figure 2.17) seem to lack any intermediate stage in between the marine and terrestrial lineages.
3. *Intertidal*. The terrestrializations of woodlice, land hoppers and land crabs (Figure 2.4 and Figure 2.11), suggest a scenario in which invertebrates made their way to the land after having lived for an appreciable time in brackish intertidal environments, e.g., floodplains, coastal slacks, beaches and muddy shores like mangroves. Also the terrestrialization of ellobioid snails (Figure 2.26) might have involved an intertidal phase Under this scenario it is expected that the fully terrestrial lineages show traits that can be considered heritages of their intertidal past. The jumping legs of terrestrial amphipods may indeed be interpreted in this manner.
4. *Cavernicolous*. In several phylogenetic reconstructions, a cave-living lineage is positioned at the base of the tree, in between a marine ancestral lineage and the terrestrial lineages. This could suggest that coastal caves, formed during sea level changes and isolated from the marine environment, could have been an intermediate stage. While the present cave-dwelling lineages stayed in the caves and evolved into true troglobionts, other lineages made their way to the land, from cavernicolous ancestors. The phylogeny of pancrustaceans (Figure 2.4, Figure 2.5) suggests this type of scenario for the terrestrialization of hexapods. Also in the phylogeny of Tricladida a cave-dwelling group is present at the base of the tree, however, the terrestrial flatworms are nested in a freshwater cluster (Figure 2.21).
5. *Phreatic*. In principle, the groundwater could also be an intermediate medium in the marine-terrestrial transition, however, there is no phylogenetic evidence for this scenario. As we have seen in Section 2.2, groundwater invertebrates, which are especially prominent among crustaceans, descend either from freshwater ancestors or directly from marine ancestors. They should be regarded as ecological specializations of freshwater lineages. Their position in the phylogeny of crustaceans is at the tips of the tree, nowhere in the main splits.

There is hardly any evidence that groundwater invertebrates can be considered ancestral to a soil-living lineage (one example was mentioned in the paragraph on groundwater crustaceans).

6. *Freshwater*. This seems to be the most common scenario and it clearly holds for earthworms and potworms (Figure 2.24), as well as flatworms (Figure 2.21), stylommatophoran snails (Figure 2.26) and caenogastropods (Figure 2.27). These animals are all soft-bodied. In the aquatic environment, the problem of osmoregulation in a medium with low salinity was solved first, before these animals ventured onto the land. The next step, air breathing and drought resistance, were difficult to realize with the worm body plan and so these animals

Table 2.1. Summary of putative evolutionary terrestrialization scenarios for soil invertebrates.

Scenario	Pathway	Evidence from phylogeny	Examples
Commuting	Frequent transitions between marine, freshwater and soil environments, in both directions	No clearly separate marine, freshwater or soil clades in the phylogeny; habitat use with irregular distribution over tree	Nematodes
Direct	Invasion of the land from marine habitats without notable intermediate adaptations to intertidal or freshwater environments	Terrestrial lineages positioned directly in a cluster of marine representatives	Eutardigrades Onychophorans Myriapods Chelicerates
Intertidal	Intertidal habitats (muddy shores, beaches, floodplains) colonized from the marine environment intermediate; adaptations to the intertidal habitat that help terrestrial colonization	Terrestrial lineages with many intertidal representatives, transitional series of marine-terrestrial species	Isopods Amphipods Land crabs Echiniscoid tardigrades Ellobioid snails
Cavernicolous	Marine lineages isolated from the open sea in coastal caves formed by geological processes, Terrestrial lineages splitting from cave-dwelling invertebrates	Cave-living lineage in stemgroup position ancestral to a terrestrial species cluster	Hexapods
Phreatic	Marine lineages become isolated in groundwater habitats and from there colonize the soil	Groundwater lineage in stemgroup position ancestral to a terrestrial species cluster	No evidence as yet among Metazoa for this scenario
Freshwater	Marine lineages become isolated in swamps, pools or lakes that gradually lose their marine character, soil colonized from freshwater habitats	Terrestrial lineages clustering inside a group of freshwater species, which in turn are in a derived position relative to marine lineages.	Terrestrial flatworms Earthworms Potworms Panpulmonate snails Caenogastropod snails

remained tied to the groundwater or pore water. The gastropods were able to use a specific ancestral structure, the mantle cavity, for respiration, but in general also remained restricted to moist habitats.

These six scenarios are summarized in Table 2.1, including the evidence from the phylogenetic analyses reviewed above. It must be emphasized that all reconstructions are hypothetical; nobody has seen arthropods crawling out of the sea onto the land in the Ordovician. In addition to all kind of technical issues that come with deep phylogenetic analysis of DNA, such as *heterotachy* and *heteropecily* (Roure and Philippe 2011) there is also the problem of missing lineages due to extinction. Statistically underpinned ancestral state reconstructions still have to be made for many groups of soil invertebrates and a wider selection of taxa is necessary in order to distinguish the exceptions among the general pattern. The choice of outgroups (stem groups) is often critical to the analysis as well (Grandcolas et al. 2004).

As a way forward, comparative functional genomics may help. The scenarios proposed here are testable by analysing functional genes in the genomes of present terrestrial lineages and comparing them across a given phylogeny. For example, a significant crustacean signature can be found in the genome of the springtail *Folsomia candida* (Faddeeva-Vakhrusheva et al. 2017), providing support for the pancrustacean origin of hexapods. In groups with many independent terrestrializations, convergent evolutionary signatures can provide cues to the genomic adaptations to life in soil, as is illustrated by work on panpulmonate snails (Romero et al. 2016b). These molecular explorations will be further discussed in Chapter 6.

CHAPTER 3

Populations in Space and Time

Help us!

That Genesis-thing is beyond our memory
- how when, why, did who, what, which us create -
But once we 're with two, we start to variate
and for millions of years we colonized the sea.

When it becomes overcrowded, we migrate
in an explosive Cambrian scenery
to wet ground, hot sand, rotting plant debris.
But the sun, the wind, the cold aggravate fate

and ruthless peasants usurped what was free.
Ploughed, raked, pest controlled, we suffocate,
our diversity now threatened in highest degree!

We want better, yeah: we want to ..... emancipate.
Nico! Don't observe us, but HELP us, that's our plea!
Show how to become like you - a vertebrate.

Jasper Aertsz

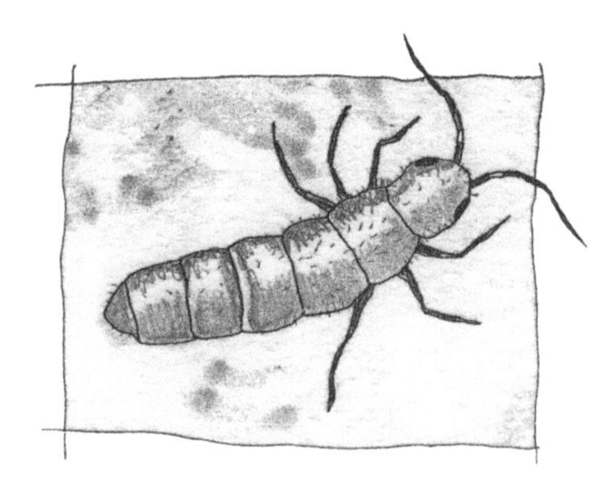

CHAPTER 3

Populations in Space and Time

"The pageant of the river bank has marched steadily along, unfolding itself in scene-pictures that succeeded each other in stately procession."

Kenneth Grahame, The Wind in the Willows

Populations are the units in which natural selection, favouring certain phenotypes over others, causes evolutionary change. An evolutionary biologist's definition runs as follows: "A population is a group of individuals of the same species that exchange genetic material and jointly live in a specific habitat". The size of a population determines how effective natural selection can be, relative to *genetic drift* (neutral evolution). In small populations the fate of new mutations tends to be dominated by drift. Even favourable alleles run a risk of getting lost by accident if the population is small. In large populations favourable genetic variants can steadily increase in frequency through selection, depending on the benefit they confer to the phenotype. Therefore, in this Chapter the first question is: how large are populations of soil animals?

Size and structure of a population are the result of dynamic processes that go on within it: recruitment, mortality, turnover, migration, dispersal, gene flow, etc. In this Chapter we will explore and review the many population studies conducted on soil invertebrates, including their life histories, dispersal mechanisms and genetic structure. Soil invertebrates are ideal animals for population studies in the first place, because their densities are easily measured, in comparison with mobile animals or animals with discontinuous distributions such as butterflies, lizards, birds and wildlife.

3.1 Population density and population size

Soil ecologists characterize populations usually not by their size but by their density. Population density is estimated by delineating a plot in the field, taking a certain number of random samples from that plot, often in a stratified scheme, extracting the fauna from these samples, counting the animals and extrapolating these counts, using the dimensions of the sample, to estimate the average number per unit area. Statistical techniques are applied to quantify the confidence limits of such density estimates. An overview of sampling designs is provided in Jiménez et al. (2020).

A great variety of methods is available for reliable extraction of animals from soil samples. A classical method is to drive all mobile animals out of a soil core by heat and gradual drying. A funnel below each sample concentrates the animals in a vial with preservative. This method is due to the Swedish entomologist Albert Tullgren

(1874–1958); a device with several replicate funnels is called a *Tullgren extractor*. Depending on the species, type of soil and design of the extractor, efficiencies of 80% and higher can be reached with this method (Van Straalen and Rijninks 1982).

The principle of Tullgren is often equated with an extraction method described by the Italian entomologist Antonio Berlese (1863–1927), but this is not correct. The apparatus of Berlese was actually quite different: it used a water bath to heat the sample from below, while Tullgren used a lamp to heat the sample from above (Berlese 1905, Tullgren 1918). The Tullgren method works best for small mobile arthropods (springtails, mites, beetles, spiders); inactive stages are not extracted. Larger springtails, spiders and beetles may be collected by hand-sorting, after sieving the litter above a white tray and using a battery-driven aspirator to collect them (Van der Laan 1963).

For nematodes, two methods are available, one relying on motility, the other on the physical properties of the animals (size and density). A common method based on motility is due to the German medical doctor Gustav Baermann (1877–1950), working on human-parasitic soil nematodes in the Dutch East Indies (Baermann 1917). The apparatus consists of a glass funnel filled with water upon which a sample in a gauze is held floating. Such "*Baermann funnels*" are also used for tardigrades and enchytraeids. Mobile animals escape from the sample into the water. The other method uses the so-called "*Oostenbrink elutriator*", which applies a two-step extraction. First the sample is gently flushed to wash-out nematodes and small soil particles; then nematodes are separated from soil particles by differential flotation in an upward flow (Oostenbrink 1954). The apparatus is named after the Dutch nematologist Michiel Oostenbrink (1921–1979), working at the Institute of Plant Pathology in Wageningen. Recovery of nematodes in the flotation step can be very high, up to 100%, but when using sieves to get rid of small soil particles, a significant fraction of the smaller animals may be lost (Verschoor and De Goede 2000).

For earthworms, a common method is to drive the worms to the surface by pouring a 0.5% formalin solution over the soil. This method, although cheap and popular, is not to be recommended since it has severe side-effects on other invertebrate populations and the recovery of worms is still modest, especially for the deeper burrowing species (Eichinger et al. 2007). Earthworms are best separated from the soil by (tedious) hand-sorting, after spreading samples on a white tray and carefully separating roots from soil.

The densities of microarthropods and earthworms are usually expressed per m^2, but the abundances of nematodes and other microscopic invertebrates are often expressed in numbers per g of soil, or numbers per g of habitat unit, such as moss cushions. The latter is especially popular among students of tardigrades (Glime 2017). The use of mass or habitat units makes it difficult to compare the abundance with invertebrate densities expressed per surface area.

Another common approach to soil invertebrate population studies is to set out pitfall traps in the field and count the number of animals caught over a defined period. This works very well with surface-active arthropods such as ground beetles and cursorial spiders. Obviously, the catches will depend not only on density but also on locomotor activity of the species, and so in turn on climatic factors. The design of the trap (diameter, rims, cover, preservative) can also have a large influence on the catches (Boetzl et al. 2018). These biases are recognized by all soil ecologists (Engel et al. 2017). Pitfall trap data can never be translated easily to a true density. Nevertheless, observations accumulated over a complete activity period, at least in some cases, show a good correlation with density (Baars 1979a), especially when combined with a temperature-driven locomotion model (Engel et al. 2017).

Using these methods and many others, a huge body of data on soil invertebrate densities and biomass has been collected in the context of the *International Biological Program* (IBP) that ran from 1964 to 1974. One of the seven IBP themes was entitled "*Productivity of Terrestrial Communities*". The methodologies developed at the time, and the data collected (Petrusewicz and Macfadyen 1970, Phillipson 1971, Petersen and Luxton 1982), are still valuable, while long-term population studies have become much rarer in the recent literature.

Figure 3.1 gives an overview of densities for most of the soil invertebrate groups presented in Chapter 2, compiled from the literature. Only estimates of annual mean density, expressed as numbers per m^2, were selected. Pitfall trap data (common for carabid beetles and spiders) or data collected by standardized habitat searching (applied, e.g., in snail and flatworm surveys) were excluded. The data are from many different habitats: tundra, pine forests, temperate grassland and woodlands, agricultural fields, savannas and tropical rain forests. The great variety of soil types and climatic conditions obviously introduces a lot of variation. Still some general patterns may be extracted from the overview.

First of all, Figure 3.1 shows that the densities of soil invertebrates as a whole vary by eight orders of magnitude. They are typically high (median density > 10^6 individuals per m^2) for small animals such as nematodes; they are intermediate (10^3 to 10^6 per m^2) for meiofauna such as springtails, pauropods, mites, and enchytraeid worms and lowest for the macrofauna, e.g., earthworms, snails and spiders. Indeed, a negative correlation between density and body size is often observed (e.g., Osler and Beattie 1999). However, when considering all soil invertebrates jointly, the correlation with body-size is not very strong. For example, tardigrades are consistently present at lower densities (by a factor of 100) than nematodes, although their body sizes do not differ too much and their habitats are much the same. It is also obvious that some predatory groups (pseudoscorpions, spiders, harvestmen) have lower densities than saprotrophic invertebrates of comparable body size.

Another interesting phenomenon seen in Figure 3.1 is that, when expressed on a log scale, many densities are skewed to the low range (e.g., dipteran larvae, ants, springtails, mites, enchytraeids). It seems that the interquartile band indicates a typical, taxon-specific density range, and the lower extremes represent populations of low density in less suitable habitats where the animals can nevertheless survive. Similar extremes in the high-density range are less common (in this graph only shown by tardigrades). The tailing towards low densities can also be observed in some of the frequency distributions of soil invertebrate densities presented in Petersen and Luxton (1982).

The densities of any soil invertebrate taxon differ greatly between habitats, but at the same time show a marked temporal constancy in one habitat when expressed as annual means. Many soil invertebrates go through high and low densities due to life-history phenomena or seasonal cycles. Still several soil invertebrates have annual

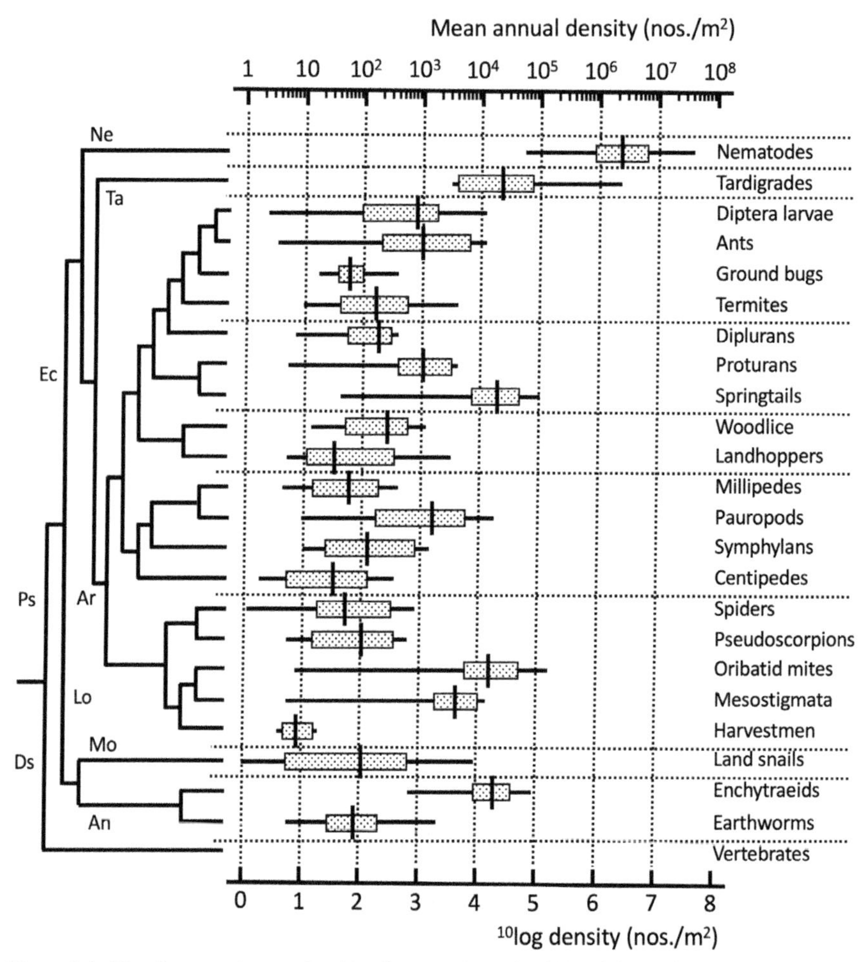

Figure 3.1. Showing annual mean densities (log-transformed) of 23 soil invertebrate taxa. The median density across studies is indicated by a vertical line, and the interquartile range (comprising 50% of the data) is shown as a dotted box. The horizontal lines link the lowest and highest values reported. The evolutionary relationships between the taxa are indicated on the left by a simplified phylogenetic tree of bilaterian animals. Ds: Deuterostomia, Ps: Protostomia, Ec: Ecdysozoa, Lo: Lophotrochoza, Ne: Nematoda, Ta: Tardigrada, Ar: Arthropoda, Mo: Mollusca, An: Annelida. The data comprise 349 density estimates from 146 publications. Literature sources and summary data for this graph are given in Annex 8.1.

mean densities characteristic for a specific ecosystem (pine forest, grassland), that are stable for many years, even though the seasonal cycles may be considerable. This is nicely illustrated by long-term studies summarized in Bengtsson (1994), Takeda (1995b), Wolters (1998) and Kampichler and Geissen (2005). One explanation may be that soil invertebrates, after a rapid growth period each year, quickly reach their carrying capacity, set by habitat properties such as resource availability, humus type and soil moisture. The maximum reached in the growing season does not seem to depend much on the number surviving the unfavourable season. We will revisit the possible population regulatory factors in Section 3.4.

A comparison with ecological data on other animals shows that the densities reached by soil invertebrates are much higher, up to three orders of magnitude, even when correcting for the effects of body size. One of the few general rules of ecology holds that a primary determinant of population density is the body size of the animal (Damuth 1981, Peters 1983, Hendriks 1999, Mulder et al. 2005). The relationship is often presented to be almost linear with slope 1, such that the product of population density (in numbers per km^2) and body size (in kg) is a constant. Halving the body size approximately doubles the expected population density.

These allometric relationships predict much lower densities than the observed data in Figure 3.1. For example, using the scaling relationship reported in Peters (1983, Figure 10.3), established for vertebrates and invertebrates jointly, one would expect, for an animal of 1 g (the approximate mass of an earthworm) a population density of 2613 individuals per km^2, i.e., around 0.003 per m^2. The regression in Damuth (1981) established for mammals, predicts 0.017 per m^2 for a small animal of 1 g. Actually, the median density of earthworms is 80 per m^2 (Figure 3.1). That the densities of soil invertebrate populations are orders of magnitude larger than recognized by "*current ecological wisdom*" is hardly ever recognized.

As a caveat to this conclusion we need to add that the data in Figure 3.1 are for the total community belonging to a specific taxon, not for individual species. We know that communities of soil invertebrates are characterized by a high degree of dominance; the first most abundant species often covers already more than 50% of the total density. Therefore, our conclusion that the densities of soil invertebrates are much higher than predicted by allometric body-size relationships still holds for the most abundant species, although maybe not for the rare species.

Can density data such as summarized in Figure 3.1 be used to estimate absolute population size? To do this we need to know more about the dispersal distances or home ranges of the animals. In the short-term concept of local population size a useful measure is the average distance between the spot where an animal hatches from an egg, and the spot where it deposits its own eggs when mature. For some species dispersal distances have been estimated by introducing animals in a field plot devoid of such animals and monitoring the gradual spread of their distribution as a function of time (Marinissen and Van den Bosch 1992, Sjögren et al. 1995, Ojala and Huhta 2001, Eijsackers 2011). These data were used in Table 3.1 to derive a rough estimate of local population size from the densities in Figure 3.1. Obviously, such estimations ignore the effects of long-range dispersal, phoresis, migration, etc., which are important phenomena for several soil invertebrates, as we will see in Section 3.9.

The data in Table 3.1 show that local (short-term) population sizes may not differ so much between the different soil invertebrate taxa, at least not by orders of magnitude. Despite the large differences in density, acknowledging for dispersal distances brings the estimates in the range of 100 to 2000 for four main invertebrate groups, earthworms, enchytraeids, springtails and oribatids. These numbers are in line with estimates for other animals, although on the low side. In conservation biology, where the estimation of population size has a long tradition (Traill et al. 2007), the so-called 50/500 rule is often used as a guideline (Franklin and Allendorf 2014). This rule says that an effective size of 50 is a short-term minimum to prevent local extinction of wildlife and an effective size of 500 is the minimal viable long-

Table 3.1. Estimates of local population size for some soil invertebrate taxa. derived from median density and average dispersal distance.

Group	Median density (numbers per m^2)	Average dispersal area within a generation (m^2)	Short-term local population size
Earthworms	79.4	3.1 – 18	245 – 1.430
Enchytraeids	21,400	0.13	2780
Collembola	19,950	0.005 – 0.13	100 – 2.500
Oribatid mites	15,100	0.005 – 0.13	75 – 2.000

Median densities are from Figure 3.1. Average dispersal distances from Marinissen and van den Bosch (1992) and Eijsackers (2011) (earthworms), Sjögren et al. (1995) (enchytraeids) and Ojala and Huhta (2001) (Collembola and oribatids), were used to calculate average dispersal area (m^2) as a circle with radius equal to the average dispersal distance.

term population size to avoid inbreeding depression. The minimum viable population sizes for insects and marine invertebrates were estimated as 2000 and 2500, *resp.* (Traill et al. 2007).

It may be concluded that soil invertebrate populations are characterized by very high densities, but without long-range dispersal would have quite low effective population sizes, comparable to protected species in mammalian conservation programmes. So, the effects of genetic drift in local populations of soil invertebrates are expected to be appreciable, like they are in protected species with small populations. The analysis also points out, that, everything else being equal, dispersal and genetic exchange on a scale exceeding the home range, are of crucial importance for the maintenance of genetic variation and the evolutionary potential of soil invertebrates.

3.2 Production and turnover

The ecological study of soil invertebrates should not be limited to simple descriptive dynamics, sometimes qualified as "*bugs go up, bugs go down*". The significance of a soil invertebrate in the food-web is determined not only by its numbers, but also by its biomass and the flux of energy, matter and elements through the population (Mulder et al. 2005, Mulder 2006). Biomass density, also indicated with the botanical term "*standing crop*", can be estimated by multiplying the density in numbers by the average mass of the individuals. A more precise measure is obtained by classifying the population according to size classes, multiply the numbers in each size class by the average mass of that class, and cumulate this over all sizes. Thus, a massive amount of biomass data has been collected in the context of the IBP, as mentioned above. The reader is referred to Petersen and Luxton (1982) for an overview of these data.

Biomass is generated in a population by growth and reproduction, and leaves the population through mortality. In a population of constant size and in the absence of migration these two components should balance each other: biomass generated by growth and reproduction equals the biomass loss through mortality. The flux through the population is designated as "*production*", or since it is heterotrophic,

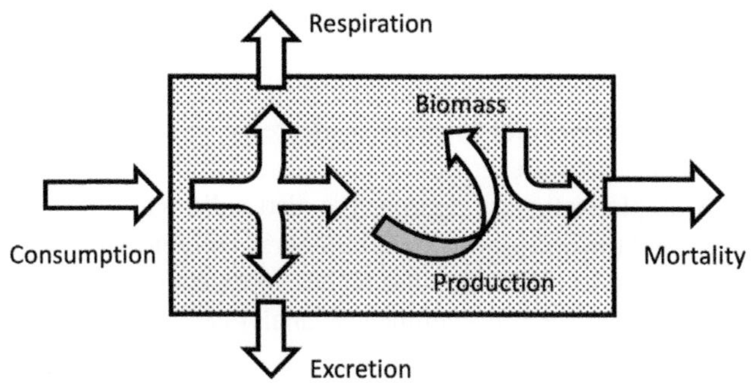

Figure 3.2. The concept of production on the population level. The energetic equivalent of total consumption (by all members of the population jointly) leaves the population by respiration and excretion (the latter including defaecation, urine and other excreta such as mucus) while a part is used to generate biomass, both in the form of growth of existing individuals and by adding new individuals to the population (reproduction). If the population is stable, biomass leaves the population by mortality at the same rate as it is generated. The biomass flux through the population can be estimated either from total growth and reproduction or from mortality, as shown in Annex 8.2.

"*secondary production*". Figure 3.2 provides a schematic illustration of the concept of biomass flux through a stationary population.

The ratio of production (P) to biomass (B) is known as *biomass turnover* (P/B ratio). Biomass turnover is one of the most important functional characteristics of a soil invertebrate population, since it defines the efficiency by which matter extracted from soil resources (litter, microbial communities) is transferred to other organisms (decomposers and predators). Productivity was therefore the main target in the IBP program in the 1970s and it is equally important in the development of soil wood-webs, which began in the 1990s.

Production of soil invertebrate field populations can be estimated in various ways. One of the more reliable methods is to estimate it directly from field data. The biomass leaving the population through mortality may be estimated from the frequency distribution of body sizes (Van Straalen 1985b). Another approach is to measure growth rates and metabolic rates of the species in laboratory cultures as a function of temperature, and use soil temperature registrations to predict growth and respiration in the field, which are then summed over all size classes to obtain the total rate of biomass generation at any point in time. This method is less reliable due to uncertainty on the precise food conditions in laboratory cultures and the extrapolation on the basis of temperature.

Biomass turnover has the dimension time^{-1} (usually per year). The reciprocal of biomass turnover, with the dimension time, is the average residence time of biomass in the population. This is comparable to the average lifetime of individuals, also called *life expectancy at birth* or average residence time of individuals. Turnover of individuals is exactly equal to turnover of biomass if mortality rate is constant, i.e., not size-dependent (Van Straalen 1985b). If mortality rate decreases with size (the younger, smaller individuals run a higher risk than older, larger individuals) the turnover of biomass is smaller than the turnover of individuals. Likewise, the turnover of substances in the biomass (e.g., heavy metals) may be larger than the turnover of biomass if the substances accumulate during life, i.e., if older individuals have a higher body burden than younger individuals (Van Straalen 1987). These formalities are further explained in Annex 8.2.

The calculation of production is relatively straightforward if the population is stationary, that is, if it does not increase or decrease in size and has a stable age distribution. This will hardly ever be the case in real soil invertebrate populations. Fortunately, the simple stationary approach is also valid when the animals emerge in a single cohort and the decrease of their numbers is monitored through time (Hale 1980, Van Straalen 1983a). If the population has an extended recruitment period, or overlapping generations, a more complicated analytical approach must be applied (Van Straalen 1982).

An important conclusion shown in Annex 8.2, is the formal proof that both growth and reproduction must be included in the production term, for it to equal the biomass lost through mortality. Moreover, the allocation of production, on the individual level, between growth and reproduction is of no relevance to total population production; only their sum matters. The fact that not only body growth but also reproduction must be included in the productivity of a population is ignored in some common methods for calculating secondary production in the aquatic literature (Ricker 1946, Allen 1951, Hynes and Coleman 1968, Hamilton 1969).

To explore patterns of productivity of soil invertebrates, data were compiled from published papers and the P/B ratios plotted as a function of body-size (Figure 3.3). Estimates for populations under extreme (e.g., arctic) conditions were not included. The data for soil invertebrates are compared with a regression established by Banse and Mosher (1980) for a variety of invertebrates (mostly aquatic), living between 5°C and 20°C. According to their regression, P/B decreases with increasing body mass

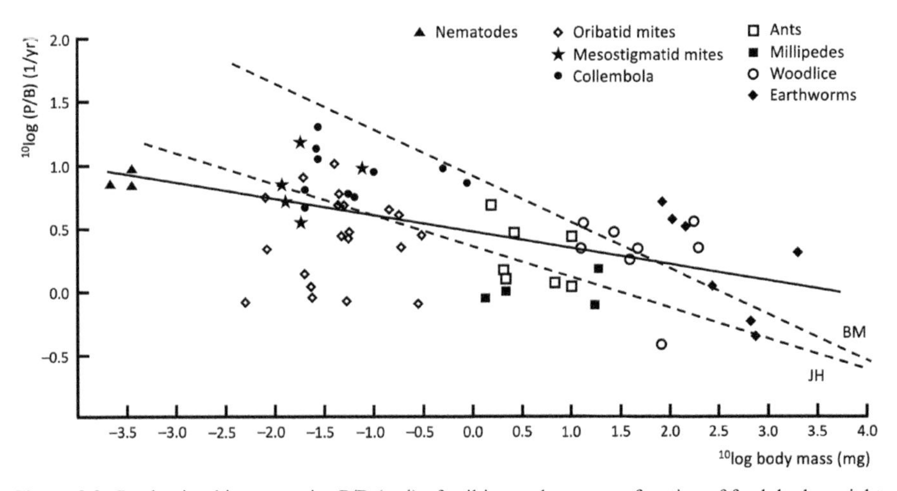

Figure 3.3. Production-biomass ratios P/B (yr^{-1}) of soil invertebrates as a function of fresh body weight W (mg), with both variables plotted on a logarithmic scale. The data are from 34 different publications, specified in Annex 8.3. The line indicated BM is a regression result derived from 33 invertebrate populations living at annual mean temperatures between 5 and 20°C (Banse and Mosher 1980). The line indicated JH is a similar regression with slope fixed at −0.25 as suggested by Hendriks (2007). The dotted line is the best fitting line through the data, reading $^{10}\log(P/B) = 0.460 - 0.128\ ^{10}\log W$ (95% confidence interval for the slope parameter: −0.176 to −0.080).

according to an allometric relationship with exponent –0.37. This exponent can be derived as the slope of the regression line in a double-logarithmic plot (Figure 3.3).

In a more extensive survey of productivity data including no less than 230 allometric regressions, the Dutch ecologist Jan Hendriks found a universal scaling exponent of –0.25 (Hendriks 2007). This regression is also plotted in Figure 3.3. A scaling coefficient of ¼ is consistent with Kleiber's "*Fire of Life*" theory for metabolic rate, which scales with body-size to the power ¾, a well-established "law" in comparative animal physiology (Kleiber 1961, Lavigne 1982, Schmidt-Nielsen 1998). An exponent of ¾ is theoretically expected when productivity is limited by internal transport systems with a lower boundary for the diameter of the smallest branches (West et al. 1997).

Figure 3.3 reveals that P/B ratios for many soil invertebrates are significantly below the regression line from Banse and Mosher (1980). In fact, the body-size dependence of P/B is much shallower than expected. The best fitting line through the data has a slope of –0.13 over the range considered here, significantly different from –0.37 (Banse and Mosher 1980) and also from –0.25 (Lavigne 1982, Hendriks 2007). Especially nematodes, small Collembola and many oribatid and mesostigmatid mites are less productive than expected based on their small body-size. The larger soil invertebrates: earthworms, isopods and some superficially living Collembola fall more or less within the range predicted by the allometric models, although the variation is large.

Why would the productivity of small soil invertebrates be so low? It may be considered an effect of adaptation to soil conditions, particularly to the decrease of energy intensity with depth, as predicted by the GKGB hypothesis introduced in Chapter 1. While living in the soil relaxes the selective forces due to predation and drought, it diminishes productivity due to resource limitation. According to this argument, the lower resource availability in the soil would affect small animals more than large, burrowing animals. The smaller animals have a body size in the range of the soil's pore size distribution; they are therefore much more limited in the exploration of the soil than animals than can dig their way through the soil. In fact, several of the larger invertebrates, such as earthworms, have a productivity higher than expected from their body size (Figure 3.3). Isopods, roaming the surface of the soil more than crawling through the soil, are also less restricted by resource availability and most species have a production turnover above the allometric expectation (Figure 3.3).

The relatively low productivity of ants (Figure 3.3) might have another reason. These animals are supposed to bear very high maintenance costs stemming from their intensive locomotory and nest-building activities. Relative running speed of ant species is correlated with colony size and whole-colony respiration (Mason et al. 2015). Ants respire more than 70% of their assimilated energy (Jensen 1978b). This might contribute to a lower rate of production relative to biomass.

The productivity patterns seen across taxa are also observed across species of the same taxon that live at different positions in the soil profile. For example, the two collembolans *Orchesella cincta* and *Tomocerus minor* are common species that can be found sympatrically in many forests of Western Europe. *O. cincta* is a superficially active, drought-tolerant species, with a morphology typical for the epigeon (Figure 1.1), while *T. minor* is a hemiedaphic species, drought sensitive and associated with wet spots in the forest floor. The P/B ratios estimated for these two species (9.4 and 6.8, *resp.*) correlate with their habitat use: the higher P/B is associated with a surface-active life-form (Van Straalen 1989). A similar trend is seen in oribatid mites. Luxton (1981c) distinguished two groups of oribatids: larger, more superficially living species with a P/B ratio greater than 2.5, versus the smaller, less fecund and deeper living species with a P/B ratio lower than 2.5 per year.

In conclusion, the environmental gradient of the soil profile is shown to have a dual selective effect: it imposes a lower level of productivity upon animals living deeper in the soil profile, and this affects the smaller, non-digging, animals more than the larger animals. The adaptations to soil living are body-size dependent and bend the classical allometric relationships valid for other invertebrates. That small soil invertebrates have a lower productivity than is expected on the basis of their body-size is an observation which, surprisingly, is lacking among the conclusions of the IBP (Petersen and Luxton 1982).

3.3 The importance of biogeographic history

Several groups of soil invertebrates show a clear geographic pattern in their distribution across the globe. We already noted in Chapter 2 the influence of latitude in shaping earthworm communities (Lavelle 1983). There is an obvious split between the families of the Northern and Southern Hemisphere which can be traced back to the break-up of Pangea, 175 million years ago. Another example is the case of terrestrial flatworms (Terricola), which have their origin and largest distribution in the Southern Hemisphere. That these signatures of the evolutionary past can still be recognized today is due to the limited dispersal capacity of many soil invertebrates. However, historical patterns have often been blurred, or even erased, by anthropogenic introductions, especially the transport of soil, leading to, e.g., Australian flatworms in Western Europe and originally European earthworm species in North America.

Among soil invertebrates the case of velvet worms (Onychophora) presents one of the finest examples of a historical biogeographic legacy. Present Onychophora show a fragmented distribution in tropical regions and the Southern Hemisphere. Two families are commonly distinguished and these show a remarkable non-overlapping distribution: Peripatidae are present in Meso-America, Central Africa and South-East Asia, and Peripatopsidae in South America, South Africa, the eastern Indonesian islands, parts of Australia and New Zealand (Monge-Najera 1995, Smith 2016, Oliveira et al. 2016), see Figure 3.4. This pattern suggests that their distribution was once continuous and has been influenced by continental drift, a process referred to as *vicariance*: the separation of an evolutionary lineage by development of geographic barriers, allowing *allopatric speciation*.

Their biogeographic distribution suggests that Onychophora evolved on the ancient supercontinent Gondwana after it separated from Laurasia about 100 Ma BP. This explains why there are no velvet worms in North America, Europe and Central Asia. An evolutionary split then gave rise to the family Peripatopsidae in the south of Gondwana and Peripatidae in the northern parts. Then Gondwana broke up, South America separated from Africa and Australia drifted from its southern position towards South-East Asia. India separated from Africa, crossed the Tethys ocean and

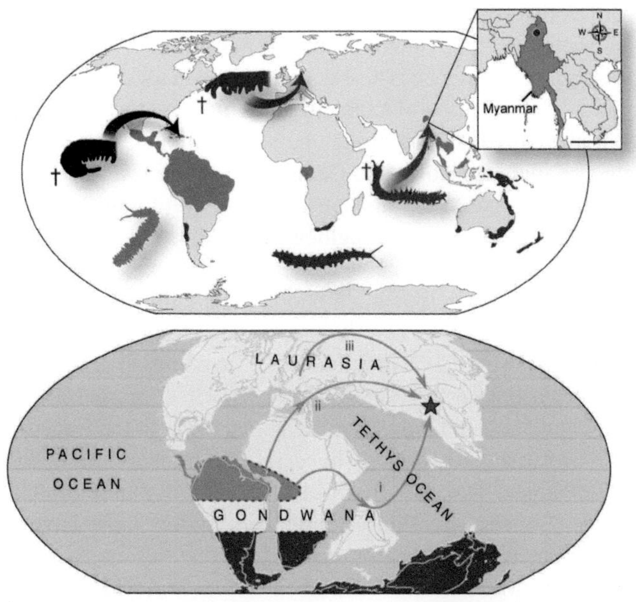

Figure 3.4. Above: present distribution of velvet worms, distinguishing the families Peripatidae (circumtropical) and Peripatopsidae (fragmented ranges in the Southern Hemisphere). Also shown are fossil onychophorans in the Caribbean and north-western Europe. A recent find of a fossil peripatid onychophoran in Myanmar is highlighted in the inset. Below: reconstructed distribution of velvet worms during the Cretaceous (100 Ma BP). Three scenarios (i, ii, iii) are shown to explain the peculiar presence of peripatid velvet worms in South-East Asia. Reproduced from Oliveira et al. (2016) (upper image) and Smith (2016) (lower image), with permission from Elsevier, Ivo de Sena Oliveira and Martin Smith.

joined Asia. The onychophorans moved with the continents. After subsequent range contractions the present fragmented pattern resulted (Figure 3.4). Onychophorans tell the "*living fossil tale*" of large-scale tectonic processes and there is no need to assume oceanic dispersal to explain their distribution (Allwood et al. 2010, Murienne et al. 2014).

However, the presence of peripatid onychophorans in South-East Asia west of the Wallace line does not fit in this scenario because the western Malay Archipelago derives from Laurasia. There are three explanations for this phenomenon (Smith 2016):

1. The South-East Asian velvet worms arrived there through "hitch-hiking" on the Indian continent (the "*Out of India*" scenario),
2. They migrated from Gondwana to Laurasia through a land connection before the complete break up of Pangea, and
3. They represent a remnant of an ancestral distribution on Laurasia.

The latter scenario is supported by peripatid fossils in Burmese amber, found in 2002 and re-analysed by X-ray microtomography by Oliveira et al. (2016). These fossils are 100 million years old and therefore contradict the Out of India hypothesis. Further fossil evidence is necessary to draw a final conclusion, but it is obvious that continental drift is the prime factor explaining the evolutionary relatedness among Peripatidae on South America, South Africa and Australia.

The importance of vicariance is also demonstrated in a study by Xu et al. (2015a) on the origin and diversification of Liphistiidae. This is a group of primitive spiders (89 species described in 2015), that build tube-shaped funnels with a trap-door. Their present distribution includes North and Central China, the southern islands of Japan and parts of South-East Asia. They are considered "living fossils" because they have typical ancestral characters such as a segmented opisthosoma (Xu et al. 2015b).

A phylogenetic reconstruction using nucleotide sequences of five genes (Xu et al. 2015a,b) indicated that Liphistiidae is a monophyletic family that has its origin early in the evolution of spiders, in the primitive suborder Mesothelae, which in addition to Liphistiidae includes several extinct families. The common ancestor of liphistiids must have lived even before the Permian, 298 Ma BP (Figure 3.5). However, the radiation of the family is dated only 50 million years ago, in the Eocene.

Mesothelae are assumed to have lived in Euramerica, before this continent got integrated in Pangea. The only fossil of this group is found in a Carboniferous deposit in France. So, the ancestors of liphistiid spiders must have dispersed in an extremely long eastward move from central Europe to East Asia, either via the "*Silk Road*" or via the Middle East. Upon their arrival in Asia liphistiids underwent a remarkable radiation, resulting in the present nearly non-overlapping distributions of genera (Figure 3.5). The long-term separation from their ancestors, another case of vicariance, promoted allopatric speciation of the group.

The third example of biogeographic principles in soil invertebrate evolution is inspired by work of the great American Collembola taxonomist and biospeleologist Kenneth A. Christiansen (1924–2017). Christiansen was famous for his work on cave adaptations. He described nearly 50 new species of cave Collembola and contributed

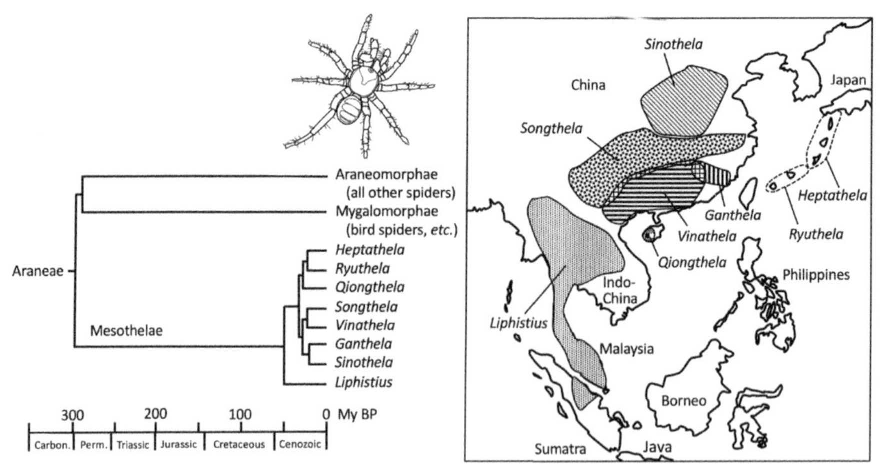

Figure 3.5. Left: Dated phylogeny of spiders (Araneae), showing the splits in three suborders. Suborder Mesothelae includes one extant family, Liphistiidae, presently living in China, Japan and parts of South-East Asia. Only the splits between liphistiid genera are shown in the phylogeny. Right: present distribution of liphistiids in China and South-East Asia. The spider image shows *Heptathela yanbaruensis*, a species from the Japanese island Okinawa. Note the segmented opisthosoma, unique for Mesothelae. Redrawn from Xu et al. (2015a,b).

greatly to a Darwinian interpretation of cave biology, promoting the "*modern synthesis*" version of evolution (Culver 2017). He has also done important work on biogeography of springtails and is considered, together with the French entomologist Paul Cassagnau (1932–2016), as the founding father of Collembola biogeography.

Analysing the global distribution of Collembola, Christiansen and Bellinger (1995) noted that it is much dominated by geographic proximity. Statistical analysis of the faunal composition allocated to biogeographic provinces revealed that any one province is most similar to an adjacent province. For example, the Collembola fauna of India is similar to continental South-East Asia and Malaysia, not to old Gondwana territories (Africa and South America). Similarly, the fauna of Madagascar is more similar to three nearby African regions, than to India. African regions generally show little similarity to South American ones. So, it seems that the present distribution of Collembola is due to relatively recent geological events in the Cenozoic, not to the great continental movements following the break-up of Pangea. Any patterns resulting from ancient tectonic events such as continental drift must have been erased by more recent dispersal events. The biogeography of Collembola reflects their much greater dispersal capacity compared to earthworms, flatworms and velvet worms (*cf.* Figure 3.4).

This conclusion of Kenneth Christiansen has been confirmed by several recent phylogeographic analyses. A number of interesting studies has been done on the Antarctic and Arctic Collembola fauna (Frati et al. 2001, Fanciulli et al. 2001, McGaughran et al. 2011, Ávila-Jiménez and Coulson 2011, Collins et al. 2019, 2020). Collembola are one of the few animal taxa that extend their distribution to both Arctic and Antarctic ecosystems and so their occurrence in these extreme environments is specifically suited to answer questions on dispersal ability. Antarctic populations of Collembola show a high degree of differentiation due to glaciers, which act as severe barriers to dispersal (Frati et al. 2001, Fanciulli et al. 2001). Populations only 16 km from each other, on either side of the Drygalski Ice Tongue on Victoria Land, showed a large degree of genetic divergence (Collins et al. 2019).

The polar environment also presents an ideal platform for studying the dispersal of soil invertebrates. Inhabitable sites in the high Arctic may be considered niches that opened after the last glacial maximum. This "*opening of the niche*" is very well dated for most of the area. In addition, a near-complete catalogue of Arctic Collembola was developed by two taxonomic experts, the Norwegian and Russian entomologists Arne Fjellberg and Anatoly Babenko.

In an analysis of 358 species living in the high Arctic, Ávila-Jiménez and Coulson (2011) distinguished nine different clusters, of which eight were geographically restricted and one occurred everywhere in the Arctic region. Interestingly, the species in each cluster were not related to local climate variables such as temperature and precipitation. The cluster composition was rather related to the nearest Atlantic (continental) region: North-Eastern Europe, Western Siberia, Eastern Siberia, Beringia, Alaska and Northern Canada (Figure 3.6). An influence of ocean currents was also noted. The geographic pattern can be described as a northerly dispersal from continental sites after the last glacial maximum. The ice sheet started to decline

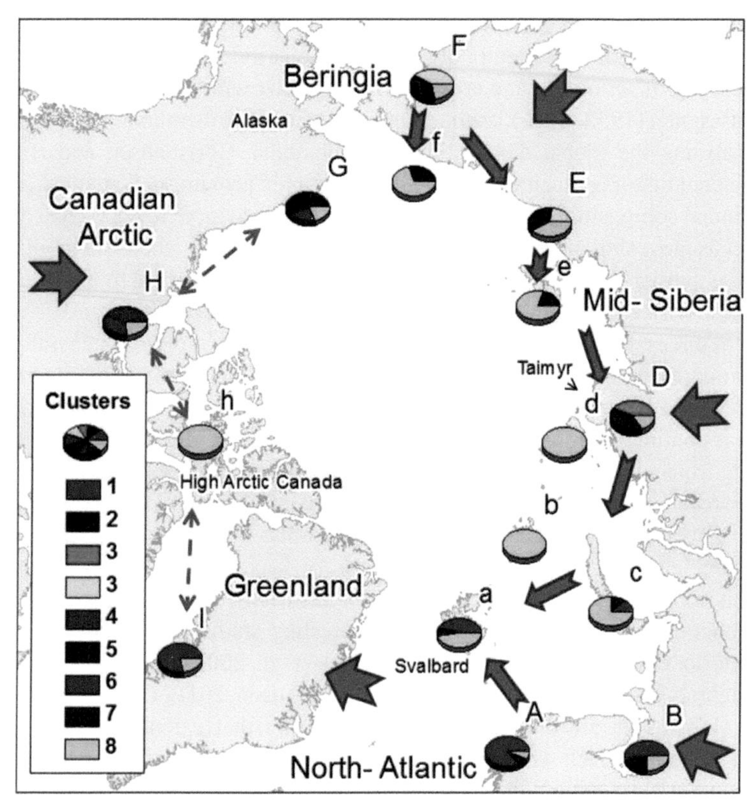

Figure 3.6. Clusters of Collembola associations in geographic areas of the Arctic, showing the similarity between high Arctic communities (indicated with lower case letters) and those of nearby continental sites (capitals). The pie diagrams express the composition of each area by eight clusters (numbered 1–8). Possible dispersal routes revealed by Gaussian mixture modelling are indicated by arrows. Reproduced from Ávila-Jiménez and Coulson (2011), with permission from Malu Ávila-Jiménez.

14,500 years ago, so Collembola colonization of the Arctic is a relatively recent phenomenon, in accordance with Christiansen's conclusion.

The three examples of biogeographical studies of soil invertebrates in this section (velvet worms, spiders and Collembola) all highlight the importance of the evolutionary past when interpreting the distribution of soil invertebrates. Historical factors such as vicariance, isolation and dispersal cannot be ignored. In large-scale inventories, the abundance of soil animals is often found to be highly correlated with climatic factors, but similarity due to proximity is ignored in such analyses. The examples presented here have illustrated that climatic factors may be confounded by spatial patterns with a historical background. To state it bluntly: temperature and precipitation cannot explain the presence of flatworms in New Zealand and their absence in Western Europe, nor the presence of velvet worms in Brazil and their absence in India.

The distribution range is also important as a template for natural selection. It is often argued that natural selection should be different in the centre of the range compared to the margins. In marginal habitats the animals are assumed to suffer more often from stress factors that limit their distribution. By *peripatric speciation* even new species could arise on the borders of a distribution range. Accordingly, we will see in Chapter 4 that variation of reproductive mode is associated with the distribution range, a phenomenon known as "*geographic parthenogenesis*".

3.4 Limiting and density-regulating factors

The density data of soil invertebrates displayed in Figure 3.1, were collected from within the distribution range of a species, but still show considerable variation across ecosystems. Some habitats that could be colonized are only marginally suitable or completely unsuitable for specific soil invertebrates. Which are the main factors that allow soil invertebrates to establish themselves at a site, and which factors prevent them from reaching a high density in others?

The literature on soil invertebrates provides an overwhelming amount of information on this question, because it is usually one of the first things a soil ecologist will do: relate abundance to soil factors. The patterns revealed by these studies are summarized in numerous soil ecology textbooks and review articles and will only be very shortly recapitulated here. We first visit the main abiotic factors that are known to limit the occurrence and population density of soil invertebrates.

- *Drought*. As all soil invertebrate lineages descend from marine or freshwater ancestors, the evolution of a certain degree of drought resistance was a crucial adaptation to live in the aerial porosphere of the soil. Not all species within a specific lineage have developed drought resistance to the same extent, that is why soil humidity is still a limiting factor for the great majority of soil invertebrates. In many lineages the species can be ranked for drought resistance and this often correlates with their position in the soil profile or the use of wet and dry microhabitats. The overriding influence of soil moisture is obvious in every large-scale inventory and usually appears as a relation between precipitation and soil invertebrate abundance (e.g., Phillips et al. 2019, Johnston 2019). However, the real (proximate) limiting factor is not precipitation but the relative air humidity of soil pores, as we have seen in Chapter 1.
- *Water-logging*. In addition to lack of moisture, excessive water may also be a limiting factor. Invertebrates that live in soil pores such as springtails and mites often need spots to evaporate their water surplus, in addition to wet spots to take up water. Many earthworms will not withstand continued inundation and flee across the surface to drier places. On the other hand, larvae of flies and midges often thrive in very wet soils, such in as marshes and tropical rain forests, partly because they deal reasonably well with the anaerobic conditions that result from waterlogging.
- *Frost*. Below-zero temperature is major limiting factor for animals with small body sizes, since they quickly assume the temperature of the environment (*cf.* Section 1.2). Many soil invertebrates rely on behavioural mechanisms to escape freezing, others have evolved different degrees of cold-hardiness, as we will see in Section 6.3. Some even profit from frozen soil or snow to roam the surface and prevent freezing by intercepting solar radiation. Still in temperate

climates winter frost is a major limiting factor for many soil invertebrates, especially insects, earthworms and snails.

- *Heat*. Like cold, excessive heat can also be limiting. Heat interacts very much with air humidity, so usually it is the drought that comes with heat that is the proximate limiting factor. But also heat itself is a life-threatening factor and is usually avoided by living in holes or burrows. The options for animals to physiologically adapt to heat are limited, as we will see in Section 6.6.
- *Acidity*. Soil pH often comes out as a determining factor in soil invertebrate surveys. This is partly due to the fact that it is so easily measured and summarizes many other soil factors. As we have seen in Chapter 1, soil pH is correlated with humus type, pore water concentrations of cations, etc., so it is often difficult to decide what is the proximate factor affecting soil invertebrate life at low or high pH.
- *Organic matter*. Several aspects of soil organic matter (SOM) may limit soil invertebrate populations. SOM, and the microbial communities associated with it, is a significant food source for many invertebrates and so low-SOM soils (deserts, dunes, bare rock) are prohibitive to the development of a soil invertebrate community. Secondly, SOM, by its association with clay minerals, determines soil porosity, water infiltration and the moisture regime, and thirdly, the degradation of SOM influences the concentrations of soluble organics, pH and element concentrations. These three aspects jointly ensure that SOM is one of the most determining factors for the occurrence of soil invertebrates.
- *Humus type*. As we have seen in Chapter 1, the mor-moder-mull gradient is another determining factor for soil invertebrates. Humus type is given by the type of vegetation growing on the soil, in a complex set of interactions involving decomposition rate, soluble phenolics, humic acids and pH. Many microarthropods (Collembola, oribatid mites) and enchytraeid worms reach high densities in mor soils while earthworms are more commonly associated with mull soils.
- *Mineral composition*. The mineral nutrition of soil invertebrates is badly known, but the scant data available suggest that some invertebrates may have unexpected requirements for certain trace metals. For example, the oribatid *Platynothrus peltifer* has markedly high body concentrations of manganese and zinc. Manganese is an element easily leaching under high acidity and this may limit the occurrence of *P. peltifer* in acid soils (Van Straalen et al. 1988).

It must be emphasized that the drivers of soil invertebrate populations very much depend on the scale of the study. For example, when viewed on a global scale, earthworm species richness is usually found to be positively correlated with precipitation while earthworm abundance is correlated with annual mean temperature. Often, no effects of soil type and land management can be demonstrated on a global scale (Phillips et al. 2019, Johnston 2019). However, these global surveys hardly have any predictive value on a local or regional scale where numerous studies have demonstrated effects of soil type (clay versus sand), humus type (mull versus mor)

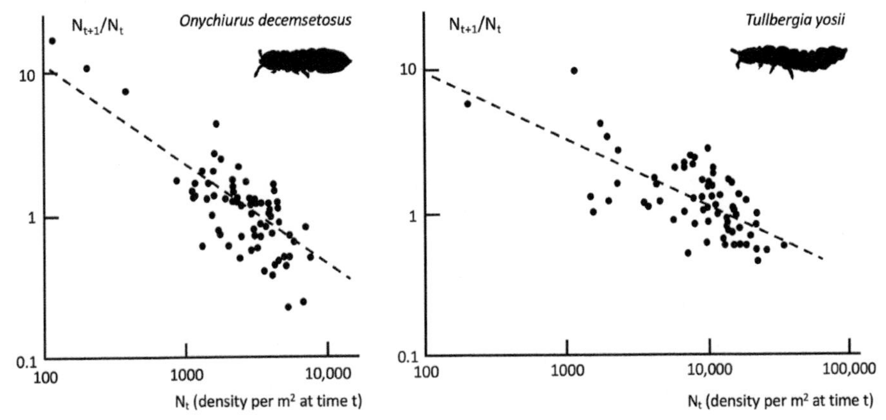

Figure 3.7. Density ratios between two subsequent sampling times versus density, over a six-year study period of two Collembola species in Japanese pine forest soil. The negative relationship between N_{t+1}/N_t and N_t suggests a density-dependent regulatory mechanism. Redrawn from Takeda (1983), with modifications.

and land use (grassland versus arable soils) (e.g., Lavelle and Spain 2005, James et al. 2021).

In addition to the abiotic factors listed above, soil invertebrate populations may also be limited by biotic factors, such as food resources, competition and predation. More than abiotic drivers, these biotic factors may act in a density-dependent fashion and so are potentially able to regulate populations rather than just limiting their occurrence. This could explain why populations of soil invertebrates, as we have seen above, show a remarkably good temporal predictability.

Density-dependent regulation was nicely demonstrated in a six-year study by the Japanese soil ecologist Hiroshi Takeda on Collembola populations in coniferous forest (Takeda 1983). Despite large seasonal fluctuations, the mean annual densities of the dominant species turned out to be quite constant, returning to more or less the same density every year. The regulatory effect was illustrated in a plot of the density ratio between two subsequent sampling times (N_{t+1}/N_t) versus density (Figure 3.7). This ratio varies from 10 (at low densities) to 0.1 (at high densities). This implies that a large population will decrease and a small population will increase. The negative correlation stabilizes the population around 3,000 individuals per m² for *Onychiurus decemsetosus* and around 15,000 per m² for *Tullbergia yosii* (the densities for which $N_{t+1}/N_t = 1$, see Figure 3.7).

The precise mechanism for density-dependent population dynamics in the studies of Takeda remains unknown. Autocorrelation due to life-history phenomena may have played a role and this was not included in the analysis. In general, the demonstration of density-dependence by correlation analysis of population data is not without methodological difficulties (Wolda and Dennis 1993, Holyak and Lawton 1993, Hanski et al. 1993). More extensive time series, preferably over 20–40 generations, are needed to prove the case, however, such data are not available for soil invertebrates at the moment.

That food resources may contribute to density regulation is also suggested in studies by Al-Assiuty et al. (1993) and Khalil et al. (2011). These authors counted

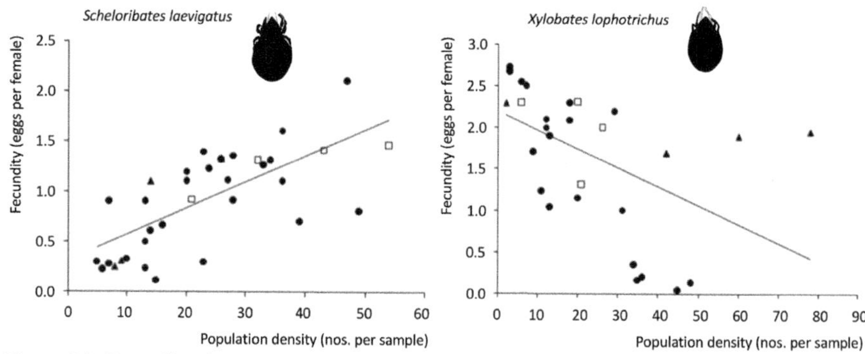

Figure 3.8. Fecundity of two species of oribatid mite, estimated by egg number, correlated with density, over different sites in the province of Al Gharbia, Egypt. Closed circles: mango plantations, triangles: orange plantations, open boxes: palm plantations. Redrawn from Khalil et al. (2011), with modifications.

the number of eggs in the bodies of oribatid mites sampled in a variety of habitats in Egypt. In many oribatid mites the eggs carried by females are visible through the integument after clearing the animals. The number of eggs can be used as an indicator of fecundity, although this cannot be compared across species since fecundity also depends on the residence time of the eggs in the body. However, it can be compared in the same species across sites. In the species *Scheloribates laevigatus* egg number turned out to be positively correlated with density, while in *Xylobates lophotrichus* egg number was negatively correlated with density (Figure 3.8).

The difference between the species was related to their feeding habits (established by gut content analysis): *S. laevigatus* is an opportunistic feeder with a wide trophic breadth, that can quickly establish a population from low densities by positive feed-back from fecundity on density. *X. lophotrichus*, however, is a specialized feeder (only one type of food was found in the gut) that is inhibited by its own density. That food resources regulate oribatid mite populations was also suggested in experimental studies by Wehner et al. (2014) and Chen and Wise (1999).

More evidence for density regulation is reported in a study on the woodlouse, *Armadillidium vulgare,* in a grazed grass heath (Hassall and Dangerfield 1990). The growth rate of every new cohort, expressed as relative weight gain per time unit, was strongly negatively correlated with the temporal mean density of that cohort (Figure 3.9). This is most likely due to crowding and food shortage at high densities. A similar negative effect of crowding is observed when earthworms are cultured at high stocking rates, as demonstrated in numerous experiments (e.g., Reinecke and Viljoen 1990, Butt et al. 1994, Domínguez and Edwards 1997, Karmegam and Daniel 2009). Finally, density regulation was suggested for populations of the ground beetle *Pterostichus oblongopunctatus,* which converged to the initial density within 1–2 years, after being experimentally reduced or enhanced (Brunsting and Heessen 1984).

These data combined suggest that regulation by food resources is important for soil invertebrate populations, although it is certainly not the only factor. There is also evidence for top-down regulation by predation. This is suggested by studies in which small prey organisms were experimentally protected against (larger) predators, such

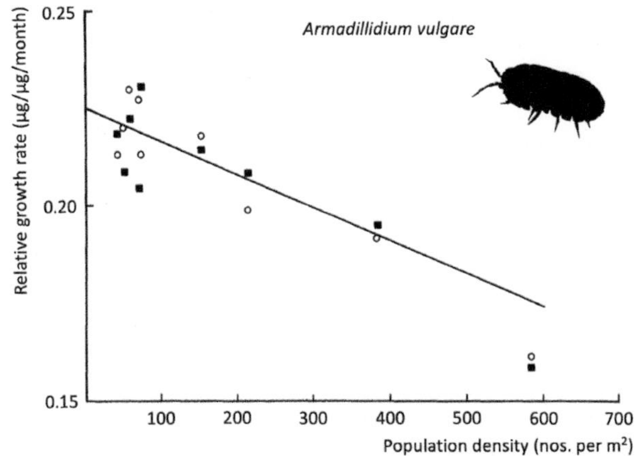

Figure 3.9. Relative body growth of different cohorts of the isopod *Armadillidium vulgare* as a function of cohort density, in a grass heathland in East Anglia, UK. The negative correlation suggests a density-dependent regulatory mechanism. Redrawn from Hassall and Dangerfield (1990), with modifications.

as centipedes (e.g., Poser 1988). In pesticide studies, eradication of predators is often followed by an upsurge of prey animals, such as springtails, which suggests that prey populations are regulated by their predators (Edwards and Thompson 1973).

In food-web models of the soil community, the presence of top-down regulation at the base of the web (and bottom-up regulation in the top of the web) is an important property contributing to stability of ecosystems (De Ruiter et al. 1995). The evidence reviewed in the present section suggests, however, that bottom-up regulation at the food-web base cannot be ignored and for several soil invertebrates may be more important than top-down regulation.

The lesson for our evolutionary analysis is that natural selection for efficient use of soil resources (food choice, food acquisition, competitive ability for resource use) is expected to be particularly strong for soil invertebrates. The adaptive responses to these selective forces will be explored in Chapter 6.

3.5 Iteroparity and indeterminate growth

As discussed in the section on productivity, resources acquired by an animal are allocated either to growth or reproduction (Figure 3.2). All soil invertebrates have to deal with this allocation but they do so in many different ways. An important distinction is between animals in which growth and reproduction are separated in different parts of the life-cycle (e.g., larva and adult) and animals that continue to grow even after reaching sexual maturity. The latter group is said to have "*indeterminate growth*", that is, a growth pattern that does not end in a fixed body size at maturity (like in many insects).

Theoretical arguments have shown that indeterminate growth promotes repeated reproduction (*iteroparity*) (Kozłowski 1992, 1996, Roff 1981, 2002). The reason is that indeterminate growers can increase their future reproductive output by growing larger. This option is not open to determinate growers: if the next reproductive output is just the same, future reproduction cannot balance mortality and it does not make sense to stage another effort. Therefore, according to theory, *semelparity* (breeding once in a life-time) should be more frequent among animals with determinate growth. This argument is often used to explain the commonness of semelparity among aboveground terrestrial insects. Soil invertebrates, however, are almost always iteroparous. Does reproductive output accordingly increase with body size in these animals?

In Figure 3.10 a typical fecundity schedule is plotted for the parthenogenetic collembolan *Folsomia candida*, a model species for this type of work (Fountain and Hopkin 2005). The example is drawn from the work of the American entomologists Renate and Richard Snider at Michigan State University, who conducted many experiments with microarthropods under laboratory conditions. Oviposition was recorded over no less than 40 moulting intervals (*instars*) (Snider 1973). The graph shows that egg laying starts early, then increases with age (and size), followed by a gradual decrease. While the decrease in later life may be attributed to senescence, the increase in early reproductive life is most relevant since in the field the great majority of springtails will die before they become old. A similar schedule of age-dependent reproduction is seen in other Collembola (Sharma and Kevan 1963, Hale 1965, Joosse and Veltkamp 1970, Gist et al. 1974, Snider 1983, Janssen and Joosse 1987).

Collembola obviously differ from insects in that they continue growing after reaching sexual maturity. This fits with the commonness of iteroparity among Collembola relative to insects. A typical aspect of collembolan reproduction, at least in the majority of species (Poduromorpha and Entomobryomorpha), is that the cycle of the ovary synchronizes with two moulting cycles. The animals go through subsequent reproductive and non-reproductive instars. During the non-reproductive instar, eggs are matured in the ovary, while in the subsequent reproductive instar, the female searches for a spermatophore, is fertilized and lays her eggs. She has to be fertilized again for every new clutch. Since feeding is suspended around the moult and during oviposition, body growth is mostly concentrated in the

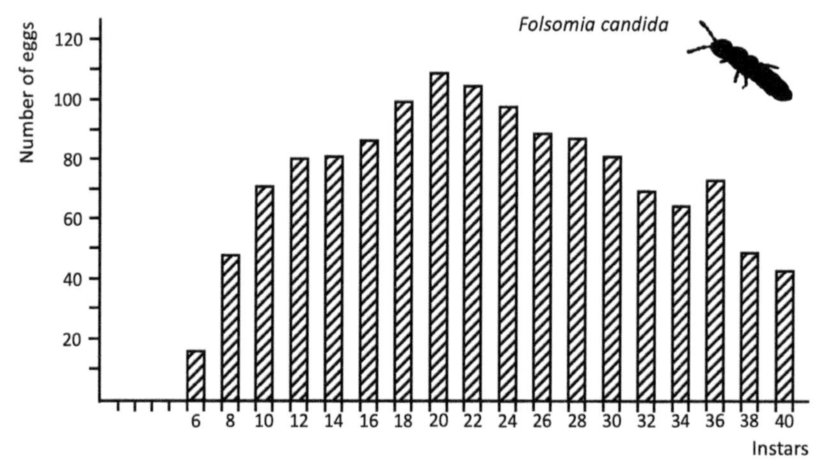

Figure 3.10. Oviposition pattern for the collembolan *Folsomia candida*, cultured at 21°C. Note that batches of eggs are deposited every other instar (moulting interval). Redrawn from Snider (1973), with modifications.

non-reproductive instars (Ernsting and Isaaks 2002). However, in species of the supergroup Symphypleona (*cf.* Figure 2.7), the animals stop moulting when mature and there is only one reproductive instar (Blanquaert et al. 1981b). So, these species (of which the majority live on top of the soil and in the vegetation) resemble the true insects in that they have become more or less semelparous. This divergence within the Collembola is perfectly in line with the association between iteroparity and indeterminate growth.

The ability to increase reproductive output by growing larger also holds for other soil invertebrates, for example isopods. Female isopods incubate their eggs and *mancas* in a brood pouch (*marsupium*) below their bodies, formed by *oostegites* extending from the leg base to the midline. Due to this morphological organization, the size of the mother is of immediate relevance to the capacity of the pouch. Accordingly, there is a strong correlation between body size and reproductive output (Sunderland et al. 1976, Antoł and Czarnoleski 2018, Figure 3.11). In *Porcellio scaber*, only 20% of the ultimate size of a female is reached at maturation, the rest of the weight is gained while the animal is already reproductively active. The body weight of females carrying eggs may differ by a factor of 6 (Figure 3.11).

A positive relationship between body size and clutch size is also observed in spiders. Wolf spiders (Lycosidae) are often used for reproduction studies as the females carry an egg sac attached to the abdomen and so the number of spiderlings crawling from the sac under contained conditions can be correlated easily with condition parameters of the female. Figure 3.12 shows an example from work done on tundra spiders in northern Canada. In three different species and at three different sites, similar relationships held (Bowden and Buddle 2012). The mass of the egg sac was directly proportional to the mass of the female, such that the ratio of egg-sac mass to female mass was more or less constant.

It seems that a positive correlation between body size and reproductive output is quite common among soil invertebrates. In addition to the examples mentioned above, it is also typical for earthworms (Lavelle 1981) and land snails (Madec et al. 1998). That most soil invertebrates have indeterminate growth, and can gain reproductive output by growing larger, may explain the wide occurrence of iteroparity.

The importance of growth as a fitness trait in soil invertebrates is confirmed by studies on the heritability of life-history traits. Only few quantitative genetics studies have been done with soil invertebrates, but some data are available for the springtail *Orchesella cincta*, the land snail *Cornu aspersum* and the woodlouse *Porcellio laevis* (Table 3.2). These data suggest that heritabilities of weight and age at maturity are generally higher than those of reproductive parameters (clutch size). It seems that, in animals with indeterminate growth, the primary genetic control is with growth and development, while clutch size is a more or less a consequence of body size.

In addition to additive genetic effects, expressed as h^2, there may be significant maternal effects on reproductive parameters, especially in isopods where the female carries the young in a brood pouch. These maternal effects may bias the estimates for heritability, as the maternal influence on the young is modulated by the diet of the mother (Carter et al. 2004). In springtails a significant negative maternal effect for development time was noted (Janssen et al. 1988), but this was not reproduced in a

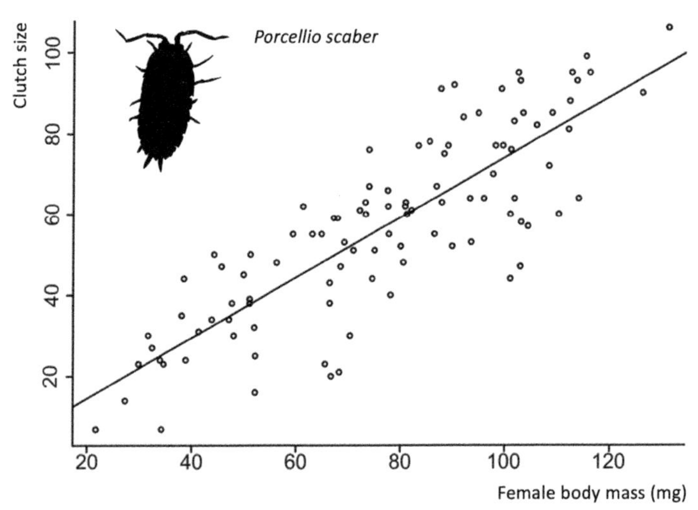

Figure 3.11. Correlation between female body mass and number of offspring (clutch size) in the woodlouse *Porcellio scaber*, sampled from an old yard in Poland. Modified from Antoł and Czarnoleski (2018).

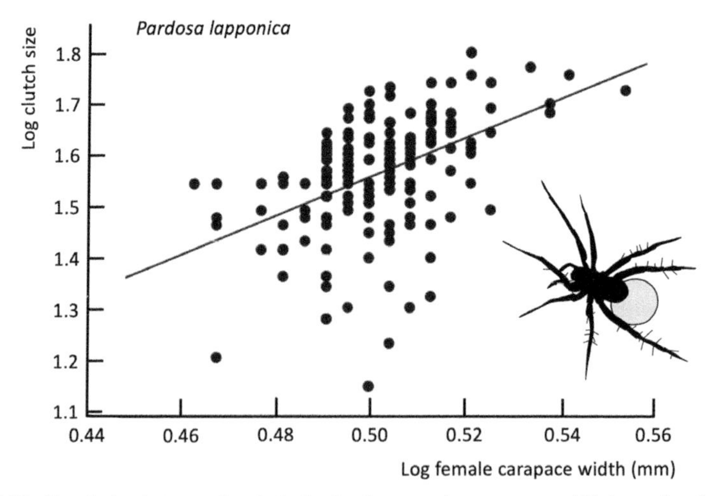

Figure 3.12. Correlation between female body size (measured as carapace width in mm) and number of offspring hatching from the egg sac in female tundra spider *Pardosa lapponica* in the Tombstone range of northern Yukon territory, Canada. The spider in the image is shown carrying an egg sac. Redrawn, with modifications, from Bowden and Buddle (2012).

later study (Stam et al. 1998a). The work on quantitative genetics of life-history traits in soil invertebrates is left fairly incomplete at the moment.

A tight relationship between egg production and body size does not hold for oribatid mites. These animals are iteroparous and continuously deposit eggs throughout their adult life, but they cannot increase in size once they reach the adult stage. In field populations many oribatids carry a single large egg in their body, but some carry 3 to 10 (Luxton 1981a, Kaneko 1989, Al-Assiuty et al. 1993). From an

Table 3.2. Estimates for heritability of life-history traits in populations of *Orchesella cincta* (Collembola). *Cornu aspersum* (= Helix aspersa) (Gastropoda) and *Porcellio laevis* (Isopoda). Estimates for the additive genetic variance relative to total phenotypic variance (h^2), derived from half-sib analyses and parent offspring regressions, are shown with their standard errors.

Species	Weight at maturity	Age at maturity	Clutch size	References
Orchesella cincta	0.15 ± 0.19	0.36 ± 0.24	0.24 ± 0.19	Posthuma and Janssen (1995)*
	0.52 ± 0.19	0.43 ± 0.18	0.16 ± 0.12	Stam et al. (1998)**
Cornu aspersum	0.36 ± 0.16	0.21 ± 0.01	0.12 ± 0.01	Dupont-Nivet et al. (1997, 1998)
Porcellio laevis			0.06 ± 0.12	Carter *et al.* (2004)***

*RBB and SCH populations pooled. **RBB population. ***With diet as a fixed effect.

overview of 24 species by Luxton (1981a) it is obvious that the greatest number of eggs are not carried by those oribatids with the largest body size.

Thus, the iteroparity of oribatid mites seems to be in conflict with their inability to grow larger. Parthenogenesis can hardly be accepted as a reason, since it is difficult to see why the relationship between growth and reproduction would be different for parthenogenetic compared to gonochoristic or hermaphroditic species. One explanation may be that the costs of reproduction are low in oribatids: their egg production does not lead to increased mortality. Iteroparity is promoted not only by indeterminate growth but also by low costs of reproduction (Pianka 1976, 1978).

There are also exceptions among isopods. The sea slater *Ligia oceanica* is semelparous in spite of a positive relationship between brood size and body size. Antoł and Czarnoleski (2018) argued that seasonal unpredictability might be an additional factor calling for iteroparity in terrestrial isopods. Whether this holds for oribatid mites is doubtful. In conclusion, there is not a single explanation for iteroparity in soil invertebrates. The three factors promoting iteroparity (indeterminate growth, low energetic costs of reproduction and bet-hedging against an unpredictable environment) may work jointly, in interaction with each other.

3.6 Life-histories in the soil profile

As we have seen above, the variation of reproductive output among soil invertebrates is great: isopods may breed up to 120 juveniles in their pouch, while many earthworms and oribatids produce one egg at a time. This variation touches the core question of life-history theory, a popular branch of ecology which arose around 1970 with influential papers by the American ecologists Eric Pianka, Stephen Stearns and Graham Bell (Pianka 1970, Stearns 1976, Bell 1976). The basic idea is that reproduction should not be isolated from the rest of the life-history, rather it is the complex of age-specific reproduction, mortality, and body growth that is optimized as a whole. The different life history traits, including their mutual trade-offs, should be combined into one measure of fitness, the *intrinsic rate of population increase*, which can be calculated from fertility and mortality schedules by demographic methods.

In line with the GKGB hypothesis formulated in Chapter 1 we may expect that life-histories of soil invertebrates vary with their position in the soil profile. We have seen in Chapter 1 that Bouché (1977) classified earthworms according to three ecological categories, surface-living species (epigés), soil-living species (endogés) and species which feed on the surface but live in deep burrows (anéciques) (Figure 1.3). The British soil ecologist John Satchell linked Bouché's scheme to life-history theory by arguing that the contrasting behaviours, morphologies and physiologies distinguished by Bouché, arise from a gradient of *r versus K selection* across the soil profile (Satchell 1980). At about the same time, this link was also recognized by the French soil ecologist Patrick Lavelle (Lavelle 1981).

At the time, the r and K selection theory had gained wide attention after the publication of a book on island biogeography by MacArthur and Wilson (1967) and two extremely influential papers by Pianka (1970, 1972). The genetic basis for the theory was developed earlier by MacArthur (1962) and Roughgarden (1971). Basically, the theory argues that in an environment that is empty, with low densities and plenty of resources, a life-history is selected that emphasizes early reproduction, high reproductive output, semelparity, high productivity and a short life. In an environment that is crowded, where densities are high and resources are limited, a life-history is selected with postponed reproduction, iteroparity, modest reproductive output, low productivity and a long life. The letters r and K derive from the classical notation of the logistic growth equation, where r is the per capita rate of increase at low density and K is the maximum density reached after a long time, also called the *carrying capacity* of the environment.

To explore the validity of the r-K selection theory, Satchell (1980) and Lavelle (1981) summarized earlier data collected on earthworm reproduction by the German zoologist and soil scientist Otto Graf (1917–2014), and others. We recapitulate here the fecundities of Lumbricidae (Lavelle also summarized data on other earthworm families). Table 3.3 shows that surface-feeding species, such as the compost worm

Table 3.3. Reproductive output of twelve species of earthworms of the family *Lumbricidae*. Classified by ecological categories, according to Satchell (1980) and Lavelle (1981).

Ecological category	Species	Cocoons per year per worm	Place in r-K continuum
Surface living and litter feeding (épigés)	*Eisenia fetida*	17 – 550	r-selected
	Dendrobaena subrudicunda	42 – 95	
	Lumbricus rubellus	92 – 94	
	Lumbricus castaneus	65	
	Dendrobaena mammalis	17	
Subsurface living, surface feeding (anéciques)	*Lumbricus terrestris*	41	Both r and K properties
	Allolobophora longa	8	
Soil living, subsurface feeding (endogés)	*Aporrectodea caliginosa*	27 – 42	K-selected
	Aporrectodea rosea	8 – 35	
	Allolobophora chlorotica	26 – 31	
	Octolasion cyaneum	13	

Eisenia fetida, have the highest rate of cocoon production. The typical production rate varies from 50 to well over 100 cocoons per worm per year. The subsurface feeders, e.g., *Aporrectodea rosea*, have a lower rate, typically between 20 and 50 cocoons per worm per year. Worms feeding at the surface but living in deep burrows, such as *Lumbricus terrestris* (the anécique category of Bouché) seem to be more similar to the soil-living species than to the surface-living species as regards their fecundity.

The reproduction data on earthworms fit into the theory that soil-living species are limited by food resources more often than surface-living species. Thus, the soil represents a K-selecting environment, calling for a typical K-selected life-history: low reproduction and a long life. Living at the surface favours an r-selected life-history with high reproductive output, but a shorter lifetime.

A similar pattern may be expected in other soil invertebrate lineages. Table 3.4 shows fecundity data extracted from laboratory studies of Collembola at 15°C, performed by various authors (where necessary recalculated to obtain rates per week). The species are classified according to the life-forms of Gisin (*cf.* Figure 1.2). The trend expected under the GKGB hypothesis is that epigeic species show r-selected traits (high reproduction) and euedaphic species show K-selected traits. However, this tendency is less clear than in the case of earthworms. Among the epigeon, *Sminthurus viridis* has a high rate of egg production, but highly productive species are also present in the hemiedaphon (*Arrhopalites sericus*) and the euedaphon (*Folsomia candida*).

Collembola lay their eggs in clutches, typically between 20 to 60 eggs, usually smaller for younger individuals (*cf.* Figure 3.10). Table 3.4 shows that the rate of egg production (eggs per female per week) varies more than clutch size. This observation fits into the idea that for indeterminate growers like Collembola, growth and development are the main traits under genetic and selective control, while clutch size variation is a consequence of differences in body size, as suggested in the previous section.

The variability within species is nevertheless striking in some cases, such as *Folsomia candida*, the "standard soil arthropod" (Fountain and Hopkin 2005). Snider and Butcher (1973) reported an extraordinarily high productivity of their *Folsomia* population with clutches up to 160 or even 200 eggs in one instar. However, Green (1964) reported an average clutch size of 43, a figure more in line with other Collembola. Part of the variation may be due to different authors working on different parthenogenetic clones. It is known that clonal variation is a significant issue in *F. candida* (Grimnes and Snider 1981, Crommentuijn et al. 1995, Simonsen and Christensen 2001). Another issue is that *F. candida* should not be considered a true soil species. It is actually rare in forest and grassland ecosystems and is more typical of ephemeral, rich habitats such as compost heaps. So, while it has the morphology of the euedaphon, at least some clonal lines of *F. candida* might be r-selected, rather than K-selected.

As a third example of life-history evolution in the soil profile, we point out the case of egg number in oribatid mites. In laboratory cultures oribatids show a low but steady rate of egg production. The parthenogenetic species *Platynothus peltifer*

Table 3.4. Reproductive output of various species of Collembola cultured under laboratory conditions at 15°C, as reported in the literature. Reproduction is expressed both as the number of eggs laid per female per week (over the adult life) and as clutch size (number of eggs laid in a batch during one reproductive instar).

Morphological category	Species	Fecundity (eggs per female per week)	Clutch size	Source
Epigeon (species with colour patterns, well developed furca, eyes)	*Sminthurus viridis*	84.0	30	Maclagan (1932)
	Orchesella cincta	14.4	34 – 46	Van Straalen and Joosse (1985), Janssen and Joosse 1987
Hemiedaphon (species with uniformly distributed pigment, with furca and eyes)	*Arrhopalites sericus*	30.5	17	Blancquaert et al. (1981b)
	Sinella curviseta	11.8 – 13.0	10 – 59	Niijima (1973), Waldorf (1971), Gist et al. (1974)
	Tomocerus minor	4.3	32 – 42	Van Straalen and Joosse (1985), Janssen and Joosse (1987)
	Folsomia quadrioculata	2.3	10	Grégoire-Wibo (1974), Sengupta et al. (2016)
	Isotoma trispinata	3.6	11	Tanaka (1970)
	Desoria olivacea	1.4	35	Hale (1965)
	Ceratophysella denticulata	1.4 – 8.3	29 – 46	Hale (1965), Thibaud (1970)
	Hypogastrura purpurescens	5.8	35	Thibaud (1970)
Euedaphon (species without pigment, reduced furca, short legs and antennae, eyeless)	*Megalothorax minimus*	4.9	10	Blancquaert and Mertens (1979)
	Folsomia candida	39.0 – 52.5	43 – 180	Green (1964), Snider (1973), Snider and Butcher (1973), Grimnes and Snider (1981), Itoh et al. (1995)
	Onychiurus folsomi	20.0	30	Snider (1974)
	Protaphorura armata	2.5	14	Snider (1983)
	Schaefferia coeca	1.7	8	Thibaud (1970)
	Mesogastrura ojcoviensis	2.5	10	Thibaud (1970)

produces on the average 2.0–2.5 eggs per female per week at 20°C (Van Straalen et al. 1989, Denneman and Van Straalen 1991). Another parthenogenetic species, *Oppia nitens*, produces an average of 0.2 to 0.6 eggs per female per week (Keshavarz Jamshidian et al. 2017, De Lima e Silva et al. 2017). As both species continue egg laying for a considerable time (up to 10 weeks in laboratory culture and throughout the growing season in the field), their lifetime reproductive output is nevertheless substantial.

Table 3.5. Egg numbers of oribatid mite species according to Luxton (1981a), with mites classified by functional groups on the basis of growth efficiency (P/A) as in Luxton (1981c).

Functional category	Species or family	Eggs
P/A I litter dwellers highly productive species P/B > 2.5 adult body mass > 10 μg growth efficiency 55–65% r-selected	*Damaeus claviceps*	3.7
	Belba corynopus	3.9
	Steganacarus spinosus	2.5
	Steganacarus magnus	5.1
	Phthiracarus anonymus	1.6
	Phthiracarus nitens	2.8
	Xenyllus tegeocranus	10.1
	Nothrus palustris	2.7
	Nothrus sylvestris	1.8
	Galumna lanceolata	3.0
P/A II both soil and litter species P/B around 2.5 growth efficiency 50–52% r or K selected	*Adoristes ovatus*	6.3
	Hypochthonius rufulus	1.0
	Ceratozetes gracilis	2.1
	Achipteria coleoptrata	2.5
P/A III soil dwellers P/B < 2.5 adult body mass < 10 μg growth efficiency 26–30% K-selected	*Oppia arnata*	1.4
	Oppia subpectinata	1.6
	Oppiella nova	1.2
	Suctobelbidae	1.0
	Chamobates cuspidatus	2.8
	Tectocepheus velatus	1.1
	Oribella paolii	1.7
	Hemileius initialis	3.2

Note that the nomenclature here differs from Luxton (1981c). P/A I is Luxton's group I. P/A II is Luxton's group III and P/A III is Luxton's group II. Also note that in his paper Luxton calls groups I and III K-selected and group II r-selected (p. 98), while we adhere to Pianka's original meaning where K selection operates under conditions of resource limitation, so is typical of Luxton's group II (P/A III).

Mite reproduction studies in the laboratory are quite scarce, but a good deal is known about the number of eggs contained in the body of field specimens, and these data may be taken as an indication of fecundity. Table 3.5 summarizes egg numbers from the monumental work of the British soil biologist Malcolm Luxton, at the University of Liverpool, who did his mite studies at the Mols Laboratory at Ebeltoft, Denmark, with Henning Petersen, in the IBP programme (Luxton 1975, Luxton 1981a,b,c, Petersen and Luxton 1982).

Luxton (1981c) divided the oribatid mites that he investigated into three groups, based on growth efficiency (production to assimilation ratio, abbreviated P/A). When we correlate this with the egg numbers reported in Luxton (1981a) a picture emerges which is in good agreement with the GKGB hypothesis: species ovipositing in the litter layer have a higher average egg number than species laying their eggs in the 0–3 cm layer of the soil (Table 3.5). We have seen in Section 3.2 that litter dwellers also have a higher biomass productivity as measured by the P/B ratio. Thus, the r and K selection regimes that grade across the soil profile, lead to a gradient of reproductive output similar to the one in earthworms (Table 3.3).

However, as noted by Luxton, this does not correlate with body size. The trend in body size runs mostly in the other direction: the litter-dwellers, with the highest egg numbers, have the largest body sizes. The r-K theory usually states that r selection, due to the emphasis on reproduction, leads to a lower allocation to growth and a smaller body size. In the case of oribatid mites this is counteracted by selection for small body size due to the physical restrictions of the soil. We have seen the same trend in the body-size dependence of P/B: the productivity of small soil invertebrates tends to be low due to selection for small bodies acting in concert with K-selection on the life history.

To complete the argument from life-history theory, the patterns of fecundity must be combined with mortality schedules, so as to cover the whole life-history. This is conveniently done in a *life-table*, as illustrated in Table 3.6. By way of example, fecundity and mortality schedules were derived from a study by Gist et al. (1974) on the springtail *Sinella curviseta*. From the life-table data ("*vital rates*") the intrinsic rate of natural increase can be calculated using standard demographic methods, as explained in ecology textbooks, e.g., Krebs (1972) and illustrated in Table 3.6. The analysis shows that the average lifetime of *S. curviseta* under the conditions of the experiment was 57 days, life-time reproductive output per female ("*gross reproduction rate*") was 449 eggs, *net reproduction rate* (reproductive output corrected for mortality) was 208, *generation time* was 61 days and the intrinsic rate of natural increase of this population was 0.0874 per day (Table 3.6).

The intrinsic rate of natural increase measures how fast the population can grow when mortality and fertility schedules would stay as they are. This refers to the central theorem of demographic theory which says that any population with age-dependent (but time invariant) mortality and fertility schedules, independent of the founding population, will approach a stable age distribution and then grow exponentially at rate *r*, according to the equation: $N(t) = N_0e^{rt}$. A positive r indicates growth, a negative r implies exponential decrease.

The intrinsic rate of natural increase has attracted a good deal of attention since its introduction into ecology by the famous Australian geneticist (also theologian) Louis Charles Birch (1918–2009) (Birch 1948). In Birch's view *r* was a measure of *biotic potential*, especially of insect populations, in an unlimited environment where the effects of density (crowding) do not need to be considered and population growth is exponential. In life history theory, *r* is considered an index of fitness which is maximized by natural selection, given certain constraints among the vital rates. A large body of theoretical and experimental work is based upon this principle, summarized in textbooks such as Charlesworth (1980), Stearns (1992) and Roff (2002). In ecotoxicology, *r* is taken as a measure of performance under toxic stress, translating toxicity at the individual level to the population level (Kammenga and Laskowski 2000).

The value of *r* calculated from the life-history of *S. curviseta* in Table 3.6 implies that the population can multiply itself every day by a factor of $e^{0.0875} = 1.09$, so an increase of 9% per day. Similar high values for the growth potential of Collembola

Table 3.6. Showing how the intrinsic rate of natural increase is calculated from life-table data. Survival and fecundity figures are for the collembolan *Sinella curviseta* cultured at 30°C in the laboratory (Gist et al. 1974). Age includes the egg stage (10 days egg development with 84.4% hatching).

Age (days)	**Survival up to age x**		**Fecundity (eggs per female per 10 days)**		
x	l_x	$x\,l_x$	m_x	$l_x m_x$	$x\,l_x m_x$
0	1.000	0.0	0	0.0	0.0
10	0.844	8.4	0	0.0	0.0
20	0.716	14.3	0	0.0	0.0
30	0.608	18.2	37	22.5	674.9
40	0.537	21.5	63	33.8	1353.2
50	0.490	24.5	85	41.7	2082.5
60	0.476	28.6	85	40.5	2427.6
70	0.462	32.3	44	20.3	1423.0
80	0.443	35.4	42	18.6	1488.5
90	0.382	34.4	19	7.3	653.2
100	0.349	34.9	24	8.4	837.6
110	0.316	34.8	22	7.0	764.7
120	0.307	36.8	20	6.1	736.8
130	0.264	34.3	6	1.6	205.9
140	0.250	35.0	0	0.0	0.0
150	0.184	27.6	2	0.4	55.2
160	0.099	15.8	0	0.0	0.0
170	0.018	3.1	0	0.0	0.0
180	0.000	0.0	0	0.0	0.0
Total	**7.745**	**440.0**	**449**	**208.0**	**12703.1**

Calculation of demographic parameters			
Life expectancy (average lifetime)	$e_0 = \Sigma x l_x / \Sigma l_x$	= 440.0/7.745	= 57 days
Gross reproduction rate	$GRR = \Sigma m_x$		= 449
Net reproduction rate	$R_0 = \Sigma l_x m_x$		= 208
Generation time	$T = \Sigma x l_x m_x / \Sigma l_x m_x$	= 12703.1/208	= 61 days
Intrinsic rate of natural increase	$r = \ln(R_0)/T$	= ln(208)/61	= 0.0875 day^{-1}

were calculated by the Belgian soil ecologist Colette Grégoire-Wibo for *Folsomia candida* (16% per day), and *Protaphorura armata* (7% per day) (Grégoire-Wibo and Snider 1977, 1983). These high potential growth rates will never be realized in the field because survivorship observed in the laboratory ignores predation, which is the dominant source of mortality for many soil invertebrates. The high values of *r* nevertheless indicate that Collembola populations can increase rapidly from low densities in a favourable season, e.g., in early spring. This is confirmed by many field population studies as reviewed in Sections 3.1 and 3.4.

Life-table studies have often been done with Collembola, because these animals are eminently suitable: many species are easy to culture while all life stages, including the eggs, can be observed and counted. Similar work, including nematodes, oribatid mites, earthworms, isopods, centipedes, slugs, etc., has been conducted, often in the context of soil pollution (Bengtsson et al. 1985b, Van Straalen et al. 1989, Crommentuijn et al. 1993, Kammenga et al. 1996, Laskowski 1997, Klok et al. 1997, Chen et al. 2017, De Lima e Silva et al. 2021). Evolutionary studies, focusing on demographic adaptation in response to environmental conditions or geographic factors are less common (but see Sengupta et al. 2016).

In conclusion, this section has revealed an association of soil invertebrate life-histories with the soil profile, as predicted by the GKGB hypothesis. The trend is most obvious for earthworms and oribatids, and blurrier for springtails. And, interestingly, soil invertebrates present a syndrome of life-history traits that does not exist in above-ground communities: small body size, low productivity and a long life in a resource-limited, K-selected environment. This is a consequence of the soil depth gradient presenting a two-pronged selective challenge, favouring lower productivity due to resource limitation and smaller body size due to the physical limitations of the soil environment.

3.7 Environmental tuning

A soil invertebrate must not only have an optimal life-history, it must also ensure that the life-history is tuned to environmental changes. It must be active in favourable periods and avoid activity under conditions posing a threat to survival. The seasonal occurrence of an invertebrate is called *phenology*. Life-history theory tells us that a highly tuned phenology with temporally concentrated reproduction is selected for in a seasonally varying environment that affects age classes differently (MacArthur 1968). Do we see these phenological adaptations in soil invertebrates?

The simplest way of environmental tuning derives from the temperature response of life-history processes. Since all soil invertebrates are ectotherms, they all experience higher metabolic rates at higher temperatures, although the responsiveness to temperatures varies greatly between species. The simple increase of activity with temperature will already ensure that such activities become clustered in time and take place mostly when temperatures are high. The responsiveness to temperature, i.e., the slope of the temperature response, determines the degree of phenological synchronisation (Van Straalen 1983b). We will explore in Chapter 6 how temperature responses of soil invertebrates have evolved in relation to the soil profile.

The simple direct response to temperature can already explain a good deal of soil invertebrate phenology. This is illustrated convincingly in a population study of the springtail *Parisotoma notabilis* by Diekkrüger and Röske (1995). The authors developed a model driven by temperature dependence of egg development, juvenile growth and fertility. After calibrating the model on a springtail population in pasture land, the dynamics at other sites could be predicted remarkably well, only by measuring the temperature in the field (Figure 3.13).

Models like this will fail when soil moisture becomes limiting at temperatures that would still allow ongoing development. This was recognized by

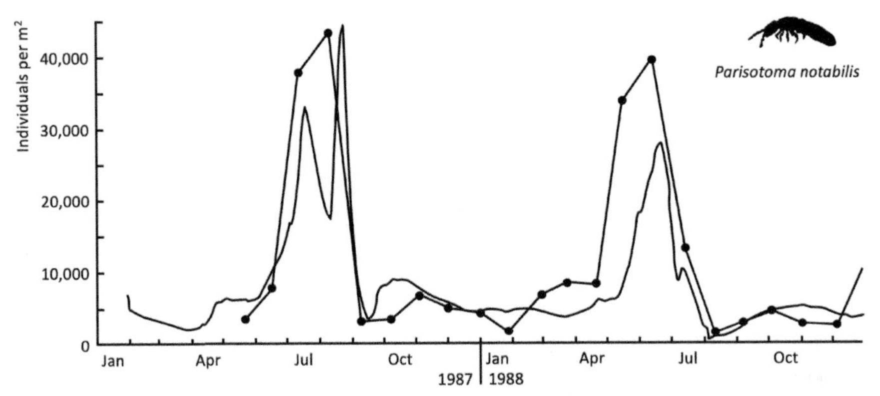

Figure 3.13. Measured (dots) and simulated (continuous line) population changes of *Parisotoma notabilis* (Collembola: Isotomidae) in an agricultural field on sandy soil with winter wheat (1987) and winter barley (1988). The model included temperature dependence of the major life-history processes, calibrated on a pasture population and driven by temperature registrations in the field. Redrawn, with modifications, from Diekkrüger and Röske (1995).

Choi and Ryoo (2003) who developed a similar model for the springtail species *Paronychiurus kimi*. A version of the model that was driven by temperature only worked reasonably well but overestimated the seasonal peaks. Inclusion of a soil moisture response in the model improved the fit with field data significantly (Choi and Ryoo 2003, Choi et al. 2006).

The generation structure of populations in seasonal environments is designated as "*voltinism*". For soil invertebrates that directly respond to the ambient temperature, the number of generations in a year is not well defined. Many soil invertebrates are multivoltine in quite a flexible way. For example, several Collembola have two overlapping generations in a year (Van Straalen 1985a), but depending on the temperature regime, a third generation is possible. Pseudoscorpions usually have one generation a year, but the species *Neobisium muscorum* has two in some places (Gabbutt 1969, Goddard 1976).

A special type of voltinism, called *parsivoltinism*, or *cohort-splitting*, is seen isopods, millipedes and spiders with a dual life-cycle. The tropical isopod *Burmoniscus ocellatus* breeds in April and the same generation breeds again in August, while the offspring of the first breeding period grow quickly and join the summer breeders (Ma et al. 1991b); so, this species has two generations per year for part of the population. A similar split of cohorts is happening in the temperate isopod *Philoscia muscorum* (Sunderland et al. 1976), in the millipede *Polydesmus angustus* (David et al. 1999, David 2009) and the dune wolf spider *Pardosa monticola* (Bonte and Maelfait 2001).

Life-cycles longer than 2 years are rare among soil invertebrates, however, exceptions do occur. A spectacular example is the 8-year cycle of the "train millipede", *Parafontaria laminata armigera* (family Xystodesmidae). This species, endemic to Japan, has larval stages living a secluded life belowground while feeding on soil. The millipede becomes adult only after completing the seventh instar, and makes its way to the surface to reproduce. Surface activity is synchronized over a large area while the millipedes form swarms of millions of animals. When a swarm passes a railway line the train schedules must be interrupted due to slippery rails (hence the name "*train millipede*"). The periodicity of *Parafontaria* outbreaks was revealed only in 2021 with a reconstruction of historical records since 1920 (Niijima et al. 2021).

Soil invertebrates with bivoltine or parsivoltine life-cycles experience contrasting selection regimes in successive generations. Janssen et al. (1988) have suggested that this could explain the observed negative maternal effect of juvenile period across generations in the collembolan species *Orchesella cincta*. In their study, parents with a short maturation time gave rise to offspring with a long maturation time and the other way around. However, the genetic or epigenetic mechanism behind this effect has remained unclear up to now (Stam et al. 1998a). In the millipede *Polydesmus angustus* no evidence for maternal effects on the life-cycle was found (David and Geoffroy 2011).

While many soil invertebrates do nothing more than just responding to the ambient temperature, there are also several examples of species that have some kind of programmed *dormancy*, e.g., arrested development, suppressed reproduction, slow growth, and activation of specific resistance mechanisms, such as cold hardiness and drought resistance. The various types of dormancy are classified by different authors in different ways, usually along an axis of increasing intensity: *quiescence*, *facultative diapause*, *paradiapause*, and *obligatory diapause*. Quiescence is a "light" dormancy that can be readily reversed by the same factors that induced it, while obligatory diapause is a phase in the life-cycle that is regulated by daylength-triggered hormonal secretions, and is induced in a stage before the diapause itself.

However, the classification of dormancies along this axis does not acknowledge the great diversity among soil invertebrates. An overview is given in Table 3.7 (ignoring insect diapause, which is a field of its own). The responses include both *aestivation* (to survive dry and hot periods) and *hibernation* (to survive cold periods and frost). In addition to the classical photoperiod-regulated diapause of insects, all kind of responses regulated by daylength, temperature, humidity and food are found in soil invertebrates. Sometimes different mechanisms occur in the same species (e.g., daylength-regulated diapause in winter and aestivation quiescence in summer).

An extreme form of environmental tuning is *cyclomorphosis*: the seasonal occurrence of divergent morphotypes of the same species. It is known as seasonal *polyphenism* in butterflies and daphnids, but it also occurs in some species of Collembola. For example, an undescribed "species" of isotomid with specialized hairs, allocated to the genus *Vertopagus*, turned out to be the winter morph of the already known *Isotoma nivea* (Fjellberg 1978). Also, in the species *Ceratophysella sigillata* the *chionophile* (snow-active) winter morph has a distinct morphology (Zettel and Zettel 1994).

Some soil invertebrate groups include species that are known for their ability to lose almost all their water and survive in a dehydrated state, known as *anhydrobiosis*, a form of *cryptobiosis*. Tardigrades are best known for this, but anhydrobiosis also occurs in nematodes and Collembola. We will return to this remarkable phenotype in Chapter 6, where we discuss the biochemical mechanisms involved.

Table 3.7. Types of dormancy observed in soil invertebrates (excluding insects), mentioning examples of species in which a dormancy effect has been demonstrated.

Type of dormancy	Induced by	Phenomenology	Examples	References
Reproductive quiescence in relation to hibernation	Low temperature	Suspended reproduction, extended reproductive latency	*Orchesella cincta*, *Tomocerus minor* (Collembola) *Alaskozetes antarcticus* (Oribatida)	Van der Woude and Verhoef (1988), Convey (1994,1996)
Reproductive quiescence in relation to seasonal synchronisation	Long daylength, short daylength	Synchronized initiation of parturial moult, temporally restricted breeding	*Armadillidium vulgare* (Isopoda)	Souty-Grosset et al. (1988a,b), Mocquard et al. (1989), Nasri-Ammar et al. (2001)
Reproductive quiescence in relation to food shortage	Food deprivation	Suspended reproduction, hypometabolism	*Caenorhabditis elegans* (Nematoda)	Gerisch et al. (2020)
Heat-induced aestivation response	Heat, low air humidity	Escape from surface heat, sealing of shell opening, inactivity, hypometabolism	*Helix pomatia*, *Cernuella virgata*, *Theba pisana* (Stylommatophora)	Nowakowska et al. (2010), Schweizer et al. (2019)
Humidity-induced aestivation response	Drop in soil water potential	Inactivity in specialized burrow, hypometabolism, regression of sexual traits, segment addition	*Aporrectodea caliginosa*, *Hormogaster elisae*, *Martiodrilus carimaguensis*, *Glossoscolex paulistes* (earthworms)	Jiménez (2000), Bayley et al. (2010), Holmstrup (2001), Díaz Cosin et al. (2006)
Egg, larval, pupal or adult diapause in relation to hibernation or aestivation, as an obligatory part of the life-cycle	Daylength, temperature, maternal imprinting, induction during a sensitive stage in advance of the stress period	Developmental arrest, inactivity, hypometabolism	*Anurida maritima*, *Sminthurus viridis*, *Sphaeridia pumilis*, *Sminthurinus aureus*, *Lepidocyrtus lignorum*, *Hypogastrura tullbergi*, *Ceratophysella sigillata* (Collembola), *Polydesmus angustus* (Diplopoda)	Witteveen et al. (1988), Wallace (1968), Roberts et al. (2011), Blancquaert *et al.* 1981a). Leinaas and Bleken (1983), Birkemoe and Leinaas (2009), Zettel and Zettel (1994), David et al. (2003)

Table 3.7 contd. ...

...Table 3.7 contd.

Type of dormancy	Induced by	Phenomenology	Examples	References
Cyclomorphosis in relation to stress survival	Cold, winter conditions	Appearance of alternative morphotype, adapted to seasonal conditions	*Isotoma nivea, Hypogastrura lapponica* (Collembola)	Fjellberg (1978), Leinaas (1981)
Dauer larva aimed for dispersal	Food shortage	Nictation, increased stress tolerance, adapted metabolism	*Caenorhabditis elegans* (Nematoda)	Riddle (1988)
Cysts as part of infectious cycle (transition between hosts)	Plant host	Formation of a cyst with eggs, detached from plant tissue, suspended development	*Globodera rostochiensis* (Nematoda)	Evans and Stone (1977)
Anhydrobiosis	Dryness	Almost complete shut-down of metabolism, genome fragmentation, extreme resistance against dehydration, cold and vacuum	*Milnesium tardigradum* (Tardigrada), *Folsomides variabilis* (Collembola)	Sømme (1996), Mali et al. (2010), Wharton (2015), Barra and Poinsot-Balaguer (1977), Poinsot-Balaguer and Barra (1991)

The diversity of dormancy responses, summarized only crudely in Table 3.7, suggests that different animal groups have found different solutions to the environmental tuning problem and that dormancy mechanisms evolved only after the terrestrialization of that lineage (although parts of the molecular machinery deployed could be universal). Since soil invertebrates represent so many independent terrestrializations (see Chapter 2) the variation of dormancy phenomena should not surprise us. We will explore the various mechanisms of temperature, drought and other resistances further in Chapter 6.

In several cases the entry of a dormancy is accompanied by behavioural responses, aimed at finding a secluded spot suitable for a prolonged period of inactivity. Some land snails will crawl in the vegetation, to escape the scorching heat at ground level in Mediterranean climates (Schweizer et al. 2019), but most animals seek shelter in the soil. The soil provides many hiding opportunities for small invertebrates, especially if it has a cover of differently-shaped structural elements (branches, dead wood, bark, stones, plant roots, etc.). The spots used by overwintering animals are called *hibernacula*. Some diapausing insect larvae, spiders and pseudoscorpions make a hibernaculum of their own, by spinning a silken chamber (Wood and Gabbutt 1979). Several earthworms make a specialized burrow called *aestivation burrow*, in which they curl up and can stay inactive for months (Figure 3.14).

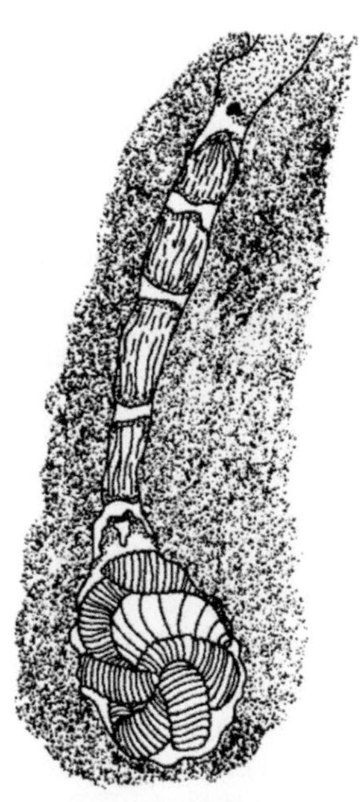

Figure 3.14. Aestivation dormancy by the tropical earthworm *Martiodrilus carimaguensis* (Crassiclitellata: Glossoscolecidae) from the Eastern Plain savannas in Colombia. At the end of the rainy season, the worm empties its gut and curls up in a deep burrow, sealed with loosely packed, brick-like, castings, to survive the 4-months dry season. Drawing by A.G. Moreno. From Jiménez et al. (2000), reproduced by permission from Elsevier and Juan Jiménez.

In conclusion, we have seen that not all, but many, soil invertebrates have some kind of dormancy response to survive adverse conditions, be it cold, heat, drought or food shortage. It is also obvious that strong dormancy mechanisms such as true diapause are present mostly among the surface-living representatives of a lineage. Diapause is never observed in euedaphic Collembola, only in epigeic and hemiedaphic species. Also, Collembola exploring the water surface (Blancquaert et al. 1981a), and those living on open intertidal mudflats (Witteveen et al. 1988) have a true diapause. This perfectly fits in GKGB hypothesis introduced in Chapter 1: diapause is promoted in "energy-intensive" habitats.

For earthworms, Bouché (1977) argued that true diapause is characteristic for epigeic species while the less pronounced forms of dormancy (quiescence) are seen in éndogés and anéciques. Despite obvious exceptions (Díaz Cosín et al. 2006) this seems to be a general trend. In mites, diapause is never observed among soil-living oribatids; they rather rely on flexibility of the life cycle and can hibernate in any life stage (Convey 1996, Belozerov 2008). However, diapause is quite common among mites living in the vegetation, such as spider mites and phytoseiids (Veerman 1992).

So, the need for a diapause stage in the life-cycle, which adds to a synchronized and environmentally tuned phenology, seems to be greater in surface-active invertebrates than in soil-living species. This is especially true for seasonally varying temperate and arctic environments (Convey 1996, Leinaas 1981, Leinaas and Bleken 1983, Birkemoe and Leinaas 1999). It may be speculated that diapause evolved in terrestrial animals when ancestral quiescence mechanisms (that we see in many soil-living species) were brought under control of neurosecretory pathways and so became more tightly regulated. This provided an advantage in seasonal climates and exposed habitats. A molecular analysis of the homologies among dormancy mechanisms is needed to test this suggestion.

3.8 Spatial distribution and aggregation

Every soil ecologist knows that soil invertebrates are not uniformly distributed over their habitat. When taking soil cores from the field and extracting animals from them, some cores deliver hundreds of individuals, while others deliver none, to the frustration of the investigator. The number of sampling units required to estimate the mean density with any acceptable confidence may amount to 50 to 100, even when the habitat looks quite homogeneous.

The variation across samples is an obvious consequence of heterogenous spatial distribution or *aggregation.* Soil ecologists have devoted quite some effort to describing aggregations by means of statistical distributions, such as lognormal, Poisson and negative binomial distributions (Taylor et al. 1979). A variety of indices has been used to measure the degree of aggregation, for example the slope of variance to mean in a power plot (Taylor 1961), Morisita's *index of dispersion* (Morisita 1962) and Lloyd's *index of mean crowding* (Lloyd 1967). In almost all cases investigated, soil invertebrates show a clumped distribution, that is, they are more aggregated than expected from a random allocation of numbers to sampling units. What are the biological reasons and possible evolutionary consequences of this nearly universal tendency to aggregate?

As a starting hypothesis one may assume that soil invertebrates just respond to the non-random distribution of abiotic factors such as soil moisture. We have seen in Section 3.4 that soil moisture is a crucial factor for almost all soil invertebrates, therefore we expect aggregation to be particularly strong for drought-sensitive species. This was demonstrated in a classical study by the Dutch soil ecologist Els Joosse (Joosse 1970, 1971, Joosse and Verhoef 1974). She showed that drought-sensitive Collembola such as *Tomocerus minor* and *Isotoma viridis* had a more pronounced aggregation (measured by Taylor's index) than *Orchesella cincta* and *Entomobrya nivalis*, two species with a fair degree of drought resistance. This pattern has been confirmed by many other authors, both in Collembola (Usher 1969, Usher and Hider 1975, Mertens and Bourgoignie 1977, Verhoef and Nagelkerke 1977) and in oribatid mites (Gérard and Berthet 1966, Usher 1975).

In these older studies, the tendency of a species to aggregate was described by statistical indices. However, this approach does not address the spatial structure explicitly. It was not until the introduction of *geostatistics* in soil ecology that this problem was fully recognized. The Dutch ecologist Christien Ettema was one of the first scientists to develop the new field, called "*spatial soil ecology*". At the time she worked with David C. Coleman at the University of Georgia (Ettema et al. 1998, Ettema and Wardle 2002).

A standard approach in geostatistics is to consider the spatial autocorrelation of the data, that is, the degree to which similarity between sample units is a function of the distance between them. This is usually expressed as a *semivariance* function which is estimated from spatially explicit sampling data by the following equation (Fortin and Dale 2005):

$$\gamma(h) = \frac{1}{2N(h)} \sum_{i=1}^{n} [z(x_i) - z(x_i + h)]^2$$

where:
γ(h) is semivariance,
h is the geographic distance between two sample locations,
N(h) is the number of sampled locations separated by distance h,
$z(x_i)$ is the number of animals present at location x_i,
$z(x_i+h)$ is the number of animals present at location x_i + h, and
n is the total number of locations sampled.

The expression under the sum is actually a covariance between two spatially separated locations. The factor ½ is applied because in the summation every pair of sampling locations is visited twice. This is why the outcome is called semivariance (half a variance).

The function γ increases monotonically with *h* because the variance across samples will always be larger if they are taken further apart (except in special cases, e.g., regular grid distributions). The intercept of the function (at zero distance) equals the variance obtained when (theoretically) sampling the same location multiple times (called the *nugget* of the *semivariogram*, indicated by C_0). At great distance the autocorrelation does not increase any further: the variance reaches a maximum (called *sill*, C_0+C). The value of *C* relative to the sill ($C/(C+C_0)$) at any distance is the fraction of the total variance that is explained by spatial structure over that distance.

These terms are illustrated in Figure 3.15, which shows a selection of semivariance functions from Ettema et al. (1998). The shape of the curves suggests that semivariance as a function of distance can be described by an exponential function, which starts at the nugget and approaches the sill asymptotically. The average distance at which spatial structure extends is 1/α where α is the shape

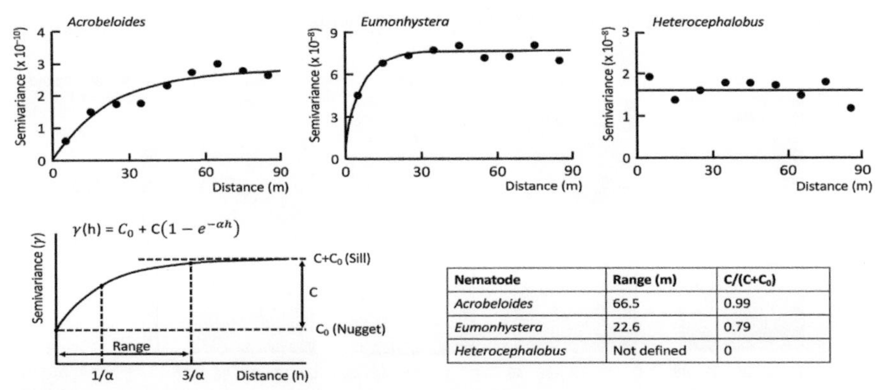

Nematode	Range (m)	C/(C+C₀)
Acrobeloides	66.5	0.99
Eumonhystera	22.6	0.79
Heterocephalobus	Not defined	0

Figure 3.15. The principle of spatial analysis illustrated by semivariograms for three nematode genera in a riparian wetland. The lower (theoretical) graph explains the parameters of the semivariogram, when modelled using an exponential function. The table provides parameter estimates for the three nematodes. Based on data from Ettema et al. (1998).

parameter of the exponential function. Three times this value (3/α) is usually taken as the "*range*" of spatial autocorrelation.

The semivariograms show that *Acrobeloides* is spatially structured over quite a large range (autocorrelation extending up to 66.5 m), while *Eumonhystera* densities vary on a smaller scale (no further than 22.6 m), and *Heterocephalobus* shows no spatial structure at all. It is obvious that these nematodes differ significantly in their habitat use and respond to different soil factors. However, there were only weak correlations with soil factors such as moisture and nitrate.

As a next step, the spatial autocorrelation can be used to create "smoothed" maps of the density. A statistical technique, called "*kriging*", was developed for this purpose, named after Danie G. Krige, a South-African mining engineer who used spatial interpolation to locate the most likely sites of gold ores. The technique is computationally complicated, but the details don't need bothering us here; the reader is referred to textbooks on geostatistics (e.g., Fortin and Dale 2005).

In Figure 3.16 smoothed density distributions over the habitat are shown for the nematode *Acrobeloides*, sampled four times in a year. Interestingly, the spatial distribution is not constant in time. In November, a cluster of high densities is found in the north-west corner of the plot that gradually moves south, then dissolves, while in summer a new cluster is forming in the eastern side of the plot (Figure 3.16). These temporal changes were not correlated with spatial shifts in soil factors. Obviously, spatial structure changes with time, but what exactly is happening within these nematode communities is not clear.

The geostatistical approaches, applied to nematodes and other soil invertebrates (e.g., in earthworms by Cannavacciuolo et al. 1998 and in spiders by Birkhofer et al. 2006) have provided great insights into the spatial distribution, but like the single indices of aggregation, do not shed light on the biological processes that underly the formation of such aggregations. Several biological factors have been proposed to explain how aggregates are formed in soil invertebrates:

1. *Hydrokinesis*. This is the tendency to move or sit still depending on the local moisture level of the substrate. In a kinesis (in contrast to a *taxis*) there is no direction in the movement; the simple rule "walk when dry and sit still when

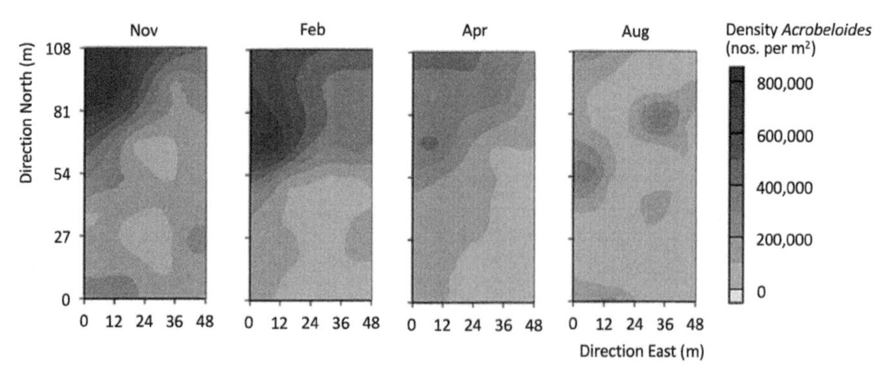

Figure 3.16. Temporal changes in the spatial density distribution of the nematode *Acrobeloides* in a 50 x 100 m plot in a riparian wetland. Reproduced from Ettema et al. (1998). Copyright by the Ecological Society of America.

moist" will nevertheless always cause the animals to aggregate in moist spots (if these are present in the habitat available).

2. *Photokinesis*. The tendency to move in the light and to sit still in the dark will, like in hydrokinesis, result in aggregations formed in dark spots.
3. *Thigmokinesis*. This is the tendency to reduce locomotion as long as there is lateral contact with the substrate, especially three-dimensional structures such as reliefs, grooves, crevices, ridges, etc. Like in the response to humidity and light, this will cause aggregations in sheltered places.
4. *Orientation by excreta*. Many animals respond to faecal pellets or droppings excreted by conspecifics to feed upon (e.g., juvenile woodlice do so abundantly). This mechanism is not very specific, but it effectively causes aggregations to form at places where other animals are or have been.
5. *Orientation by pheromones*. Many soil animals use specific chemical compounds, emitted to the air, excreted on the substrate or shed with the faeces, depending on their volatility. If these compounds can be sensed in low concentrations by conspecifics they act as aggregation pheromones.

So, aggregations can be formed by many different mechanisms, and they may also have several different biological functions. This may explain why they are so widespread. A famous model that illustrates many of these points is the gregarious behaviour of woodlice. Already in 1926, the American ecologist and zoologist Warder Clyde Allee (1855–1955) published a paper on the aggregation behaviour (called "*bunching*" by him) of land isopods (Allee 1926). The "*Allee effect*", a decrease of population growth at low densities, causing populations to go extinct below a critical size, is named after him. In fact, Allee is one of the few famous biologists who developed a major biological principle from work on soil invertebrates (in addition to Darwin and his earthworms, of course).

Experiments with *Porcellio scaber* have shed interesting light on the Allee effect in woodlice. *Porcellio* and other woodlice tend to form multi-layered bunches, in which the animals crawl on top of each other (Figure 3.17). Both soil moisture and light avoidance may act as initiating events. Small aggregations then tend to grow because, when the number of animals in the aggregate is low, the likelihood of joining an aggregate is larger than the likelihood of leaving. The consequence is that bunches increase in size, up to 50–100 animals. That the dynamics of aggregation are density dependent (Figure 3.17), can be considered a kind of *quorum sensing*, as is also known in bacterial biofilms. It may even be considered a form of subsocial behaviour, constituting a first step towards eusocial behaviour in other arthropods (Allee 1926, Broly et al. 2012).

Broly et al. (2014) showed that the gregarious behaviour of isopods has clear advantages to every member of the aggregate: average water loss per animal is reduced by almost a factor of 2. The tendency to form aggregates, which is also present in aquatic isopods, might even have helped the evolutionary terrestrialization of Isopoda (Broly et al. 2013b). Respiration is also somewhat lower in an aggregate; individual growth rate is higher and the animals survive better. But the need to form aggregates causes an Allee effect in the population of woodlice, because at

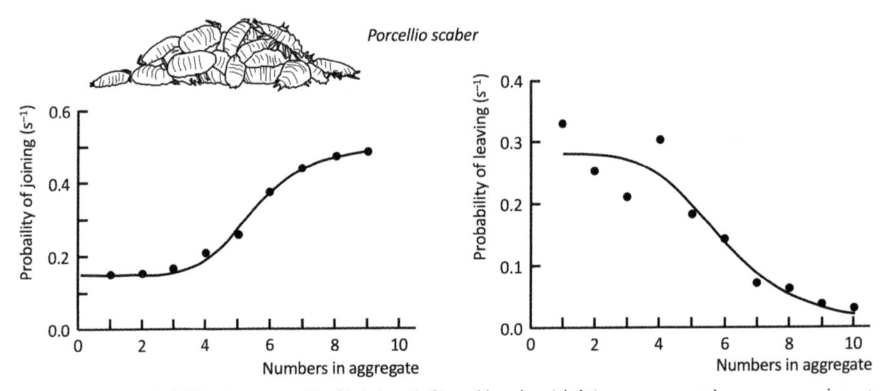

Figure 3.17. Probability (per second) of joining (left) and leaving (right) an aggregate in arena experiments with *Porcellio scaber*. Aggregates tend to grow because joining increases and leaving decreases with aggregate size. Redrawn from Broly et al. (2013b, 2016). The image shows an aggregate of woodlice, drawn from a photo by Pierre Broly.

low densities, the animals spend a considerable time moving around searching for clusters to join (Brockett and Hassall 2005).

The fact that isopods, and other soil invertebrates, show complex aggregative behaviours suggests the existence of specific pheromones that elicit the response. That such pheromones exist has been demonstrated by many researchers in several soil animal groups (Verhoef and Nagelkerke 1977, Verhoef et al. 1977, Joosse and Koelman 1979, Leinaas 1983, Takeda 1984, Verhoef 1984, Pfander and Zettel 2004, Salmon et al. 2019). However, the precise nature of the compounds involved remains unclear. Nilsson and Bengtsson (2004) obtained evidence that fatty acids, endogenous to the cuticle of springtails, act as attractants and repellents to *Onychiurus armatus*. Linoleic acid from dead animals had a repellent action and palmitic acid from conspecifics acted as an attractant. However, electrophysiological analysis in *Folsomia candida* showed that the chemosensory "*organ III*" in the antenna is not sensitive to fatty acids (Liu and Wu 2017). To really understand the action of aggregation pheromones in soil invertebrates, identification of both the active compounds and their sensory receptors is necessary (Salmon et al. 2019).

This short excursion into spatial soil ecology has shown that the methodology for describing spatial patterns is now well established. Also, we know a good deal about the evolutionary advantages of aggregative behaviours. We may even recognize the origin of social behaviour in some soil invertebrates. However, how aggregates develop, and what pheromones, releasing or priming signals, are involved, remains a mystery to date. A lot of work is still to be done.

3.9 Movement and dispersal

In the earlier sections of this Chapter we have seen that dispersal is a crucial phenomenon for soil invertebrates. Without dispersal beyond the home range effective population sizes of many soil invertebrates would be quite small, despite the high densities. Without dispersal we would expect highly differentiated local populations and a dominating influence of genetic drift. Several soil invertebrates indeed show this kind of genetic structure, as we will see below, but many others have evolved effective dispersal mechanisms.

Dispersal is difficult to measure in any animal and even more so in small soil invertebrates. It is no wonder that dispersal was an underrated issue in population ecology for a long time. Population ecologists realized its significance but due to the lack of a data just assumed that dispersal was a neutral factor in the dynamics of a population. Evolutionary biologists, however, have recognized the overriding influence of "*gene flow*" upon population genetic structure for a long time. Estimates of gene flow derived from genetics are however, difficult to phrase in terms of a dispersal mechanism.

The case for dispersal as an ecological mechanism changed in the beginning of this century with the formulation of the "*movement ecology*" framework (Nathan et al. 2008). This paradigm aims to explain the displacement of animals from an interaction between external factors (temperature, soil factors, physical objects) and the internal state (developmental stage, sex drive, hunger, etc.). These two categories of drivers combine with *motion capacity* (how to move?) and *navigation capacity* (where to move?) to realize a specific movement path. Superimposed upon this framework is an argument of evolutionary optimization, in which the costs (energy expenses) and benefits (fitness advantage) of movement are taken into account (Nathan et al. 2008).

The movement ecology paradigm has not yet penetrated soil ecology much, although several groups of soil invertebrates are eminently suitable for movement studies, in particular animals that can be marked or located by telemetry such as carabid beetles, spiders and isopods. Also for earthworms specific tagging techniques have been developed (Mathieu et al. 2018). Still, in a review of movement patterns of forest invertebrates, Brouwers and Newton (2009) recovered data on only 30 species in total, of which 17 were carabid beetles and no other soil invertebrates.

A typical movement pattern of carabid beetles is shown in Figure 3.18 (Baars 1979b) It is characterized by two elements:

- Random short walks, undirected, apparently exploring the local habitat in search of food
- Directed movements, more or less in straight line, at higher speed, apparently in search of new habitat.

However, the length of the short and long distances overlapped considerably and they were performed in more or less random order without link to sex or age. The data can be analysed using the "*distance bearing*" method, which assumes a Markov chain mechanism, each move characterized by a step length (distance) and a compass direction ("bearing"). This method is superior to the continuous registration of GPS coordinates, which for small distances is not accurate enough (Růžičková and Veselý 2016, Růžičková and Elek 2021).

These movement recordings have revealed remarkable facts about the private life of carabid beetles in heterogenous landscapes. For example, the movements of *Carabus ullrichii* turned out to be sex-specific. Males tend to be active at the border of the forest where they seem to wait for females, which prefer to deposit their eggs inside the forest. The forest edge could be a mating site for these beetles

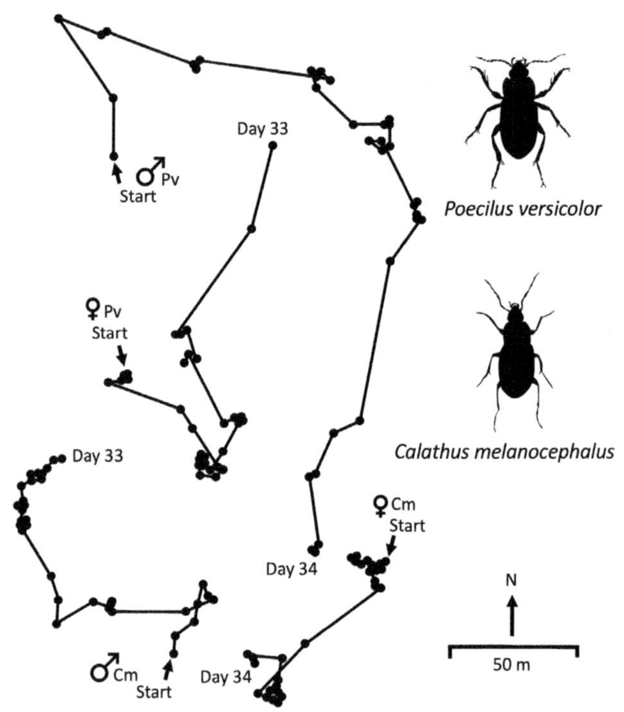

Figure 3.18. Daily movement pattern of two species of carabid beetle in a heathland, analysed using specimens marked with radioactive iridium-192. Reproduced and modified after Baars (1979b).

(Růžičková and Veselý 2018). This research also underlines the importance of small-scale landscape elements such as forest edges, hedgerows and river banks in the life of surface-active soil invertebrates.

Less detailed movement studies, focussing on overall dispersal rates without consideration of the spatial pattern, have been done with spiders (Bonte et al. 2004), earthworms (Marinissen and Van den Bosch 1992, Zorn et al. 2005, Eijsackers 2011), enchytraeids (Sjögren et al. 1995), Collembola (Dunger et al. 2002) and oribatids (Ojala and Huhta 2001). Dispersal distances in conjunction with population density can be used to estimate short-term population size (Table 3.1).

By means of dispersal, different populations of the same species are dynamically connected to each other. This was recognized in the famous concept of "*stabilization of numbers by spreading of risk*" formulated by the Dutch ecologist Piet den Boer (1926–2016). From a unique set of long-term term monitoring observations on carabid beetles in Dutch heathlands, Den Boer noted that some species tended to fluctuate in numbers much more than others. Species with large fluctuations synchronized across the landscape (e.g., *Calathus melanocephalus*), run a larger risk of local extinction than species in which the local populations are connected by intensive dispersal (e.g., *Poecilus versicolor*). So, dispersal protects the population from local extinction (Den Boer 1968, 1970, 1981). This idea was developed further in the theory of metapopulation dynamics (Hanski and Gilpin 1991) and it also became common in nature conservation programs aimed at connecting habitats.

In addition to active movement, as studied in the movement ecology paradigm, passive or facilitated movement may actually be more important for most soil-living animals. An informative review of non-volant dispersal in terrestrial arthropods is given by Reynolds et al. (2014). In soil invertebrates we may recognize three types of facilitated movement:

- *Anthropogenic transport with soil, plants or seeds.* This is the most important dispersal mechanism for earthworms. Their natural dispersal capacity is so small that without (usually unintended) anthropogenic transport the distribution ranges of earthworms would still reflect their speciation in the Jurassic (see Section 2.8). Other soil invertebrates are also mainly transported by humans, e.g., flatworms (Jones et al. 2001b) and Collembola (Roberts and Weeks 2011). This type of dispersal leads to erratic and fragmented distributions without any distance relationships; the network rather reflects trading routes and economic connections (Helmus et al. 2014).
- *Wind dispersal.* This is important for small invertebrates that are light enough to be blown away, such as nematodes, tardigrades and microarthropods. Wind dispersal is actively exploited by ballooning dwarf spiders of the families Erigonidae and Linyphiidae. These spiders, on certain occasions, depending on the weather, crawl to a high point, direct their abdomen to the air and secrete very fine silken threads that, when long enough, lift the animal into the air. Collembola and mites are also often transported by wind and some species actively move into the trees, apparently to increase the likelihood of being blown by the wind (Bowden et al. 1976). The *dauer* larva of *Caenorhabditis elegans* (an arrested and highly resistant third larval stage) shows a behaviour called "*nictation*", in which it climbs up to the tips of grass blades and waves its body so as to increase the likelihood of being blown away (Riddle 1988). Several soil insects that were often assumed to be non-volant can sometimes be found in massive numbers at 200 m height in the air (Chapman et al. 2005). Collembola can even be sampled at 10,000 m height using samplers attached to airplanes (Freeman 1952). Although the ecological relevance of dispersal at these high altitudes is unclear, wind dispersal at ground level is probably very significant for many arthropods (Hawes et al. 2007).
- *Phoresis.* This is a specialized way of transport practiced mainly by mites and pseudoscorpions, in which the phoretic animal attaches itself to a flying or fast-walking larger animal. In several cases phoresis is an obligatory part of the life-cycle. The deutonymph of some prostigmatid and uropodine mites is entirely specialized in phoresis and has a deviating morphology. Such a stage is called *heteromorph* or *hypopus*. The phoretic stage may have adapted legs, a specialized perianal organ with suckers, or a stalk ("*pedicel*") excreted from perianal glands, to attach itself to the host (Figure 3.19). In most cases the mite has a preferred site on the host to which it clings (e.g., lower leg, upper leg, head, underside of elytra, etc.). Phoresis is considered a neutral phenomenon, although sometimes the host is hampered in its flight. Some phoretic mites develop into parasites or predators after phoresis has brought them to the host nest (e.g., *Varroa* mites in bee hives). Harvestmen are a popular victim for phoretic and parasitic soil mites, as the morphology of their legs makes it very difficult to get rid of hitchhikers.

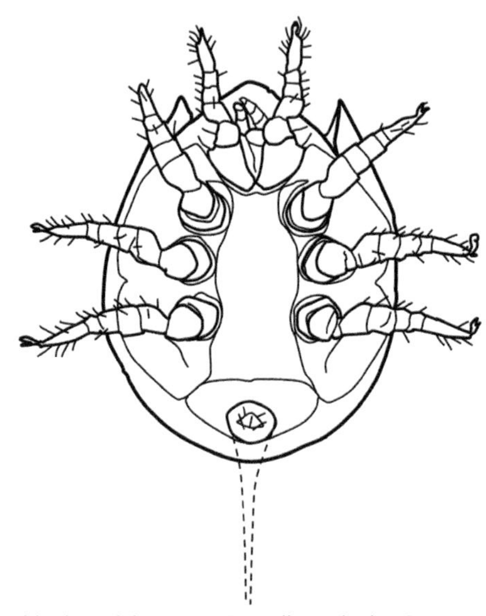

Figure 3.19. Mesostigmatid mites of the group Uropodina colonize European badger burrows and may be collected from nest debris heaped near the entrance of the burrow (Kurek et al. 2020). The deutonymph of *Trematura patavina* is phoretic and attaches to its insect host using a pedicel (indicated by dashed lines) excreted from perianal glands. Which host the mite uses to colonize badger nests is however not known. The image shows a *Trematura* deutonymph in ventral view, redrawn from a photograph on Wikimedia Commons.

One may expect that dispersal will be more common among species living at the soil surface in comparison to euedaphic species. Soil-living species must survive adverse conditions by phenotypic plasticity or local adaptation, while surface-living species have the option to move away. This trend seems to hold at least for Collembola (Ponge 2020). Also, in earthworms, the peregrine species are épigés or anéciques. The dichotomy "*move or change*" adds to the many life-history strategies that show a gradient across the soil profile.

In conclusion, dispersal can result from random or directed walking, in case it will be a local phenomenon, or it can involve morphological, physiological and behavioural adaptations that allow long-range, facilitated, transport by wind or by hitchhiking on other animals. Dispersal is more common among epigeic species than among true soil species. In all cases dispersal has great consequences for the genetic composition of local populations. It counteracts the effects of genetic drift and inbreeding in small populations as it introduces new genetic variants in populations living distant from each other.

3.10 Genetic population structure and isolation by distance

The genetic composition of a population is measured by means of polymorphic genetic markers. In the 1970s population genetics was mostly based on allozyme polymorphisms, but in the 1990s various DNA markers became fashionable, such as RAPDs (*random amplified polymorphic DNA*), *microsatellites* and *single nucleotide polymorphisms* (SNPs). In addition, DNA sequences of marker genes like *cytochrome c oxidase subunit 1* (COI) or *subunit II* (COII), *16S ribosomal RNA* (16S rRNA) and ITS1 (*internal transcribed spacer 1*) were used to analyse population structure. The latter markers also find application in phylogenetics and taxonomy (*cf.* Chapter 2). A special marker is the *DNA barcode*, a designated region of COII or ITS1, that is agreed among taxonomic specialists as correctly identifying the species (Hogg and Hebert 2004, Huang et al. 2007, Jeratthitikul et al. 2017). Table 3.8 lists the most common techniques and describes the sources of polymorphism (Van Straalen and Roelofs 2012).

Among the various markers, microsatellites stand out as the most useful for population analysis, since they are *codominant* (both alleles can be recognized in a diploid individual) and allow estimation of all classical population genetic variables such as *allele frequencies*, *heterozygosity* and *linkage* (Goldstein and Schlötterer 1999). In the development of primers for microsatellite loci, it is recommended to check for null alleles (alleles that don't give a PCR product) and confirm Mendelian inheritance using controlled crosses. The latter issue is often ignored, as not all soil invertebrates can be easily crossed in the laboratory.

Table 3.8. Some of the most common genetic markers used to investigate population structure of soil invertebrates (Van Straalen and Roelofs 2012).

Type of marker	Detection method	Source of polymorphism
Allozymes	Separation of allelic protein variants on electrophoresis gel and detection by enzymatic activity	Substitutions of amino acids changing the electrophoretic mobility of a protein
Random amplified polymorphic DNA (RAPD)	Polymerase chain reaction using short primers with arbitrary sequence; band profile visualized on electrophoresis gel	Variation in annealing sites for PCR primers
Amplified fragment length polymorphism (AFLP)	Selective PCR amplification of DNA restriction fragments, visualization of profile on electrophoresis gel	Variation in restriction sites and in flanking regions of the amplicon
Microsatellites	Polymerase chain reaction using primers annealing to flanking regions of a tandem repeat, separation of DNA alleles on gel	Variability of the number of repeated core sequences, indels in the core sequence
Single nucleotide polymorphisms (SNPs)	SNP detection oligonucleotide microarray, genome-wide DNA sequencing, bioinformatic analysis	Point mutations
Cytochrome c oxidase subunit I or II haplotypes	DNA sequencing of a selected PCR-amplified fragment of the mitochondrial COI or COII gene	Sequence variation, giving rise to different haplotypes
Ribosomal RNA genes	DNA sequencing of a selected PCR-amplified fragment of an rRNA gene (16S or 18S)	Sequence variation in variable regions of the gene
Internal transcribed spacer 1 or 2	DNA sequencing after PCR amplification of one of the non-coding regions of the rRNA gene cluster	DNA sequence variation of the spacer

Population genetic surveys have been done in various soil invertebrates, most often in earthworms, Collembola, isopods and nematodes. We concentrate here on the work conducted with earthworms and Collembola, as these studies jointly illustrate how geographic scale interacts with genetic differentiation between populations. An overview of selected studies is given in Table 3.9.

An important concept used in these studies is "*isolation by distance*" (IBD), that is, the degree to which genetic differences between subpopulations of a species increase with the geographic distance between them. As a measure of genetic distance, the fixation index, F_{ST}, is most often used. This index, introduced in 1921 by the American geneticist Sewall Wright (1889–1988), is defined as:

$$F_{ST} = \frac{H_T - H_S}{H_T}$$

where H_T is the total heterozygosity of all subpopulations combined (frequency of heterozygotes observed when considering the set of subpopulations as one large population), and H_S is the average heterozygosity within a subpopulation (Hartl and Clark 1997). Obviously, if there is no genetic structure, the allele frequencies of any subpopulation are equal to those of the combined population, $H_S = H_T$ and $F_{ST} = 0$. Conversely, if the subpopulations are all fixed for different alleles $H_S = 0$ and $F_{ST} = 1$. Thus defined, F_{ST} refers to a single locus, but to estimate genetic distance reliably it is taken over a number of unlinked loci across the genome, e.g., for microsatellites preferably 10–20. There are several other statistics to measure genetic distance, one of the more popular being the *chord distance*, D_C, proposed in 1967 by the Italian population geneticist Luigi Cavalli-Sforza (1922-2018).

There is a general relationship between F_{ST} and the rate of dispersal (usually called migration in population genetics), however, this relationship depends on the model assumed for the dispersal mechanism. Under the so-called *island model*, it is assumed that the population is split into many subpopulations dispersed geographically like islands in an archipelago. In that case one may derive that, by approximation:

$$m = \frac{1}{4N}\left(\frac{1}{F_{ST}} - 1\right)$$

where m is the number of migrants between subpopulations per generation and N is the size of a subpopulation. So, in principle, the rate of migration m may be estimated from the degree of differentiation of allele frequencies between subpopulations (F_{ST}).

Dispersal rates estimated from genetics tend to be higher than the rates derived from observational studies in the field (summarized in Table 3.1). For example, in a study on fine-scale genetic differentiation in the earthworm *Allolobophora chlorotica*, Dupont et al. (2015) derived a dispersal distance of 3.41 to 3.78 m within one generation, which would imply a dispersal area of 36.5 to 44.9 m^2, while in experimental field studies the one-generation dispersal area varies from of 3.1 to 18 m^2 (Table 3.1). The suggestion emerging, which is open to further testing, is that the measurement of dispersal rates in the field, in many cases is an underestimate of the "realized" dispersal as estimated from genetic differentiation.

Table 3.9. Selection of population genetic studies of earthworms and Collembola, illustrating the geographic scale of genetic differentiation.

	Markers	**Genetic structure**	**Region**	**References**
Earthworms				
Aporrectodea rosea	4 allozyme loci	No evidence for IBD over 200 km, but one population stands out	Southern Finland	Terhivuo and Saura (1993)
Lumbricus terrestris	RAPDs	Genetic variation within populations as large as across populations; no evidence for IBD up to 310 km	Germany	Kautenburger (2006)
Dendrobaena octaedra	609 bp of COI	Large divergences between populations but no IBD, long-range dispersal mediated by human transport	Alberta, Canada	Cameron et al. (2008)
Hormogaster elisae	4 microsatellites	Low genetic differentiation in plot of 64 m^2	El Molar, Spain	Novo et al. (2010)
Lumbricus rubellus	COII and AFLP	High local differentiation associated with heterogenous lead pollution	Cwmystwyth Valley, Wales	Andre et al. (2010)
Aporrectodea icterica	COI and 7 microsatellites	No evidence of isolation by distance over 11 km, suggesting bottlenecks	Normandy, France	Torres-Leguizamon *et al.* (2014)
Allolobophora chlorotica	8 microsatellites	Weak isolation by distance relationship over 90 m, isolation by resistance over 11 km	Normandy, France	Dupont et al. (2015, 2017)
Lumbricus castaneus	8 microsatellites	Weak differentiation over 2 km, but no IBD, evidence for founder effects	Paris region, France	Dupont et al. (2019)
Springtails				
Bilobella auriantiaca	15 allozyme loci	No differentiation over 184 km, but Naples population different from others due to founder effect	Italy	Dallai et al. (1983)
Tetrodontophora bielanesis	14 allozyme loci	No differentiation of Central European populations over 444 km; strong divergence with transalpine populations	Poland to northern Italy	Fanciulli et al. (1991)
Tomocerus vulgaris	10 allozyme loci	Sardinian, coastal and Apennine population clusters distinct from each other with low gene flow between clusters	Northeastern Italy and Sardinia	Fanciulli et al. (2000)

Table 3.9 contd. ...

...Table 3.9 contd.

	Markers	Genetic structure	Region	References
Springtails				
Desoria klovstadi	678 bp COII	Populations isolated from each other by harsh environmental conditions and glaciers	Victoria Land, Antarctica	Frati et al. (2001)
Gressittacantha terranova	5 allozyme loci	22 populations clustering in three groups isolated from each other by glaciers	Victoria Land, Antarctica	Fanciulli et al. (2001)
Orchesella cincta	563 bp of COII, AFLPs and 6 microsatellites	No differentiation at local scale, IBD at distances larger than 60 km; strong subdivision between north-western, Central European and Italian populations	Europe	Van der Wurff et al. (2003, 2005), Timmermans et al. (2005)
Folsomia candida	Five ISSR loci	Weak IBD over 200 m distance between wells of aquifer	Michigan, USA	Sullivan et al. (2009)
Lepidocyrtus curvicollis	Complete COII and partial EF1α sequence	Deeply divergent genetic lineages with no apparent gene flow even among mainland populations at 10 km distance	Thyrrhenian islands and facing Italian and French coast	Cicconardi et al. (2010)
Sminthurus viridis	38 allozyme loci and 8 microsatellites	Populations differentiated but no clear relationship with distance over 2500 km, evidence for human-mediated long-distance dispersal	Southern Australia	Roberts and Weeks (2011)

The work by the French population ecologist Lise Dupont represents a particularly nice illustration of how population genetics can enlighten the spatial dynamics of soil invertebrates (Torres-Leguizamon et al. 2014, Dupont et al. 2015, 2017, 2019). Dupont and her group investigated populations of *Aporrectodea icterica* and *Allolobophora chlorotica* in grasslands of Normandy, France, at different spatial scales. Using a topographic map, they included a variety of landscape elements (grasslands, agricultural fields, roads) that were given weights in proportion to their "resistance" against earthworm dispersal. For neither species there was a significant isolation by distance effect, but for *A. chlorotica* there was a significant correlation with cost-weighted geographic distance (Figure 3.20). In this scenario roads were considered as barriers (high costs) and grasslands as corridors (low costs).

Despite these examples of small-scale differentiation, the work on earthworms as a whole (see Table 3.10) illustrates an apparent absence of genetic distance effects at the larger scale, even up to several hundreds of km. This may seem unexpected, given the low rate of horizontal locomotion in earthworms. But we need to remember that anthropogenic transport is the main mechanism of long-range dispersal in earthworms, causing many bottlenecks and founder effects. These completely

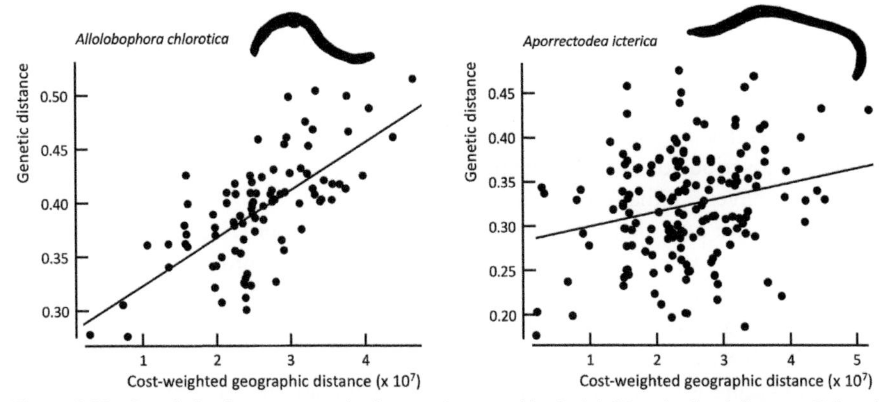

Figure 3.20. Correlation between genetic distance (measured by Luigi-Sforza's chord distance, D_C) and geographic distance (weighted by costs associated with the resistance of habitat units in the landscape), for subpopulations of two earthworm species in grasslands of Normandy, France. In the scenario giving the highest correlation for *A. chlorotica*, grasslands had low resistance costs while roads and agricultural fields had high costs. For *A. icterica* no significant correlation with cost-weighted geographic distance scenarios could be established. Redrawn, with modifications, from Dupont et al. (2017).

override the effects of "natural" dispersal, especially in the man-made landscapes of Europe and the ecosystems receiving many human introductions in North America (Cameron et al. 2008).

The situation in springtails is slightly different. There are several studies showing a correlation between genetic distance and geographic distance. This evidence was reviewed by Costa et al. (2013) and is reproduced partly in Figure 3.21. Wind dispersal seems to be relatively significant in at least some Collembola. However, the variability is large and sometimes distorted by outliers due to anthropogenic transport, sharp divides by, e.g., glaciers (Fanciulli et al. 2001), mountain ranges (Fanciulli et al. 1991), and historical colonization processes from glacial refugia

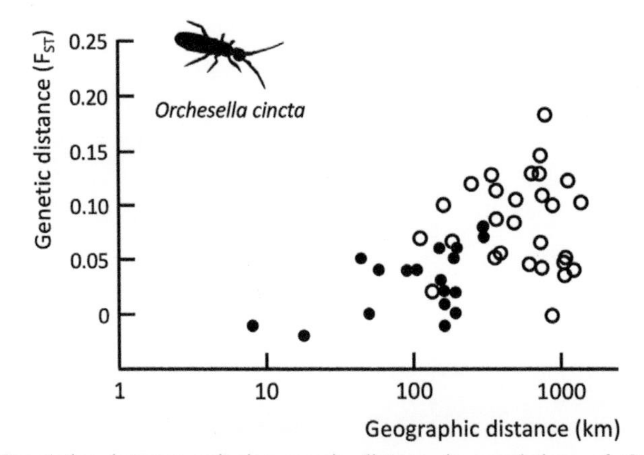

Figure 3.21. Correlation between pairwise genetic distance in populations of *Orchesella cincta* (Collembola) at forest sites across Europe (Netherlands to Poland, Denmark to northern Italy). The data suggest that isolation by distance begins at around 60 km. Graph redrawn from Costa et al. (2013), based upon data from Timmermans et al. (2005) (open circles) and Van der Wurff et al. (2005) (closed circles).

(Timmermans et al. 2005). Studies on populations of *Lepidocyrtus* in northern Italy show a very deep divergence between populations, over distances as small as 10 km (Cicconardi et al. 2010). It seems that isolation by distance may hold for Collembola populations in the relatively flat landscapes of North and Central Europe (where wind dispersal is the main mechanism), while vicariance due to mountain ranges and islands is a serious obstacle for dispersal in Mediterranean landscapes. Glaciers in Antarctica have a similar barrier function.

Finally, we note that the genetic population studies provide some support for a correlation between the position of a species in the soil profile and its apparent dispersal rate (the GKGB hypothesis introduced in Chapter 1 of this book). The higher dispersal rate of *A. allolobophora* compared to *A. icterica* as noted by Dupont et al. (2017), is explained by the smaller size and surface activity of the former species, making it more prone to passive dispersal than the deep-burrowing, larger *A. icterica*. The IBD effect in *Orchesella cincta* is likewise explained by its epigeic lifestyle and its habit to climb trees and being blown away (Van der Wurff et al. 2003). These conclusions are, however, based upon little evidence at the moment and stand out to be investigated in more detail.

Chapter 4

Reproduction and Development

The Underground Harmony Orchestra

Drop out: A life on grass grants you a reset.
Inhale wet earth, fresh manure adds more zest.
No? No glow of the mind - simply stressed?
Tune in an ear to the ground. Hear: a trumpet.

Nematode is doing her beastly best!
Larva plays the flute, Woodlice castanet.
Snail rams the timpani, Spider the spinet.
A slave choir chants whole notes from the ants nest.

Springtail elicits thrills from a cornet.
Mite raises the horn, Beetle the clarinet:
the high C is the ultimate test!

And on trombone Worm blows the heart out his chest.
Feeling addressed - makes the presence manifest
of this nonesuch underground nonet.

Jasper Aertsz

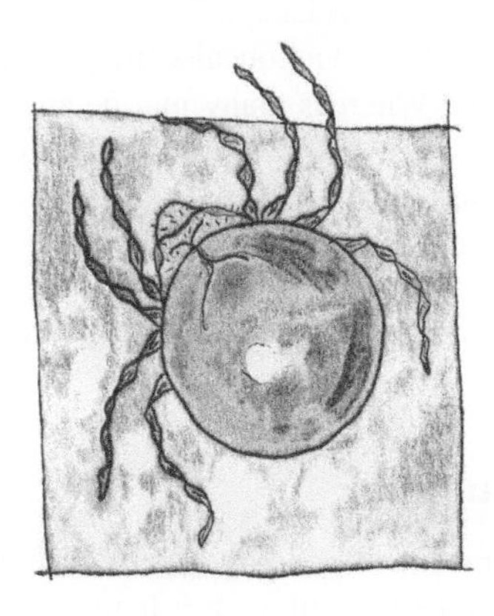

Chapter 4
Reproduction and Development

"O, I have girls", said Toad lightly: "twenty girls or thereabouts, always at work. But you know what girls are, ma'am! Nasty little hussies, that's what I call 'em!"

Kenneth Grahame, The Wind in the Willows

Soil invertebrates show a great variety of reproductive habits. The most familiar system is the one with separate sexes, males by definition producing many small, motile, gametes and females producing fewer, larger and largely immobile gametes. Such a system is called *gonochorism*, as distinct from *hermaphroditism* (two sexes expressed in one individual) and *parthenogenesis* (females producing eggs without intervention by males). In addition, *asexual reproduction* by means of fission and fragmentation is found in several invertebrates. Sometimes gonochorism and hermaphroditism are combined under the term *amphimixis*, to emphasize that sexual reproduction involves a fusion of two gametes, sperm and egg, which are either produced by the same individual or by separate males and females.

The marine lineages from which soil invertebrates evolved are predominantly gonochoristic. It seems most logical to assume that both parthenogenesis and hermaphroditism are derived traits that evolved after the terrestrialization of the group, or maybe in association with it. The question we are asking in this Chapter is: if soil invertebrates started out as gonoristic, what are the conditions under which hermaphroditic and parthenogenetic lineages evolved and are these transitions in line with the GKGB hypothesis set out in Chapter 1?

In addition to the mode of reproduction, also the early development of land-living animals changed profoundly. Whereas many marine ancestors of soil invertebrates have free-swimming larvae, almost all soil invertebrates (except insects) have *direct development*: the juvenile stage resembles the adult and there is no metamorphosis. What happened with the developmental programmes responsible for the production of a free-swimming larva from a fertilized egg? These and other questions of developmental adaptation to life in soil are addressed in the last three sections of this chapter.

4.1 Asexual reproduction

Asexual reproduction, by means of budding, fission or fragmentation of the body is relatively rare in soil invertebrates, although it is the sole mode of reproduction in some enchytraeids. This was discovered in 1959 by both Bell (1959) and Christensen (1959) and it is now known to occur in eight species: *Enchytraeus fragmentosus*, *E. bigeminus*, *E. variatus*, *E. japonensis*, *E. dudichi*, *Buchholzia appendiculata*,

Cognettia glandulosa, and *C. sphagnetorum* (Dósza-Farkas 1995, Christensen et al. 2002, Collado et al. 2012).

In laboratory cultures, fragmenting enchytraeids spontaneously fall apart into three to eleven pieces at regular intervals. For example, *E. japonensis* fragments about every 14 days, after it has reached a body size of 10 mm. Each fragment, even when comprising only two intermediate segments, regenerates a complete worm, including a head region. The fragmentation process requires muscle contractions to separate the pieces and so is best done on a solid substrate. In liquid media the ability to fragment is impaired.

Inomata et al. (2000) showed that fragmentation in *E. japonensis* is under control of the head region, most likely the cerebral and subesophageal ganglia. Decapitation stimulates fragmentation, even when the critical body size has not yet been reached. This suggests that the brain produces an inhibitory signal, which is gradually lifted when the worm matures and reaches a critical body size after which fragmentation is no longer suppressed.

In *Cognettia sphagnetorum* mature females occur rarely, and these individuals seem to lay eggs by parthenogenesis. However, the few eggs produced never hatch and decay within a few days (Christensen 1959), making *C. sphagnetorum* completely dependent on asexual reproduction. Other species, e.g., *Enchytraeus bigeminus*, *Cognettia glandulosa* and *Buchholzia appendiculata*, have not completely lost the ability to reproduce sexually, although they mainly reproduce by fragmentation. In *B. appendiculata* the sexual lineages are hermaphroditic, the most common mode of reproduction in enchytraeids, while in *C. glandulosa* the sexual animals produce eggs by parthenogenesis.

Fragmentation of the body not only occurs spontaneously, but is also induced by physical injury and environmental stress. Sjögren et al. (1995) described how the rate of fragmentation in *C. sphagnetorum* increased in soils with high concentrations of copper and zinc. Rundgren and Augustsson (1998) suggested that the fragmentation response could be used as an endpoint in laboratory toxicity tests with *C. sphagnetorum*. The fact that fragmentation is sensitive to environmental conditions is consistent with its being a regulated process, under control of the brain, as described above for *E. japonensis*.

The existence of asexual propagation in enchytraeids causes a large degree of clonal variation and cryptic diversity within species complexes. This diversity is amplified by the fact that several lineages are triploid and reproduce solely by parthenogenesis. In addition, hybridizations between species, causing *allopolyploidy*, add additional complexity. Martinsson and Erséus (2014) showed that *Cognettia sphagnetorum* from Northern Europe consists of at least four different molecular operational taxonomic units that could be raised to species level on the basis of sequence divergence of marker genes. Even worse, two lineages of another fragmenting species, *C. glandulosa*, cluster inside the four *C. sphagnetorum* lineages, and the same holds for two other *Cognettia*s (Figure 4.1).

The analysis of Martinsson and Erséus (2014) leaves only the genus *Cognettia* monophyletic, while several presently delimited morphotypes are polyphyletic. Obviously, the species complex requires thorough taxonomic revision. It is recommended that all material designated as *C. sphagnetorum* be characterized by

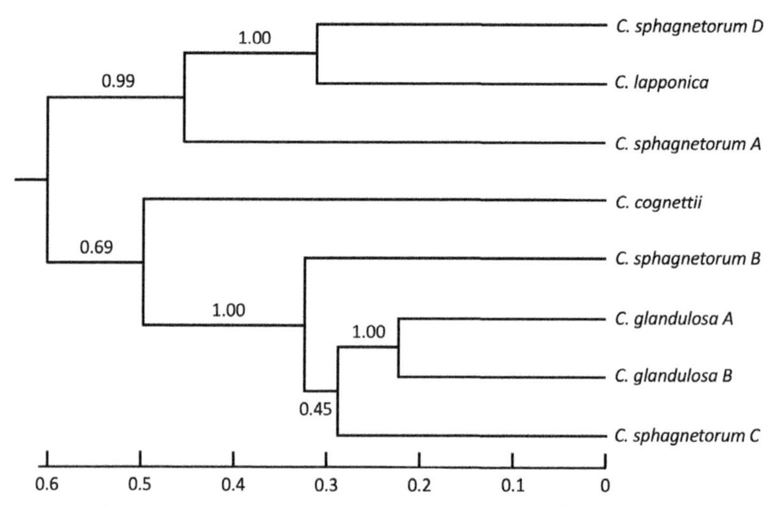

Figure 4.1. Species tree constructed from 32 Northern European isolates of *Cognettia*, established using a Bayesian multispecies coalescence method, applied to sequence variation in four marker genes. The horizontal axis shows the expected number of changes per site in cytochrome oxidase subunit 1. Reproduced from Martinsson and Erséus (2014), with permission from Elsevier Science Publishers.

molecular methods. Until taxonomy has provided a solution, it seems best to opt for a "*sensu lato*" approach in delineating the fragmenting enchytraeid morphospecies.

Asexual reproduction is also known to occur in terrestrial flatworms (Platyhelminthes: Tricladida: Terricola). Winsor (1998) aptly describes how flatworms tend to fragment easily when one tries to pick off animals that tenaciously stick to the substrate. Flatworms even show a phenomenon known as "*worm melting,*" the complete and rapid degradation of the animal when it is disturbed, surely a collector's nightmare!

Most likely, the fragmentation response of Terricola rests on the same mechanism as the one in their better-investigated sister group, the aquatic planarians. For a long time, *Planaria* was a model species for regeneration research; however, this position is now taken by *Schmidtea mediterranea*, due to its more stable diploid genome and the availability of genetic tools (Newmark and Sánchez Alvarado 2002).

Planarians are famous for their ability to regenerate their bodies from even small parts of tissue. This ability depends crucially on the presence of *neoblasts*: somatic stem cells distributed throughout the body, whose primary function seems to be the regeneration and repair of worn tissues. However, neoblasts can also act as a source of germ cells (Figure 4.2). The number of fragments into which an animal can be divided such that each fragment still generates a complete individual depends on the number and the tissue distribution of neoblasts (Bely and Sikes 2010, Baguña 2012). The role of neoblasts in regeneration of Terricola is unknown at the moment, but is likely to follow similar principles. Maybe the "worm melting" described by Winsor (1998) is just a falling apart to unicellularity, while new worms will regenerate from surviving neoblasts.

Enchytraeids and Terricola are about the only soil invertebrates in which body fission and fragmentation are important reproductive processes. Other

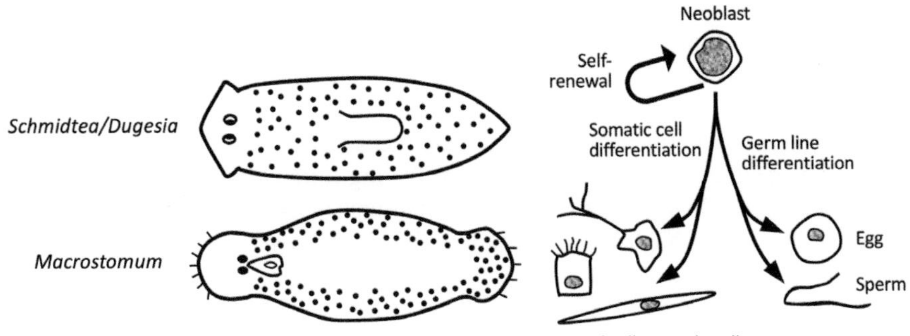

Figure 4.2. Flatworm neoblasts, which are primarily somatic stem cells, can generate various differentiated tissue cells as well as germ-line cells (right). They are characterized by a very large nucleus, which almost fills the whole cell. The distribution over the body is species-specific (left). Redrawn, with modifications, from Bely and Sikes (2010).

soil invertebrates only show a capacity to regenerate parts of their bodies. For example, several earthworms can to some extent regenerate lost segments and many arthropods can regenerate limbs and antennae, after being attacked by a predator. Often the regenerating parts can be recognized due to deviating segmentation, and the frequency of such damage can be used to obtain an estimate of predation pressure in the field (Ernsting and Fokkema 1983).

In summary, asexual reproduction is an important aspect of the life-cycle for a limited number of soil invertebrates. It is not to be considered an escape from sexual reproduction, e.g., to overcome cytoplasmic incompatibilities due to polyploidy, rather, it is reproductive mechanism of its own. The ability to propagate by fission or fragmentation seems to be an ancient capacity of metazoans, which has been retained in flatworms and to some extent in enchytraeids, but was lost in higher animals. However, even in enchytraeids it is not the most common mode of reproduction. Asexual reproduction is much more frequent in Naididae and Aelosomatidae, aquatic clitellate families holding an ancestral position to enchytraeids (see Figure 2.24). The soil environment may actually have selected against asexual reproduction.

4.2 Parthenogenesis

If there is one thing true about the reproduction of soil invertebrates, it is that so many of them are parthenogenetic. We will see in this Chapter that, in contrast to asexual reproduction, parthenogenesis has evolved to become widespread among soil invertebrates: tardigrades, nematodes, enchytraeids, earthworms, woodlice, collembolans, and mites. All these groups contain parthenogenetic species; in some cases, complete evolutionary lineages consist of only parthenogens.

Often parthenogenesis is qualified as "asexual", however, it is actually quite different from asexual reproduction. The production of eggs has nothing to do with tissue fragmentation or body fission. The only similarity is that the offspring are often (not always) a clone of the mother. From a biological point of view, parthenogenesis could be considered a form of sexual reproduction with one sex. Some authors have explicitly argued against equating parthenogenesis with asexuality (Boyden 1950), but this has remained a minority of point of view (cf. Schön et al. 2009). Table 4.1 provides an overview of the different terms used in reproductive biology.

The term parthenogenesis covers quite a heterogeneous collection of phenomena on the cellular level, with different genetic consequences (De Meeûs et al. 2007). Since parthenogenesis is derived from sexual reproduction with two haploid gametes, the lack of fertilization would lead to haploid offspring if no compensation mechanism would be in place. In some groups haploid offspring is indeed the result. Best known is the production of haploid males from unfertilized eggs in Hymenoptera, which have a haploid-diploid sex determination system. Among soil invertebrates, ants are the main representatives of this mode of reproduction.

Table 4.1. Terminology used in describing the various modes of reproduction in animals. From various sources.

Term	Description
Asexual reproduction	Fragmentation or fission of the body, followed by regeneration of a new individual from a part
Gonochorism	Form of sexuality with two sexes in two different individuals, one with small and motile gametes (male) and one with large and relatively immobile gametes (female)
Hermaphroditism	Form of sexuality with two sexual functions, as in gonochorism, but with the two sexes combined in a single individual, either simultaneously or sequentially
Aphally	Loss of male copulatory organs in a hermaphrodite
Sequential hermaphroditism	Form of sexuality in which an individual can switch from one sex to the other in the course of its development or in response to an environmental or social stimulus
Amphimixis	Form of sexual reproduction in which a zygote is formed by fusion of two different gametes (as in gonochorism and hermaphroditism)
Parthenogenesis	Form of reproduction in which females lay viable (female) diploid eggs without being fertilized by males
Androgenesis	Form of reproduction in which the male parent is the sole source of the offspring's nuclear genome
Automixis	Form of parthenogenesis in which haploid gametes fuse directly after being formed by meiosis, to form a diploid zygote
Apomixis	Form of parthenogenesis in which meiosis is suppressed and female eggs are produced directly from diploid germ cells (mitotic parthenogenesis)
Thelytoky	Parthenogenesis in the strict sense, in which females lay diploid eggs produced either by automixis or apomixis (as different from arrhenotoky)
Arrhenotoky	Form of sexual reproduction in which males develop from unfertilized haploid eggs and females from fertilized diploid eggs
Facultative parthenogenesis	Form of parthenogenesis that can be induced by environmental cues, e.g., food shortage
Spanandry	The occurrence of rare males in populations of parthenogenetic or facultative parthenogenetic females
Cyclical parthenogenesis	The occurrence of parthenogenetic generations during part of the life-cycle, in relation to seasonal changes in the environment

The production of haploid males from unfertilized eggs is often considered a special form of parthenogenesis, called *arrhenotoky*. Parthenogenesis in the stricter sense, i.e., the production of diploid female offspring by females is called *thelytoky*, to distinguish it from arrhenotoky. The male counterpart of parthenogenesis, *androgenesis*, is seen in some species of ants but not in other soil invertebrates. In this weird form of sexual reproduction the female nuclear genome is eliminated before or after fertlization. The offspring's nuclear genome is wholly paternal, while the mitochondrial genome remains maternal (Schwander and Oldroyd 2016).

In thelytokous parthenogenesis, the issue arises how the somatic chromosome number is restored or maintained in the embryo. In addition, since fertilization of the oocyte is usually a necessary trigger for development of the egg, an alternative trigger for development must be in place. To assure diploidy in the embryo, there are four basic solutions:

- Pre-meiotic doubling of all chromosomes without cell division. This assures that meiosis starts with tetraploid cells, which automatically leads to diploid eggs. Parthenogenetic earthworms use this mechanism.
- Fusion of two haploid oocytes directly after the second meiotic division, e.g., the second polar body fuses with the female pronucleus. This is a common mechanism in enchytraeids, nematodes and arthropods.
- Fusion of cells after the first zygotic cleavage, e.g., fusing the first two blastomeres. This is known to occur in sea urchins and fish, but whether it happens in soil invertebrates is unknown.
- Suppressing the reductive division (meiosis II) altogether (*ameiotic* or *mitotic parthenogenesis*). This type is assumed to have evolved from meiotic parthenogenesis. It is the most common mode of parthenogenesis in nematodes and arthropods.

The variety of mechanisms in different animal groups confirms the fact that parthenogenesis has emerged multiple times independently in different evolutionary lineages, a prime example of *convergent evolution*. In addition, parthenogenesis can evolve along various pathways, both from gonochorism and from hermaphroditism (Simon et al. 2003); in rare cases reverse evolution (from parthenogenesis back to gonochorism) has been documented as well.

If the diploid chromosome number is restored while retaining a complete meiosis this is called *automixis*. Depending on the precise cytological course of events, automixis allows a certain degree of recombination and in its purest form may be considered equivalent to self-fertilization. It may easily lead to complete homozygosity; a population of automicts can be considered a collection of (genetically different) highly inbred lines. On the other hand, in the case of mitotic parthenogenesis, also called *apomixis*, the offspring are genetic clones of the mother. Heterozygosity, if present in the mother, is maintained in the offspring, which all have the same heterozygosity. Which cytological form of parthenogenesis prevails in soil invertebrates, is only known for a few species, as we will see below.

In many species that reproduce by parthenogenesis, males are found in very low numbers, a situation called *spanandry*. For example, in the oribatid mite *Oppiella nova* living in forest soil, Wehner et al. (2014) found 2% males, despite the fact that *O. nova* is considered truly parthenogenetic. Such rarely occurring males are assumed to be non-functional, however, occasional return of male functionality in parthenogenetic populations cannot be excluded. The fraction of males in *O. nova* depends on environmental conditions, the number of males being higher under conditions of resource scarcity (Wehner et al. 2014). The presence of *spanandric males* in principle allows occasional production of sexual eggs, in addition to parthenogenetic eggs, a situation described as '*facultative parthenogenesis*' by Goto (1960) and Kurup and Prabhoo (1977) in Collembola. A system of *cyclical parthenogenesis* as seen in cladocerans and aphids is not known to occur among soil invertebrates.

When parthenogenesis coexists with male-female sexuality in the same species, the parthenogenetic strains tend to have specific geographical ranges, different from the main range. This was first described by the French zoologist Albert Vandel in the isopod *Trichoniscus provisorius*. Vandel (1928) observed that in the region of Toulouse, *T. provisorius* populations consisted of two 'races', one parthenogenetic and triploid, with spanandric males, the other diploid, with a normal sex ratio. In northern and eastern France, however, only parthenogenetic populations were found. Vandel noted that a similar phenomenon was present in other invertebrates, e.g., the millipede *Polyxenus lagurus*. He called it '*parthénogenèse géographique*'.

The parthenogenetic triploid form of *Trichoniscus* is now considered a separate species, *Trichoniscus pusillus*. At the same time the original interpretation by Vandel (1928) has been questioned. Fussey (1984) found no evidence of geographic parthenogenesis for *Trichoniscus* in the British Isles. He noted an association of *T. provisorius* with calcareous soils and rocks, mostly likely in relation to the greater energy intensity of such habitats. In the Netherlands, however, a geographic separation of the two forms seems to occur: *T. pusillus* is very common, occurs all over the country, except in saltmarshes along the coast, while *T. provisorius* has a more restricted distribution and is found only in a few places in the West of the country (Berg et al. 2008).

Geographic parthenogenesis also holds for the common parthenogenetic springtail *Folsomia candida*, a well-known laboratory model. While all *F. candida* cultured for experiments in laboratories worldwide are parthenogenetic, occasional male/female populations do occur in the wild. Frati et al. (2004) discovered an isolated sexual population in northern Italy (Figure 4.3).

The phenomenon of geographic parthenogenesis proves to be quite common and has attracted a lot of attention, since evolutionary biologists have hoped it would shed more light on the "*paradox of sex*": why sexuality is so common despite the many costs (Lehtonen et al. 2012). According to several authors, parthenogenetic populations are biased towards high latitudes, high altitudes, islands and dry environments, compared to their closest sexual relatives. This is often explained by greater colonization capacity of parthenogens after the last glaciation or by other factors, e.g., genetic drift in small marginal populations (Haag and Ebert 2004). In addition, marginal and empty habitats would offer a suitable niche for parthenogenetic forms because biotic interactions in such habitats are weak. This argument follows the *Red Queen hypothesis* (Van Valen 1973): sexuality provides a selective advantage in

Figure 4.3. Male (below) and female (above) *Folsomia candida* from a population at Cremona, Italy. The male is somewhat smaller, has a more slender abdomen, longer antennae and – quite remarkable – a more "inquisitive" behaviour. Populations with males are extremely rare in *F. candida*. Photograph by Viola Tanganelli, University of Bologna.

habitats in which pressing biotic interactions, such as parasites, require an on-going adaptive capacity, achieved by recombination. Such habitats are more abundant in the centre than in the borders of a distribution range. The absence of competitive interactions in the range margins would allow a switch to parthenogenesis.

While this "*marginal habitat theory*" for the advantage of parthenogenesis may be well accepted in ecology, it does certainly not hold for soil invertebrates. Actually, the data rather indicate the reverse: sexual populations of otherwise parthenogenetic species are the ones that occupy marginal habitats, deal with stressful environments and face fluctuating conditions. For example, in a study of Maraun et al. (2013), parthenogenetic oribatid mites were more common at lower, rather than at higher altitudes. Petersen (1978) noted a sexual population of the otherwise parthenogenetic collembolan *Mesaphorura* (*Tullbergia*) *macrochaeta* in a wind-exposed sand-dune habitat in Jutland, Denmark, clearly a marginal habitat. Likewise, Christensen et al. (1992) noted that the sexual form of *Fredericia galba* (Enchytraeidae) occupied the higher altitude and marginal habitats among thirteen sampling stations in Denmark, while the more common parthenogenetic forms were dominant at sites of intermediate elevation with high population densities and a rich soil fauna.

The data suggest that the *Red Queen hypothesis* and the *marginal habitat theory* may not be the right explanations for the predominance of parthenogenesis in soil invertebrates. Scheu and Drossel (2007) developed another argument, using a model inspired by work on oribatid mites. According to their model, a switch to sexuality in an otherwise parthenogenetic population, is advantageous when resources are limiting and vary in an unpredictable manner. This fits in the "*Tangled Bank hypothesis*", formulated by the British evolutionary biologist Graham Bell (Bell 1982), and named after Darwin's famous image evoked in the *Origin of Species* (Darwin 1859). Bell's hypothesis argues that parthenogenesis is promoted in an environment with constant, structurally available resources while sexual reproduction is associated with a "tangled bank", i.e., an environment characterized by complexity and unpredictability. The soil may not have an abundance of resources, but they are structurally available and constant. The tangled bank hypothesis, rejected by many ecologists including Bell himself, may have been dismissed too early (Song et al. 2011).

With so many parthenogenetic representatives, in different evolutionary lineages, soil invertebrates present a unique model to test the various theories aiming to explain the advantages and disadvantages of parthenogenesis versus sexual reproduction. It is therefore instructive to review the occurrence of parthenogenesis in the various lineages of soil invertebrates in some detail.

4.3 Parthenogenetic worms

Starting with nematodes, we note that the majority of species is gonochoristic, however, transitions to parthenogenetic reproduction have occurred numerous times. A review by Denver et al. (2011) showed that parthenogenesis is especially common in the families Cephalobidae and Panagrolaimidae, and rare in the Rhabditidae (to which the well-known genetic model species *Caenorhabditis elegans* belongs). It is also common in plant-parasitic root-knot nematodes of the genus *Meloidogyne*. A phylogeny of this genus shows that gonochorism is the likely ancestral state. In each of three clades several transitions to meiotic and further to mitotic parthenogenesis must have happened, in total at least five times in this genus (Figure 4.4).

A factor that may potentially influence the evolution of parthenogenesis in nematodes is the extent to which sperm and other mating factors are required for

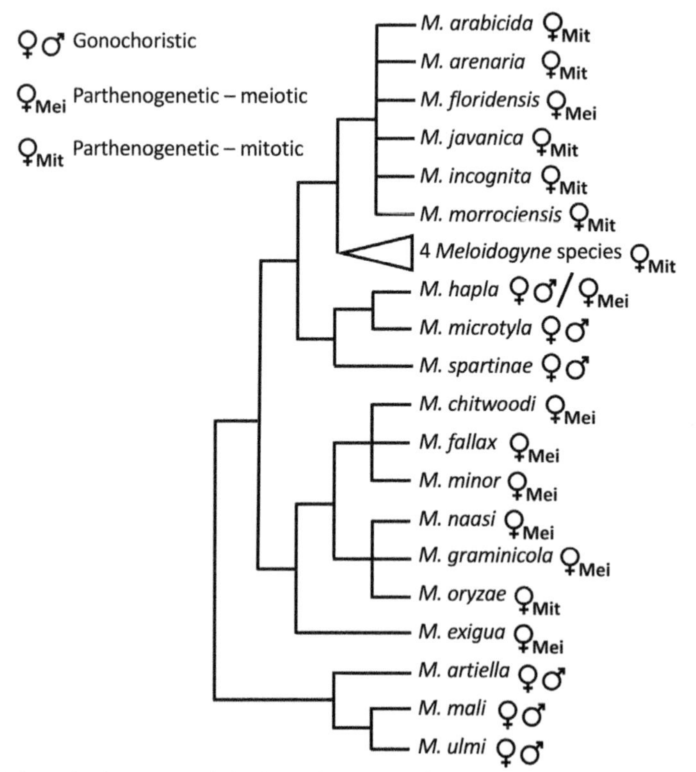

Figure 4.4. Reproductive mode variation in root-knot nematodes, the genus *Meloidogyne*. The distribution of gonochorism and two forms of parthenogenesis, indicated by icons, suggests at least five evolutionary transitions. Redrawn and modified from Denver et al. (2011).

initiating egg development. In rhabditid nematodes, including *C. elegans*, this requirement is known to be effective, but activation by sperm is not required in Cephalobidae and Panagrolaimidae. This may explain why parthenogenesis has evolved more easily in these latter families. Denver et al. (2011) also point out that in parasitic nematodes very small effective population sizes are part of the life-cycle, e.g., when a single individual penetrates a host; under these conditions, parthenogenesis might have a selective advantage.

In annelid worms, parthenogenesis seems to be linked to *polyploidy*. Many earthworms have polyploid populations, almost in equal frequency as the normal diploid types (Viktorov 1997). From data provided in Christensen (1961), the frequency of parthenogens in species with a polyploid karyotype is 34% for Enchytraeidae and 67% for Lumbricidae, while the respective frequencies under amphimictic species are only 2% and 3%. Enchytraeids differ from lumbricids, however, in that many polyploids can still reproduce sexually. This is explained by the occurrence of *achiasmatic meiosis*, i.e., the chromosomes do not engage in crossing-over sites (chiasmata).

An interesting case of parthenogenesis is present in *Lumbricillus lineatus*, a widespread enchytraeid of the upper littoral zone, living in decaying seaweed and other organic material. This species consists of two lineages with different cytological types, one is diploid and hermaphroditic, the other is triploid and parthenogenetic. The triploid form does not produce mature spermatozoa itself. It copulates with the hermaphroditic diploid to obtain viable sperm. However, these spermatozoa are only needed to initiate egg development and do not genetically contribute to the offspring (Christensen and O'Connor 1958). This "*parthenogenesis with pseudofertilization*" can be seen as an intermediate step in a sympatric speciation process leading to a separate lineage of parthenogens, but still constrained by developmental requirements. A similar situation occurs in some species of earthworms and flatworms.

In earthworms, parthenogenesis is quite common in the family Lumbricidae (several species in the genera *Aporrectodea*, *Dendrobaena* and *Octolasion*) and it also occurs in Megascolecidae (*Amynthas*, *Megascolides*), but not in Glossoscolecidae or other earthworm families (Díaz Cosin et al. 2011). Among North-American lumbricids more than thirty species are reported to be parthenogenetic, which would constitute nearly 20% of the earthworms on this continent.

Almost all parthenogenetic earthworms are polyploid (usually triploid), and they apply pre-meiotic chromosome doubling in the primary oocyte to maintain the somatic chromosome number in the offspring after meiosis. However, an exception occurs in *Dendrobaena octaedra*, in which no premeiotic doubling takes place, but meiosis 2 is suppressed, the system applied by most parthenogenetic arthropods, as discussed above.

Because all parthenogenetic earthworms evolved from hermaphrodites, demonstration of parthenogenesis is more complicated compared to arthropods, where gonochorism is the ancestral condition. Absence of males is not a criterion and culture in isolation cannot exclude self-fertilization. Indications for parthenogenesis can be the absence of sperm in the spermatheca or degenerated male sexual organs. However, in some parthenogenetic earthworm species such as *Octolasion tyrtaeum*, male structures are not reduced and copulation is even necessary to trigger parthenogenetic reproduction. Such "*pseudogamic*" copulations do not lead to sperm transfer; their sole function is to trigger egg production using mechanical stimulation or chemical clues associated with the copulation. Parthenogenetic egg production is very difficult to demonstrate in such cases. The most conclusive evidence for pseudogamic parthenogenesis comes from genotyping parents and offspring as applied by Shen et al. (2012) in two *Amynthas* species from Taiwan.

The best-investigated, but still puzzling parthenogenetic earthworm system is the *Aporrectodea caliginosa* complex (Fernández et al. 2012). This complex contains the widespread and very abundant parthenogenetic species, *A. trapezoides*, in fact one of the most abundant earthworms in the world. Using four different marker genes sequenced in worms from more than thirty locations, mostly in Europe, Fernández et al. (2012) revealed an astoundingly complex pattern of evolutionary branching (Figure 4.5). Two different clades are present in the complex, a Mediterranean and a Eurosiberian, and a transition to parthenogenesis occurred in each of them, first in the Mediterranean clade, and much later also in the Eurosiberian clade. Both parthenogenetic lineages are commonly identified as *A. trapezoides*, but the Eurosiberian *A. trapezoides* is actually more related to *A. longa*, while the Mediterranean worms are clustered with *Nicodrilus carochensis* and *N. monticola*. By applying a molecular clock analysis, Fernández et al. (2012) were able to correlate the two origins of parthenogenesis with paleoclimatic events, a dry period in the Pliocene, and glaciations in the Pleistocene (Figure 4.5).

Because the occurrence of parthenogenesis is scattered across the Lumbricidae and evolved many times independently, a comparison of the ecological preferences of parthenogenetic species versus hermaphroditic species may shed light on the conditions that favour either strategy. In such an analysis of North-American lumbricids, Jaenike and Selander (1979) noted that parthenogenetic species tend to occupy ephemeral and fluctuating habitats such as the litter layer and rotting

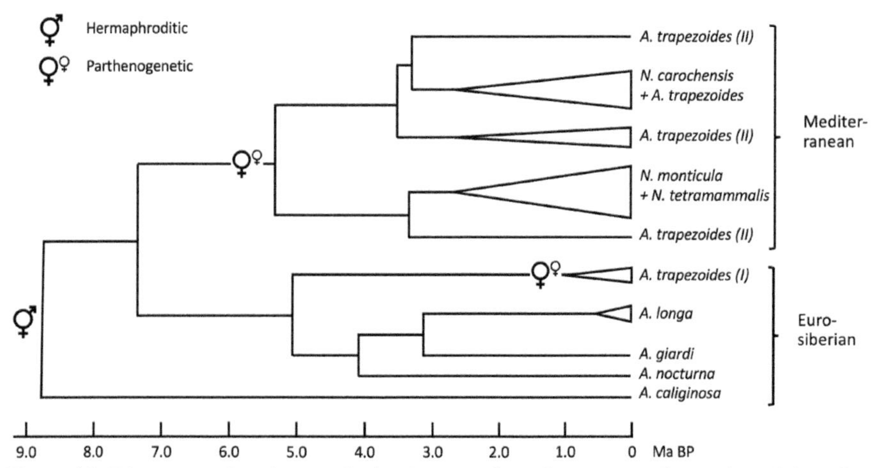

Figure 4.5. Divergence of earthworms in the *Aporrectodea caliginosa* complex, estimated from four marker genes, showing the position of parthenogenetic lineages in the two main clades (Mediterranean and Eurosiberian) of the genus. The origin of parthenogenesis from a hermaphroditic ancestor was estimated to have occurred 3.85–6.41 Ma BP in the *A. trapezoides* II clade and 1.05–3.48 Ma BP in the *A. trapezoides* I clade. N. = *Nicodrilus*. Adapted from Fernández et al. (2012).

logs, while hermaphroditic species tend to be more common among the worms that inhabit the stable environment of the deeper soil. This analysis emphasizes the greater colonization capacity of surface-active parthenogens. The association of parthenogenesis with surface-active species is contrary to the situation in Collembola, as we will see below.

Summarizing the situation in nematodes, potworms and earthworms, we conclude that each group shows little correlation between parthenogenesis and certain ecological conditions, apart from a possible association with plant-parasitism in nematodes. Bottlenecks associated with colonization seem to be the key issue for all three lineages. A scenario is suggested in which parthenogenesis is a rare, naturally occurring phenomenon in most populations, but because of their greater colonization ability, parthenogenetic mutants gain an advantage in the colonization of new habitats arising under changing environmental conditions. The consequence is that the present distribution of parthenogenetic worms carries a significant historical signature.

4.4 Parthenogenesis in Collembola

The rate of parthenogenesis in springtails is even higher than in earthworms, but like in earthworms and nematodes, there is a strong tendency for parthenogens to cluster in specific families, while other families are nearly devoid of them. For example, only one parthenogenetic species is known for the family Entomobryidae, while in Tullbergidae, Arrhopalitida and Neelidae more than one half of the species are reported parthenogenetic. In addition, families with parthenogenetic representatives are scattered across the whole phylogenetic tree of Collembola (Chernova et al. 2010).

There is however, one obvious pattern in Collembola: parthenogenesis is tightly linked to the average vertical position of a species in the soil profile. All species living at the soil surface and in the vegetation have two sexes, while almost all species living in the deeper layers of the soil profile are parthenogenetic. This divergence reflects the so-called *Lebensformen* ("life-forms") of Gisin (1943), depicted in Figure 1.2. The occurrence of parthenogenesis in Collembola is one of the strongest illustrations of the GKGB hypothesis.

Several authors have shown that the depth distribution of parthenogenetic Collembola differs from the one in sexual species. Chernova et al. (2010) found that parthenogenetic species are common at depths between 5 and 30 cm, while the greatest abundance of sexual species is found above 20 cm depth. The depth distribution depends on the type of soil, the moisture gradient and the use of soil cultivation, however, the general pattern still holds. Figure 4.6 illustrates the point. It holds in all families of Collembola that have both parthenogenetic and sexual species (*cf.* Figure 4.6), so it is not an issue of phylogenetic constraint, but a true example of convergent evolution.

Some of the cytological changes that underly the origin of parthenogenesis have been elucidated in a few species of Collembola (Riparbelli et al. 2006). In normal cell division, the organization of a spindle requires the activity of a *centrosome* in either pole of the cell. Centrosomes contain essential structures called *centrioles*; two of them are present in one centrosome. In an unfertilized egg of a sexual species

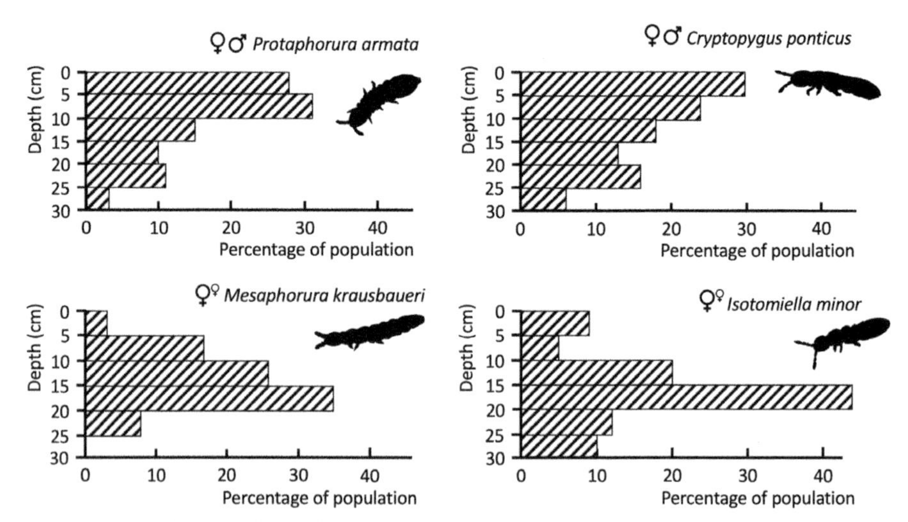

Figure 4.6. Distribution of four springtail species over the profile of a chernozem soil in European Russia. *Protaphorura armata* and *Cryptopygus ponticus* are bisexual species, while *Mesaphorura krausbaueri* and *Isotomiella minor* are parthenogenetic. Redrawn from Chernova et al. (2010), with modifications.

the maternal centrosome is inactivated or absent, and the male gamete provides the centrioles during fertilization. This is the reason why in so many species fertilization is needed not only to restore diploidy but also to initiate development of the egg.

In parthenogenetic eggs the male centriole is not provided, so the question becomes, how do such eggs start their development? Riparbelli et al. (2006), in a study of the parthenogenetic springtail *Folsomia candida*, showed that the parthenogenetic egg is able to self-assemble astral arrays of microtubules, without the need of maternal or paternal centrosomes. In eggs of the sexual species *Folsomia fimetaria*, a close sister species of *F. candida*, unfertilized eggs will never develop and self-organizing cytoplasmic asters have not been observed. Only parthenogenetic oocytes seem to have evolved the capacity to self-organize a spindle. Given the widespread occurrence of parthenogenesis among arthropods, the basic materials for assembly of microtubule asters are apparently available in the cytoplasm of many eggs, but are either not used, or the assembly process is switched off during normal development.

Another condition that may explain the predisposition of Collembola to develop parthenogenesis may be related to the peculiarities of sex determination, at least in the group Symphypleona, as revealed by Dallai et al. (2000, 2001) in six species of the families Sminthuridae, Bourletiellidae, Dicyrtomidae and Katiannidae. The females of these Collembola have a chromosome number of 2n = 12, which includes 2n = 8 autosomes and 2n = 4 sex chromosomes. The males have a diploid chromosome number of 2n = 8 autosomes, plus n = 2 sex chromosomes, so the male sex is defined by the absence of two sex chromosomes. The peculiar situation is that these two sex chromosomes are eliminated after fertilization, in the first cleavage of the zygote.

In addition, spermatogenesis of these Collembola is aberrant because the 2n = 10 males are able to produce sperm cells with n = 6 chromosomes. This is arranged during the first meiotic division, when the chromosomes separate into a set of 6 and a set of 4, while the latter do not proceed to the second meiotic division and

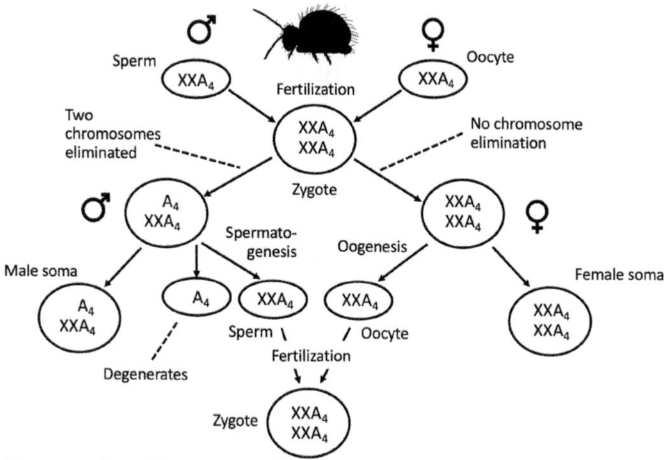

Figure 4.7. Reconstruction of the sex determination system in two symphypleone springtails, *Sminthurus viridis* and *Allacma fusca*, showing post-zygotic elimination of two sex chromosomes in males and aberrant spermatogenesis. A_4 = four autosomes, XX = two sex chromosomes. Image inspired by the work of Dallai et al. (2000).

degenerate. So the male produces half the number of spermatids (Figure 4.7). There is no recombination in the male lineage.

This system in principle brings the sex ratio of the offspring under control of the female. This might explain why collembolan sex ratios are quite variable and often far from unity, even in male/female populations. In other insects with a similar mechanism of sex determination it is known that females can add a signal to the oocyte, which predetermines it to become male by removing one of the sex chromosomes (two in the case of Collembola). It is always the paternal chromosome that is removed in these cases.

Dallai et al. (2000) speculate that this mechanism may act as a predisposition for facultative parthenogenesis: if the female is able to withhold the male-determining signal, she can easily produce all-female offspring. In addition, reproductive manipulators such as *Wolbachia* could also easily control the sex ratio by suppressing the male signal (see Section 4.6). On the other hand, chromosome elimination as a sex determining mechanism is not known in the other two main Collembola groups, Neelipleona and Arthropleona, which also have many parthenogenetic species.

Summarizing the situation in Collembola, we see that parthenogenesis is really common and is clearly associated with life in soil. This contrasts with the situation in earthworms, where parthenogens are more common among surface-active species than among endogeic ones (Section 4.3). Some cytological conditions, such as the capacity to initiate egg development without the need for male centrioles, and the mechanism of post-zygotic sex determination, may predispose some groups of Collembola to develop parthenogenetic reproduction relatively easily.

4.5 Parthenogenesis in oribatid mites

The other soil invertebrate group that includes many parthenogenetic species consists of oribatid mites. Norton and Palmer (1991) estimated that more than 600 of the approximately 7000 known species are parthenogens. Like in Collembola, earthworms and nematodes, there is a tendency for parthenogenesis to be clustered in specific lineages. Among oribatid mites many parthenogens are found in the group Desmonomata, an early-derived lineage of Oribatida with seven families and some extremely abundant and widespread species like *Platynothrus pelitifer* and *Nothrus silvestris*. Palmer and Norton (1991) discerned three patterns among the six families of Desmonomata:

- Families with males absent or very rare: Camisiidae, Trhypochthoniidae, Malaconothridae and Nanhermannidae
- Families with a significant number of both sexes, with sex ratios varying from 23% to 77% males: Crotoniidae and Hermanniidae
- Families with a mixture of all-female species and evenly proportioned male-female populations, according to genus, as in Nothridae.

Molecular phylogenetic analysis has shown that several of the completely parthenogenetic lineages of Oribatida are monophyletic (Maraun et al. 2004, Heethoff et al. 2009). This implies that these groups have undergone diversification despite being parthenogenetic. Speciation of complete parthenogenetic lineages presents somewhat of an embarrassment to evolutionary biologists, because parthenogenetic reproduction is assumed to quickly lead to loss of genetic variation, which according to the theory of natural selection nullifies the capacity to adapt to new environments. Consequently, some ancient parthenogenetic lineages such as bdelloid rotifers and darwinulid ostracods have been called "*ancient asexual scandals*" (Maynard Smith 1986, Judson and Normark 1996): speciose lineages that have reproduced by parthenogenesis for a long stretch of evolutionary time. Theoreticians have argued that such lineages should not persist, let alone that they could evolve and give rise to new species.

In spite of traditional evolutionary theory, Oribatida demonstrate that parthenogenesis does not exclude diversification. A nice example is found in the work by Heethoff et al. (2007, 2009) who investigated populations of the cosmopolitan parthenogenetic species *Platynothrus peltife*r from sixteen sites across the world. Phylogenetic analysis of the cytochrome oxidase I gene (Figure 4.8) revealed that *P. peltifer* consists of seven geographically distinct lineages with ancient splits. The species as a whole could perhaps be more than a hundred million years old, and most likely this represents the time since parthenogenesis in this taxon exists. The divergence between the Eurasian and American clades dates back to the separation between Europe and North America, and the divergence between Southern European and Northern European populations coincides with the uplifting of the Alps (Figure 4.8).

The DNA sequence of *P. peltifer* COI also demonstrated a preponderance of nonsynonymous over synonymous substitutions, a commonly accepted signal of purifying selection (Van Straalen and Roelofs 2012). So, most likely, *P. peltifer*, over millions of years, has seen not only mutation and genetic drift but also natural selection, without apparent changes in morphology. A similar situation holds for oribatids of the genus *Tectocepheus* (Laumann et al. 2007). Heethoff et al. (2007)

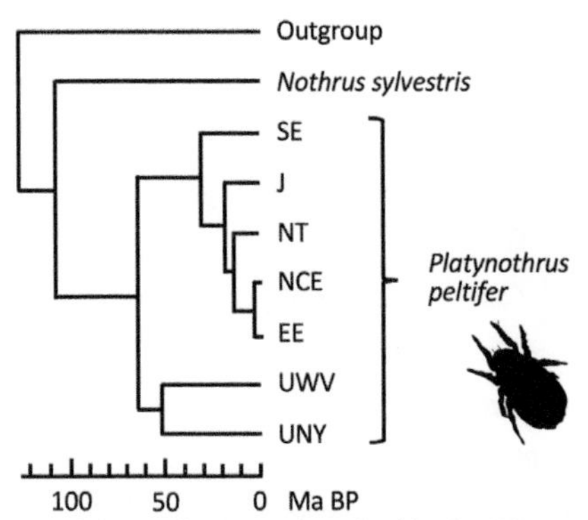

Figure 4.8. Phylogenetic tree of the parthenogenetic oribatid mite *Platynothrus peltifer* sampled from seven locations in Europe, Asia and North America. The tree was dated using a molecular clock derived from cytochrome c oxidase subunit I sequences. SE Southern Europe, J Japan, NT Northern Tyrolia (Austria), NCE Northern/Central Europe, EE Eastern Europe, UWV West Virginia, USA, UNY New York, USA. Adapted from Heethoff et al. (2007).

argued that the case of oribatid mites represents a third "ancient asexual scandal". Alternatively, one might argue that coexistence of parthenogenesis and evolution is not to be considered a scandal at all. Evolution is possible with mutation and selection only. Strictly speaking it does not require recombination and the slower rate of evolution is not to be considered a great drawback.

Another challenge that Oribatida pose to common evolutionary theory consists of evidence for reversal to gonochorism from parthenogenesis. In phylogenetic analysis, the so-called *Dollo's rule* says that complex characters cannot evolve back along the same pathway, or, as it is commonly expressed, that the evolution of complex characters is irreversible. This rule, named after the Belgian paleontologist Louis Dollo, imposes constraints on the cladistic analysis of character states: irreversible characters can evolve in only one direction (so-called *Dollo irreversibility*).

However, a phylogenetic analysis of Desmonomata demonstrated that two sexual species, *Crotonia brachyrostrum* and *C. caudata* evolved from a parthenogenetic ancestor (Domes et al. 2007b). This remarkable analysis, well underpinned with ancestral state reconstructions, proves that male/female sexuality could re-evolve after it had been lost early in the evolution of the Desmonomata (Figure 4.9). So, Dollo's law is violated by these oribatids. Alternatively, one might conclude that parthenogenesis should not be considered a complex character at all. Maybe the genetic changes needed to switch between male/female sexuality and parthenogenesis are relatively few or simple.

Like in Collembola, parthenogenesis in Oribatida is not observed in species that inhabit exposed environments like tree bark and vegetation. Although there are many sexual soil-living oribatids, sharing the environment with parthenogens, the species living in off-ground habitats are all sexual. For example, Karasawa and Hiiji (2008) found 28 species of Brachypylina oribatids in a subtropical forest in Japan, of which thirteen found in canopy leaves and tree bark were all sexual. Among soil-living mites no males were found for seven out of fifteen species. A similar pattern was observed by Fischer et al. (2010) in a study of oribatid mite communities of grasslands and forests in the Central Alps.

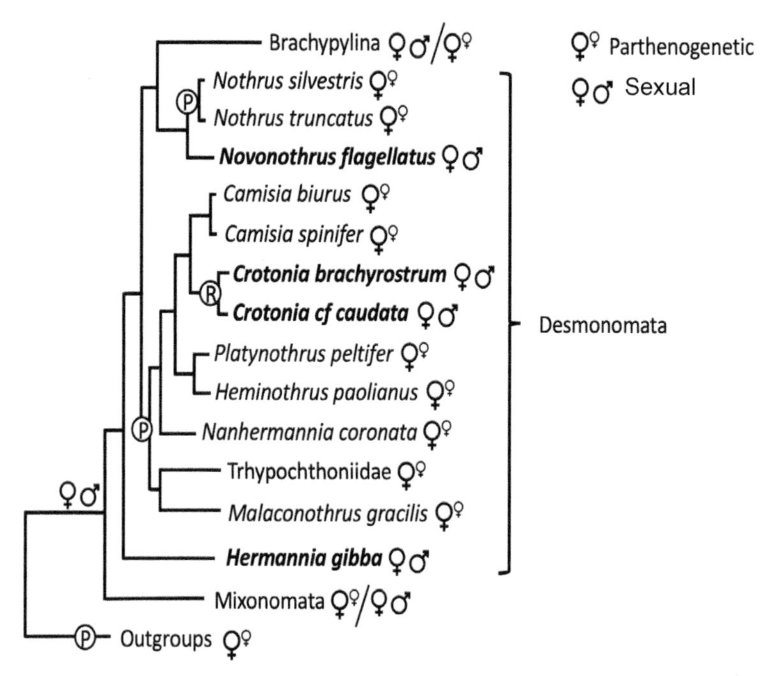

Figure 4.9. Simplified phylogenetic tree of Desmonomata, an early branching clade of Oribatida, with many parthenogenetic species. The tree combines partial DNA sequences from 18S ribosomal RNA, heat shock protein 82, and elongation factor 1 alpha for thirty taxa, of which 14 lineages are shown here (Trhypochthoniidae, Brachypylina, Mixonomata and outgroups condensed). Species with males and females are printed in bold. Ancestral state reconstruction shows that parthenogenesis evolved after the origin of Desmonomata (indicated by P), while male/female sexuality re-evolved from parthenogenesis in the clade of *Crotonia* (indicated by R). Modified from Domes et al. (2007).

Interestingly, the *Crotonia* species which regained sexuality, frequently colonize the bark of trees, although they are accidentally found also in soil samples. This perfectly fits in the idea that parthenogenesis in oribatids is a soil-bound strategy, and that there is selection against parthenogenesis in above-ground habitats. This again is very well in line with the GKGB hypothesis (*cf.* Chapter 1).

In summary, parthenogenesis in Oribatida is widespread, with a clear phylogenetic signature, like in Collembola. Importantly, the distribution of parthenogenesis in Oribatida challenges a number of widely held tenets in evolutionary biology. First, parthenogenesis does not exclude genetic divergence; second, complete lineages can evolve and give rise to new species while parthenogenetic; and third, sexuality can re-evolve in parthenogenetic lineages. The emphasis in evolutionary biology on "sex" as a necessary condition for evolution seems to be misplaced. All in all, oribatids are a fascinating group of soil invertebrates to unravel the conditions under which parthenogenesis might present an evolutionary advantage over bisexuality.

4.6 Reproductive manipulators

A treatment of the reproductive biology of soil invertebrates cannot be complete without considering the role of intracellular microorganisms acting as reproductive manipulators. *Wolbachia*, an alphaproteobacterium from the order Rickettsiales is the best-known agent. Other bacteria in the same order, e.g., *Rickettsia* and *Anaplasma,* are parasites of mammals; their life-cycle often includes an intermediate host which explains their occurrence in arthropods. *Wolbachia* itself does not regularly infect mammals, but influences—manipulates—the biology of many invertebrates, especially arthropods and nematodes (Werren et al. 2008).

Recently, the diversity of microbial intracellular symbionts in invertebrates has been expanded to include bacteria outside the Rickettsiales. Bacteria of the genus *Cardinium* are found in several insects (Zchori-Fein and Perlman 2004), and reach a particularly high incidence in spiders; Duron et al. (2008) showed that no less than 37% of spider species, in all major families, may be infected by *Cardinium*. Whether *Cardinium* manipulates the sex ratio or other aspects of reproduction in spiders is not yet clear (Vanthournout et al. 2011). Interestingly, *Cardinium* belongs to Bacteriodetes, a phylum quite remote from Alphaproteobacteria to which *Wolbachia* belongs. Other intracellular microorganisms in arthropods are *Blattobacterium* (Bacteriodetes), a symbiont of cockroaches, termites, and some beetles, *Arsenophonus* (Gammaproteobacteria, Enterobacteriales), a bacterium causing male-killing in parasitoid wasps, and *Spiroplasma* (Tenericutes, Entomoplasmatales), a bacterium that may restore fertility in fruit flies infected with female-sterilizing nematodes.
In total four main manipulative effects are known for these microbes:

- *Cytoplasmic incompatibility* between infected and non-infected hosts, e.g., the inability of *Wolbachia*-free oocytes to undergo the first mitotic cleavage after being fertilized by sperm from *Wolbachia*-infected males; this effect is common among higher insects and is also found in isopods.
- *Feminization* of male hosts; this is seen in amphipod and isopod crustaceans, and in some insects; male hosts are converted by *Wolbachia* to functional females, causing sex ratio distortion and intersex phenotypes.
- *Male killing*: infected males die in the egg stage or during early larval development, a phenomenon observed in beetles, flies, butterflies and pseudoscorpions. The sex-ratio distorting effect of *Wolbachia* in spiders may also rely on killing male embryos.
- *Parthenogenesis*: *Wolbachia* induces changes in the cell cycle of the oocyte allowing the production of eggs without fertilization but with the somatic chromosome number; this is well known in species with arrhenotoky (ants and wasps) and it is also assumed to underlie the occurrence of parthenogenesis in Collembola.

In an evolutionary sense all four effects can be understood from the fact that intracellular bacteria reside in the cytoplasm, so will normally not be vertically transmitted by males. Consequently, the fitness of such parasites is promoted by enlarging the female participation in a population.

Wolbachia was first discovered in the midge *Culex pipiens*, hence was initially designated *Wolbachia pipientis (wPip)*, however, the many different *Wolbachia* lineages that have been discovered since are now indicated by letters (supergroups A to H, Figure 4.10). With every new species of arthropod investigated new strains of

Wolbachia are added. There is no agreement yet about the question what constitutes a species of *Wolbachia*. Genome-wide phylogenetic analysis has provided conclusive evidence for the existence of at least seven supergroups, each of them a monophyletic lineage separated from the others with high statistical support (Figure 4.10). Only supergroup H, found in termites, is not yet positioned beyond doubt (Gerth et al. 2014). Other authors have proposed more supergroups (e.g., Ros et al. 2009).

Although *Wolbachia* infection is best known among higher insects (some 40% of terrestrial arthropods may be infected) it is also very common in soil invertebrates, including nematodes, termites, crickets, pseudoscorpions, spiders, isopods and springtails. *Wolbachia* is not known to occur in free-living nematodes of the soil, but filarial (parasitic) nematodes of the group Onchocercidae and plant-parasites of the family Pratylenchidae harbour *Wolbachias* of supergroups C and D (Figure 4.10). The mutualism in these parasites is obligate, that is, *Wolbachia* produces essential factors for metabolic energy generation by the nematode. Such symbiotic mutualisms have not been described for soil arthropods.

The various strains differ markedly in their host-specificity. Supergroups A and B are found in several insects, sometimes even the same strain is found in different hosts, but supergroup E is found only in Collembola and supergroup H only in termites. Some strains, e.g., *wAu*, naturally found in *Drosophila simulans*, can be transfected to other dipterans not normally harbouring *Wolbachia*, e.g., to *Aedes aegypti* (vector of several pathogenic viruses), for the purpose of mosquito control.

When the *Wolbachia* phylogenetic tree (Figure 4.10) is rooted using sequences of the veterinary pathogens *Ehrlichia* and *Anaplasma* (Rickettsiales), the *Folsomia* strain (*wFol*) turns out to be ancestral to all other strains (Gerth et al. 2014). So, it seems that *Wolbachia*'s life as a symbiont of insects started by infecting early hexapods, the ancestors of Collembola. In addition, the rooted tree shows that

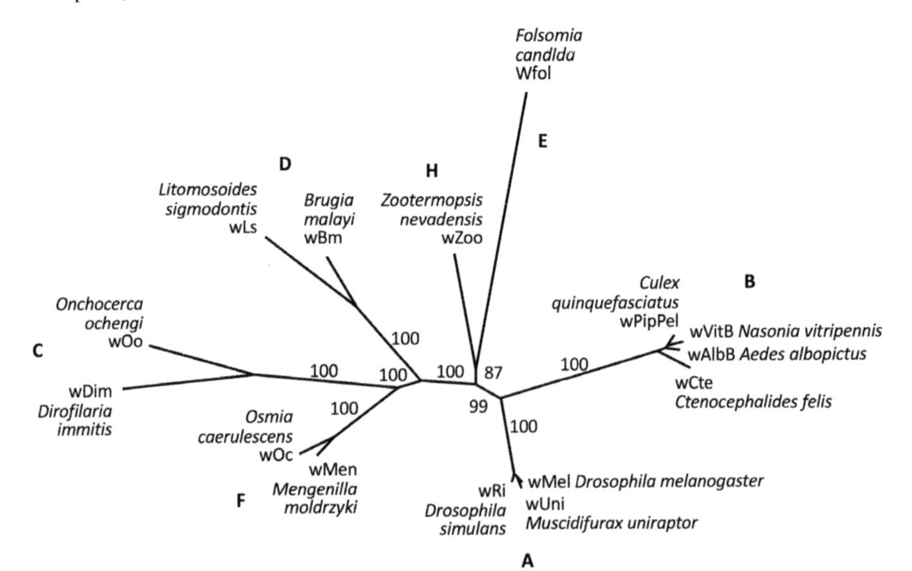

Figure 4.10. Unrooted phylogeny of *Wolbachias* in arthropods and nematodes, based on 90 orthologous loci across the genome. The *Wolbachia* strains, here indicated with the name of their main host, cluster in seven supergroups (A to H). Adapted from Gerth et al. (2014), with permission from Springer Nature.

nematodes received their *Wolbachia* from an arthropod lineage, at least twice in evolutionary history (Gerth et al. 2014).

Interestingly, *Wolbachia* has never been found in soil-living oribatid mites. Extensive attempts to detect them have been negative (Perrot-Minnot and Norton 1997). However, Cordaux et al. (2001) reported the presence of a *Wolbachia* living in an unidentified oribatid mite living phoretically on the woodlouse *Armadillidium vulgare*. This observation indicates that oribatids may not be completely free of *Wolbachia*, although *Wolbachia* has certainly nothing to do with the widespread occurrence of parthenogenesis in soil-living mites.

Why *Wolbachia* and *Cardinium* infect some lineages of arthropods much more than others is not known. Close proximity to a potential source is an important factor for horizontal transfer, as demonstrated by the presence of *Wolbachia* in predators, parasites and phoretics of infected arthropods (Cordaux et al. 2001). In addition, *Wolbachia* cells may persist in the soil for a short time and may be taken up by all kind of soil invertebrates. In this way, *Wolbachia* may live as a commensal in many species but persists as a reproductive parasite or true symbiont only in some. It is likely that the soil played a crucial role in the early evolution of *Wolbachia*.

Horizontal transfer of *Wolbachia* between species is shown by the presence of diverging *Wolbachia* lineages in related lineages of hosts, in other words, by the incongruence of *Wolbachia* phylogeny with host phylogeny. An example is given in Figure 4.11. Michel-Salzat et al. (2001) analysed nine different populations of the isopod species *Porcellionides pruinosus* from France, Greece, Tunesia and Israel, and compared the phylogeny of the host with the phylogeny of the endosymbionts.

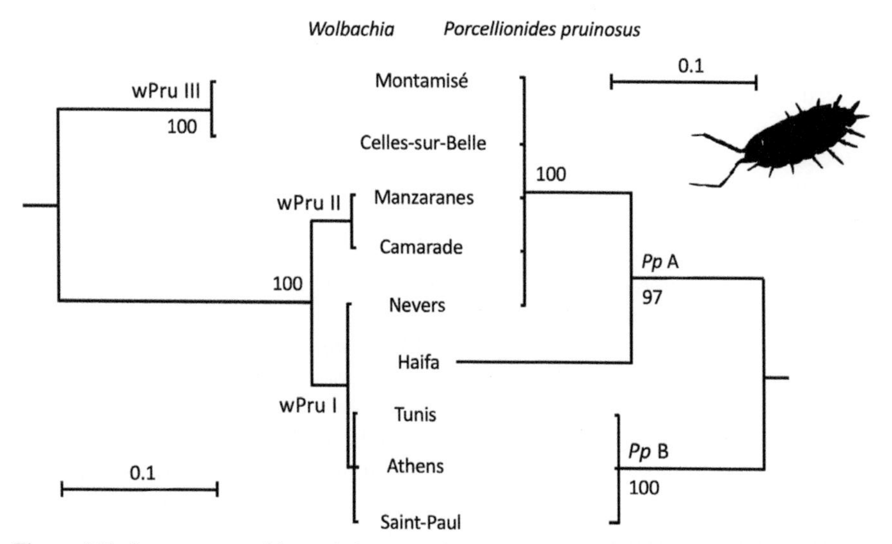

Figure 4.11. Incongruence of host phylogeny and endosymbiont phylogeny demonstrated by a wsp *Wolbachia* tree (left) and a 16S rRNA tree of its host, the woodlouse *Porcellionides pruinosus* (right). The names in the middle are sampling sites in the Mediterranean region. wPru *Wolbachia* in *P. pruinosus* (three lineages indicated I, II and III are shown), Pp *P. pruinosus* (two lineages indicated A and B). Reproduced from Michel-Salzat et al. (2001) with permission from Springer Nature.

The data revealed two different lineages of *P. pruinosus*, and three lineages of *Wolbachia*, however, there was no conjunction between the two, and so no co-evolution. Obviously, *P. pruinosus* has been infected more than once, on different occasions, by different *Wolbachia* lineages.

The sex-distorting effects of *Wolbachia* may have extremely complex effects on the genetic structure of a host population. This is apparent from a number of detailed studies on the pill bug *Armadillidium vulgare*, in which a *Wolbachia* induces feminization of genetic males (Verne et al. 2012, Cordaux et al. 2001, 2004, Rigaud and Juchault 1993). In isopods, females have karyotype *WZ* and males are *ZZ*. However, when infected by *Wolbachia*, genetic males are converted into functional females. In addition, another feminizing factor (*f*) is present in the genome of *A. vulgare*, most likely a fragment from *Wolbachia* inserted in the pill bug genome by horizontal gene transfer. The *f* factor acts as a female sex-determining gene, in addition to the normal female sex gene *W*. To complicate things further, an autosomal masculinizing gene, called *M*, is able to counteract the feminizing effect; this factor acts against *f*, but it cannot override the feminizing effect of *Wolbachia*. Finally, a complex of resistance genes *R* is able to prevent vertical transmission of *Wolbachia*, but not *f*. The result of all this genetic warfare is that almost all females in an *A. vulgare* population are actually feminized males (*ZZ* with sex ratio distorters), while true genetic females are rare. Genetic males (*ZZ* free from sex ratio distorters or carrying *M*) remain present in low numbers; they actually have a fitness advantage when rare. So, in *A. vulgare*, the normal *ZZ*/*WZ* system is not functional anymore, the *W* gene is driven to extinction and the sex determination is completely taken over by parasitic feminizing factors and their counteracting genes, acting in males.

Another population genetic consequence of the sex ratio distorters in *A. vulgare* is a separation between nuclear and mitochondrial markers. Population structure derived from mitochondrial markers in the host is closely correlated with the genetic variation in *Wolbachia*, due to the joint maternal inheritance of *Wolbachia* and mitochondria. Nuclear markers such as microsatellites, however, show the normal population structure of decreasing genetic similarity with increasing distance. No such isolation by distance pattern can be seen in the mitochondrial markers (Verne et al. 2012). The unusual pattern of mitochondrial polymorphism found in this species and also in *Porcellionides pruinosus* (Marcadé et al. 1999) is ultimately due to infection by *Wolbachia*.

Among Collembola, the presence of *Wolbachia* was first discovered in the parthenogenetic species *Folsomia candida* by Vandekerckhove et al. (1999). This *Wolbachia* appeared to belong to a then new supergroup, called E; until now this supergroup seems to be specific for Collembola. Czarnetzki and Tebbe (2004) showed that the *Wolbachia* cells were only present in the ovaries and the brain. Their absence from the gut is understood from the fact that the gut of springtails undergoes a regeneration process tied to the moulting cycle. The significance of their presence in the brain is not known. In fruit flies, *Wolbachia* also resides in the brain, which is assumed to enable the manipulation of mating behaviour.

Czarnetzki and Tebbe (2004) also demonstrated the presence of *Wolbachia* in three other parthenogenetic species, *Mesaphorura macrochaeta*, *M. italica* and *Paratullbergia callipygos*, while no *Wolbachia* could be found in the nonparthenogenetic species *Protaphorura fimata* and *Isotoma viridis*. Ma et al. (2017a) added *Wolbachias* in *Megalothorax incertus*, *Thalassiphorura houtanensis*, *Mesaphorura yosii* and *Folsomides parvulus*, all parthenogenetic. It seems that, up to now every parthenogenetic springtail investigated is positive for *Wolbachia*. *Wolbachia* strains may even occur occasionally in sexual species such as *Orchesella cincta* (Timmermans et al. 2004).

The nearly one-to-one correlation between parthenogenesis and *Wolbachia* is a strong indication that *Wolbachia* is actually the cause of parthenogenesis in Collembola. However, the evidence is not conclusive since one might argue (*cf.* Riparbelli et al. 2006) that *Wolbachia* is simply profiting from already existing parthenogenesis, by hitchhiking on the purely maternal transmission in such species. To prove causality, one would have to show that removal of *Wolbachia* from a parthenogenetic species restores sexual reproduction with males. In some invertebrates these experiments have been successful, e.g., Vanthournout et al. (2011) were able to remove *Wolbachia* from the dwarf spider *Oedothorax gibbosus* using tetracycline. This restored the female-biased sex ratio of the offspring to a normal value of 1.

In Collembola, such experiments are complicated by the fact that the development of eggs appears to be *Wolbachia*-dependent. Timmermans and Ellers (2009) cured parthenogenetic *Folsomia candida* from *Wolbachia* using a diet of 1% rifampicin and observed that despite normal egg production all eggs laid by *Wolbachia*-free *F. candida* failed to develop. Hatching success recovered in consecutive clutches and this was correlated with recovery of bacterial abundance.

These observations suggest that in some way or another *Wolbachia* has become indispensable for egg development in parthenogenetic Collembola. For example, *Wolbachias* could produce some factor necessary for spontaneous spindle formation in the first embryonic cell division, replacing the action of the male centrioles. However, the cytology of oocytes in *Wolbachia*-cured females has not been studied yet and definitive conclusions must await further research.

The collembolan *Wolbachias* are not only ancestral to all other *Wolbachias*, they also have—by far—the largest genomes (Kampfraath et al. 2019, Figure 4.12). The genome of wFol is 1.8 Mbp which is in the range of an average free-living bacterium. This is remarkable for an endosymbiont, because they usually have strongly miniaturized genomes. One of the factors contributing to large genomes in animals is invasion by mobile genetic elements and this seems to be the case also in *Wolbachia*. There is a very strong correlation between genome size and bacteriophage DNA across the various strains (Figure 4.12). This has been linked to a lack of purifying selection due to many bottlenecks (Wu et al. 2004). In addition, large *Wolbachia* genomes also have a large share of DNA repair genes, another indication of phage invasion (Figure 4.12).

All this genome sequencing has not yet resolved the genes involved with reproductive distortion or parthenogenesis induction in the host. One avenue of research is to identify strictly orthologous genes across *Wolbachias* with similar reproductive effects. Such genes can indeed be found (Kampfraath et al. 2019), but it is unclear whether parthenogenesis induction by *Wolbachia* relies on a common mechanism in different strains. Another approach is to conduct functional studies

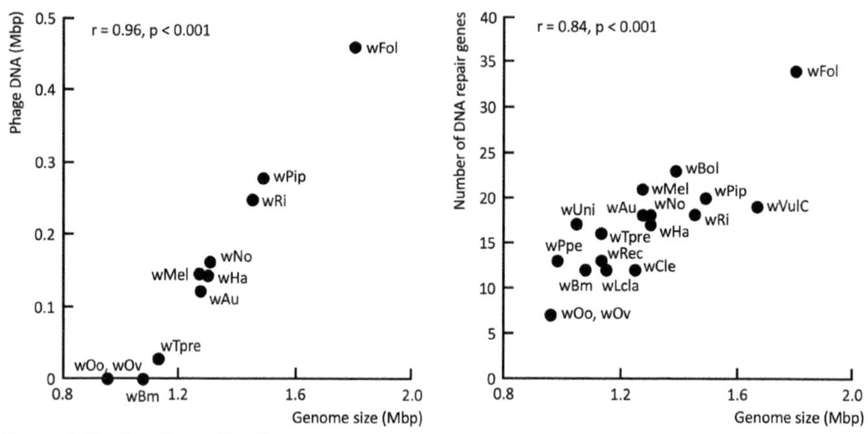

Figure 4.12. Correlates of *Wolbachia* genome size: larger genomes have more bacteriophage DNA (left) and more DNA repair genes (right). The various *Wolbachia* strains are indicated by code names; their original hosts are: Wau, wRi, wHa and wNo from *Drosophila simulans* (Diptera), wBm from *Brugia malayi* (Nematoda), wBol from *Hypolimnas bolina* (Lepidoptera), wCle from *Cimex lectularius* (Hemiptera), wFol from *Folsomia candida* (Collembola), wLcLa from *Leptopilina clavipes* (Hymenoptera), wMel from *Drosophila melanogaster* (Diptera), wOo from *Onchocerca ochengi* (Nematoda), wOv from *Onchocerca volvulus* (Nematoda), Wpip from *Culex pipiens* (Diptera), wPpe from *Pratylenchus penetrans* (Nematoda), wTpre from *Trichogramma pretiosum* (Hymenoptera), wVulC from *Armadillidium vulgare* (Isopoda), wUni from *Muscidivorax uniraptor* (Diptera). Modified after Kampfraath et al. (2019).

with *Wolbachia* outside the host. Nowadays several *Wolbachia* strains can be cultured in insect cell lines, which enables genetic engineering (Fallon 2021). We are still only beginning to understand the biological mechanisms by which *Wolbachia* interacts with its host.

4.7 Hermaphroditism

Like parthenogenesis, hermaphroditism is not uncommon in soil invertebrates but it is subject to significant phylogenetic constraints. Approximately 5–6% of all animal species are hermaphroditic, however, when arthropods are removed from this comparison, the fraction may increase to 30% (Jarne and Auld 2006, Grober and Rodgers 2007). The most striking phylogenetic aspect of hermaphroditism is its complete absence from insects. This also holds for soil invertebrates: none of the common soil arthropod groups (Collembola, beetles, ants, termites, woodlice, millipedes, centipedes, mites, spiders) is hermaphroditic. However, hermaphroditism is a common condition in all terrestrial clitellates (earthworms and enchytraeids) as well as terrestrial gastropods (land snails). Outside these main groups, hermaphroditism is observed in nematodes, terrestrial flatworms and sparsely in tardigrades.

Hermaphroditism shows a remarkable association with the great evolutionary transitions in habitat use. Both in annelids and molluscs the vast majority of marine species is gonochoristic, while hermaphroditism is the dominating mode of reproduction for species inhabiting freshwater and soil. Likewise, in tardigrades, none of the marine species is hermaphroditic but hermaphrodites are found in freshwater, moss, leaf litter, and soil habitats (Bertolani 2001). Why were freshwater and terrestrial habitats conducive to the evolution of hermaphroditism in annelids and molluscs?

The evolution of hermaphroditism has been the subject of many theories and models, but an all-encompassing answer is still difficult to give. Ghiselin (1969), in his classical overview of hermaphroditism throughout the animal kingdom, suggested three factors:

- *Low density*: when mates are relatively rare, a strategy in which an individual can mate with every conspecific it encounters, and can also self-fertilize, is obviously advantageous. According to Ghiselin (1969) this explains the relatively high incidence of hermaphroditism in sessile animals, deep-sea animals and parasites. However, whether this is applicable to earthworms and land snails may be questioned, since these animals can hardly be said to live under low densities. In fact, as we have seen in Chapter 2, their densities are actually higher than expected.
- *Size advantage*: sequential hermaphroditism may be advantageous when the male function is best performed at small size and the female function at larger size. This may explain the occurrence of hermaphroditism in nematodes, but not simultaneous hermaphroditism as found in earthworms and land snails.
- *Gene dispersal*: animals with restricted motility or with small and isolated populations may benefit from hermaphroditism, especially when it is sequential, to avoid the negative effects of inbreeding and genetic drift. The applicability of this explanation to hermaphroditic soil invertebrates is doubtful.

Charnov et al. (1976) added to this list their '*resource allocation*' theory, in which the mode of reproduction of a species is determined by selection acting on the relative allocation of resources to male and female functions. Hermaphroditism is predicted when the number of offspring produced by combining male and female functions is greater than the number of offspring for pure males and females. This model emphasizes *low mobility* as a factor promoting hermaphroditism. Puurtinen and Kaitala (2002) reiterated this interpretation. They developed a simulation model showing that under conditions of poor mate-search efficiency and high costs of searching, hermaphroditic reproduction is stabilized. This became known as the "*moving to mate*" model: if moving to mate is costly, hermaphroditism is favoured.

One of the few rigorous empirical tests of the "moving-to-mate" hypothesis is due to Eppley and Jesson (2008). These authors built a database for mode of locomotion, used as a proxy for adult mate search efficiency, and correlated this with breeding system (separate sex versus simultaneous hermaphroditism) in a phylogeny of 122 species of plants and animals. The data showed a statistically significant association between these two traits. Consequently, breeding system and mate search efficiency did not evolve independently from each other.

This conclusion is in line with the classical allometric relationship, established by comparative animal physiology, on the costs of locomotion across the animal kingdom (Figure 4.13). The cost of locomotion, expressed as energy expenditure per kg body mass per distance travelled, decreases linearly with body size when both axes are expressed as logarithms. Different intercepts are found for different

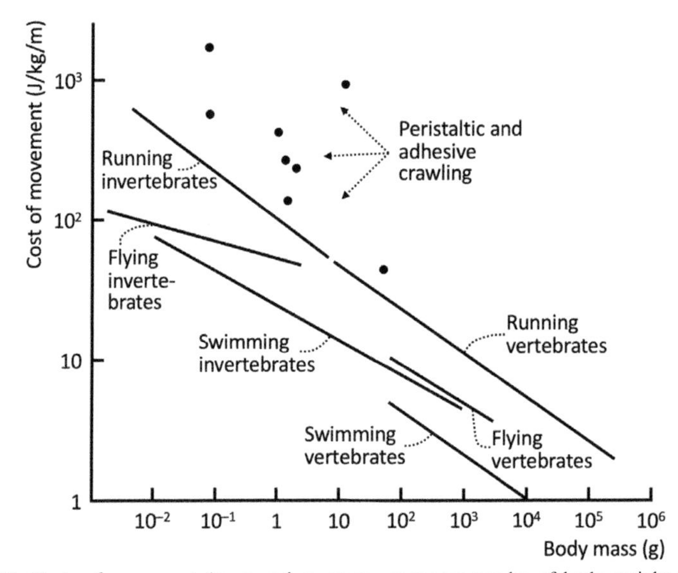

Figure 4.13. Costs of movement (measured as energy expenses per kg of body weight per distance travelled), as a function of body mass, across the animal kingdom. Regression lines are shown for flying, swimming and running vertebrates and invertebrates, derived from numerous studies. The dots refer to individual measurements of crawling invertebrates (slugs, snails, insect larvae). Redrawn from Eppley and Jesson (2008).

modes of locomotion. Swimming is most efficient, followed by flying and running. For animals relying on peristaltic and adhesive crawling on surfaces, locomotion costs are considerably above the average predicted by the allometric relationship for running animals (Figure 4.13).

While hermaphroditism is obviously plesiomorphic in earthworms and enchytraeids this has not prevented them from becoming secondarily parthenogenetic in several cases, as we have seen in Section 4.3. Thus, in earthworms, parthenogenesis evolved from hermaphroditism, while in tardigrades, nematodes, and all arthropods it evolved from gonochorism. Conversely, while we have seen at least one case of reverse evolution, from parthenogenesis back to gonochorism (Figure 4.9), I do not know any example for the re-evolution of gonochorism from hermaphroditism (but never exclude anything in biology). Still, it seems that hermaphroditism, once evolved, does not easily allow separation of the two sexes, although elimination of one sex (parthenogenesis) does happen.

To find out more about the biological conditions that promote the evolution of hermaphroditism, it may be useful to consider a group of species that do not share a hermaphroditic ancestor. Nematodes provide an interesting case. On numerous occasions, in various lineages, nematodes have switched from gonochorism to hermaphroditism.

The genus *Panagrolaimus* covers a group of nematodes with extremely diverse ecologies. Many of them are free-living and associated with decomposing organic matter, but others are associated with bark beetles and a few of them are parasites on mammals. The genus also includes three modes of reproduction:

gonochorism, parthenogenesis and hermaphroditism. Three hermaphroditic species form a monophyletic group within the genus (Lewis et al. 2009). Among them is *P. detritophagus*, a common species of agricultural soils. Whether the three hermaphrodites share an ecological niche that is different from the other species in the genus is not yet known.

In the genus *Pristionchus* a similar diversification of reproductive modes is present. Nematodes of this genus live in association with various species of scarab beetles (cockchafers, dung beetles) and chrysomelids (Colorado potato beetle). Eighteen species of *Pristionchus* have been described, of which 13 are gonochoristic and five are hermaphroditic. In contrast to the situation in *Panagrolaimus*, however, the hermaphroditic species are not monophyletic; they constitute five independent lineages (Mayer et al. 2007). So, in this genus alone hermaphroditism seems to have evolved no less than five times!

A similar situation holds for *Caenorhabditis* (Figure 4.14). This genus includes at least 20 different species (probably much more), of which until now three have been found to be hermaphroditic and self-fertilizing (*C. briggsae*, *C. tropicalis* and *C. elegans*). Molecular phylogenies of the genus (Kiontke et al. 2011, Ellis and Lin 2014) show that, like in *Pristionchus*, hermaphroditism evolved independently in each species.

Because *Caenorhabditis* includes the famous genetic model *C. elegans*, a lot is known about the mechanisms of nematode sex determination and the molecular changes that underlie the conversion of gonochorism into hermaphroditism. In the first place we should note that hermaphroditism in *C. elegans* and most likely in all other nematodes, is *sequential*, that is, sperm is produced during the last larval stage; this is carried over to the adult stage when it is used to self-fertilize the eggs. In addition to hermaphrodites, *C. elegans* also has "real" males, which occur on occasion and may mate with hermaphrodites. Pure males are never found in *simultaneous hermaphrodites* such as earthworms and land snails, which are always male and female at the same time.

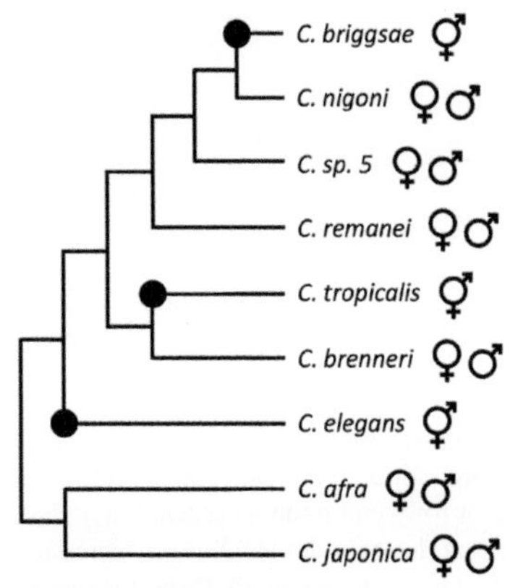

Figure 4.14. Simplified phylogenetic tree of the nematode genus *Caenorhabditis*. Only the species with sequenced genomes are shown. Hermaphroditism of *C. briggsae*, *C. tropicalis* and *C. elegans* evolved on three independent occasions (indicated by dots). From Ellis and Lin (2014).

Sex determination in *C. elegans* is chromosomal; females have karyotype XX and males are X0. Normally the twofold X dosage in females inhibits a gene *xol-1*, the master sex-switch gene of *C. elegans*. In males, however, the absence of a double X dose allows *xol-1* to be active, which triggers a cascade ultimately leading to repression of a crucial gene called *tra-2* followed by activation of the male programme in germline cells (Figure 4.15).

Hermaphrodites are actually genetic females in which spermatogenesis is temporarily turned on. Initiating factors, activated during larval stage 4, repress the key gene *tra-2* and this activates, via a number of steps shown in Figure 4.15, the

Gonochoristic ancestor

XO
XX
xol-1
sdc-1
sdc-2
sdc-3
her-1
tra-2
tra-3
cul-2
fem-1
fem-2
fem-3
trr-1
tra-1
fog-3
fog-1
sperm

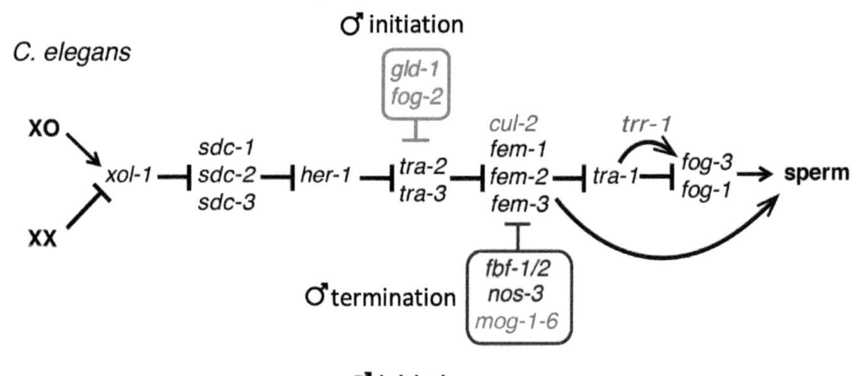

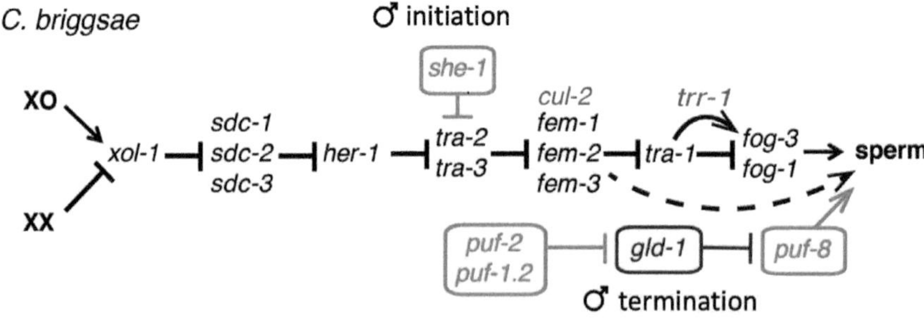

Figure 4.15. Comparison of the genetic regulatory pathway allowing sperm production in the hermaphroditic nematodes *C. elegans* and *C. briggsae*, compared to their gonochoristic ancestor (phylogenetic outgroups). Genes are indicated by abbreviated code names. The genes printed in grey are pleiotropic accessory factors. The arrows ending in a vertical line indicate repression. In genetic females (XX) the male pathway is repressed due to a double X-dose repressing xol-1. However, male function can be initiated in larval stage 4 by factors acting on tra-2; spermatogenesis is terminated by repression of fem-3. In *C. briggsae* the core pathway is the same but sperm initiation and termination are regulated by different factors acting upon different genes in the pathway, with the same final result. Reproduced from Haag et al. (2018), with permission from Oxford University Press and Eric S. Haag.

production of sperm, despite the inhibiting signal from *xol-1*. Spermatogenesis is terminated by repression of a gene called *fem-3*, downstream of *tra-2* (Baldi et al. 2009, Cook Hill et al. 2006, Thomas et al. 2012, Ellis and Lin 2014, Haag et al. 2018). After the moult to adulthood, the regulating factors lose their effect and the female programme is on.

Interestingly, sperm production can be experimentally induced in females of nematode species that are normally gonochoristic (e.g., *C. remanei*). This is done by repressing *tra-2* with RNAi (Baldi et al. 2009). A second mutation is required to activate the sperm cells. This remarkable result shows that just two mutations, in different pathways, are sufficient to create a hermaphroditic nematode from a gonochoristic ancestor.

As we have seen above (Figure 4.14), *C. briggsae* acquired hermaphroditism independently from *C. elegans*. This is also apparent on the molecular level. The factors that regulate the expression of the male pathway in *C. briggsae* appear to be different from *C. elegans* and act more downstream in the cascade (Figure 4.15). The apical effect is, however, just the same. *C. elegans* and *C. briggsae* use different regulating factors to achieve the same phenotype, a magnificent example of *convergent evolution.*

Despite the fact that the origin of self-fertilization in *C. elegans* and *C. briggsae* is relatively recent, mating-related traits are already degraded. The occasional males that occur in cultures are no reliable mating partners anymore and mating is only successful under artificial conditions. For example, the sex pheromone normally excreted by females is not excreted by hermaphrodites or is less efficient to attract males; also, hermaphrodites are not sensitive to the '*soporific factor*' excreted by the male that aims to sedate the female during copulation. In addition, hermaphroditic sperm outcompetes allosperm from males. In the wild, successful matings must be extremely rare, and continued self-fertilization leads to a high degree of inbreeding. So, while there may be short-term advantages to the sequential hermaphroditism of nematodes, on the long run such self-fertilizing lineages are likely to go extinct. This interpretation is consistent with the fact that hermaphroditism in nematodes is only seen in terminal branches of the phylogeny. There are no ancient clades of hermaphroditic nematodes, comparable to parthenogenetic oribatids.

It may be fair to say that hermaphroditism in nematodes is of a different nature than hermaphroditism in earthworms and land snails. In the latter case, the male and female programmes are active simultaneously and there is a strong preference for outbreeding, even when self-fertilization is perfectly possible. So, the ecological conditions under which hermaphroditism evolved and is maintained in nematodes might not shed much light on the evolution of hermaphroditism in earthworms and snails. However, the molecular mechanisms may. The work on sex determination in nematodes suggests that mutations that cause a switch from gonochorism to hermaphroditism are not that difficult to envisage. Simple mutations could also underlie the evolution of hermaphroditism in molluscs and annelids.

In annelids, hermaphroditism is wholly connected with the origin of Clitellata (*cf.* Figure 2.24). Since this lineage began its evolution in freshwater (soil-living clitellates evolved only later), hermaphroditism cannot be considered an adaptation to life in soil. Exactly the same pattern we see in the evolution of hermaphroditic land snails (Figures 2.26 and 2.27). Hermaphroditism evolved in freshwater representatives of Panpulmonata and Caenogastropoda long before they made their way to the land.

So, in the end, hermaphroditism on land may be seen as a developmental heritage from the past, a "*frozen accident*" or an "*exaptation*" in the sense of Gould and Vrba (1982). It arose, maybe as a neutral mutation, in freshwater and, millions of years later, turned out to be useful on land. In contrast to the forward switch (from gonochorism to hermaphroditism), the reverse switch seems to be more difficult. This may be explained by pointing out that during evolution of hermaphrodites, the male and female reproductive systems became intertwined, even partly fused (*cf.* Figure 5.18). Any mutation eliminating the developmental programme of one sex would also affect the other and so it was no longer possible to separate the sexes.

Despite this suggestion there are some species of aquatic hermaphroditic snails in which the male copulatory organs are missing. This condition, called *aphally*, is also seen in land snails, e.g., in the tropical stylommatophoran species *Phaedusa ramelauensis* (Köhler and Mayer 2016). Aphally is often polymorphic in a population and induced by environmental conditions, especially high temperature. It could be a step towards the evolution of parthenogenesis, as suggested for the terrestrial slug *Deroceras leave* (Nicklas and Hoffmann 1981). Like in earthworms and enchytraeids, hermaphroditism in snails can give rise to parthenogenesis by dysfunction or loss of male organs, but never to complete restoration of separate sexes (gonochorism).

For terrestrial hermaphrodites there could also be ecological reasons disfavouring back-mutations to gonochorism. Such mutations would force the animal in a "moving to mate" scenario, which is costly without limbs, and so would decrease fitness. That is why hermaphroditism persisted. Had the snails had limbs, things might have been different. Further comparative research is needed to decide whether this interpretation makes any sense.

4.8 The great switch to direct development

All marine ancestors of soil invertebrates spawn in seawater and have free-swimming larvae. These larvae often have a body plan that typifies the phylum to which they belong. All members of the phylum go through that particular larval stage and you can recognize the phylum from the larva. This is summarized in the concept of *phylotypic stage*, that is, a stage in early development that is shared by all members of a phylum. The concept goes back to the Estonian biologist Carl Ernst Von Baer who formulated his embryological laws in 1928. According to his third law the embryos of different species are similar in the early stages of development but come to differ from each other more and more in the course of development. The laws of Von Baer are the foundation for the relation between phylogeny and development, as elaborated in the magnificent book of Stephen J. Gould, *Ontogeny and Phylogeny* (Gould 1977).

According to the "ontogeny view" of evolutionary change, major transitions commence with changes in early development. For example, Vargas et al. (2021) argued that terrestrialization of hexapods began with the synthesis of a serosal cuticle which protects the egg against desiccation. The Danish zoologist Claus Nielsen argued

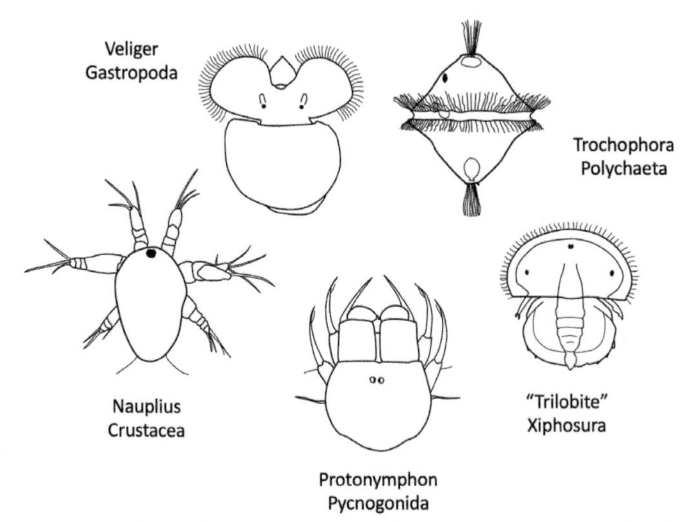

Figure 4.16. Free-swimming planktonic larvae of animal lineages assumed ancestral to freshwater and soil-living invertebrates. Nauplius redrawn from Wikimedia Commons, protonymphon from Brenneis et al. (2017), "trilobite" from Harzsch et al. (2006), veliger from Croll (2009), trochophora from Nielsen (2018).

that there are good reasons to assume that the ancestral metazoan had a morphology similar to the free-swimming larvae of present marine invertebrates (Nielsen 2013). According to Nielsen's theory, the present larval stages, with ciliated bands as adaptations to swimming, represent the ancestral morphology of the Metazoa better than the adults. In Figure 4.16 some of these free-swimming larval stages are shown.

Marine crustaceans, after their hatching, start their life as a planktonic *nauplius*. This common larval type, shared by many groups, has three pairs of appendages (homologous to *antennula*, *antenna* and *mandibula* of the adult) and a single median eye, which disappears later. Many crustaceans have a second larval stage following the nauplius, e.g., the *cypris* larva of barnacles and the *zoea* larva of crabs. The free-swimming larva of sea spiders (Pycnogonidae), called *protonymphon*, is somewhat similar to a nauplius; it also has a median eye complex, but the first appendage already shows the typical characteristics of a chelicera. Sea spiders are considered ancestral to terrestrial Arachnida together with horseshoe crabs (Xiphosura), the direct sister groups of Arachnida (*cf.* Figure 2.17). However, the free-swimming larva of horseshoe crabs, called "*trilobite*" has a more derived morphology. This is due to the fact that horseshoe crabs mate in the intertidal zone and lay their eggs on land, where they develop in a "nest". The seabound larvae that hatch from the eggs have already gone through the earlier stages. In fact, the on-shore breeding of horseshoe crabs can be said to represent a primordial stage of terrestrialization, that wasn't continued, but yet shows the beginning of direct development.

The planktonic larva of marine bivalves and gastropods is a *veliger*, named after the (often bilobed) ciliated velum, used for swimming. Eyes are present at the base of the primordial tentacles and a simple shell is usually already recognizable (Figure 4.16). The common larval stage of annelid worms is a *trochophora*, which in fact consists of the first two segments of the adult animal, but equipped with ciliated tufts and bands for swimming. Several authors see a morphological continuity between veliger and trochophora (e.g., Nielsen 1995), which reinforces the phylogenetic link between Mollusca and Annelida in the superphylum Lophotrochozoa.

The biphasic life cycle with planktonic larvae, called *indirect development*, was the common mode of early development for all marine metazoan lineages (Arenas-Mena 2010). When crustaceans, chelicerates, gastropods and clitellates colonized the terrestrial environment and gave rise to soil invertebrates, they had to get rid of these larval stages. Discharging gametes in the medium became impossible and a separate larval stage adapted to swimming did not make sense anymore. They all had to switch to *direct development*, in which the juvenile is a miniature version of the adult. We know that in metamorphosis, larval tissues undergo substantial transplantation and replacement. Likewise, in the evolutionary switch from indirect to direct development a substantial reprogramming of developmental pathways was necessary. The switch to direct development may be considered one of the greatest evolutionary changes associated with terrestrialization. But how was it done?

An indication of the developmental reprogramming is provided by studies on the polychaete-clitellate transition, as summarized by Kuo (2017). The early development of polychaete worms is similar to other lophotrochozoan lineages (bivalves, flatworms) and is characterized by *spiral cleavage*, i.e., a system of asymmetric cleavages, producing small and large *blastomeres*, while the plain of cell division is inclined with respect to the animal-vegetative pole, such that the small cells form a spiral on the large blastomeres (Figure 4.17). This basic pattern is also seen in clitellates. In polychaete marine worms, the embryo continues with a trochophora with its swimming adaptations (Figure 4.16). Then, after a while, the trochophora undergoes axial elongation, in which segments are added. The clitellate development, however, proceeds to a *gastrula*: the gut is formed by movement

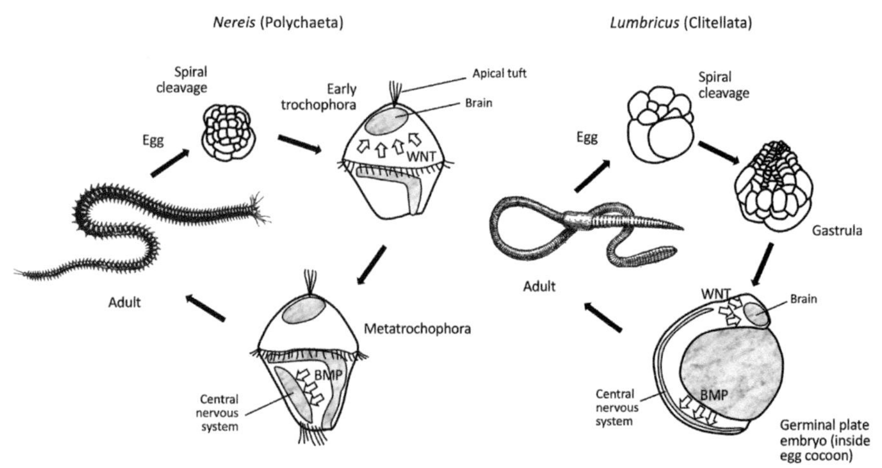

Figure 4.17. Scheme of the life-cycle of a polychaete and a clitellate worm, showing indirect development in the former and direct development in the latter. WNT and BMP are two signalling cascades contributing to prostomium identity and ventral axis identity, resp. In the ancestral, indirect, development the two patterning signals are separated in time but in the direct developmental mode, they are active simultaneously. Composed from images in Kuo (2017).

of cells to the inside, while at the same time the embryo starts to elongate (axial growth). Gastrulation and axial growth occur simultaneously.

On the molecular level, the great difference between the two modes of development is that in polychaetes, the *WNT* signalling cascade, which patterns the *prostomium*, the brain and the apical tuft, is active early in development, while *BMP* signalling, which induces the formation of the ventral nerve cord, becomes active in a later trochophora stage (*metatrochophora*) (Figure 4.17). In the clitellate embryo, however, *WNT* signalling and *BMP* signalling are active at the same time, thus causing axial elongation in synchrony with prostomium patterning.

So, the evolution of direct development was associated with a temporal advance of *BMP* signalling or a delay of *WNT* signalling. These two signalling cascades operate as independent genetic modules, so it is perfectly possible to shift their temporal expressions without affecting each other. This type of evolutionary change is called *heterochrony*, an extremely important principle in evolutionary developmental biology. It is made possible by the modular organization of developmental pathways.

The evolution of direct development is also an example of *evolutionary tinkering*, a concept introduced by the great French molecular biologist François Jacob (1920–2013). Jacob (1977) argued that evolution rarely invents completely new structures. Rather, it tends to combine, expand or modify existing structures to make something new. The heterochronic shift underlying the polychaete-clitellate transition is fine example of this principle.

The shift to direct development in annelid worms is not to be considered an adaptation to life in soil, since the first clitellates lived in freshwater; earthworms and enchytraeids evolved only later (*cf.* Figure 2.24). In fact, it would seem that life in freshwater did not strictly require the abolishment of planktonic larvae. Several freshwater groups have actually maintained planktonic larvae, e.g., freshwater copepods still have a nauplius. However, direct development was certainly a preadaptation, or rather an exaptation (*sensu* Gould and Vrba 1982), that allowed the evolution of soil-living lineages in those groups that had made the developmental switch before.

How the evolution of direct development took place in the crustacean-hexapod transition and the pycnogonid-arachnid transition is not known, but it wouldn't surprise this author if it would turn out to involve heterochronic shifts in early development as well.

4.9 Endless forms of limbs, most beautiful

Locomotion in the terrestrial environment is most efficient when the body is lifted from the ground and supported by limbs. Crawling over the substrate comes with substantial energetic costs (Figure 4.13). However, not all soil invertebrates have developed limbs. Their absence among nematodes, flatworms, earthworms, enchytraeids and gastropods is remarkable, because the basic genetic machinery to make appendages seems to be present among all metazoans.

For example, the homeobox protein *Distalless*, discovered in *Drosophila*, is active not only in all arthropod appendages but also in those of velvet worms (Onychophora), polychaete worms (Annelida), sea urchins (Echinodermata) and sea squirts (Urochordata). In all these animals the protein contributes to distal outgrowth (Panganiban et al. 1997). Even cuttlefish tentacles are patterned by the same genes as arthropod legs and vertebrate appendages (Tarazona et al. 2019, Prpic 2019).

This should not be taken to imply that arthropod legs, parapodia of ragworms, cephalopod arms, tubular feet of sea urchins and siphons of sea squirts are all homologous to each other, but rather that a homologous developmental mechanism underlies the parallel origin of these structures. This follows the *evo-devo* perspective in which homology is related to similarity in *character identity networks* (Wagner 2014). The character identity network for making appendages was already present in the ancestor of all bilaterians and was deployed several times in different contexts.

The evolution of limbs is best illustrated by the phylum Arthropoda. The ancestral arthropod limb is assumed to consist of a proximal *protopodite* and two distal branches called *endopodite* and *exopodite* (Figure 4.18). In addition, the *coxa* (*coxopodite*) of the protopodite often has a dorsal extension called *epipodite*. The coxa may also bear serrated plates or lobes; the ones on the inside are called *endites* and the ones on the outside *exites*. Endites are often involved in food processing; the median part of the protopodite when bearing teeth is called *gnathobase* (Figure 4.18).

All these structures of the arthropod limb have been modified in endless variations. The biramous plan is still seen in fossil and several extant crustaceans, including the pleopods of isopods (Figure 2.10). The pereiopods of isopods and the walking legs of all millipedes, arachnids and hexapods are, however, uniramous: they lack an endopodite. In these cases, the protopodite cannot easily be distinguished from the exopodite as they form one structure. The distal ramus is then called *telopodite*. In female amphipods and isopods, endites of walking legs may be transformed to *oostegites*, forming the floor of a brood pouch below the thorax, but these structures are lacking in most other arthropods. In decapod crustaceans (crabs and lobsters) the epipodites of the walking legs form a gill; they point upwards from the leg base and are protected in a gill chamber by a downward fold of the carapace.

As the legs of arthropods show such an enormous variety of structure, with teeth, plates, hairs, flagella, etc., attached to any of the substructures, they are a goldmine for comparative morphology (*cf.* Boxshall 2004). The morphological variation is paralleled by an equally wide variation in functions. Arthropod legs are not only made for walking but also used to swim, process food, capture prey, mate, sense vibrations, pump water and even support respiration, as we will see below. The evolution of arthropods in fact boils down to the evolution of their legs.

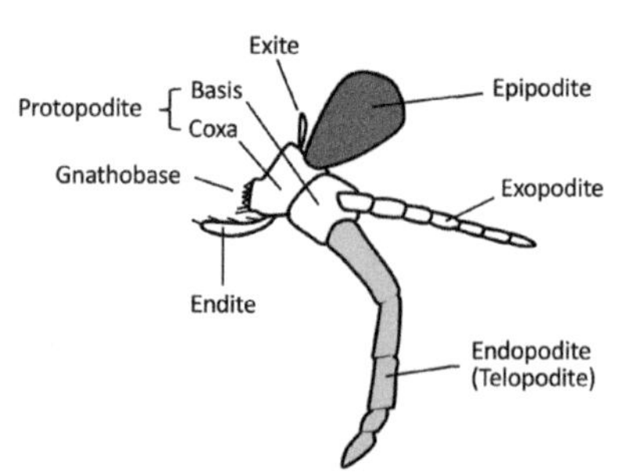

Figure 4.18. Organization of the arthropod biramous limb. Not all structures and appendages shown are present on every limb.

The study of limb homologies gained a new impetus when evolutionary developmental biologists discovered the genetic networks underlying their formation. Averof and Cohen (1997) were the first to suggest that the wings of insects were derived from an epipodite on the limb of an ancestral crustacean. Their conclusion was based upon the observation that two regulatory genes, *nubbin* (*pdm1*) and *apterous* (*ap*) had similar expressions in the wing primordia of the fruit fly, *Drosophila*, and the epipodites of the brine shrimp, *Artemia*. As the epipodite is the ancestor of the crustacean gill this allowed the remarkable suggestion that the wings of insects might be homologous to the gills of crustaceans. However, rather than calling the structures homologous, it is the character identity network that is homologous. Expressing it differently, Carroll et al. (2005) argued that the deployment of *nubbin* and *apterous* in insect wing development is a case of *co-option*, i.e., a secondary recruitment of an existing developmental network to develop a similar structure in another context.

Molecular research on the regulatory genes involved in the "*limb identity network*" has revealed some surprising insights. Damen et al. (2002) suggested that not only the wings of insects are derived from an ancestral epipodite, but also several structures in the opisthosoma of chelicerates, including the book gills of horseshoe crabs, the book lungs of spiders, as well as the tubular trachaea and spinnerets of spiders. However, more recent research has shown that the tubular trachaea of chelicerates are not exactly homologous with the trachaea of insects. In insects, the trachaea are not affected by homeotic mutations of the legs; this is consistent with their origin from the epipodite. However, in arachnids, such mutations affect legs and book lungs alike. Work on harvestmen (Sharma 2017) has confirmed that the opisthosomal appendages of chelicerates must be considered serial homologs of the telopodite, rather than of the epipodite. These new insights are summarized in Figure 4.19.

It thus seems that arachnids and insects have found different solutions for the problem of air breathing when they colonized the land. To make tracheal tubules,

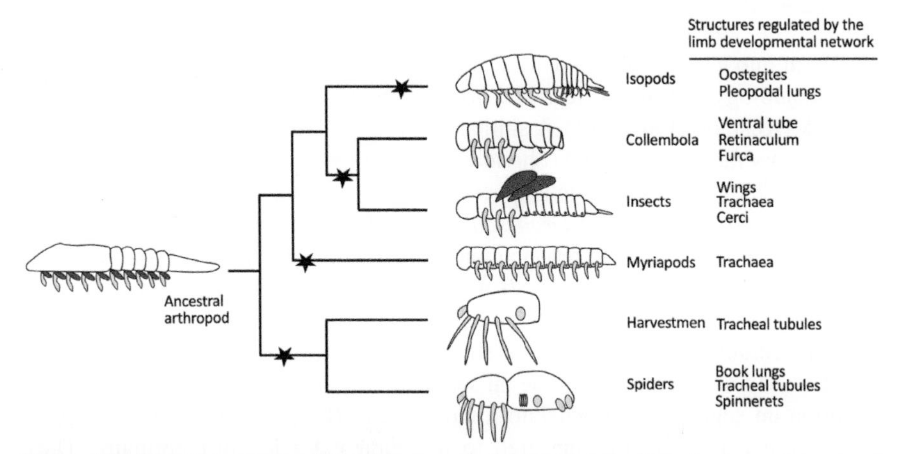

Figure 4.19. Evolution of soil arthropod limbs and structures of the limb developmental network (ignoring appendages of the head). Structures derived from telopodite primordia are shaded light grey, structures derived from epipodites are shaded dark grey. Terrestrializations are indicated by asterisks. The divergences in the phylogenetic tree are not to scale. Modified from Damen et al. (2002).

chelicerates recruited the telopodite developmental pathway while insects relied on the epipodite pathway (Sharma 2017). The fascinating conclusion is that the ancestral programme for making legs, which is so typical for all arthropods, turned out to be extremely useful in their terrestrialization, since by relatively small modifications, it allowed the formation of respiratory structures (trachaea, tracheal tubules, book lungs). The landfalling success of the arthropod phylum may be ascribed to their legs, not so much because they allowed walking, but because they allowed air breathing.

The strongest experimental proof for homologies between body structures is obtained by homeotic mutations. If an antenna can be mutated to a leg (like in the *Antennapedia* mutant of *Drosophila*) or a haltere in a wing (the *Ultrabithorax* mutant) this proves that the two structures are regulated by a similar genetic network. Most likely the ancestral arthropod had legs on all segments (Figure 4.19). The lack of appendages on the abdomen of insects is due to active repression of limb formation. Knock-down of the suppressor genes will cause the formation of legs in segments where they shouldn't be (*ectopic* legs). Such work has been done mostly on the genetic model species *Drosophila*, but there is also evidence among spiders and Collembola.

In a landmark study, Khadjeh et al. (2012) analysed the genetic regulation of limb formation in the spider *Achaearanea tepidariorum*, a species often invading houses and glasshouses (family Theridiidae). In insects the *Hox* gene *Ultrabithorax* suppresses limb formation in the first abdominal segment while *abdominal A* has a similar effect in the second and other abdominal segments. However, in spiders, the primary limb-suppressor is *Antennapedia,* a gene which in insects stimulates limb formation in the thorax. Its role in the opisthosoma of chelicerates is assumed to be ancestral. In insects this gene has been recruited to act in the thorax, leaving the limb suppression in the abdomen to *Ubx* and *abd-A* (Angelini and Kaufman 2005).

In spiders, the first opisthosomal segment (O1) is normally without appendages, but suppression of *Antp* by RNA interference (*RNAi*) leads to the appearance of ten-legged spiderlings with a perfectly functional leg on O1. In addition, repression of both *Antp* and *Ubx* causes the book lung, which is normally present on O2, to transform into a small limb bud. This study convincingly shows that book lungs are serial homologs of walking legs and are regulated by the same limb development network (Khadjeh et al. 2012). A similar story holds for the tracheal tubules of the harvestman *Phalangium opilio* (Sharma 2017).

The homology of abdominal appendages and walking legs was also illustrated for Collembola, in the species *Orchesella cincta* (Konopova and Akam 2014). These authors showed that furca, retinaculum and ventral tube can be experimentally modified by RNAi addressing the *Hox* genes *Ubx* and *abd-A*. Suppression of *Ubx* converted the ventral tube on abdominal segment 1 to a pair of walking legs and the retinaculum on segment 2 to a structure similar to a furca. Knock-down of *abd-A* caused an extra ventral tube on segment 2 (replacing the retinaculum) while the furca on segment 4 turned into a pair of legs (Figure 4.20). The four-legged *Ubx*-suppressed springtails appeared to use their extra leg just normally. These remarkable experiments reveal that the system of suppressing appendage formation in the abdomen, which was first discovered in *Drosophila*, already began in the first

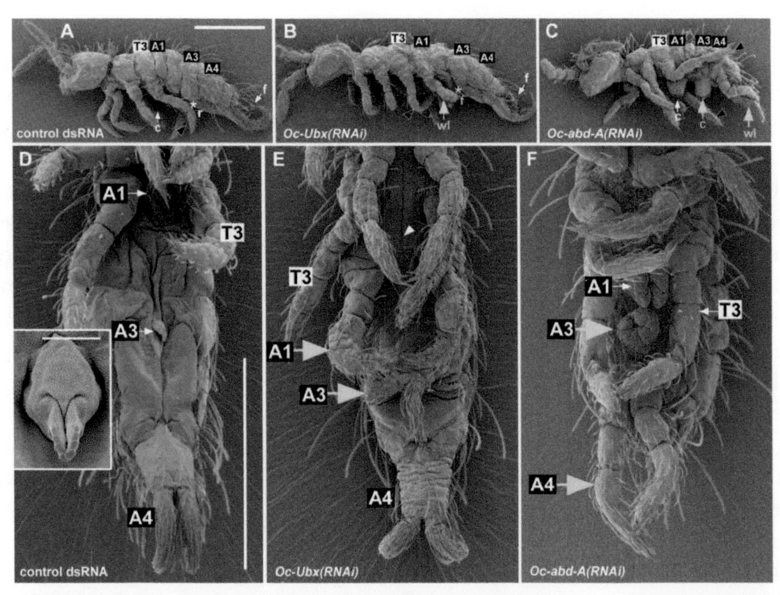

Figure 4.20. Homeotic transformations in the springtail *Orchesella cincta*. Double-stranded RNAs against Ultrabithorax (Ubx) and abdominal-A (abd-A) were prepared from *O. cincta* cDNAs and injected into females carrying eggs. The hatchlings raised from eggs laid by control females showed normal abdominal morphology (Figures A,D). Suppression of Ubx transformed the ventral tube in a pair of legs (A1) while the retinaculum was transformed into a furca (A3) (Figures B,E). Suppression of abd-A transformed the retinaculum into a ventral tube (A3) and the furca into a pair of legs (A4) (Figures C,F). f furca, c collophore (ventral tube), r retinaculum, wl walking leg, T3 thorax 3, A1 to A4 abdominal segments. The inset in Figure. D shows a close-up of the retinaculum. From Konopova and Akam (2014), with permission from Barbora Konopova.

hexapods, and most likely was part of the repertoire of developmental changes that came with terrestrialization. The same holds for terrestrial chelicerates.

The lesson from these evolutionary developmental studies is that modification of the legs, to change their function from swimming to walking, was maybe not the crucial innovation associated with terrestrialization. Walking was a question of adapting muscles and nerves rather than making something quite new. The real innovation was the use of limb developmental networks to make specific structures that allowed air-breathing: book lungs, trachea and trachaeal tubules. The "*endless forms, most beautiful*" in Charles Darwin's famous last sentence of the *Origin of Species* (Darwin 1859), were realized in arthropods by an endless process of co-option.

4.10 The mantle cavity of snails

If arthropods conquered the land with their legs, gastropods did so with their mantle cavity. As we have seen in Section 2.9, all molluscs avail of a *pallial cavity*, in which one or two gills (often in the form of lamellated comb-like structures, *ctenidia*) are suspended (Simone 2021). To support efficient respiration, the mantle cavity is actively flushed with water, which is often sucked up and pumped out through tubular openings called *siphons*. This system is seen in all bivalves and cephalopods, but not in all gastropods. Those gastropod lineages that made the transition to land have lost their gills and use the internal walls of the mantle cavity as a respiratory surface. And, as we have seen in Chapter 2, this evolutionary change already began in freshwater. The presence of lungs in land snails should not be considered an adaptation to terrestriality.

In several gastropod groups, species can be ranked in a transitional series between aquatic and terrestrial life. Can such comparisons tell us something about the origin of the lung? An interesting model to study the question is found in the family Ampullariidae (Caenogastropoda), or apple snails (for an extensive review of their biology see Hayes et al. 2015). This group includes several species that are more or less amphibious. Some lay their eggs in the water, others in the sediment, but the best-known species, *Pomacea canaliculata* (golden apple snail), is known to deposit its eggs in the vegetation. It regularly crawls out of the water and is an obligate air breather. When kept immersed, eggs cannot develop properly. Although often considered a freshwater snail (also in Figure 2.27), *P. canaliculata* is actually a limnoterrestrial species. Due to its voracious herbivory it is a serious pest in wet-field agriculture, e.g., paddies.

Apple snails are known to have both a lung and a gill. The lung is composed of a dorsal fold of the mantle cavity, which has displaced the gill somewhat, although both are present and functional. A study of the vascularization of the pallial complex of *P. canaliculata* by Rodriguez et al. (2019, 2021) showed that haemolymph from the body is passed first through the lung, then through the gill, to be passed over to the heart, which pumps it into the body cavity (Figure 4.21).

This system, with the lung and the gill connected to one another in series, implies that the blood is already oxygenated when arriving in the gill. This situation presumably has altered the function of the gill. Histological studies of the gill epithelium suggest that it has an osmoregulatory function, in addition to a gas exchange function (Rodriguez et al. 2019, 2021). Because the gill is less efficient in its respiratory function, the apple snail became an obligate air breather. The dual respiratory system of apple snails shows that the transition from gill-breathing to aerial breathing might have been relatively easy and could be made gradually, by switching emphasis from one organ to another.

Similar scenarios are seen in other snail lineages, such as in the panpulmonate group Ellobioidea, which shows terrestrializations in two genera, *Pythia* and *Carychium* (Romero et al. 2016a,b, see also Section 2.9). In each of these lineages the invasion of the land was accompanied by loss of the gills and use of the mantle cavity as a lung. Because of multiple land invasions (another one took place in the Stylommatophora), panpulmonate snails are an excellent model for studying parallel adaptations associated with the terrestrial life-style.

A most peculiar situation is valid for intertidal snails of the genus *Siphonaria*, so called "*false limpets*". They are externally similar to limpets, but are classified as a basal lineage of Panpulmonata, quite remote from "true limpets" of the genus *Patella* (Kocot et al. 2013, see also Figure 2.26). In agreement with this position, *Siphonaria* is hermaphroditic, like other panpulmonates, but unlike *Patella*. Figure 4.22 shows the morphology of the pallial cavity of *S. pectinata* to illustrate their dual terrestrial/aquatic adaptations (Simone and Seabra 2017).

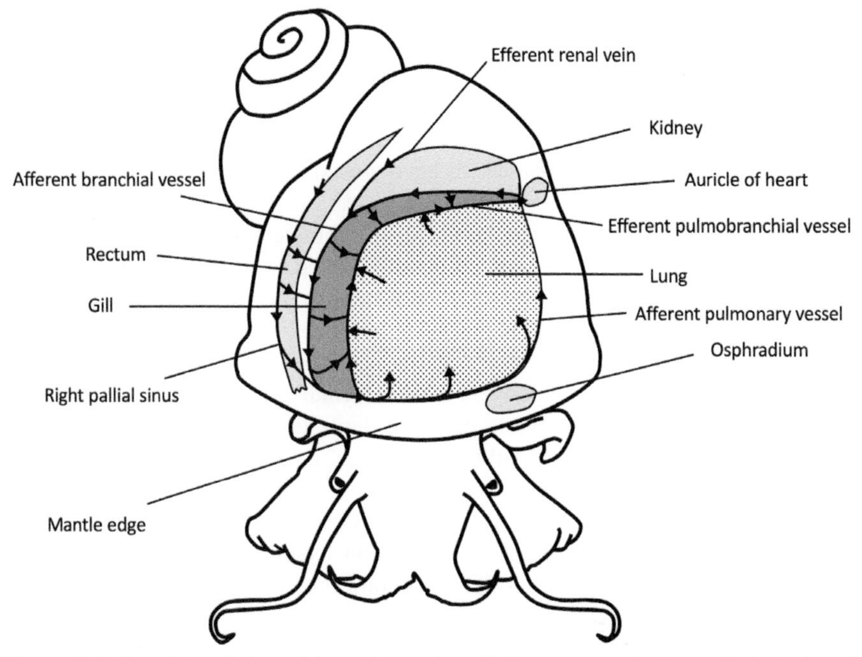

Figure 4.21. Frontal-dorsal view of the golden apple snail, *Pomacea canaliculata*, with the body wall partially removed to show the position of lung, gill, kidney and heart. The arrows indicate the direction of haemolymph flow. Redrawn from Rodriguez et al. (2019).

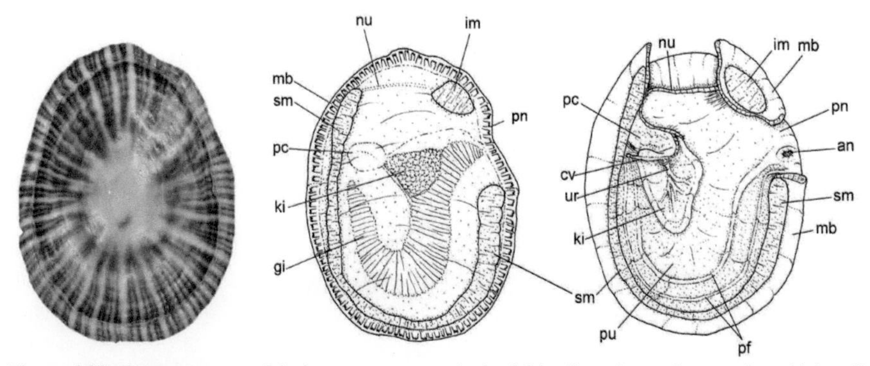

Figure 4.22. Gross anatomy of *Siphonaria pectinata* (striped false limpet), a pulmonate intertidal snail from the south coast of Portugal, showing morphologies for both gill-supported aquatic respiration and lung-supported aerial respiration. Left: dorsal view of the shell, middle figure: dorsal view of the whole animal with shell removed, right figure: view of the pallial cavity with its dorsal wall removed. Abbreviations of anatomical structures: an anus, cv ctenidial vein (efferent gill vein), gi gill, im shell muscle, ki kidney, mb mantle edge, nu nuchal connection of mantle, pc pericardium, pf pulmonary folds, pn pneumostome, pu pulmonary cavity (lung), sm shell muscle, ur urethra. From Simone and Seabra (2017), reproduced with permission from Luiz Ricardo L. Simone.

Like apple snails, siphonariids have both a lung and a gill. In water they respire with the gill and on land the lung is used. The gill is quite large and also the lung covers a considerable part of the pallial cavity (Figure 4.22). The pneumostome is however, not contractile as it is in land snails. The phylogenetic relationship of Siphonariidae to the non-marine Hygrophila and Stylommatophora (Holznagel et al. 2010, Kocot et al. 2013), suggests that their intertidal life-style does not reflect a marine ancestry but should be considered a secondary adaptation, the beginning of a marine regression. It is possible that they descended from a terrrestrial ancestor without a gill. This would imply that siphonariids have moved to the seashore from the land, not from the sea.

Like Ampullaroidea, Cyclophoroidea, Ellobioidae and Stylommatophora, Siphonariidae illustrate the repeated co-option of the pallial cavity to accommodate aerial respiration. The flexibility of this system, a heritage from their marine ancestors, may explain the many independent terrestrialization events in gastropods, as discussed in Section 2.9.

Molecular mechanisms underlying developmental reprogramming of the mantle cavity in snails remain unknown to date. We don't know which genetic networks define the character identity of pallial complexes. Evo-devo approaches in snails have come off the ground only recently (Lesoway and Henry 2021). Still, because of their multiple terrestrial invasions, in many different lineages, land snails present eminent models for the study of adaptive (and exaptive) evolutionary changes associated with the life-style of a soil invertebrate.

CHAPTER 5

Mate Choice, Brood Care and Predatory Behaviour

Sex & Slugs (& Rock & Roll)

Let's talk. Talk about sex as between male
and female 'pussy riots' are to deplore:
No, she isn't either a madonna or a whore.
And yes, he must learn: butthole surfers prevail.

Why must mankind play everything hardcore?
Why are queers, libertines objects to assail?
Why so many rappers scrap and derail?
Do like us, we strongly like to implore:

1. Join the roaming comrades' caravan trail.
2. Inflame. Wave your willies with ardour.
3. Then, like flagellants, let your rhythm sticks flail.

4. Enter the deep purple, adore, explore...
5. Lend some chromosomes of us, slug and snail:
6. with sex pistols you too enjoy much more!

Jasper Aertsz

CHAPTER 5

Mate Choice, Brood Care and Predatory Behaviour

"O, Mole! The beauty of it! The merry bubble and joy, the thin, clear, happy call of the distant piping! Such music I never dreamed of, and the call in it is stronger even than the music is sweet!"

Kenneth Grahame, The Wind in the Willows

One might think that the soil environment leaves little room for animal behaviour, but this is quite a misconception. Many soil invertebrates show intricate behaviours in relation to sexual activity, social interactions and prey capture. Moreover, these behaviours show interspecific variation that is related to the soil profile. The GKGB hypothesis outlined in Chapter 1 is very well applicable as a framework for explaining the evolution of soil invertebrate behaviours.

In this Chapter we visit the main trends in courtship, mate selection and mating, and compare them across species in the various evolutionary lineages. Like in Chapter 4, we will see that the steep abiotic gradients across the upper layers of the soil profile act as a strong template for the diversification of reproductive behaviour. In addition, like all other sexually reproducing animals, soil invertebrates are subject to sexual selection, which has led to a wonderful variety of courtship and conflict behaviours, varying from nuptial gifts to love darts.

Many animals exert some degree of control over the paternity of their offspring. Such control may come from preferences in courtship and mating (*pre-copulatory mate choice*) or it may consist of differential treatment of sperm inside the female genital tract (*cryptic female choice*). A huge body of research in mammals, birds and insects has shown that female choice is often driven by fitness optimization, and may lead to the well-known process of *sexual selection*, already described by Darwin (1871). Do soil invertebrates also show female choice, male competition and sexual selection? We discuss a number of model systems that have been probed to answer this question.

Soil invertebrates also show social behaviour at several levels. We will see that in addition to the advanced social structure of ant and termite communities, a great variety of soil-living insects and other soil invertebrates engage in social interactions that may be considered the beginning of social behaviour. The study of subsocial behaviour in species that do not have the peculiar genetic system of Hymenoptera may shed a broader light on the evolution of sociality. Finally, predatory behaviour and predator escape reactions are another fruitful area of comparative behavioural research.

5.1 Indirect sperm transfer

A *spermatophore* is a coated structure containing sperm cells, which is transferred as an intact unit from male to female and releases sperm when in touch with the female gonopore, or inside the female genital tract, sometimes in a special receiving organ. Spermatophores take a great variety of shapes, depending on the species, but the common purpose of them all is to enhance the success of fertilization. An important aspect contributing to fertilization success is that a spermatophore will keep the sperm together, often in an inactive state, until 100% of it is put inside the female genital tract.

Spermatophores are present throughout the animal kingdom: marine, aquatic and terrestrial, and they evolved many times independently in marine lineages, e.g., polychaetes, cephalopods and crustaceans. Obviously, spermatophores are not to be seen as an adaptation to terrestrial habitats or to low humidity (Proctor 1998). Rather, packaged sperm transfer evolved to enhance fertilization success in marine animals of which the majority relies on spawning freely in the water. The very same structures proved to be extremely useful when animals colonized the land, and spawning in the water became impossible.

Many soil invertebrates have a spermatophore of some kind, but the structure is best known in the arthropod lineages employing *indirect sperm transfer*: a spermatophore is deposited on the substrate, after which the female takes it up to fertilize herself. This indirect sperm transfer may or may not be accompanied with courtship behaviours. In this section we review the evolutionary trends and transitions of sperm packaging and transfer, including the behavioural changes that come with them.

Patterns of indirect sperm transfer in arthropods are discussed in two classical review papers: Schaller (1971) and Proctor (1998). The latter author distinguished four different categories to classify modes of sperm transfer:

- *Direct transfer*: the sperm or spermatophore is placed directly inside the female tract or in a sperm-receiving organ during copulation.
- *Pairing with indirect transfer*: male and female engage in courtship behaviour during which a spermatophore is placed on the substrate and taken up by the female.
- *Incompletely dissociated transfer*: males deposit spermatophores when triggered by chemical or other cues from the female, but there is no physical contact between the sexes.
- *Completely dissociated transfer*: males and females never meet; spermatophores are deposited without intersexual communication.

Proctor (1998) adds that the distinction between these categories is not absolute and all kind of transitions occur. For example, male pseudoscorpions may initially deposit a spermatophore on the substrate, but during a later stage of courtship push the sperm package inside the female genital opening with their legs (Weygoldt 1969).

An impression of the morphological variety of arthropod spermatophores may be obtained from Figure 5.1. Structures the shape of a droplet attached to threads or webs are found in centipedes, millipedes, Archaeognatha and Zygentoma; stalked packages are seen in springtails and many mites, and intricate contraptions involving valves and expelling mechanisms are observed in scorpions. In addition to packaging and releasing devices, a spermatophore may also carry structures used as a *nuptial gift*, which are consumed by the female during uptake of sperm. For example, the spermatophores of bush crickets (Tettigoniidae) and some crickets (Gryllidae) are equipped with a bulbous structure called *spermatophylax*, which is eaten by the female during or directly after copulation and possibly indicates male quality (Simmons et al. 1999).

Assuming that the first arthropods that went on land already had a sperm packaging structure of some kind, the most likely plesiomorphic condition for the spermatophore of all terrestrial arthropods is a relatively large sac-like structure with a rigid sheath that was placed on the substrate, while the reproductive behaviour was dissociative (Witte and Döring 1999). In centipedes and several millipedes, this structure was subsequently placed on a thread- or web-like support produced by spinnerets. In Archaeognatha and Zygentoma, the sperm is also placed on threads, but these are not produced by spinnerets but by specialized glands in the male genital tract, so called *paramera*; the use of threads in association with the spermatophore in Archaeognatha and myriapods is due to convergent evolution. In Collembola and Diplura, the spermatophore evolved further to become equipped with a stalk and a viscous sheath that allowed water exchange with the air. In addition, a process of miniaturization took place, allowing production of multiple spermatophores, while sperm cells became more tightly packed and immobilized inside the capsule.

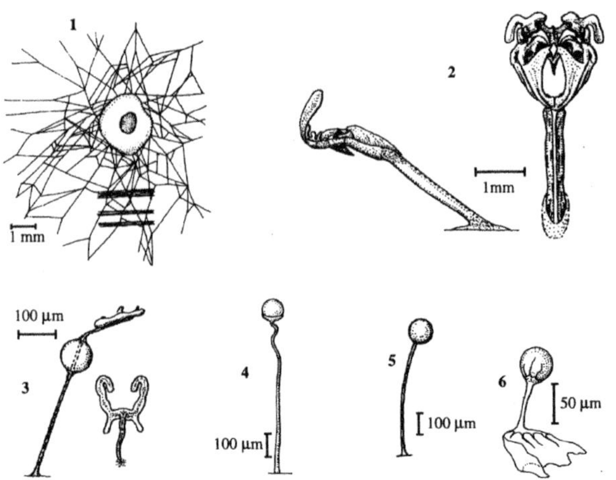

Figure 5.1. Showing the variety of sperm packaging structures in soil arthropods. 1. *Lithobius forficatus* (centipede), 2. *Tarantula marginemaculata* (spider), 3. *Dinocheirus tumidus* (pseudoscorpion), 4. *Belba geniculosa* (oribatid mite), 5. *Orchesella flavescens* (springtail), 6. *Podura aquatica* (springtail). Compilation by Stam (1998b), redrawn from Schaller (1979), published by Van Nostrand Reinhold Open Library.

A similar miniaturization took place in the chelicerate lineage, leading to the small spermatophores of oribatid and prostigmatid mites.

In this view, dissociative, indirect sperm transfer is the ancestral condition for all terrestrial arthropods, while associative sperm transfer, including copulatory behaviours, evolved several times independently, at least four times among soil-associated arthropods: in woodlice, insects, harvestmen and astigmatid mites. The evolution of sperm transfer systems in arthropods nicely illustrates the interplay between phylogenetic constraint and natural selection (Witte and Döring 1999).

Which ecological conditions favoured a transition from dissociative to associative sperm transfer is difficult to say. In general, dissociation is much more common among organisms living close to the soil surface or in the soil, while association is seen in lineages colonizing the aboveground vegetation. However, isopods and harvestmen are an exception to this rule.

Given the evolutionary persistence of indirect sperm transfer and its widespread occurrence, there must be clear advantages to it. Stam and Hoogendoorn (1999) developed an interesting argument, based on functional response theory. From their model follows that, factors favouring dissociative sperm transfer are:

- a high chance of finding a mate
- a short 'handling time' (time needed for courtship, copulation and mate guarding), and
- a high density.

So as long as these conditions prevailed among soil arthropods (and they certainly hold for many species), indirect sperm transfer was the best strategy.

5.2 Mating dances of pseudoscorpions

Among arthropods, pseudoscorpions were the first group for which the mechanism of indirect sperm transfer was described; the discovery dates back to a publication by H.W. Kew in 1912. There is not much recent work, however, on these small chelicerates. Much of what we know about their reproductive biology is due to the detailed observations by Peter Weygoldt at the Marine Laboratory of Duke University, summarized in his 1969 book (Weygoldt 1969).

The diversity of mating and reproductive behaviours in pseudoscorpions is remarkably large. It seems like pseudoscorpions have mainly diversified in their behaviour, rather than in their ecology or morphology. The behaviours vary from purely dissociative to various forms of associative behaviours. Weygoldt (1969) classified them as follows:

- *Complete dissociation*: a large number of spermatophores is deposited without communication between the sexes; females are attracted by chemotactical means. This holds for the majority of species, especially those that are gregarious and live in humid environments such as forest soils (Neobisiidae, Chthoniidae).
- *Partially dissociated*: spermatophores are only deposited in the presence of females. Very interesting is the behaviour of *Serianus carolinensis*, a species found under debris along the coast of South Carolina. In a crevice with an

overhead cover the male spins a series of vertical threads that mark a 'hallway' tapering towards his spermatophore. The female is attracted by this and when approaching is pulled over the spermatophore by the male.

- *Associated*: sperm is transferred during a courtship behaviour. Usually a series of ritual movements are conducted by the couple, during which a spermatophore is deposited and taken up. The rituals are often accompanied by head-to-head contact, but they may also remain without physical contact. Such elaborate courtship behaviours have only been observed in the superfamily Cheliferoidea.

As an example of the wonderful 'mating dances' performed by pseudoscorpions I reproduce here an illustration on the behaviour of *Dinocheirus tumidus*, a species found in large numbers under pieces of wood in the beach drift line in Florida (Weygoldt 1966), see Figure 5.2. Two aspects of this behaviour are remarkable and not seen in other soil invertebrates. First, the male clasps the female with only one pedipalp, leaving the other hand free to be used for quick poking and tapping movements on the female body. Second, after uptake of sperm through a duct extending from the spermatophore capsule, the remains of the spermatophore are destroyed by both partners while the male keeps on holding the female. Then the dancing ritual is repeated and the male produces another spermatophore, which is also taken up (Figure 5.2). This may continue up to six times.

It is assumed that chemical signalling supports the formation of a pseudoscorpion couple. A special structure, the so-called *ram's horn organ* may play a role in this. The lateral wall of the male's genital atrium has two sac-like invaginations that may be greatly extended and form long tubes stretching out of the genital operculum under the body forward like a ram's horn. It is assumed that these structures are

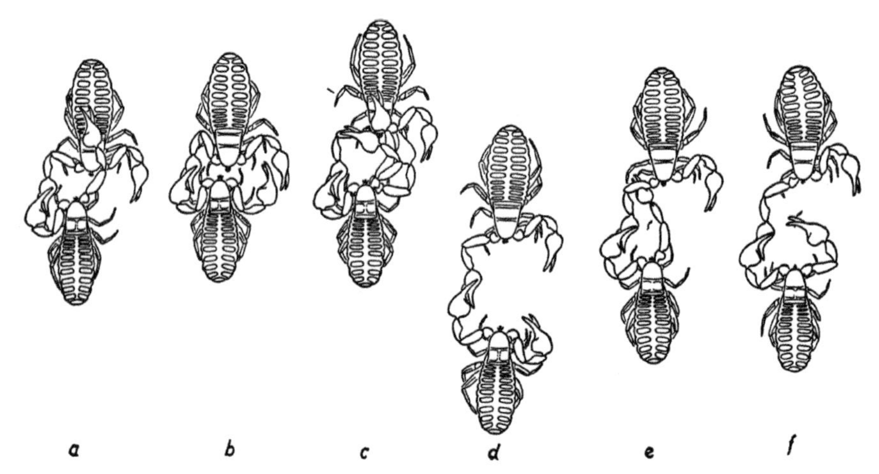

Figure 5.2. Mating dance of *Dinocheirus tumidus*, male below, female above. In (a) and (b) the partners make contact with their pedipalps. In (c) the male deposits his spermatophore, holding the female with his left pedipalp and touching her excitedly with his right pedipalp. In (d) the male pushes the female backwards towards the spermatophore and fertilization takes place. In (e) the male shakes the female with a series of short movements; between mates the stalk of the spermatophore is visible. In (f) the remains of the spermatophore are removed. Reproduced from Weygoldt (1966) with permission from The Marine Biological Laboratory.

covered with odorous material from the genital chamber that stimulates or guides the behaviour of the partner.

The courtship behaviours of the different species reveal a strong phylogenetic signature. Specific combinations of mating behaviours characterize a family and are not seen in other families. For example, Proctor (1993) argued that the four families of Cheliferoidea (Cheliferidae, Chernetidae, Atemnidae and Withiidae) can be discerned unequivocally on the basis of four behavioural traits: "*presence of spermatophore droplet*", "*male pulls female over spermatophore*", "*presence of ram's horn organs*", and "*male pushes sperm in female genital opening*". While morphological characters do not provide good synapomorphies to resolve the cladistic relationships between the four cheliferoid families, behavioural characteristics provide a solution (Proctor 1993).

While pseudoscorpions display the full range of disssociative to associative behaviours, the dissociative behaviours are seen only in species living in protected soil environments. Weygoldt (1969) noted an evolutionary trend ranging from the "primitive", mostly soil-living, families Chthoniidae, Neobisiidae, Garypidae and Cheiridiidae, which have completely dissociative behaviours, towards the more advanced families Chernetidae and Cheliferidae with their complex mating dances, which live in more open habitats (sand dunes). However, a rigorous test of this association, by mapping habitat use and reproductive behaviour on the phylogenetic tree of pseudoscorpions (e.g., the one developed by Harvey 1992) has not yet been done.

5.3 Mate choice in spiders and mites

Fascinating courtship behaviours may be observed in wolf spiders (Lycosidae). Thomas H. Montgomery, at the University of Pennsylvania described the various elements of it already in 1903 (Montgomery 1903). The male spins a small web on the soil, places an amount of sperm on it and sucks this into the sperm bulbs of his pedipalps. These palpal bulbs are complicated modifications of the tarsus, able to store sperm and pump it in and out (Figure 5.3).

After locating a female, the male then shows a series of characteristic behaviours, involving raising and trembling motions of the left and right pedipalps or the first legs, before he mounts the female, or in some species, crawls under her. Sperm is transferred in copula position by inserting the tip of the palpal bulb, called *embolus*, into one of the two genital openings in the female *epigynum* (also spelled *epigyne*) on the ventral side of the abdomen (Figure 5.3). The male embolus and the female epigynum fit tightly. Due to their rapid evolution, the male palp and the female epigynum differ between species much more than the rest of their morphology, so are among the most reliable characters for species identification.

Based on the complexity of the epigyne, spiders can be classified as *Haplogynae* (with simple epigynes) and *Entelegynae* (with complex epigynes). In haplogyne spiders, the epigynum has a single opening. The two spermathecas open in a central oviduct, through which also the eggs pass. In Entelegynae each spermatheca has its own distal opening, through which sperm is received during copulation. On the proximal side, the spermathecae open in the unpaired part of the oviduct, through

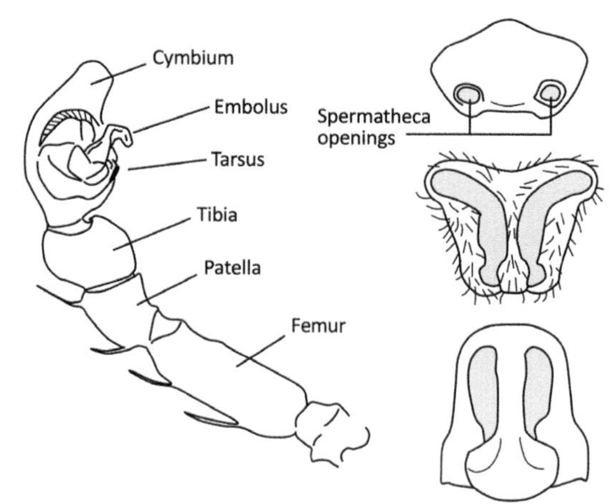

Figure 5.3. Left: pedipalp of male *Hygrolycosa rubrofasciata* (Lycosidae), showing the highly modified tarsus. Redrawn from Ahtiainen et al. (2003). Right: epigynes of female lycosid spiders, from top to bottom: *Pirata montanus*, *Trabeops aurantiacus* and *Geolycosa pikei*, redrawn and modified from Wikipedia.

which the eggs pass. It is assumed that the haplogyne organization, although not necessarily simple, is the plesiomorphic condition.

The precopulatory complexity of male courtship is often interpreted as a test on female receptivity. Since spiders show quite some aggressive behaviour towards each other, it is essential for the male to communicate his intentions from a distance and to make sure the female is ready to mate. The function of male courtship could be to suppress the predatory response of the female and to prevent cannibalism.

Another aspect of wolf spider courtship is that it contributes to evolutionary radiation. Stratton and Uetz (1983) investigated two related species, *Schizocosa ocreata* and *S. rovneri*, which have very similar morphologies but differ in their courtship behaviour and therefore do not recognize each other. Consequently, it is possible that two closely related species with nearly identical morphologies, using the same microhabitat and the same phenology, remain reproductively isolated from each other due to their differential mating behaviour. This situation is very similar to what was described above in pseudoscorpions: behaviour can be a driving factor of speciation.

Detailed insight into the nature of sexual communication in wolf spiders was obtained for *Schizocosa ocreata* (Figure 5.4). The male uses rhythmic movements of the forelegs as a sexual signal to appease the female during courtship. The forelegs have a conspicuous tufted area, contributing to the visual signalling (Figure 5.4). However, detailed experiments (Gibson and Uetz 2008, Gordon and Uetz 2011) showed that sexual signalling is in fact multimodal: it includes not only visual but also tactile, chemical, acoustic and seismic elements:

- Sensing of silk and possibly pheromones from the female
- Tapping with the legs causing vibrations in the substrate
- Jerking of the body back and forth

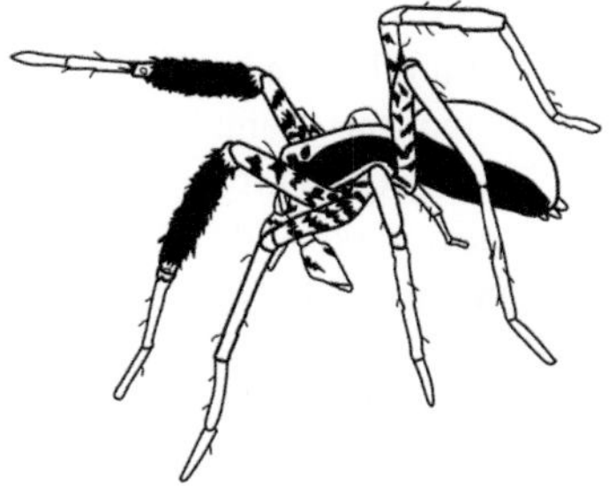

Figure 5.4. *Schizocosa ocreata* (Lycosidae), a model species for studies on multimodal sexual communication. Note the raised right foreleg used as a sexual signal during courtship Redrawn from a photograph by G.W. Uetz, University of Cincinnati.

- Vibrations of the abdomen
- Palpal raisings and vibratory movements
- Stridulation

The seismic and visual signals appeared to be redundant, that is, they convey more or less the same information and may be deployed in different environments. Both male and female prefer the surface of leaf litter, which allows good vibrations, but when placed on hard substrates such as soil or rock that do not easily transmit seismic communication, males display more intense visual signalling.

Multimodal sexual communication in wolf spiders not only serves to prevent cannibalism, it also functions in mutual mate assessment. Experiments by Rypstra et al. (2003) with another lycosid, *Pardosa milvina*, suggest that males use substrate-born cues (silk, pheromones) to assess the reproductive status of the female. Males display more intense courting before virgin females, while females use visual signals (leg raises, body shakes) to assess the quality of the male (body size and condition). This is in accordance with sexual selection theory, which emphasizes assessment of female receptivity by the male and assessment of male physical condition by the female.

Another interesting case of mate choice in spiders is seen in the nursery web spider, *Pisaura mirabilis*. Pisauridae are closely related to wolf spiders, but the females carry their egg cocoons with their palps and chelicerae, in contrast to wolf spiders, which carry the cocoon attached to the abdomen. In addition, the female pisaurid deposits the cocoon in a suitable place under low vegetation on the soil, spins a tent-shaped web over it and guards the '*nursery web*' until the spiderlings emerge. *Pisaura mirabilis* shows an additional remarkable behaviour: the offering of a *nuptial gift* as part of its courtship behaviour.

Nuptial gifts are taxonomically widespread, especially in insects, but are rare among soil invertebrates. A few species of pisaurid spiders, some scorpions and onychophorans show this behaviour, and we have mentioned above the nuptial gift of crickets, a spermatophylax attached to their spermatophore. The nuptial gift of spiders takes the form of a prey item, wrapped in silk, which is offered to the female and eaten by her during copulation.

Several hypotheses have been coined concerning the possible adaptive significance of the nuptial gift. It has been considered an investment from the male's side that increases the capacity of the female to produce eggs, or it could act as a protection from cannibalism and allow the male to mate without being eaten by the female. It is not impossible that the "cannibalism prevention" or the "paternal investment" hypotheses are valid explanations for the origin of the nuptial gift phenomenon; however, there is now good evidence that, at least in *Pisaura mirabilis*, the nuptial gift behaviour is maintained by female choice and sexual selection (Stålhandske 2001, Albo et al. 2013, Tuni et al. 2013).

Experiments have shown that female *Pisaura* do not necessarily cannibalize males if they come without gift, but the giftless copulations are significantly shorter, and less sperm is transferred. Moreover, Albo et al. (2013) showed that females are able to preferentially store sperm from copulations with a gift. This was demonstrated by experiments in which copulations with and without gift were experimentally interrupted so as to equalize the time available for sperm transfer. Still the females given a gift appeared to have stored more sperm, and the hatching success of their eggs was higher, compared to giftless copulations (Figure 5.5). In addition, by mating with several males, females gained higher oviposition rates and higher hatching success compared to imposed monogamy (Tuni et al. 2013). Apparently, females are able to have their eggs fertilized by the best male and use the nuptial gift as a token of quality.

How female pisaurids are able to control differential fertilization is unknown. However, several mechanisms are suggested in work on other spiders. In the goblin spider *Silhouettella loricata* (Arachnida, Oonopidae), a behaviour has been described called "*sperm dumping*" (Burger 2007). The female of this tiny spider, living almost unseen in leaf litter, is able to expel the ejaculate of a previous male from her receptaculum before mating with a new one. The entire ejaculate is enclosed in a sac-like package, made by glands adjoining the receptaculum, and then dumped on the ground, most likely with help from the new male.

In the dome spider *Linyphia litigiosa* (Arachnida, Linyphiidae) the first male fertilizes most of the eggs despite a high degree of polyandry (Watson 1991a). A fight between males, incited by the female, precedes mating. In addition, the male performs a large number of fake copulations with empty palps, before he finally

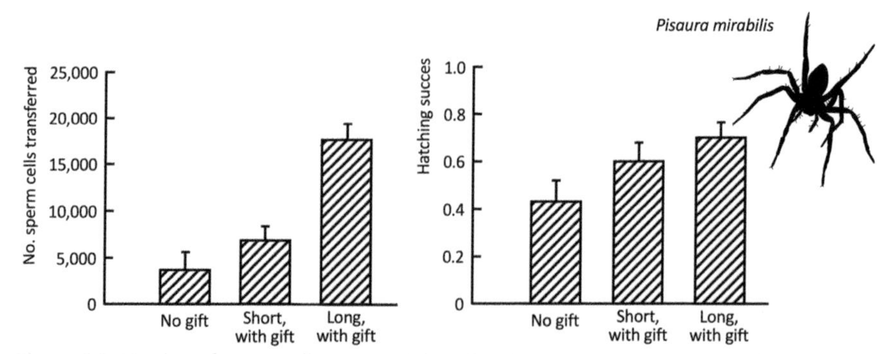

Figure 5.5. Number of sperm cells recovered from female sperm storage organs (left) and hatching success (right) of *Pisaura mirabilis* spiders involved in copulations without nuptial gift, short copulations with gift (interrupted to match the duration of non-gift spiders) and normal matings (long copulations with gift). Reproduced from Albo et al. (2013) with permission by Royal Society Publishing.

takes up sperm from his web and inseminates the female (Watson and Lighton 1994). It seems likely that the energy-demanding pre-insemination phase, in addition to the pre-courtship fights, inform the female of male quality. The females can then control fertilization through the action of valves and tubes in the pre-spermathecal duct, allowing only a small fraction of the sperm to reach the site of fertilization. Watson (1991b) speculated that this morphology might also reduce the risk of venereal infection and so there could be more than one factor contributing to the evolution of female choice. Anyway, by enforcing first mate sperm precedence, and by relying on the combat between males, females are able to select the larger and more vigorous fighter, enhancing the fitness of their offspring.

The unique and unusual features of spider sexual biology, involving a great diversity of morphological, behavioural and sensory features, make them very interesting models for the study of female choice and sexual selection (Eberhard 2004).

A similar case of condition-dependent mate choice is demonstrated by the predatory mite, *Hypoaspis aculeifer* (Mesostigmata, Laelapidae). These mites are typical polyphagous predators; they attack a great variety of prey species, including springtails, fly larvae, enchytraeid worms, nematodes and astigmatid mites. Because of its voraciousness and polyphagy, *Hypoaspis* is greatly valued as a biocontrol agent. The mites have a haploid-diploid sex determination system, in which males are produced by arrhenotokous parthenogenesis (see Table 4.1). This makes genetic analysis easy, since inbred lines can be made through mother-son matings.

The variability of prey choice in *Hypoaspis* appears to have a genetic basis: mite populations can be selected to feed preferentially on one or another prey type. Lesna and Sabelis (1999) developed isofemale lines that after four generations preferred to feed exclusively on bulb mite, *Rhizoglyphus robini*, a pest of lily bulbs, while another line preferred copra mite, *Tyrophagus putrescentiae*, a pest of stored products (both Astigmata, Acaridae).

The question is, how can such polymorphisms of prey preference be maintained in a population without leading to ecotype differentiation? The answer is that hybrids between the two lines perform better than either parental line. In a series of choice experiments Lesna and Sabelis (1999) showed that hybrid vigour is promoted by mate choice: copra-mite preferring females mated preferentially with bulb-mite preferring males. For bulb-mite preferring females the situation was slightly more complicated: they chose males of their own line when fed either bulb mites or copra mites alone, but chose mates of the copra-mite preferring line when fed a mixture of the two prey mites.

These data show that mate choice in *Hypoaspis aculeifer* varies with the diet and can be understood as a mechanism that promotes fitness of the offspring. Under dietary conditions in which hybrids are superior, mites choose to mate with a partner with opposite prey preference. How female mites may sense the prey preference of their male partner is not known. They cannot select partners for their heterozygosity since the males of *Hypoaspis aculeifer* are haploid. Most likely, the smell of prey in some way or another stays with the predator and acts as an olfactory cue determining female choice.

5.4 Selective spermatophore handling in springtails

Springtails form a species-rich group and occupy many ecological niches, and so their reproductive biology is equally diverse. Like in pseudoscorpions, three of Proctor's four categories of indirect sperm transfer are present in Collembola: "completely dissociated transfer" in the majority of species, "incompletely dissociated transfer" in some hemiedaphic species and "pairing with indirect transfer" in several others (Figure 5.6).

The distinction between these categories is nicely correlated with the habitat choice of the species. Partly associative behaviours with physical contact between the sexes are seen only in species living on the soil surface, on the vegetation or on tree bark (Types III, IV and V in Figure 5.6). The fact that this equally holds for species living on the water surface (*Podura aquatica, Sminthurides aquaticus*), demonstrates that it is not air humidity that drives these transitions. Rather, it is the openness of the environment and its energy intensity *sensu* Kennedy (1928) that apparently favour an associative mode of reproduction. In addition, the density argument of Stam and Hoogendoorn (1999), discussed above, might apply.

Interestingly, associative sperm transfer is correlated with *sexual dimorphism.* Although Collembola generally are not dimorphic, some degree of sexual dimorphism

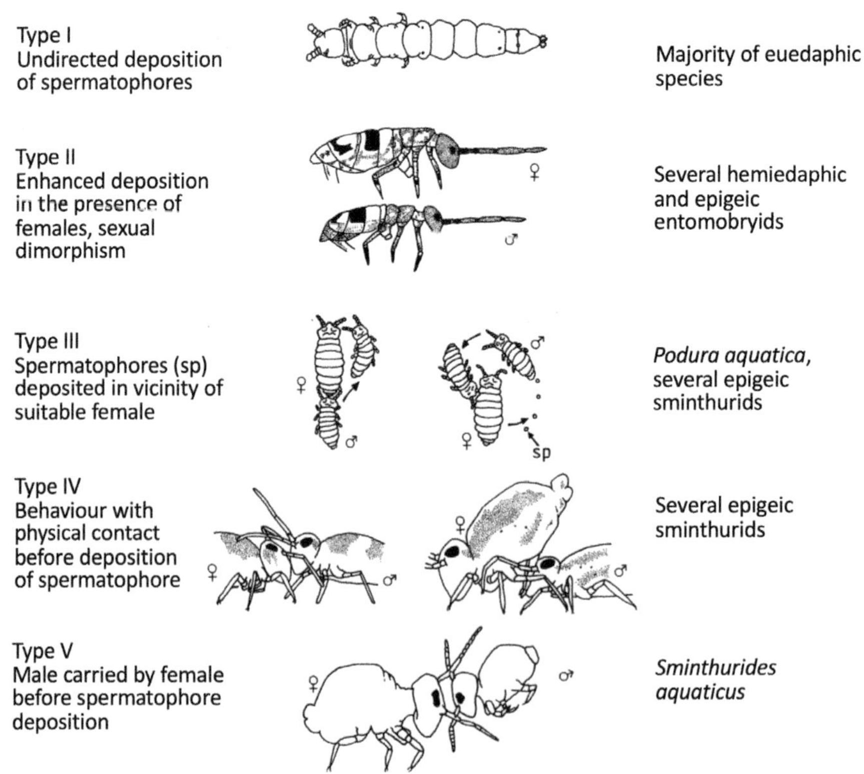

Figure 5.6. Illustrating the spectrum of dissociative and associative reproductive behaviours in Collembola. Redrawn with modifications, from various sources, including Lindenmann (1950), Gisin (1960), Bretfeld (1970) and Schaller (1971).

does occur in species living on the soil surface and in the vegetation. Males are usually smaller than females in these cases, and the pigmentation patterns may differ. For example, pigmentation of males in the genus *Orchesella* differs from females in that it shows sharper contrasts and sometimes a quite different patterning, e.g., in *O. flavescens*. In male *O. cincta*, bands on the abdomen and the second antennal segment are a shining yellow especially during the reproductive phase. This could indicate that body colouration serves a purpose in communication between the sexes.

Sexual dimorphism is strong in *Sminthurides aquaticus*, where the male is much smaller than the female and has the second and third antennal segment modified into a "*Klammerorgan*" (clasping organ). In this species the female carries the male for a considerable time before it produces a spermatophore. "Riding couples" of *Sminthurides* are often seen on the surface of aquaria and water tanks.

Research on the model species *Orchesella cincta* has shown that the male deposits spermatophores preferentially on spots that have been conditioned by conspecifics (Stam et al. 2002), including other males. In addition, video recordings revealed that males regularly return to their spermatophores to inspect them, "sniffing" them and sometimes eating them (Figure 5.7). Males guard their spermatophores like a garden, and "weed out" spermatophores from other males. Their behaviour is obviously shaped by male-male competition. When offered a choice between a spermatophore of their own making or one produced by another male, in almost all cases not their own but the strange spermatophore is eaten.

These observations show that the deposition of spermatophores is not a random process. Male *O. cincta* are able to recognize their own spermatophores, most likely using chemical cues sensed by the chemosensory organ in antenna III. All Collembola possess a sensory complex on the third antennal segment (Hopkin 1997). That the perception of spermatophore odour is located in the antenna is confirmed by the behaviour of the animal, in which it waves the proximal part of the antenna towards the spermatophore when "sniffing" (Figure 5.7).

Not only the male, also the female *O. cincta* is able to perceive individual spermatophores. Directly after moulting to a reproductive instar the female springtail shows "inquisitive" behaviour. It walks around, does not eat, is able to smell a

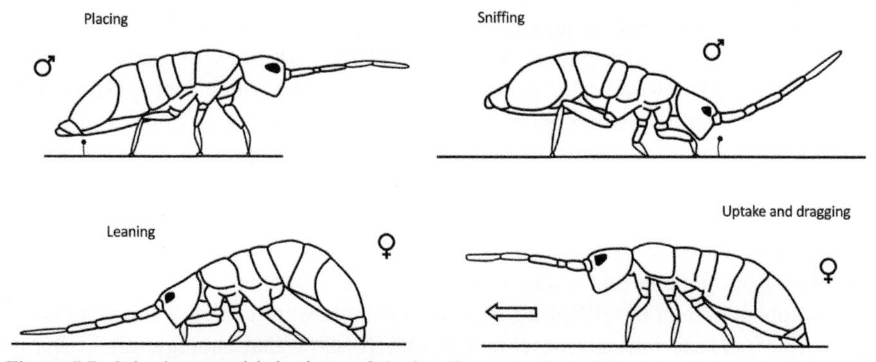

Figure 5.7. Selective sexual behaviours of *Orchesella cincta*. Top: Male depositing a spermatophore, "sniffing" it with its antennae and sometimes eating it. Below: female taking up a spermatophore followed by "dragging" behaviour and "leaning". Redrawn after Gols et al. (2004) and video recordings reported in Stam et al. (2002).

spermatophore from a distance, walks to it and moves the abdomen towards it to take it up (Figure 5.7). The female then shows a characteristic "dragging" behaviour, in which the tip of the abdomen is hauled across the substrate while the female moves forward (Gols et al. 2004). This is sometimes followed by a "leaning" posture (Figure 5.7). Within 1 h from sperm uptake the eggs are laid (varying from 10 to 100), usually in a single bout, but this may take several hours at 20°C. At lower temperatures eggs of one clutch may be laid over more than one day. The female needs only a single spermatophore to fertilize all the eggs of a clutch. Even if more spermatophores are offered, only one is taken up. There is no carry-over of sperm and so in a next receptive phase (two moults later) a new spermatophore must be found.

Given these biological constraints, sperm competition and cryptic female choice are unlikely to play a role in Collembola. Instead, the only possibility for a female to gain control over the paternity of her eggs is by selective spermatophore uptake. Such preferences do indeed exist, as shown by Hedlund et al. (1990) and Gols et al. (2004). In choice experiments followed by paternity analysis using microsatellite genotyping, Gols et al. (2004) were able to show that spermatophores from specific males were much more attractive, to more than one female, than others. The preference was not determined by the relatedness between male and female and so outbreeding avoidance does not seem to play a role.

These choice experiments strongly suggest that mate choice in *O. cincta* is mediated by a chemical signal emanating from the spermatophore, acting on both males and females. The Italian ecologist Valentina Zizzari and the Dutch entomologist Astrid Groot elucidated the nature of this chemical as *z-(14)-tricosenol*, $C_{23}H_{46}O$, a linear carbon compound with a double bond between the 14th and 15th carbon atom, belonging to the fatty alcohols (Zizzari et al. 2017). Most likely, the amount of z-(14)-tricosenol put into a spermatophore by the male is variable and is sensed by the female as an indicator of quality. However, there might be other compounds in lower quantities that contribute to the "bouquet" of a spermatophore. Up to now this is the only case in which a spermatophore-associated pheromone was identified to influence mate choice.

Zizzari et al. (2013) also showed that the attractiveness of a spermatophore is influenced by male-male competition. Males exposed to the smell of other males deposited fewer spermatophores than they did in a previous reproductive instar (Figure 5.8, left). However, these spermatophores were much more attractive to females compared to spermatophores produced in the absence of competitor males (Figure 5.8, right). Obviously, when spermatophores are present in high density, females may have difficulty choosing one; under such conditions, any investment that increases the olfactory attractiveness of a spermatophore will be greatly beneficial. In addition, females gain indirect benefits from their choosiness, as the male offspring from females that were allowed to choose a spermatophore produced more spermatophores than the offspring from females that were denied a choice (Zizzari et al. 2009). These intricate observations clearly show that strategic allocation of resources in response to male-male competition and female choice are important aspects of reproductive behaviour, even in species with dissociated sperm transfer.

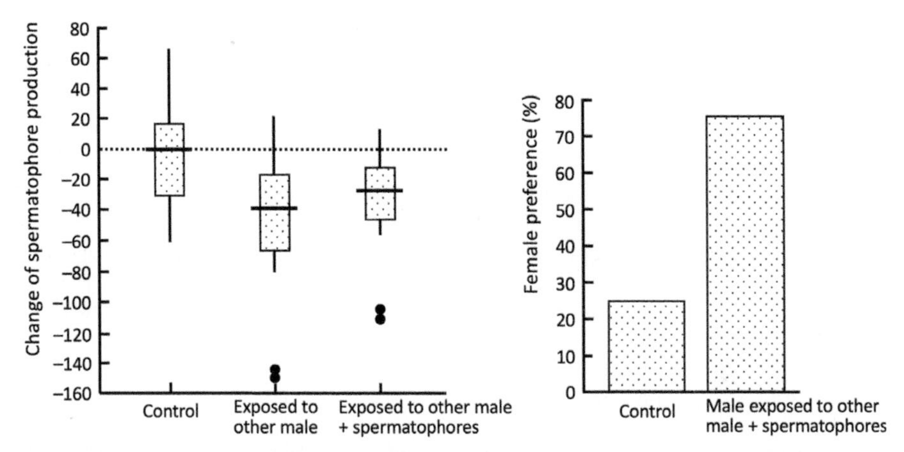

Figure 5.8. Left: Male springtails, *Orchesella cincta*, produce fewer spermatophores (relative to the previous reproductive instar) when exposed to other males with or without their spermatophores. The graph shows the median as a horizontal line, the interquartile range as a box and the full range as a vertical line; outlier observations are indicated by dots. Right: females prefer spermatophores from males that were exposed to the smell of other males and their spermatophores. From Zizzari et al. (2013), reproduced with permission from Royal Society Publishing.

Another model system for behaviours associated with indirect sperm transfer is found in various symphypleone Collembola, notably species of the families Bourletiellidae (*Bourletiella*, *Deuterosminthurus*, *Heterosminthurus*) and Sminthuridae (*Sminthurus*, *Allacma*, *Sphaeridia*). Many authors have described the fascinating and complex rituals of the courtship behaviour of these tiny animals (Bretfeld 1970, 1971, Blancquaert and Mertens 1977, Betsch-Pinot 1977, Kozlowski and Aoxiang 2006).

In *Deuterosminthurus bicinctus*, the sequence of behavioural components runs as follows (Kozlowksi and Aoxiang 2006) (see also Figure 5.6):

- Excited rushes of the male towards the front of the female
- Rubbing of the head, antennation
- “Riding” to and fro while clasped
- Quick 180° turn of the male, deposition of a spermatophore, turn back
- Male drags female backwards, “prancing” of the female, uptake of spermatophore
- Disconnection and avid consumption of the remains of the spermatophore by both male and female.

Similar behaviours are seen in other symphypleone Collembola, although the details vary per species. A common element to all is that a male will deposit a spermatophore only after a number of preparatory rituals and that he actively forces the female to take up the spermatophore very quickly after it has been produced. In one case the spermatophore is not placed on the substrate but directly on the female genital pore (*Sphaeridia pumilis*, Blancquaert and Mertens 1977). The various courtship behaviours all seem to serve two purposes: first, to make sure that the female is prepared to take up the spermatophore when produced, second to prevent that the spermatophore is eaten.

The courtship behaviour is relatively simple in *Sminthurus viridis* and *Allacma gallica*, and more elaborate in *Deuterosminthurus* and *Bourlietella* species. Since all these species are vegetation dwellers, an adaptive trend related to the ecology of the species can hardly be recognized. The main lesson seems to be that associative behaviour evolved in response to off-ground open environments, but the details of the courtship are subject to evolutionary specialization unrelated to their ecology.

5.5 Courtship in bristletails, silverfish and crickets

Within Archaeognatha (bristletails), only a few species have been studied in detail for their reproductive behaviour, but these studies have revealed an interesting grading within the group. Three different modes of sperm transfer have been described (Sturm 1978, Sturm 1992, Goldbach 2000):

- Spermatophores attached to a *support thread* in most Machilidae such as *Machilis germanica, Lepismachilis y-signata, Dilta insulicola, Pedetontus unimaculatus* and *Promesomachilis hispanica*. The support thread is excreted by lobe-shaped glands of the male genital apparatus called *paramera*.
- Production of *stalked spermatophores*: most members of the family Meinertellidae (*Machilinus rupestris, Machiloides tenuicornis* and *Neomachilellus scandens*).
- *Direct transfer* of the spermatophore to the ovipositor of the female (*Petrobius maritimus*).

In Figure 5.9 this variation is mapped on the phylogeny of Hexapoda. The ancestral condition for all hexapods is assumed to consist of a sac-like spermatophore with a rigid sheath that remains outside the female's genital opening (only sperm enters). In Protura and Diplura the spermatophore stayed close to the ground

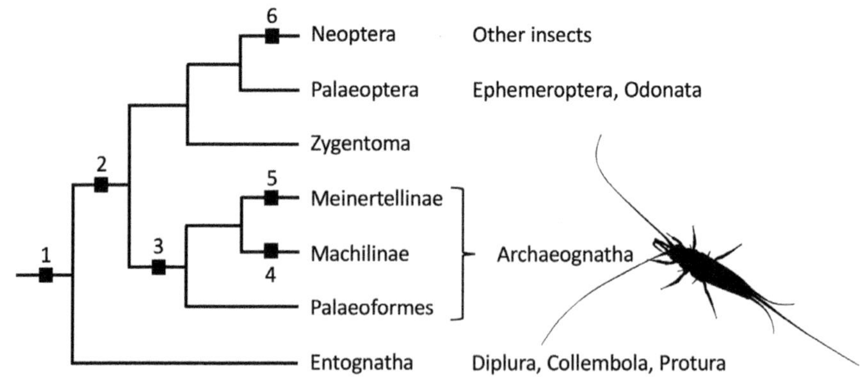

Figure 5.9. Evolutionary changes in the reproductive biology of Archaeognatha and related arthropods. 1. sac-like spermatophore, not taken up entirely, rigid sheath of spermatophore, 2. carrier threads produced by paramera, mating behaviour starts with head-to-head position, 3. viscous sheath of spermatophore, complete uptake, 4. drumming with maxillary palps, U-shaped bending of male, carrier thread stretched by male, 5. deposition of stalked spermatophores, no thread, piggyback position during mating, 6. spermatophore sheath secreted by accessory glands. Redrawn after Goldbach (2000). Note that in contrast to this rendering, Entognatha may not be monophyletic (*cf.* Figure 2.5). The silhouette shows *Petrobius maritimus* (Machilidae).

but in Collembola it was raised on a stalk while the capsule became smaller. In Archaeognatha, the sperm package was attached to a carrier thread produced by the paramera, while the spermatophore capsule became more viscous and is taken up entirely (characters states 2 and 3). In the subfamily Machilinae this is combined with a characteristic drumming of the maxillary palps on the substrate (4). A transition to stalked spermatophores and a piggyback position during mating is seen in the tropical Meinertellidae (5,6); this is quite comparable to the transition in the Collembola lineage. In the species *Petrobius maritimus*, however, male and female stand side-by-side during courtship, and only the abdominal tips are in contact. Why this species is an exception within the Machilidae remains unclear; maybe its habitat (crevices in shoreline rocks, including man-made dikes) makes the spinning of threads impossible.

The fact that the sexual behaviour of Archaeognatha matches their phylogeny quite well, suggests that reproductive behaviour could have been a driving factor for speciation in this group. We saw the same phenomenon in pseudoscorpions (Section 5.2) and spiders (Section 5.3): species with very similar morphologies are reproductively separated due to their differential sexual behaviour. Reproductive behaviours within these groups might be shaped more by interspecies competition and genetic drift than by adaptation to the habitat.

In Zygentoma (silverfish), the apterygote group of hexapods directly outside Archaeognatha (*cf.* Figure 2.5), reproductive behaviours are seen that are superficially similar, especially the use of threads and deposition of the spermatophore on a web. However, detailed comparisons have shown that the behaviours of three species of silverfish (*Thermobia domestica*, *Lepisma saccharina* and *Tricholepidion gertschi*) are more similar to each other than they are to Archaeognatha (Figure 5.10). A characteristic aspect of Zygentoma courtship behaviours is the "*by-passing*" element, in which male and female run past each other. The degree to which this behaviour is ritualized varies within Zygentoma according to the species, but it is never seen in bristletails. The absence of strong homologies between the reproductive behaviours of Zygentoma and Archaeognatha, formerly united in "Thysanura", confirms the phylogenetic separation of these two apterygote groups (*cf.* Figure 2.5).

As Archaeognatha and Zygentoma take an evolutionary position in between Collembola and pterygote insects, it is tempting to consider the reproductive behaviours of the apterygote groups as a transition from dissociative sperm transfer (in so many soil-bound invertebrates) towards the copulatory behaviours of the higher insects (Goldbach 2000). To explore this link, we discuss here the sexual behaviour of crickets, terrestrial insects at the base of the insect phylogeny (*cf.* Figure 2.9). The courtship, mating and guarding behaviours of the model species *Gryllus bimaculatus* have been described in great detail and provide an interesting comparison with apterygotes (Sakai et al. 1991, Adamo and Hoy 1994, Sakai et al. 2017, Figure 5.11).

Tactile contact between the sexes using the antennae plays an important role in the initiation of courtship. The male recognizes another cricket as a mature female from antennal contact; visual stimuli don't play a role. These contacts stimulate the preparation of a spermatophore by the male, while the couple remains in the

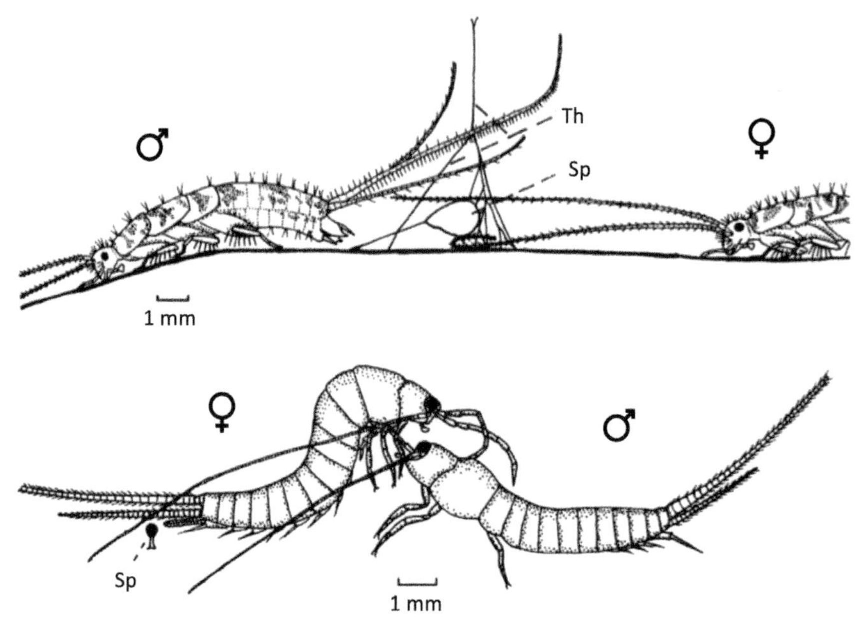

Figure 5.10. Mating behaviours associated with indirect sperm transfer in Zygentoma (*Thermobia domestica*, above) and Archaeognatha (*Neomachilellus scandens*, below). In *T. domestica*, the male has just spun a web and deposited his spermatophore. The female remains motionless until the male turns and crawls by her. In *N. scandens* the male has placed a spermatophore on the substrate and pushes his anterior under the female, to drag her abdomen onto the spermatophore. Reprinted from Sturm (1997) with permission from Elsevier.

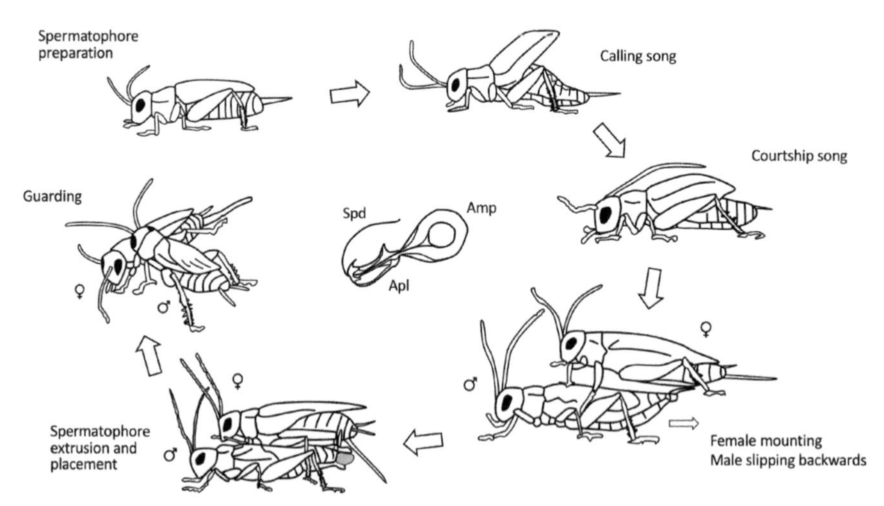

Figure 5.11. Sequence of courtship, mating and guarding behaviours shown by field crickets, *Gryllus bimaculatus*. In the centre a spermatophore is shown. Amp ampulla, Apl Attachment plate, Spd Sperm duct. During copulation the attachment plate with the sperm duct is pushed into the female genital opening. Redrawn from Sakai et al. (1991, 2017).

vicinity of each other. The spermatophore is a complex structure which has a sperm containing ampulla and a barbed attachment plate with sperm duct (Figure 5.11).

When the spermatophore is prepared, the male starts calling and singing courtship songs, repeatedly turns around the female, presents his abdomen to her, then lines up and allows her to mount him. The male then slips backwards under the female with his abdominal cerci raised as a guide. A set of small hooks (*epiphallus*) on the male genitalia attach to the female's subgenital plate. When the genitalia are connected correctly, the male extrudes the spermatophore, which is pushed up onto the genital pore of the female. The attachment plate and the sperm duct are fixed in the female opening. After demounting, the male guards the female for some time (Figure 5.11). Sperm is received in a spermatheca and the female buries the fertilized eggs in the substrate (soil or plant material) using the ovipositor.

Females tend to mate multiple times, with the same or different males. Each mating results in attachment of a spermatophore to the female genital pore, which however may be removed by the female after the male stops guarding. The multiple matings, plus the possibility to ignore courtship and remove spermatophores, provide ample opportunity for the female to gain control over the paternity of her offspring.

In a series of experiments, Simmons and co-workers (Simmons 1986, 1989, 1991, Tuni et al. 2013) have shown that female field crickets are indeed choosy. They actively evade unwanted males and if mated with them try to escape from guarding as soon as possible. The fertilization success of unpreferred matings is low because females are able to close the opening of the spermatheca, blocking the access of sperm (Tuni et al. 2013), a case of *cryptic female choice* similar to what we have seen in linyphiid spiders (Section 5.3). In addition, females tend to stay with preferred males longer and by mating multiple times allow a large amount of sperm to accumulate in their spermathecae, thereby diluting the sperm stored from previous matings. Female preference seems to be driven by characteristics of the male chirp, its body-size and its genetic relationship to the female.

That the genetic relationship comes into play was demonstrated by Simmons (1991) in experiments in which females were mated with males that were either full sib, three-quarter sib, half-sib, cousin or unrelated to the female. The duration of spermatophore attachment was short for matings with related males (Figure 5.12), and the fertilization success varied accordingly. However, the effect was most obvious when males and females courted in burrows (which they naturally do) compared to courtship in an open arena. This may be explained by the perceived exposure to predation, which necessitates shorter courting times and diminishes the capacity of the female to accurately assess the quality of a mate.

The ability to recognize kin is most likely due to male chemical cues, e.g., the mix of partly volatile cuticular hydrocarbons of which the composition is plastically adjusted to the social environment (Thomas et al. 2011). The female senses the male chemical signal during guarding and so it is the relationship to the guard, which determines female choosiness, rather than the relationship to the actual mate (Tuni et al. 2013). Avoidance of kin will reduce inbreeding depression in the offspring and so will be under positive selection. However, the population genetic consequences of female choice in field crickets have not been investigated in detail.

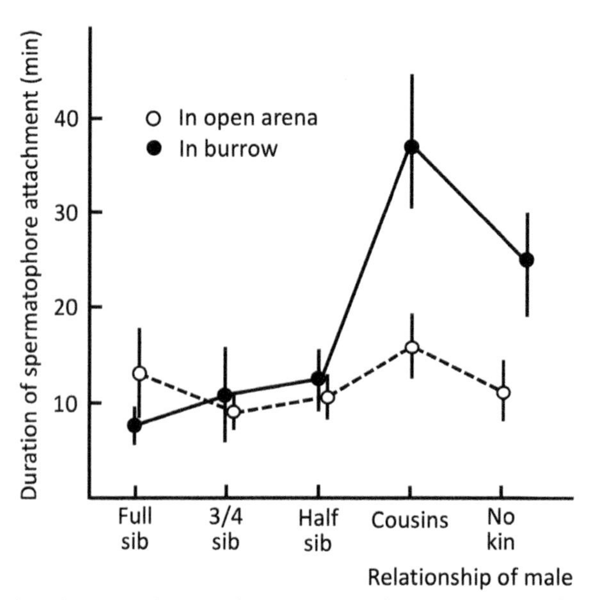

Figure 5.12. Duration of spermatophore attachment (means and standard errors) to female field crickets (*Gryllus bimaculatus*) when mated with males of different degrees of relatedness, inside burrows or in an open arena. From Simmons (1991), reproduced with permission from Elsevier.

The placement of spermatophores on the female genital pore, in crickets and some vegetation-living Collembola and bristletails, can be seen as an intermediate stage between fully dissociated sperm transfer and the copulation behaviours with direct sperm transfer shown by most higher insects. In this sense, the position of soil invertebrate reproductive biology conforms to the GKGB hypothesis outlined in Chapter 1. However, on top of this general trend, we see many specialized behaviours in every evolutionary lineage. This may be expected due to the high rate of evolution of reproductive behaviours in general. Sexual behaviour diversified enormously in each lineage and contributed to speciation.

5.6 Sexual selection in dung beetles

Sexual selection is the process by which the two sexes evolve in response to each other due to challenges associated with partner choice. It is such a widespread and important phenomenon in nature that Charles Darwin devoted most of his book "*On the Descent of Man*" not on human evolution but on sexual selection (Darwin 1871).

One of the most famous examples of sexual selection in soil invertebrates is the case of exaggerated horn growth in dung beetles. Horn-like projections on the head or thorax occur in many species of beetles (staphylinids, cerambycids, curculionids, tenebrionids, chrysomelids), but they are particularly common and often exuberant in the superfamily of Scarabaeoidea. For example, large horns or stags are seen in Lucanidae (stag beetles), Geotrupidae (earth-boring dung beetles), Scarabaeidae-Dynastinae (rhinoceros beetles) and Scarabaeidae-Coprinae (dung beetles). Throughout these groups, the presence of horns is a typical sexually dimorphic character: in almost all cases only the males have them.

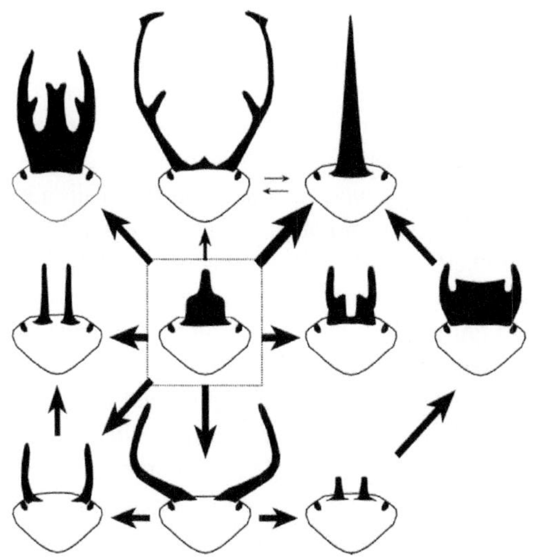

Figure 5.13. Evolutionary changes of horn morphology in the dung beetle genus *Onthophagus*. Arrows indicate pathways of evolution and arrow thickness reflects average frequency of change. The assumed plesiomorphic horn type is shown in the centre. Shape evolution was reconstructed by mapping horn morphologies on a phylogenetic tree of 48 *Onthophagus* species, of which 30 have this type of horn. Only the evolution of the most common horn type is shown: a projection of the vertex. Many species also have outgrowths of clypeus, frons or pronotum which are not shown here. Reprinted from Emlen et al. (2005).

Emlen et al. (2005) reconstructed the phylogenetic history of horn development in the dung beetle genus *Onthophagus*. This is one of the largest genera of beetles, maybe even the most speciose genus of the animal kingdom, with more than 2000 species described. The radiation of Onthophagini took place around 20–30 million years ago, in conjunction with the appearance of grasslands and the radiation of mammals. The genus includes many species with prolific outgrowths, but also many hornless species, entangled in a complicated phylogenetic pattern.

Among 48 species studied in detail, Emlen et al. (2005) identified five different main types of horns. The most common one is an outgrowth of the *vertex* (the insect's dorso-caudal head plate). Within this main type a wealth of derived horns have evolved by lateral extension and splitting. The evolutionary transitions among ten shapes, reconstructed by mapping the various morphologies on a phylogenetic tree of *Onthophagus*, are depicted in Figure 5.13. Similar radiations are seen for the other four horn types: projections of the *frons*, the *clypeus*, the centre of the *pronotum* and the sides of the pronotum. These other types may develop in the absence of a vertex horn, or they may occur in addition to the main type. For example, *Onthophagus pentacanthus* has five horns: a single horn on the vertex, a horn on either side of the pronotum, as well as a split horn on the central pronotum.

Several hypotheses have been put forward to explain the sometimes-bizarre horns of these beetles. Early authors have suggested that dung beetle horns might be an indicator of male quality and their exuberant growth would be promoted by female choice. Others have proposed that the horns might serve to protect against predators or act as a digging tool, and still others have suggested that they should be considered a neutral by-product of a large body size. However, it is now generally accepted that the horns of dung beetles are weapons of male competition. Contrary to earlier suggestions, they are not adornments sexually selected by female choice.

Horn growth in dung beetles can be understood perfectly from the beetles' behavioural ecology. There are three behavioural types, one that digs a vertical tunnel close to a fresh dung patch, down to a meter deep and places the dung at the bottom for the female to lay eggs in ("*tunnellers*"). Another type rolls a ball of dung from the dung patch to sequester it away from other insects ("*rollers*"), while a third type digs a burrow in the dung itself ("*dwellers*"). Phylogenetic analysis of these behaviours, in conjunction with the presence of horns, showed a clear association between the two traits (Emlen and Philips 2006). It turned out that the tunnelling behaviour must be considered the ancestral condition, while ball-rolling and dwelling have evolved eight times in 46 species (Figure 5.14). The phylogeny shows a clear association between tunnelling and horn evolution, confirming the interpretation of horns as weapons of sexual selection driven by male-male competition.

Emlen and Philips (2006) argued that weaponry outgrowths are generally advantageous when there is fierce competition over limited resources in physically restricted situations. Other horned beetles such as dynastids and geotrupids also use their horns in contests over some form of burrow.

Interestingly, many dung beetles show a dimorphism of horns among males, in addition to sexual dimorphism. In these species, a minor fraction of the males lacks any horn and is unable to defend a tunnel. Hornless males, which are also somewhat smaller, avoid male contests and depend on "sneaky" copulations with the females. Interestingly, the hornless condition is not a genetically determined morph but a nutritional phenotype: the amount of food available during larval development determines body size as well as horn development (Moczek and Emlen 1999). In addition, the minor males have a larger testis size than the horn-bearing morphs, which may be explained by the fact that their ejaculates always face sperm competition (Simmons and Buzatto 2014). So, while intrasexual selection forces the major males to invest in armament, possibly even at the expense of their ejaculate, the minor males, in accordance with sexual selection theory, must invest in testis size and large ejaculates, due to sperm competition.

While it is relatively easy to envisage how natural selection might have promoted exaggerated horn growth in tunnelling dung beetles, the question how such horns came to be there in the first place has been unclear for a long time. However, recent molecular analysis has elucidated this question. Warren et al. (2014) first showed that insect genes with horn-biased expression show strong evidence of positive selection (estimated from the ratio of nonsynonymous to synonymous substitutions). In addition, genes that were differentially expressed between horned versus non-horned morphs of the same species showed relaxed selection. The latter is expected on the basis of the argument that such horn-promoting genes are expressed only in a part of the population and so should be decoupled from the rest of the developmental network.

Further work on the developmental genetics of horn formation in *Onthophagus taurus* and related species has provided an interesting twist to the dung beetle story (Hu et al. 2019, Nijhout 2019). It appears that horns are formed by a network of genes which are serial homologs of wing formation genes. When genes necessary

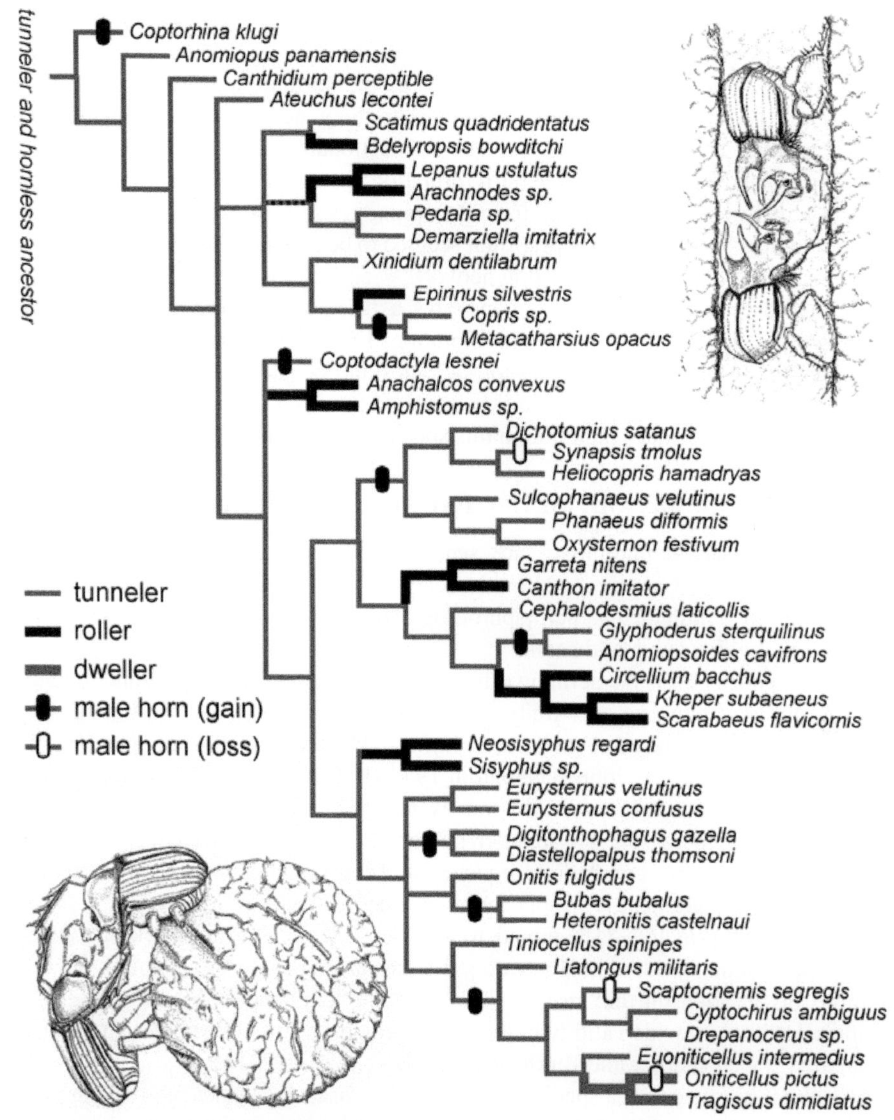

Figure 5.14. Phylogenetic association between dung sequestering behaviour ('tunnelling', 'ball rolling' or 'dwelling') and horn gain or loss in 46 species of dung beetle. Gains of horns occurred eight times but only in lineages with tunnelling behaviour. One of the three horn losses occurred on a branch of non-tunnellers. The two species shown at the bottom, *O. pictus* and *T. dimidiatus*, are scored as dwellers, the other non-tunnellers are rollers, while tunnelling is the plesiomorphic condition. Reproduced from Emlen and Philips (2006), ©2006 The Coleopterists Society.

for wing growth such as *vestigial*, *apterous* and *dishevelled* were knocked down by RNAi, not only wing formation was impeded but also horn formation (Hu et al. 2019). The medial prothoracic horn begins its development as two patches of tissue in the dorsal edge of the tergites (in the wing developmental field), that migrate to the midline and fuse to form a prothoracic horn (Hu et al. 2019, Nijhout 2019).

The discovery that homologs of the wing developmental network are responsible for horn formation on the beetle's prothorax raises questions on homology and novelty similar to what we discussed in the formation of arthropod limbs (Section 4.9). There is an obvious homology on the genetic level, but whether this can be extended to the morphological level is questionable. The deployment of genes like *nubbin* and *apterous* in beetle horn formation is another case of co-option of the network that was co-opted earlier from the epipodites of crustacean legs. The origin of prothoracic horns, which on first sight seemed to be a novelty *pur sang*, turned out to be the result of an already existing network.

The case of dung beetles demonstrates a fascinating interplay between soil ecology, developmental biology and evolution. Extreme competition for a scarce resource (dung) and the gamut of morphologies and behaviours available for male competition combined with the limited space of a burrow have contributed to the magnificent evolutionary radiation in this species-rich lineage.

5.7 Mating behaviour of hermaphroditic earthworms

Two types of hermaphroditism are seen in soil invertebrates, sequential and simultaneous (see Chapter 4). In this section and the next, we will limit the discussion to simultaneous hermaphroditism of the two main soil invertebrate groups, earthworms and land snails.

Due to *anisogamy*, the defining difference between the sexes, males and females pursue different interests over traits such as mating frequency and re-mating with novel mates, which may lead to *sexual conflict*. This was already recognized by the British geneticist Angus John Bateman in 1948 (Bateman 1948). He argued on the basis of his work on *Drosophila*, that males, more often than females, may enhance their fitness by increasing the number of copulations (with different partners). A graph of fitness versus copulation frequency will show a line known as the *Bateman gradient* and this gradient is steeper for males than for females.

Bateman's principle is assumed to hold throughout the animal kingdom as almost all animals with sexual reproduction are anisogamic. It explains Darwin's famous observation on the "*ardent male*" and the "*coy female*" (Darwin 1871), a principle with very wide validity in the animal kingdom and often applied to human sexuality as well (Parker and Birkhead 2013, Janicke et al. 2016).

Hermaphrodites take a special position in this discussion since the two partners are involved in a complicated pattern of interactions, not only inter-sexual but also intra-sexual (Leonard 2006, Anthes 2010). The diverging interests of the male and the female functions, as determined by Bateman's principle, promote a tendency in the same individual to donate sperm but not to receive it. This may cause the courtship behaviour to run out of hand (Anthes et al. 2010). Indeed, reproductive behaviours in hermaphrodites are often extremely elaborate, costly, damaging and bizarre. Model simulations confirm that under conditions in which a moderate degree of harm to the male function is associated with an advantage in fertilization, this advantage will be greater in hermaphrodites than in gonochorists. Bateman's principle may explain why hermaphrodites, more than gonochorists, are caught in costly copulations (Michiels and Koene 2006).

Considering first the sexual behaviour of hermaphroditic earthworms, we note that this is known in detail only for species that can be observed mating at the soil surface at night, for example *Lumbricus terrestris*, or can be sampled in loose substrates such as manure, in the case of *Eisenia fetida* and *E. andrei*. However, it is assumed that endogeic species, that mate unseen in their burrows, show similar behaviours.

A mating of *L. terrestris* is preceded by a "probing" phase in which the animals explore the soil surface in the immediate surroundings of their burrows, visit a neighbouring burrow and shortly touch the other worm with their front end (Nuutinen and Butt 1997a,b). Such short visits may occur repeatedly at irregular intervals. The function of this precopulatory behaviour is still somewhat vague. Experiments by Michiels et al. (2001) have shown that probing is size-selective; it could be a mechanism by which worms acquire information about the body-size of their partner. Field observations show that there is a tendency for mating earthworms to be of similar size (Monroy et al. 2005). Another function of the "neighbour probing" behaviour might be to lure the potential partner to crawl further out of its burrow. There is an advantage for both worms to remain anchored to their home burrow at all times so as to be able to withdraw immediately in case of a predatory attack. So, the visiting behaviour may be seen as a tug-of-war over copulation ground near the burrow.

During mating, earthworms align to each other anteriorly in opposite directions; the posterior end usually remains inside the burrow. The typical copula position is an S-shape in which the ventral sides between segments 10 and 40 are in close contact, while the anterior body is bent away from the partner. The contact is most close in segments 31–38 where large epidermal glands of the clitellum produce a mucus that glues the partners together. Sperm droplets are transferred in posterior direction along the body wall, through an external groove running from the male pore in segment 15 to the clitellar region of the partner, where they are collected and taken up into two paired spermathecae (Figure 5.15).

The symmetrical nature of the copulating position suggests that sperm transfer is always bidirectional, however, this does not seem to be the case. Domínguez et al. (2003) showed that in *Eisenia fetida* about 10% of copulations with sperm transfer were unidirectional. In addition, a rare case of self-copulation was observed, the animal curling its body to bring the spermathecae into contact with its own clitellum.

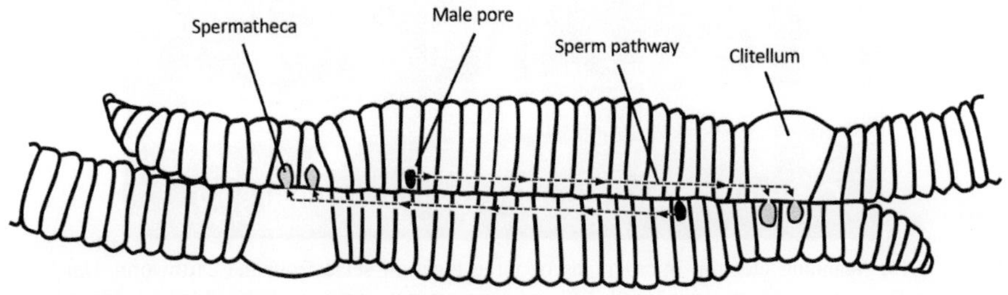

Figure 5.15. Schematic drawing of two *Lumbricus* earthworms in copula, with their anterior ends aligned in opposite directions. Sperm is transferred reciprocally through the seminal groove on the body wall, from the male pore to the spermathecae of the partner opposite the clitellum. After Domínguez and Velando (2013).

In single-worm cultures of *Eisenia fetida* some worms were able to produce cocoons, apparently by self-fertilization.

A quite bizarre aspect of earthworm copulation is the practice of body piercing. *Lumbricus terrestris* carries a double set of *copulatory setae* on the ventral side of segments 10, 26 and 31–38. These setae are nearly twice as long as the normal setae, which are used during crawling (Figure 5.16). The copulatory setae can be extended and retracted by the action of muscles in the body wall and do not show the common wear and tear of the other setae (Koene et al. 2002). The setae are driven into the body wall of the partner during copulation, deep enough to reach the blood vessels.

While early authors have suggested that copulatory setae serve to fix the partner in an alignment optimal for sperm transfer, Koene et al. (2005) discovered that their main function is to inject a biologically active substance into the partner. Copulatory setae carry four grooves along their length through which a mucus-like substance may be transported originating from the glands at the base of the setae. This substance was shown to promote the uptake of sperm and its distribution over the four receiving spermathecae. Proteomics analysis of the setal gland secretion has suggested a possible role of *ubiquitin*, a small regulatory protein, known for its function in tagging larger proteins for degradation or cellular location (König et al. 2006).

Many animals transfer biologically active substances during copulation, to enhance fertilization success or to decrease the partner's willingness to mate again. These compounds are jointly called *accessory gland products* (*Acps*). In most animals they are added to the semen itself. The body piercing of earthworms represents an alternative delivery, offering the advantage of direct action, and avoiding possible counter-measures by the partner (Zizzari et al. 2014). The transfer of bioactive substances during copulation, separate from the sperm, is also seen in land snails, as we will discuss below.

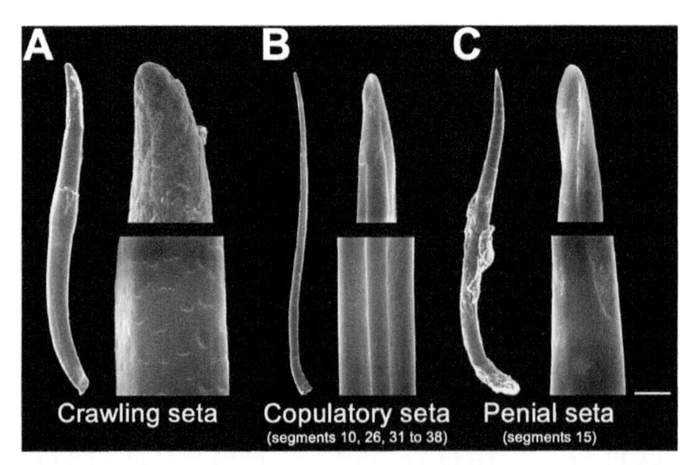

Figure 5.16. Scanning electron micrographs of three types of setae from the earthworm *Lumbricus terrestris*. Crawling setae (average length 1.06 mm) are present on the ventral and lateral sides of most segments. Copulatory setae (average length 1.81 mm) are present on segments 10, 26 and the clitellum; they have a sharp tip and four longitudinal grooves. Penial setae are present on segment 15, and have a typical spoon-shaped ending. Reprinted from Koene et al. (2002), with permission from Taylor & Francis.

In addition to harmful copulations, hermaphrodites are also known for their promiscuous mating; earthworms are no exception. Novo et al. (2010) genotyped worms plus spermathecal sperm in a natural population of the endogeic species *Hormogaster elisae* and showed that worms carry sperm from two partners on the average. In any system of promiscuous mating, sperm competition is expected to be common. Indeed, controlled mating experiments by Velando et al. (2008) showed that worms increase their ejaculate if they are the second sperm donor (Figure 5.17). How earthworms are able to assess the mating status of their partner, by chemical or tactile perception, is not known, however, the fact that they can, and act accordingly, shows that even the lowly worm obeys the laws of sexual selection.

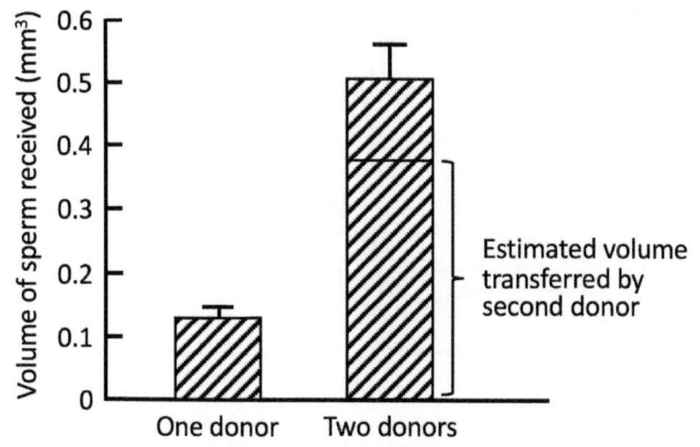

Figure 5.17. Sperm volumes recovered from the spermathecae of *Eisenia andrei* are much larger after a second mating, compared to first matings, especially at larger body sizes. Redrawn, modified, from Velando et al. (2008).

5.8 Bizarre reproductive behaviours in land snails

Sexual conflict seems to have reached a climax in land snails. Every zoologist who has dissected a snail knows how the interior of the animal is packed with reproductive organs. Figure 5.18 shows the general organization of the reproductive anatomy of a stylommatophoran gastropod. The *ovotestis* has a short hermaphroditic duct (transporting both sperm and oocytes). This duct reaches a junction called "*carrefour*", where the male and female tracts split. The sperm is directed to the *vas deferens*, of which the proximal end is fixed to the oviduct. The distal end hangs loose in the haemolymph to allow the penis to be everted during a copulation. The penis is expelled inside out, up to the point where the *penis retractor muscles* are attached. Sperm, contained in a spermatophore, enters the female tract of the partner and is received in the *bursa copulatrix*, the sperm digesting organ. In some species the spermatophore is taken up by a *diverticulum* of the bursal tract. Most of the sperm is digested in the bursa, but some of it manages to travel a long way through the female tract up to the carrefour, where it is stored in one or more *spermathecae*. The carrefour also contains a "*fertilization pouch*", where oocytes from the ovotestis are fertilized. The eggs are packed in albumen from the albumen gland, then pass

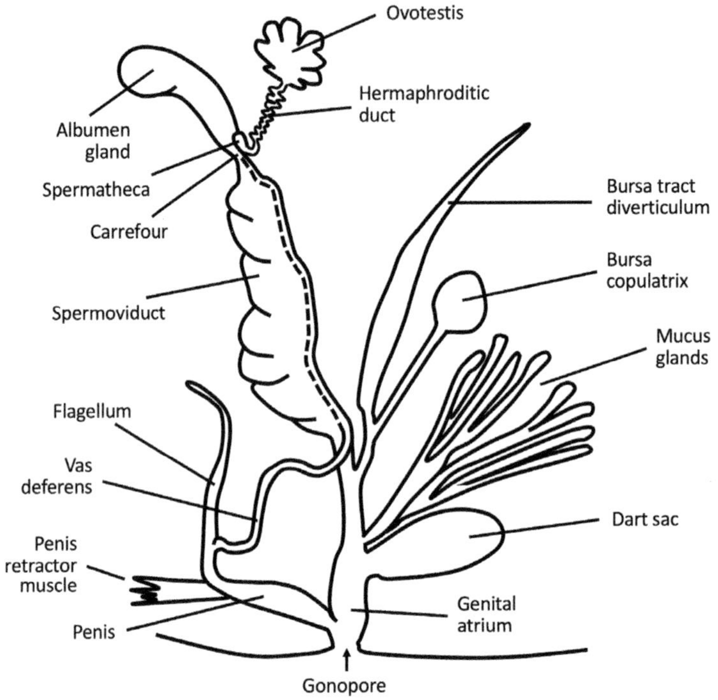

Figure 5.18. Schematic drawing of the organization of internal reproductive organs of a typical stylommatophoran snail, such as *Cornu aspersum*. Diverticulum, mucus glands, dart sac and flagellum are however, not present in all species. After Beese et al. (2008).

through the oviduct and exit through the same opening (*gonopore*) where the sperm entered.

Many land snails show reciprocal sperm transfer during copulation, that is, the two partners perform the male and female functions at the same time. Each partner has its penis inserted in the other partner's gonopore. Garden snails (*Cornu aspersum*) are commonly observed in this position; the copula may last several hours. Other snails mate unilaterally, one performing the male function, the other acting as female, and the roles can be reversed later. In the pond snail *Lymnaea stagnalis* it has been shown that sensory information from the prostate gland (most likely from stretch receptors) modulates the tendency to act in the male role (De Boer et al. 1997). This will ensure that the partner with the longest sexual deprivation time (with the fullest prostate) is most motivated to act as male. Whether a similar mechanism is present in land snails is not known.

Before penis intromission or during copulation, land snails tend to stab their partner with sharp calcareous spines called "*love darts*". These are produced by a specialized gland ("*dart sac*"), which opens in the genital atrium close to the gonopore (Figure 5.18). The dart pierces the partner's skin, up to the haemolymph. The stabbing is often seen to startle the other snail and it may cause real damage to the skin (Lodi and Koene 2016a). In *Cornu aspersum* the dart, when shot, is left in the partner (Figure 5.19). Making a new one takes around 6 days. In other species the same dart is used repeatedly to stab. Some species have more than one dart, up

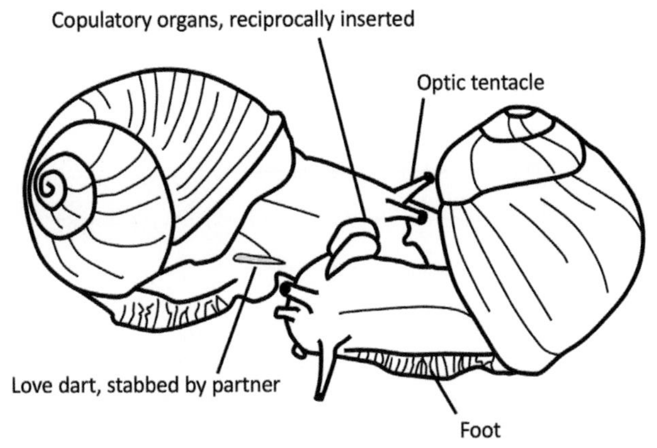

Figure 5.19. Brown garden snail (*Cornu aspersum*) pair mating, showing reciprocal penis insertion and love dart shooting.

to eight in the genus *Humboldtiana*, contained in four dart sacs that are arranged in a ring around the genital atrium (Lodi and Koene 2016a). The use of love darts is not monophyletic, it has arisen several times in the evolution of land snails (Koene and Schulenburg 2005).

The shape of the dart varies from one species to another; it may be just smooth or have four blades and it may be straight, curved or twisted. Phylogenetic analysis shows that the morphological complexity of the love dart is correlated with the complexity of the reproductive organs. This suggests that love darts and reproductive organs of land snails are subject to a sexual arms race, i.e., co-evolution (Koene and Schulenburg 2005).

Earlier naturalists have often assumed that love darts aim to increase sexual arousal in the partner (hence their name), however, this romantic conception has not stood up to scrutiny. The love dart is more like a dagger than a dart (it does not fly through the air), so the name dart is awkward anyway. Research published since 1990 has shown that the love dart is a means to transfer bioactive substances (accessory gland products, Acps) to the partner, very much like the copulatory setae of earthworms (Adamo and Chase 1990, Koene and Chase 1998, Kimura et al. 2013, Koene et al. 2013). When the dart is shot from the sac, it is covered by mucus from the *mucus glands*, also called *digitiform glands* (Figure 5.18). In *Cornu aspersum*, about one-third of the weight of a dart is due to mucus.

Active substances in the mucus cause contractions in the bursa copulatrix complex such that the bursa itself is made less accessible and the diverticulum opens wider. The net effect is that the spermatophore is taken up more easily and less of the sperm is digested. In this way a sperm donor may double its paternity in the fertilized eggs (Chase and Blanchard 2006). Because the Acps are delivered separate from the sperm, it is more difficult for the sperm-receiving partner to prevent allosperm reaching the fertilization site. The love-dart is an instrument of sexual manipulation rather than sexual arousal.

The activity of love dart mucus is widespread among land snails. Lodi and Koene (2016b) showed that mucus from *Cornu aspersum* also acted upon other species of

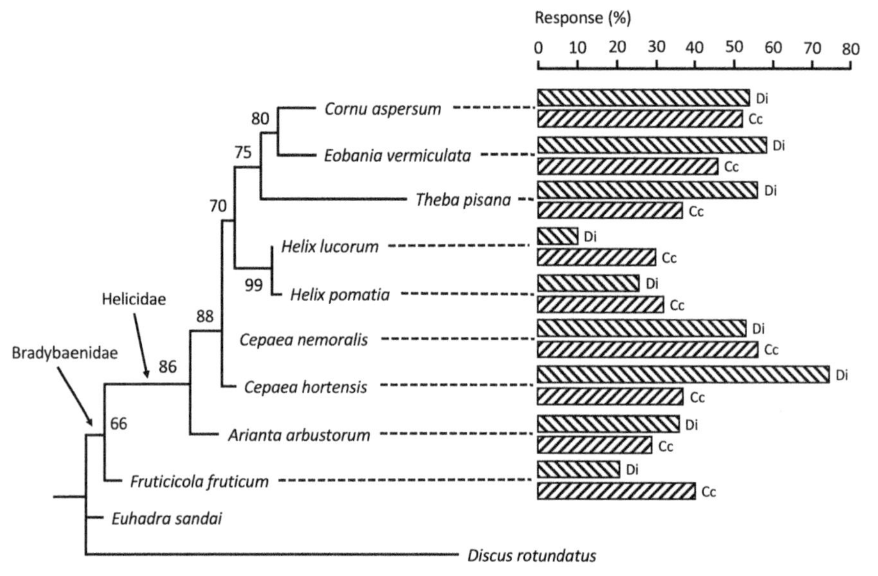

Figure 5.20. Physiological responsiveness to love dart mucus related to phylogeny of snails in the Helicoidea. Mucus from *Cornu aspersum* was applied to fresh bursa copulatrix preparations from various species. Snails were scored as responsive if there were contractions of the diverticulum (Di) or the bursa tube entry (copulatory canal, Cc). The maximum likelihood phylogenetic tree is based on a 726 bp sequence of the 28S rRNA gene. Based on Lodi and Koene (2016b).

Helicidae and even on representatives of the related helicoid family, Bradybaenidae (Figure 5.20). In physiological tests for cross-reactivity, several species responded equally strong as *C. aspersum* itself (*Eobania vermiculata, Theba pisana, Cepaea nemoralis*), but some were less responsive (*Helix lucorum, H. pomatia, Arianta arbustorum*), while others responded even stronger than the focal species (*Cepaea hortensis* diverticulum). However, not all species responding to the mucus of *C. aspersum* use love darts to the same extent. For example, in *Arianta arbustorum* dart-shooting is facultative and sperm storage is not affected by the dart. This suggests that the mucus might have more than one effect; it not only increases paternity, but also shortens the duration of courtship (Adamo and Chase 1990) and suppresses the preparedness to engage in subsequent matings (Kimura et al. 2013, 2016, Shibuya et al. 2022).

The compound responsible for the bursa contraction effect was characterized by Stewart et al. (2016). Using chromatographic fractionation of the mucus coupled to bioassays of contractility of the copulatory canal, these authors showed that the activity is due to a peptide called *Love Dart Allohormone, LDA*. LDA is part of a 235-amino acid polypeptide of the *Buccalin*-family. These genes occur throughout the metazoans and generally encode co-transmitters associated with neuromuscular transmission. The vertebrate orthologue of *Buccalin* is *Allatostatin A*. Polypeptides encoded by these genes are post-translationally cleaved into a number of smaller molecules. In *C. aspersum*, one of them is LDA (25 amino acids). The LDA of helicoid snails differs appreciably from *buccalin*-related peptides in other molluscs.

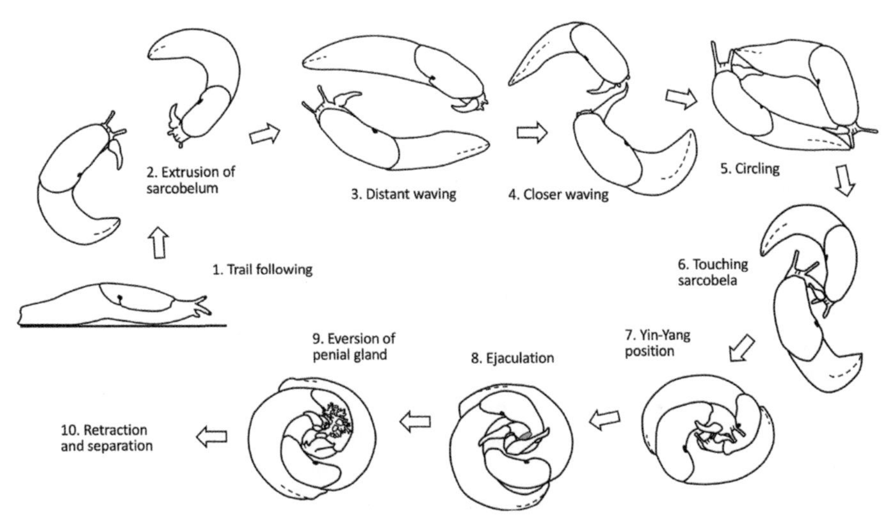

Figure 5.21. Ten steps in courtship and copulation behaviour of *Deroceras gorgonicum*, a terrestrial slug of the family Agriolimacidae. The structure extending on the slug's right side is the sarcobelum, the tip of the penial complex. In step 8 the penial sac is everted as well, and the two penises are entangled on the outside. Sperm is ejaculated onto the partner's penis. This is followed by slapping the penial gland across the partner's surface (only the right slug's gland is visible, step 9). Retraction of the penis brings allosperm inside the genital tract. Redrawn from drawings after video stills in Reise et al. (2007).

For example, the terrestrial slug *Deroceras reticulatum* has a *buccalin*-related gene, but lacks an LDA (Ahn et al. 2017). So, it seems that recruitment of the *buccalin* gene for the purpose of allohormone production has happened only in snails of the superfamily Helicoidea. Subsequently, the gene, because of its direct reproduction-related function, was subject to accelerated evolution (sexual selection) by which it came to differ from the other *buccalin*-related peptides.

Our exploration of bizarre reproductive behaviours in terrestrial gastropods cannot be complete without considering the case of *Deroceras*, a genus or around 100 species of slugs belonging to the family Agriolimacidae. These gastropods have found still another way of sexual partner manipulation. The morphology of these animals is fairly uniform and species are very difficult to identify based on their external features. About the only structure showing differentiation is the penial complex. The penis of these animals is extremely elaborate, including appendages, sacs, folds, pilasters and glands. One particularly important structure is the *sarcobelum* ("fleshy sail") at the tip of the penis, which has a sensory and excretory function. When two snails meet after crawling along a slime trail ("*trail following*"), one of the first stages in the courtship is to evert the sarcobelum (Figure 5.21).

Then follows a long courtship process, involving waving and circling and sometimes slapping each other with the sarcobelum. The slugs end up in a "*Yin-Yang position*", where both partners have their penis completely extended, including the sac-like main structure. The penises do not, however, penetrate the other's gonopore like in helicid snails, but remain entangled on the surface. The ejaculate is shed on the partner's penis and allosperm is brought inside when the penis is retracted. The final stage of copulation is the eversion of the penial gland, a structure at the base of the penis. This is slapped quickly across the partner's skin and then retracted (Figure 5.21).

In most species, the "copulation" is relatively short compared to the extremely long courtship (Reise 2007). For instance, *Deroceras gorgonicum* spends 6 to 9 hours in courtship, while sperm exchange takes one second and retraction is completed within 18–25 seconds (Reise et al. 2007). The extensive waving and circling behaviours during courtship may reflect the tendency to maximize fitness by donating sperm without receiving some, as suggested by Bateman's principle (see Section 5.7).

The penial gland is likely to have the same function as the mucus gland in helicid snails, the source of love dart allohormone. Histological investigations have shown that the gland epithelium is secretory indeed (Benke et al. 2010). The investigators suggest that the secretion includes a bioactive substance that influences the partner's sexual physiology, so represents another "alternative delivery" of accessory gland products (*sensu* Zizzari et al. 2014). However, the active principle has not yet been identified and the mode of action is not as well-known as in the love dart story of helicid snails discussed above.

As a *grand finale* of bizarre reproductive behaviours in terrestrial gastropods may serve the *apophallation* behaviour shown by slugs of the genus *Ariolimax*. This genus includes several fairly large slugs, up to 25 cm length, with conspicuous yellow colours, often called banana slugs. They are known for occasionally biting off their partner's penis. In some copulations, one of the slugs curls its body and gnaws the other's penis with its radula. The process takes considerable time, e.g., in *Ariolimax californicus* a copulation lasted 268 min and the subsequent apophallation 65 min (Leonard et al. 2002). The partner may struggle to free itself but in about 20% of the cases nevertheless loses its penis. The victim will survive the treatment but in subsequent matings can only assume the female role. In eight observed unilateral copulations, five had a partner that missed a penis (Leonard et al. 2002). This bizarre behaviour illustrates again how evolution in hermaphrodites has pushed to the limit a sexual conflict that at moderate levels exists in all animals.

5.9 Brood care and the origin of social behaviour

After fertilization many soil invertebrates lay their eggs in the soil and just leave them there to develop and hatch. Eggs are often laid in protected places, like crevices, and are usually laid deeper in the soil compared to the adult animals. Some invertebrates, e.g., grasshoppers and crickets have extended ovipositors to reach such sites, others, e.g., centipedes, cover their clutch with a little soil or leaves. The careful selection of specific microhabitats for oviposition will protect the eggs against desiccation, cannibalism or predation and so contributes to fitness, but such paternal behaviours are generally not considered a form of brood care.

In fact, brood care is virtually absent among several species-rich soil invertebrate groups, like nematodes, tardigrades, springtails, mites, earthworms and land snails, although some show behaviours that may be considered a primitive form of care. For example, the tropical megascolecid earthworm *Pontoscolex corethrurus* digs small chambers in the soil, coated with mucus, in which single cocoons are deposited

(Ortiz-Ceballos et al. 2009). These "*nesting chambers*" are provided with mucus from the mother, which contributes to the growth of the juveniles (Ortiz-Ceballos et al. 2016).

A classification of brood care behaviours is given in Table 5.1 (based on Trumbo 2012), and examples of such behaviours among soil invertebrates are listed. Inspection of the table learns that, despite the general absence of brood care in soil invertebrates, there are many groups, especially among soil-associated insects and spiders, that do show quite advanced levels of care. The most common type of parental care is designated as "*food provisioning*". Eggs laid in the soil are supplemented with food items for the hatchlings to feed upon, such as dung (in dung beetles), carrion (burying beetles) or paralysed prey (digger wasps and related families). More advanced types of brood care, in which food items are collected from the environment and provided to larvae in the nest, are found in subsocial and social invertebrates such as theridiid spiders, ants, wasps, bees and termites.

Table 5.1. Patterns of brood care behaviours according to Trumbo (2012), with examples among soil invertebrates.

Type of brood care	Description	Examples among soil invertebrates
Habitat choice	Leaving eggs in selected microhabitats favourable for egg development and offspring survival	Almost all soil invertebrates
Trophic eggs	Adding unviable eggs to a clutch, for hatchlings to feed upon	Ants (Hymenoptera: Formicidae) Burrowing bugs (Heteroptera, Cydnidae)
Guarding	Protecting eggs or hatchlings from predation or other dangers	Earwigs (Dermaptera) Centipedes (Chilopoda: Geophilomorpha)
Herding	Protecting and facilitating the feeding of gregarious mobile offspring	Fungus beetles (Coleoptera: Erotylidae)
Nest building	Construction of a secluded place for eggs to develop and young to grow through early stages	Pisaurid spiders (Arachnida: Pisauridae) Land crabs (Decapoda: Grapsidae) Some earthworms (Crassiclitellata: Megascolecidae)
Fostering	Carrying eggs in a brood pouch or cocoon in association with the mother's body	Woodlice (Isopoda: Oniscidea) Landhoppers (Amphipoda: Talitridae) Cockroaches (Dictyoptera: Blattidae) Wolf spiders (Arachnida: Lycosidae)
Provisioning	Adding nutritional materials or prey to eggs or nest for offspring to feed upon	Dung beetles (Coleoptera: Scarabaeoidea) Digger wasps (Hymenoptera: Sphecidae) Ants (Hymenoptera: Formicidae) Termites (Isoptera) Comb-footed spiders (Arachnida: Theridiidae)
Feeding	Procuring food items or prey from the environment or from the mother's body and actively providing it to young begging for it	Burying beetles (Coleoptera: Silphidae) Velvet spiders (Eresidae)

The evolution of parental care is often explained from the fitness benefits accrued to the offspring, when parents care for their young during the early, vulnerable stages of life. In this argument, brood care will evolve when the fitness gain achieved will exceed the fitness loss due to parental investment. This is assumed to hold under harsh and unpredictable environmental conditions or frequent food scarcity. It is also often associated with a semelparous life-style. However, the fact that so many soil invertebrates do not show any form of brood care, and that brood care is concentrated in specific phylogenetic lineages much more than in others (Table 5.1) suggests that intrinsic factors, related to life-history traits are at least as important (Kölliker 2007). In addition, developmental constraints and genetic factors may play a role.

One of the important genetic factors is the well-known *haplodiploid* sex determination of Hymenoptera. Because males in these insects are haploid, female offspring are more related to each other than they would be to their own offspring. For each individual female, the maximization of *inclusive fitness* (fitness including the fitness of kin, weighted by genetic relatedness), will promote the female to help her queen produce more workers, rather than to have offspring herself. This process called *kin selection* provides a commonly accepted explanation for the evolution of brood care and eusociality in Hymenoptera (Keller 2009). However, it cannot explain brood care in the many other soil invertebrates which lack the haplodiploidy genetic system (Nowak et al. 2010).

One particular case is the evolution of eusociality in termites (Isoptera). In these insects both sexes are diploid, so the genetic argument applied to Hymenoptera is not applicable. Thorne (1997) mentions several other factors that may have contributed to brood care in termites: monogamy, long development, iteroparous reproduction, high costs of founding a new nest, etc. The view that kin selection is the single comprehensive explanation for social behaviour is too limited. There may be different reasons in different evolutionary lineages. Social behaviour of termites is a continuation of the subsocial behaviour in their evolutionary ancestors, as still present in the family Cryptocercidae of cockroaches (Dictyoptera). In fact, termites are sometimes called "social cockroaches".

It is generally assumed that sociality finds its origin in parental care (Dumke 2016, Socias-Martínez and Kappeler 2019). Soil invertebrates are particularly useful models since they show so many different grades of care (Table 5.1). The comparative study of brood care behaviours may generate hypotheses about the origin of social systems. In this section we therefore explore the brood care behaviours in three groups of soil invertebrates: earwigs, centipedes and subsocial spiders.

Earwig brood care has been studied extensively in the common species *Forficula auricularia*. After fertilization the female clears a piece of soil and digs a shallow burrow where she lays her eggs. Staying under cover during several months during winter, up to early spring, she tends the clutch (Figure 5.22). The female regularly licks the eggs to remove fungi and chases away any intruders, including the male. After hatching she guides the nymphs in nocturnal food expeditions, until they leave the nest.

Experimental research has shown that brood care of earwigs is shaped by a complex of interactions between offspring, parents and environment. Of crucial importance are the trade-offs between brood size, hatchling quality, parental nutrition

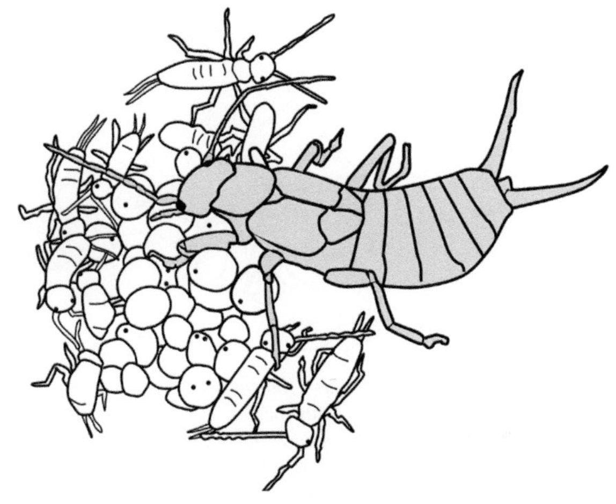

Figure 5.22. Nest of European earwig, female *Forficula auricularia* tending her eggs and hatchlings. Line drawing from a photograph by Tom Oates, 2010, published by Wikimedia Commons.

and costs of postponing the next clutch. Survival of the nestlings is enhanced (although not dramatically) by the presence of the mother (Kölliker 2007), but there are costs to the mother because the second clutch is produced earlier when care is lifted experimentally.

The trigger to caring behaviour emanates from the offspring, because as soon as eggs or nymphs are removed, the behaviour vanishes. Earwigs cannot tell the difference between their own young and those of others; they will preferentially express caring behaviour towards nymphs in good condition whether they are their own or from another nest (Wong and Kölliker 2013, Van Meyel et al. 2019). In addition, mothers in a good nutritional condition show more caring behaviour and this stimulates food-sharing among the nestlings (Kramer and Meunier 2016). Apparently, in earwigs, the various trade-offs between life-history traits are such that brood care is a favourable option. However, what exactly is the prime reason why it evolved in Dermaptera and not in related orders like grasshoppers and crickets, is difficult to say.

As a second example of brood care among soil invertebrates we consider centipedes (Chilopoda). The care behaviour of these myriapods is somewhat comparable to earwigs, but the association between the mother and her eggs and hatchlings is tighter (Figure 5.23). The female curls itself around the clutch, while hiding under cover or in a piece of decaying wood (Siriwut et al. 2014, Edgecombe et al. 2010, Mitić et al. 2012). Usually the ventral side is bended towards the eggs, but in the order Geophilomorpha the eggs are protected with the dorsal side of the mother. This seems to be related to the presence, in geophilomorphs, of ventral pores, which secrete poisonous substances that could harm the development of the eggs. Some species that lack these pores (in the family Mecistocephalidae) have reverted to ventral curling (Edgecombe *et al.* 2010).

Parental care in centipedes is concentrated in three orders: Craterostigmomorpha, Scolopendromorpha and Geophilomorpha, which jointly form the monophyletic

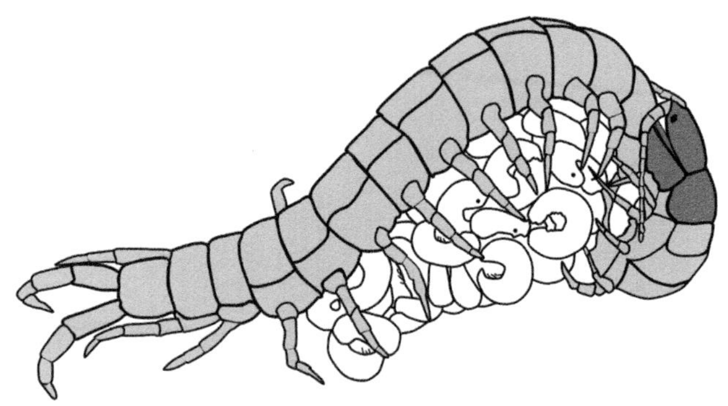

Figure 5.23. The centipede *Cormocephalus hartmeyeri* (Chilopoda: Scolopendromorpha: Scolopendridae), showing maternal brood care with maggot-like juveniles. Drawn after a photograph by Gonzalo Giribet published on the Centipede Systematics website of Greg Edgecombe at the Natural History Museum (www.nhm.ac.uk). The head capsule is seen on the right; the front end of the animal is curled backwards.

taxon Phylactometria (Figure 5.24). The name of this taxon in fact refers to the shared derived character of this group, maternal care (Edgecombe and Giribet 2007). The other two orders (Scutigeromorpha and Lithiobiomorpha), which represent about 38% of the species, show no brood care at all (Edgecombe et al. 2010, Mitić et al. 2012).

So, it seems that parental care in centipedes is due to a single evolutionary event at the base of the Phylactrometria. Following from the argument developed in the earwig example, this could be a change in life-history trade-offs, allowing a fitness gain when switching to brood care. The phylogeny also suggests that the ancestral chilopod did not have brood care and that this situation was retained in Lithobiomorpha, which we know as common surface-active centipedes. Geophilomorpha, with their elongate body, short legs and slender shape, are mostly soil-living. The sister group of geophilomorphs, Scolopendromorpha, is, however, surface-active. This latter group includes some large poisonous (and scary!) centipedes, with a body length up to 30 cm.

We must conclude that brood care in centipedes cannot be considered an adaptation to soil-living. Soil-living in Chilopoda is a derived trait. It evolved in geophilomorphs that already had brood care. The tight relationship of brood care with phylogeny in centipedes suggests that it could have been a driving force for the evolution of this group, in a manner similar to what we noted for courtship behaviour in pseudoscorpions and wolf spiders (Sections 5.2 and 5.3).

As a third example in our exploration of brood care behaviour in soil invertebrates we address the subsocial and social behaviour of spiders. Parental care is observed in many groups of spiders. For example, in Chapter 3 we have noted that female lycosid spiders carry an egg sac attached to their body (Figure 3.12), while females of the related family Pisauridae carry an egg sac with their pedipalps (Section 5.3). Almost every behavioural category of brood care is represented by a spider family (Table 5.1).

The ubiquity of brood care in spiders is considered an evolutionary predisposition to social behaviour (Whitehouse and Lubin 2005). However, there are several grades

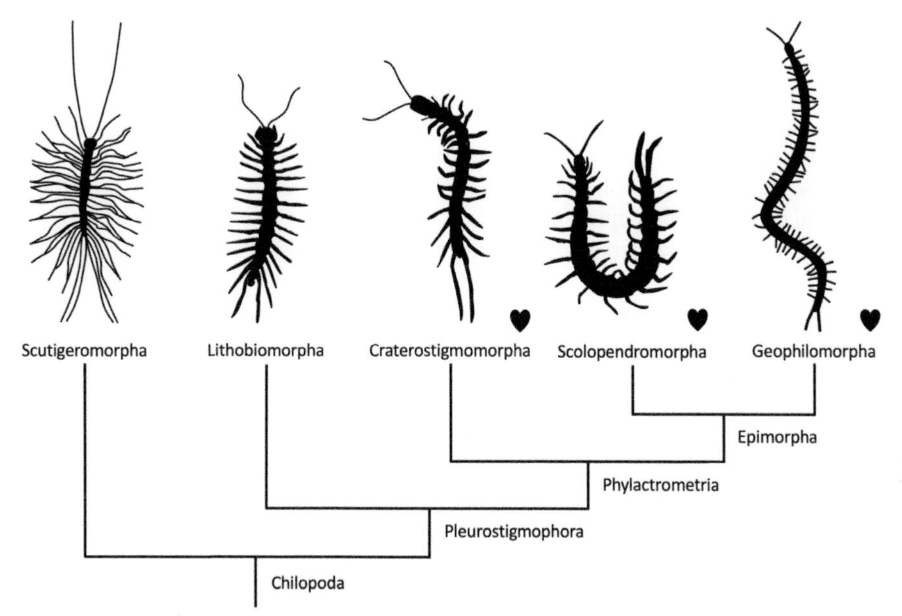

Figure 5.24. Phylogenetic relationships between the five orders of Chilopoda, including the cladistic group names. Orders showing parental care are marked with a heart. Most likely brood care is due to a single evolutionary change at the base of the Phylactrometria. After Edgecombe and Giribet (2007).

of group-living. A useful distinction is between colonial and cooperating social species. The colonial or *communal* category comprises species with different degrees of aggregation, from facultative to permanent. The colony is open to migration and there is no task division nor food sharing. The spider colony is like a flock of birds that forage in a group. In cooperating social species, however, the individuals join forces in prey capture and share food. Still such cooperating spider groups are quite egalitarian compared to the task division in ants and termites. Some authors argue therefore that "true" sociality (*eusociality*), qualified as a "*major evolutionary transition*" by Szathmáry and Maynard Smith (1995), is limited to species with worker sterility (Bernadou et al. 2021), so excludes spiders.

The different grades of social behaviour in spiders may have their origins in different ancestral behavioural patterns. One evolutionary pathway could start from the foraging function, that is, social behaviour might be promoted by individual fitness gains when foraging in groups. It is assumed that this pathway would mainly lead to colonial sociality as defined above. Another ancestral condition could be brood care. Social behaviour is seen to result from juveniles staying with the mother because of her care, decreasing dispersal, the formation of juvenile groups and finally the formation of adult groups (Whitehouse and Lubin 2005, Mougenot et al. 2012). This pathway would lead to cooperating sociality as defined above. In this view, social intolerance and aggressiveness is a consequence of dispersal (not the other way around) (Chiara et al. 2019).

Group-living spiders often live in large, self-constructed tents, made of silk and environmental materials. If spiders remain in such nests for several generations, the group develops a biased sex ratio (preponderance of females) and suffers from inbreeding depression. This type of sociality is therefore considered an "*evolutionary dead-end*" (Agnarsson et al. 2006, Avilés et al. 2006). The interpretation is supported by phylogenetic analysis: cladogenesis in social spiders is concentrated in the tips of the tree, giving it a "*spindly*" aspect. The consequence is that sociality in spiders has a high turnover: it has arisen many times (nine times in a single family, Theridiidae, Agnarsson et al. 2006), but it does not stay long (in evolutionary terms), due to the dead end of inbreeding depression.

The number of spiders characterized as "*quasisocial*", "*subsocial*" or "*social*" differs between studies and depends very much on how sociality is defined. A particular useful definition was proposed by Yip and Rayor (2014) who argued that subsocial spiders are those species in which the offspring stay together with the parent beyond the age at which they begin to feed, but disperse before they start reproducing themselves. Defined in this way, 70 subsocial species were identified. When this trait was mapped onto the phylogeny of spiders, it appeared that subsociality has evolved at least 18 times in 16 families. Table 5.2 provides a list of these 16 families.

Table 5.2. Overview of spider families in which at least one species shows subsocial behaviour, from Yip and Rayor (2014). Habitat use derived from Wikiwand webpages.

Family	Common name	Microhabitat of social species
Dipluridae	Curtain-web spiders	Funnel webs or silk-lined burrows in crevices or damp earth, under logs or leaf litter
Nemesiidae	Funnel-web trapdoor spiders	Extensive burrows in soil, with a hinged trapdoor at the surface
Theraphosidae	Bird spiders	In trees, at the soil surface or in silk-lined burrows
Filistatidae	Crevice weavers	Funnel or tube-like webs, or holes dug in soft ground
Scytodidae	Spitting spiders	Under stones, surface active
Eresidae	Velvet spiders	Large above-ground silk-woven nests, made of leaves, twigs and prey carcasses
Uloboridae	Hackled orb weavers	Cribellate orb webs weaved in above-ground vegetation
Theridiidae	Comb-footed spiders	Large tangled webs in low vegetation
Sparassidae	Huntsman spiders	Under loose bark, surface-active
Salticidae	Jumping spiders	In vegetation and surface-active
Thomisidae	Crab spiders	Silken nests spun in the vegetation
Agelenidae	Funnel weavers	Complex communal webs woven in shrubs and trees
Desidae	Intertidal spiders	Horizontal sheet webs in open vegetation
Amaurobiidae	Lace-weavers	Fine-woven funnel web under rocks, behind tree bark, or burrows in crusty soil
Miturgidae	Sac spiders	Sac woven in vegetation
Lycosidae	Wolf spiders	Funnel-shaped webs in crevices or loose soil

It turns out that subsociality, as defined by Yip and Rayor (2014) is spread across a wide phylogenetic range: from the basal mygalomorph families (trapdoor spiders, bird spiders) to the more derived araneomorph groups (jumping spiders, crab spiders, wolf spiders). It is, however, relatively rare among orb-weaving spiders (Araneidae, Tetragnathidae) and is also absent in Linyphiidae, a common family associated with low vegetation and soil.

Recalling the purpose of this book, we ask the question whether subsocial behaviour in spiders in some way or another is related to the soil environment. But Table 5.2 does not provide any indication that the behaviour is associated with a particular use of the soil. We find subsocial species among soil-burrowers, surface-active spiders and vegetation-dwellers. We are led to conclude that the evolution of social behaviour in spiders, like in centipedes, has evolved independently from their relationship to the soil. Like brood care, social behaviour of soil invertebrates cannot be considered an adaptation to soil-living, neither to living above soil.

5.10 Predatory behaviour

To close this Chapter on the behavioural evolution of soil invertebrates we consider shortly the evolutionary aspects of predatory behaviour. We have seen in Chapter 3 that predation is an important density-dependent factor in populations of soil invertebrates. Several soil invertebrate predators are known for their potential to control prey species, including agricultural pests (Marc et al. 1999, Koehler 1999, Lesna et al. 2000). What types of predatory behaviour have evolved in association with the soil profile?

The first observation is that there are so many soil predators. Predators are part of the "*enigma of soil animal species diversity*" (Anderson 1975, see also Section 2.7), not only because they are so numerous, but also because there are so many species. One finds predators in almost every soil-living group, e.g., carabid beetles, earwigs and bugs, almost all spiders, all pseudoscorpions and harvestmen, several lineages of mites, e.g., trombidiforms and mesostigmatids, and all terrestrial flatworms. Also, in the microfaunal groups predators abound, e.g., in nematodes (feeding on other nematodes and protists). The only groups more or less devoid of predators are apterygotes, earthworms, enchytraeids, terrestrial isopods and land snails.

The usual explanation for Anderson's enigma is that there are many more nutritional niches in the soil than one would think. Decomposer invertebrates do not all eat the same resources, they avail of different repertoires of digestive enzymes, they feed in different microhabitats of the soil or have their highest abundance at different times of the year. Extensive niche partitioning is the evolutionary mechanism allowing coexistence of so many species. Are predators equally divergent in their foraging strategies?

Almost all predatory invertebrates associated with the soil are *ambush predators*. They hide among soil structures, use mimicry to conceal themselves, sit still and wait, until a prey is in reach. Then they strike with a quick snap. For example, assassin bugs (family Reduviidae) stab their prey with a single thrust using their proboscis, gamasid mites pierce their prey with their chelicerae and beetles catch their prey between the basket of their mandibles. Spiders are also mostly ambush predators, since they sit and wait in their web, in a burrow, below a trapdoor or at the end of a funnel.

Ambush predation requires two specific sets of adaptations:

1. The predator should remain unseen by the prey.
2. The predator should be able to strike quickly from a standstill.

The first aspect explains why so many soil predators master the art of sitting still. Since most soil invertebrates have only limited visual acuity and respond to movement rather than to visual images, sitting still is the first thing to do for an ambush predator. It is sometimes combined with remarkable camouflage. For example, assassin bugs of the genera *Paredocla* and *Acanthaspis* carry a "back pack" of plant material and prey remnants glued to their abdomen and covered with a veil of dust. This makes them invisible to any prey passing by, although the main function may be to protect them from predation by other predators (Brandt and Mahsberg 2002, Figure 5.25). Ambush predators are also masters of sensing vibration, that is, they can note an approaching prey from subtle air and surface vibrations, using the tips of their legs, sensory hairs on the body or secondary sensory contraptions such as silken threads or webs.

The second set of adaptations implies that ambush predators should show an extreme acceleration in their strike because they have to do it from standstill. For example, Wignall and Taylor (2010) describe how the assassin bug *Stenolemus bituberus*, which specializes on spiders, hits the prey with its proboscis by quickly

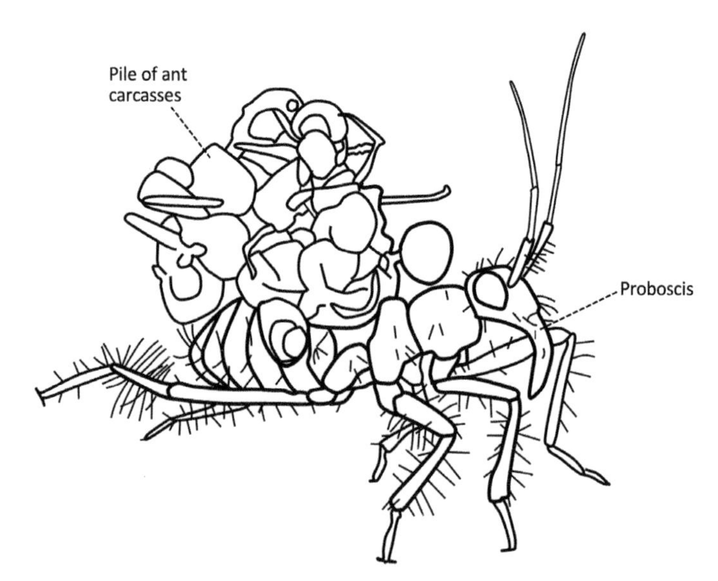

Figure 5.25. Nymph of *Acanthaspis petax* (Heteroptera: Reduviidae), an assassin bug preying on ants, with a pile of ant carcasses glued to its back by way of camouflage. Drawn after a photograph published by Wikimedia Commons, licensed under the Creative Commons Attribution-Share Alike 3.0 Unported licence.

flexing the femur-tibia and trochanter-femur joints of its first two legs and so brings the anterior of the body downwards within 0.1 s.

Despite the preponderance of ambush predation, several soil invertebrates apply hunting or pursuit. This type of predation is found among carabid beetles, rove beetles, wolf spiders and jumping spiders. They use visual information mainly to identify and stalk their prey. Consequently, all hunting invertebrate predators have large eyes. Since the eyes are immobile relative to the head both in beetles and in spiders, the animal relies on a large angle of vision to spot a prey. However, the area over which a visual predator has binocular sight is usually much smaller. A nice example of this analysis is given in a paper by Bauer (1981) who reconstructed the visual space of the carabid beetle *Notiophilus biguttatus.*

N. biguttatus is a small (5–6 mm), common carabid beetle of forest floors, heathlands and grasslands. It has large compound eyes, set on the sides of the head (Figure 5.26). The optical axes of the ommatidia point slightly more to the front, compared to the anatomical axes, due to their slanted position under the cornea. The area of the compound eye with the highest visual acuity (called *fovea*) points straight ahead. On the basis of this information, Bauer (1981) was able to reconstruct the angle of binocular vision as no more than 74°, while the total visual space covers an angle of 200° (Figure 5.26).

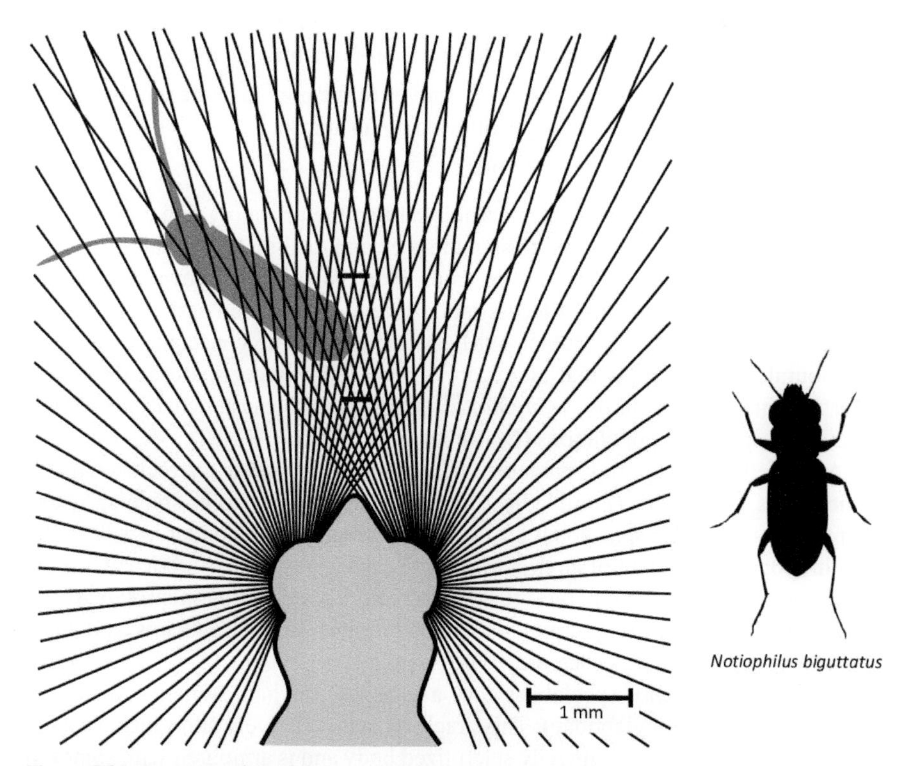

Figure 5.26. Reconstruction of the visual space for *Notiophilus biguttatus* (Coleoptera, Carabidae) in the horizontal plane. The two short lines indicate the distance, in the area of binocular sight, that releases a predatory strike. The experiments were done with the springtail *Heteromurus nitidus* as a prey. After Bauer (1981), redrawn.

N. biguttatus attacks mainly springtails and the average distance at which an attack is released measures 1.7 mm straight in front of the animal. Since springtails avail of a specialized jumping organ, the *furca* (Figure 2.6), the prey has a fair chance to escape. The furca in rest is bended under the abdomen, but can be smacked downwards to the substrate, which launches the animal into the air. The release of the furca is not brought about by muscles directly, but by a flip of the joint, due to a surge of haemolymph pushed backwards by contracting trunk muscles (Bretfeld 1963). Due to this efficient escape mechanism, an attack strike of the beetle, even if extremely quick, is not always successful. In the experiments of Bauer (1981) the success rate was 51–53%.

N. biguttatus normally roams around until it spots a potential prey. Only moving prey will trigger an attack, so springtails with high natural activity are more at risk than species with low activity. This explains the higher attack rate towards the epigeic species *Orchesella cincta* compared to the less active, drought sensitive, species *Tomocerus minor* (Ernsting and Jansen 1978).

When the beetle spots a prey, the following behavioural repertoire is performed:

- *Turning*: the beetle turns its body bringing the prey in its binocular field straight ahead.
- *Waiting*: if the prey stops moving the beetle also stands still; if the prey does not move for minutes, the beetle falls back to non-predatory behaviour.
- *Pursuit*: if the prey moves, the beetle moves to the prey to come within striking distance, usually but not always followed by an attack.
- *Attack*: the predator snaps at the prey.

The "waiting" element is a remarkable part of the predatory behaviour in this beetle. Visual information is used to spot the prey and to align the predator, but once it is close, it sits still like an ambush predator. Experiments have shown that waiting time, i.e., attack readiness and success ratio, depend on satiation and are higher for smaller prey (Ernsting and Mulder 1981, Bauer 1981). The consequence of these behavioural components is that larger preys make up a larger share of the diet at lower densities. Predation by *N. biguttatus* is predicted to cause density-dependent mortality in springtail populations (De Ruiter et al. 1988).

Another type of hunting predatory behaviour is described for recluse spiders of the genus *Loxosceles* (family Sicariidae, also known as violin spiders). These neotropical spiders (leg span about 2.5 cm) are commonly found under logs or stones in anthropogenic and natural environments of Meso- and South-America. They are known for their potent venom; bites to humans require medical treatment. *Loxosceles* species attack a wide variety of prey, including termites, isopods and ants, and apply different strategies towards each prey type (García et al. 2016).

Loxosceles gaucho specializes on a species of harvestman, *Mischonyx cuspidatus* (family Gonyleptidae). This arachnid, with relatively short legs, is about the size of its predator, has a heavily sclerotized body and is armoured with spines. It can also defend itself behaviourally, by "*intense leg tapping*" (Dias et al. 2014) and *thanatosis* (death feigning). The spider is obviously unable to bite through the heavy exoskeleton. Instead it applies a unique strategy to overcome the defences of the harvestman. It specifically attacks the weak parts of the body, the joints in the distal parts of the legs. After several small bites or a single long one, the prey is paralysed by the spider's venom and then turned over.

A breakdown of the behavioural components is shown in Figure 5.27 (Segovia et al. 2015). Transition probabilities between the various components indicate a capture sequence typically running as follows. When the spider spots a prey, it orients its body towards it, moves to it and makes contact. It then gives several bites in the weak parts of the legs, until the prey is unable to flee. It then continues to weave silk threads and turns the prey on its back. Interestingly, the behavioural components labelled "*detection from a distance*", "*orient to prey*" and "*remaining motionless*" are rather similar to the "*turning*" and "*waiting*" components of the beetle's predatory behaviour.

In summary, we may conclude that predation in the soil environment is dominated by ambush strategies, which call for "sit-and-wait" behaviours and very rapid strike responses. However, the surface of a soil is home to several visually hunting predators as well. These animals use their often-remarkable eyesight to spot potential prey and come close to them. The behaviour at close range, however, varies greatly from one predator to another and may have an element of ambush predation again. Activity is a risk factor on the prey side, since invertebrate predators usually respond to moving prey only.

In the lower layers of soil and litter, ambush predation is the only strategy possible. At the surface it is supplemented with pursuit and hunting predation. Therefore, it is likely that the total predation pressure will be higher at the surface, compared to soil-living communities. This conclusion is confirmed by the higher rate of mortality and

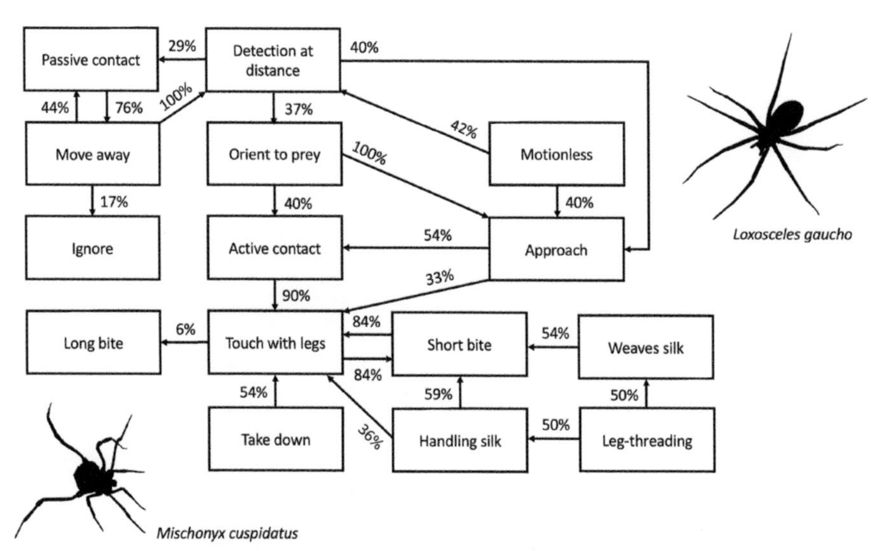

Figure 5.27. Flowchart showing transitions between behavioural components of the recluse spider *Loxosceles gaucho* attacking the harvestman *Mischonyx cuspidatus*. The arrows indicate transition probabilities (percentage of times that one behaviour was followed by another behaviour). Data from two experiments, conducted with different substrates, were combined in the present figure. After Segovia et al. (2015), redrawn.

higher biomass turnover among epigeic springtails compared to hemiedaphic species (see Section 3.2). It is also consistent with the trend towards semelparity in epigeic communities compared to iteroparity in euedaphic communities (Section 3.6). The evolution of soil communities as portrayed in the GKGB hypothesis (see Chapter 1) is likely to be associated with a higher predation pressure for those lineages that adopted a surface-active life-style.

Chapter 6

Physiological Adaptation and Microbial Interactions

Badger and Mite

I live a life of pure delight. Yes, I 'm blessed
with my burrow, cosy and watertight.
Me and Mr Badger, my familiar sight,
as I share bed and chest with my boarding guest.

But overnight the situation caused fright!
Things were fucked up and totally messed.
Mr. Badger got stressed, he was a pest,
acting like an obsessed Luddite!

But what was the subject of his quest?
Don't worry, he said, this is my annual rite!
And there's no place for protest, so don't detest

me removing dirty straw and leaves, that shite.
I said please relax, just give this a rest:
you're now spoiling my pantry and so, my appetite!

Jasper Aertsz

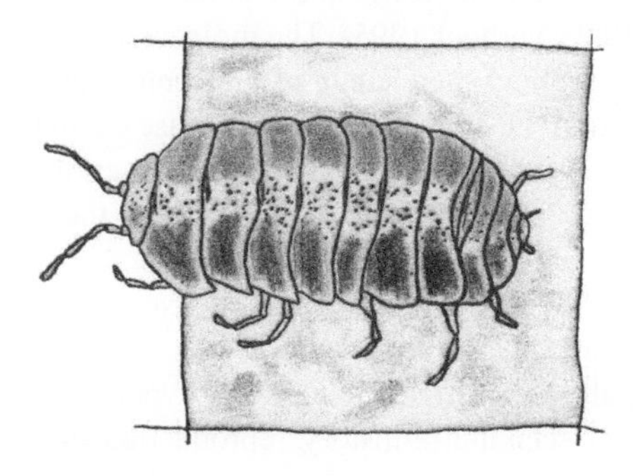

CHAPTER 6

Physiological Adaptation and Microbial Interactions

"Look baby! There goes the great Mr. Toad! And that's the gallant Water Rat, a terrible fighter, walking along o' him! And yonder comes the famous Mr. Mole, of whom you so often have heard your father tell!"

Kenneth Grahame, The Wind in the Willows

In the previous chapters we have been discussing adaptations to the soil environment at a variety of levels: life-history, mode of reproduction, development and behaviour. At each of these levels soil invertebrates show evolutionary changes in relation to soil living and above-soil life-styles. Underlying all this are mutations, new allelic variants, novel genes, changes in gene expression and physiological changes. In this chapter we will discuss the adaptation of soil invertebrates on the physiological and molecular level.

Soil invertebrates are not among the favourite model organisms in physiological and genetic research. Some Collembola come close to *Drosophila* in suitability for genetic research but they have never reached the same star status. Earthworms have an ecological relevance exceeding by far the other "worm", *C. elegans*, but that has not made them generally accepted model species. In addition, soil ecologists, with their emphasis on biodiversity and soil processes, have often rejected the idea of model species, a bare necessity for advancement in molecular science.

Physiological work on soil invertebrates, although popular in the 1980s, has therefore always remained a bit scanty and incomplete (Vannier and Verdier 1981, Joosse and Verhoef 1987, Verhoef 1995). The molecular revolution, however, which reached soil ecology by about the change of the century (Van Straalen et al. 2008) has changed this situation. DNA isolation, polymerase chain reaction, genome-wide gene expression, RNA interference, etc., are all techniques that can be easily applied to tiny soil invertebrates. The perceived great barrier of the model species has been taken away by new technology.

In this Chapter we start with some classical concepts of environmental physiology and then turn towards molecular genetics, genomics and metagenomics of soil invertebrates, all to investigate the GKGB hypothesis: are the adaptations to soil-living, that we have seen in life-history, reproduction and behaviour, also visible on the physiological and molecular level?

6.1 Patterns of metabolic rate

Whole-body *metabolic rate* expressed in energy units per time (e.g., J h^{-1}) is one of the most informative measures of animal activity since it reflects the totality of processes in the organism that deal with oxygen-fuelled energy metabolism. It is a classical parameter of animal physiology, measured numerous times in numerous animals under numerous conditions. *Respiration*, expressed as mL of oxygen consumed, or carbon dioxide produced, per time unit (mL h^{-1}) is the most common indicator of metabolic rate. Metabolic rate and respiration can be converted into each other using the *oxyenergetic coefficient* (usually assumed to equal 20.2 Joule per mL of oxygen) and the *respiratory quotient* RQ (moles of carbon dioxide produced per mole oxygen consumed, usually assumed to equal 0.82). However, the relationship between respiration and metabolic rate is not precise, since RQ depends on the substrate respired (carbohydrate, protein or lipid).

Measuring respiration in small invertebrates is no sinecure. For animals with a body mass of 5 mg and below, the most sensitive (but quite laborious!) method is *Cartesian diver respirometry* (Linderstrøm-Lang 1937, Zeuthen 1950). For larger animals, modern microelectrodes can be deployed, preferably in a manifold flow-through system. Carbon dioxide can be measured using *infra-red absorption*, a sensitive and reliable approach, especially when the inflowing air is free of CO_2. The measurement of oxygen consumption is more complicated as it relies on the difference between inflowing and outflowing levels; the amount consumed by a small invertebrate must be measured against a background of 20%. An excellent overview of respiration methods for small animals is given by Lighton (2008).

Most studies aimed at quantifying metabolic rates of soil invertebrates have been done in the context of ecological energetics. That is, the data are used to estimate energy flows through invertebrate communities organized in food-webs (Engelmann 1968, Klekowski and Duncan 1975, De Ruiter et al. 1994). Invertebrate respiration, together with microbial and plant root respiration, contributes to total soil respiration, an important component of the carbon budget of an ecosystem, as studied in the context of global change (Lei et al. 2021).

We are interested here in a more fundamental evolutionary question: how did metabolic rates change when animals colonized the soil and evolved towards fully terrestrial lineages? Our basic framework is the GKGB hypothesis presented in Chapter 1, which suggests that metabolic rate is higher among surface-living species compared to permanent soil-dwellers. Is there any evidence for this prediction from comparative physiology?

Two covariables must always be considered when comparing metabolic rates across invertebrates: body size and temperature. Larger animals have a higher metabolic rate when expressed in J h^{-1} but a lower rate when expressed per unit of body weight (J h^{-1} g^{-1}). The reason is the non-linear allometric relationship introduced in Section 3.2 (Kleiber 1961, Lavigne 1982, Peters 1983, Schmidt-Nielsen 1998). This relationship is written as: $M = aW^b$ where M is metabolic rate (J h^{-1}), W is body mass (g) and a and b are constants. According to Kleiber's law, b has a universal value of 0.75, when considering all animals great and small.

In addition to body mass, temperature must also be taken into account in metabolism studies, since all soil invertebrates are ectotherms. We adopt here the model proposed by Gillooly et al. (2001), in which the body-size-dependent metabolic rate is multiplied by the *Boltzmann factor*, a quantity from physical chemistry that measures the fraction of molecules crossing the energy barrier characteristic for thermal activation. The use of a separate multiplication factor ensures that any effect of temperature is the same for all body sizes, a requirement for the use of physiological time (Van Straalen 1983b). In a comparison of seven different metabolic models, tested against data for springtails and mites, the Gillooly model provided the best fit (Caruso et al. 2010).

The extent to which metabolism depends on temperature is determined by a crucial parameter of the model called *activation energy*. Gillooly et al. (2001) expressed this quantity in electron-Volt, as in particle physics, however, we rather measure activation energy in biochemical terms, as Joules per mol, indicated by H_A. This can be done by introducing the universal gas constant R, rather than Boltzmann's constant. The model is explained in Annex 8.4. We will study the temperature responses of soil invertebrates in more detail in Section 6.2, and here only use activation energy to normalize the metabolic rate data for temperature differences.

Ehnes et al. (2011) analysed a large collection of metabolic rates, covering 3,661 measurements of 580 taxa from 192 publications, both old data (from the IBP) and more recent work. Linear regression of the model, applied to logarithmically transformed rates, allows estimates for the three key parameters: "intercept", "allometric scaling exponent" and "activation energy". For ease of interpretation I recalculated Ehnes's parameter "intercept" to reference metabolic rate (M_{ref}) and activation energy was expressed in J mol^{-1} rather than in eV as in the original publication (Annex 8.4). Figure 6.1 shows the results.

Breaking up the data by taxonomic groups shows that reference metabolic rate does not differ much between the groups. It is somewhat lower for mites and highest

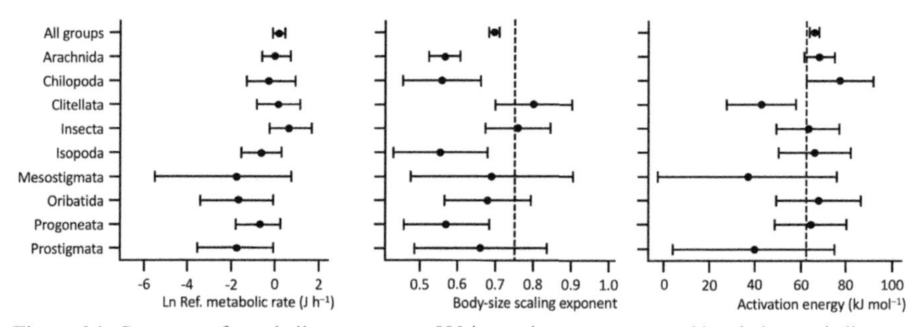

Figure 6.1. Summary of metabolic rates among 580 invertebrate taxa, grouped by phylogenetic lineages of the animal kingdom. Progoneata = all Myriapoda except Chilopoda (cf. Figure 2.15). The graphs show point estimates of three parameters with 95% confidence intervals, following linear regression of the model explained in Annex 8.4. The left diagram shows the natural logarithm of metabolic rate standardized by body weight (1 g) and temperature (15°C) (M_{ref}). The middle diagram shows the scaling exponent for body weight (b in Annex 8.4); the dotted vertical line in this diagram is the value expected by Kleiber's law. The right diagram shows the activation energies (H_A) for temperature dependence of metabolic rate (kJ/mol); the dotted vertical line indicates the overall mean (62.7 kJ mol^{-1}). The three parameters were estimated for each group from the data compiled by Ehnes et al. (2011).

for the insects, but the differences are small and not significant. The overall mean is 1.23 J/h (Figure 6.1, left diagram). However, the scaling exponent for body size does show significant variation. Almost all invertebrates, except earthworms and insects, have a value below 0.75, the value expected from Kleiber's law (Figure 6.1 middle diagram). In activation energies there are also significant differences (Figure 6.1, right diagram), which will be discussed in Section 6.2.

In contrast to Gillooly et al. (2001), the analysis of Ehnes et al. (2011) demonstrates that there is no universal metabolic model. A model in which the parameters M_{ref}, b and H_a were allowed to vary with phylogeny provided a better description than the model with a universal scaling coefficient of ¾ and a single activation energy. The average metabolic rates may be similar, but the way they depend upon body size and temperature differs between species. As Ehnes et al. (2011) pointed out quite rightly, this variation reflects the enormous diversity in body architecture, ecological life-styles and habitats among the different invertebrate lineages. Similar conclusions were reached by Chown et al. (2007) in their analysis of metabolic rates for 391 insect species and by Hoste-Danyłow et al. (2013) in a review of respiration data of forest invertebrates.

Even if the overall means are similar, do metabolic rates of soil invertebrates vary with their position in the soil profile? To answer this question, we first consider data for earthworms, as reported in Meehan et al. (2006) and Ehnes et al. (2011), and a few papers not covered by these reviews (Uvarov 1998, Uvarov and Scheu 2004, Šustr and Pižl 2009, 2010). The original metabolic rates reported in these publications were re-analysed and adjusted for body weight (1 g) and temperature (15°C) using the allometric scaling constant and activation energy determined for earthworms by Ehnes et al. (2011).

Table 6.1 summarizes these adjusted metabolic rates obtained for twelve species of earthworms from the family Lumbricidae. For each species, a range is given, showing variability in the data due to technical errors associated with the measurements, or to physiological differences between worms related to their feeding or reproductive status, which still affect the metabolic rate even when the animal is in complete rest. The differences between species are small. In fact, the metabolic rates of earthworms become similar when standardized by temperature and body size. The overall mean is 1.28 J h^{-1}; the highest and lowest values differ from the mean by about 30%. High average metabolic rates are observed for *Lumbricus rubellus* and *Dendrobaena octaedra*, which both are typical surface-active species. A low metabolic rate is noted for the anecic species *Dendrobaena mrazeki*, a central European earthworm adapted to dry ecosystems (Šustr and Pižl 2009).

The GKGB hypothesis suggests that, in association with their higher mobility and higher reproductive output, epigeic worms should have a higher metabolic rate than endogeic species. However, Table 6.1 does not reveal evidence for this. The average value for epigeic species (1.37 J h^{-1}) is only slightly higher than for endogeic species (1.27 J h^{-1}), and this difference is minor compared to the variation between species. Metabolic rate of earthworms seems to be a quite stable character which, unlike reproductive output (Table 3.4), hardly shows a trend across the soil profile.

A similar exercise can be done for springtails. Petersen (1980, 1981) already hypothesised, on the basis of respiratory measurements for eleven species of

Table 6.1. Adjusted metabolic rates of lumbricid earthworms, classified by ecological categories according to Satchell (1980) and Lavelle (1981). The original data in Meehan (2006), Ehnes et al. (2011), Uvarov (1998) and Šustr and Pižl (2009, 2010) were adjusted to a body weight of 1 g and a temperature of 15°C using the allometric scaling coefficient and activation energy for earthworms, as described in Annex 8.3.

Ecological category	Species	Metabolic rate at 1 g body weight and 15°C (J h^{-1})	
		Mean	Range
Surface living and litter feeding (épigés)	*Eisenia fetida*	1.37	0.83 – 3.03
	Eisenia rosea	1.18	-
	Eiseniella tetraedra	0.94	0.67 – 1.43
	Lumbricus rubellus	1.63	1.46 – 1.81
	Lumbricus castaneus	1.31	0.93 – 1.91
	Dendrobaena octaedra	1.62	1.28 – 2.15
	Dendrobaena veneta	1.54	0.93 – 2.37
Subsurface living, surface feeding (anéciques)	*Lumbricus terrestris*	1.04	0.69 – 1.44
	Dendrobaena mrazeki	0.87	0.82 – 0.97
Soil living, subsurface feeding (endogés)	*Aporrectodea caliginosa*	1.10	0.48 – 2.28
	Aporrectodea rosea	1.57	0.75 – 2.28
	Octolasion lacteum	1.15	1.12 – 1.17

Collembola, that metabolic rates of epigeic (surface-active) species should be higher than euedaphic (soil-living) species. Has this suggestion held up, now that more data are available?

Metabolic rates of Collembola were collected from the review of Meehan (2006), supplemented with data from Van der Woude and Joosse (1988), Šustr (1996) and Stam et al. (1996). Like in the case of earthworms, measured metabolic rates expressed in J h^{-1} were adjusted for body weight and temperature using the allometric scaling coefficient and activation energy for Collembola, derived in Meehan (2006). Antarctic and montane species as well as experiments close to freezing conditions, were ignored, because the induction of cold hardiness invalidates the universal temperature dependence correction (but see Section 6.3). The results are shown in Table 6.2.

Unfortunately, metabolic measurements on euedaphic Collembola are scanty. There are many more data on hemiedaphic and epigeic species. However, the two euedaphic species investigated (including the model species *Folsomia candida*) show a consistently lower metabolic rate than the species in the other two categories. The highest metabolic rates are found among typically epigeic species like *Sminthurus viridis* and *Allacma fusca*, springtails that are often seen in low vegetation and on tree bark, although they reproduce in the soil. The hemiedaphon shows a wide range of values, bridging the other two categories. So, the suggestion by Petersen (1980) that a correlation between metabolic rate and depth profile exists in Collembola seems to be supported by updated analyses, though weakly, given the variation.

As a third test of the GKGB hypothesis we consider metabolic rates of oribatid mites. In Section 3.3 we have introduced the classification of Luxton (1981c), which was based on indicators of energy metabolism and reproductive output, which

Table 6.2. Metabolic rates of springtails, classified by ecological categories according to Gisin (1943). Using the model explained in Annex 8.3, the original data compiled by Meehan (2006) and those reported in Van der Woude and Joosse (1988), Šustr (1996) and Stam et al. (1996) were adjusted to a body weight of 1 mg and a temperature of 15°C with an allometric scaling coefficient of 0.77 and activation energy of 55.9 kJ/mol, determined for Collembola from the data in Meehan (2006).

Ecological category	Species	Metabolic rate at 1 mg body weight and 15°C (mJ h^{-1})	
		Mean	Range
Epigeon: surface living, on top of litter, in vegetation, tree climbing	*Sminthurus viridis*	10.8	10.1 – 11.3
	Allacma fusca	9.68	9.39 – 9.97
	Dicyrtomina minuta	7.17	4.71 – 9.48
	Orchesella flavescens	5.64	-
	Orchesella cincta	7.11	6.36 – 7.94
	Orchesella villosa	3.98	-
	Pogonognathellus longicornis	4.93	4.31 – 5.55
	Pogonognathellus flavescens	7.62	5.51 – 10.37
Hemiedaphon: subsurface living, in deeper litter layers, moist microhabitats	*Sminthurinus aureus*	6.88	5.98 – 7.43
	Lepidocyrtus lignorum	6.60	5.43 – 7.95
	Tomocerus minor	6.83	4.20 – 7.76
	Tomocerus vulgaris	5.82	4.91 – 6.45
	Isotoma viridis	6.75	3.92 – 8.02
	Parisotoma notabilis	6.00	5.56 – 6.41
	Folsomia quadrioculata	4.84	4.16 – 5.96
	Tetradontophora bielanensis	3.61	3.29 – 4.04
	Neanura muscorum	2.39	-
	Protaphorura armata	5.23	3.75 – 6.28
	Protaphorura meridiata	4.70	1.13 – 7.65
	Supraphorura furcifera	5.27	4.65 – 6.02
Euedaphon: soil living, in pores	*Folsomia candida*	3.83	3.26 – 4.44
	Isotomiella minor	3.53	3.19 – 3.76

correlated with the soil profile (Table 3.5). Here we present data on metabolic rates of these oribatids. The rates compiled by Meehan (2006) were adjusted for standard body size (0.1 mg) and temperature (15°C), see Table 6.3. However, not all species could be assigned to one of Luxton's classes, due to lack of ecological knowledge.

The data reveal a relatively large variation. Most likely, technical difficulties in estimating metabolic rates play a role here. Even the Cartesian diver technology reaches it limits of precision with these tiny animals (fresh weight down to 10 μg). Moreover, the metabolic rates do not show a pattern related to the soil profile, when species are classified by Luxton's functional groups. High rates are found in the group of litter dwellers (*Belba corynopus*) but also in soil dwellers (*Chamobates cuspidatus*) and in the intermediate group (*Achiptera coleoptrata*).

Table 6.3. Adjusted metabolic rates of oribatid mites classified according to Luxton (1981c). Note that the nomenclature of P/A groups differs from Luxton (*cf.* Table 3.6). Data from Meehan (2006), adjusted using the model explained in Annexure 8.3.

Functional category	Species	Metabolic rate at 0.1 mg body weight and 15°C (mJ/h)	
		Mean	**Range**
P/A I litter dwellers highly productive species P/B > 2.5 adult body mass > 10 μg growth efficiency 55–65%	*Damaeus claviceps*	0.440	0.387 – 0.536
	Belba corynopus	0.497	0.463 – 0.563
	Steganacarus spinosus	0.305	0.262 – 0.380
	Steganacarus magnus	0.200	0.104 – 0.356
	Xenyllus tegeocranus	0.340	0.283 – 0.428
	Nothrus palustris	0.345	0.245 – 0.487
	Nothrus sylvestris	0.291	0.205 – 0.365
P/A II both soil and litter species P/B around 2.5 growth efficiency 26–30%	*Adoristes ovatus*	0.422	0.348 – 0.524
	Hypochthonius rufulus	0.387	0.256 – 0.499
	Ceratozetes gracilis	0.493	0.289 – 0.632
	Achipteria coleoptrata	0.553	0.382 – 0.728
P/A III soil dwellers P/B < 2.5 adult body mass < 10 μg growth efficiency 26–30%	*Oppia subpectinata*	0.326	0.259 – 0.412
	Chamobates cuspidatus	0.553	0.312 – 0.770
	Tectocepheus velatus	0.168	0.123 – 0.217
	Hemileius initialis	0.502	0.432 – 0.547

Comparing the three groups, earthworms, springtails and oribatids, and expressing the metabolic rates of all three groups per g of body weight, earthworms at 15°C would score 1.3 J g^{-1} h^{-1} on the average, springtails between 3 and 10 J g^{-1} h^{-1} and oribatid mites 2 – 4 J g^{-1} h^{-1}. It is sometimes argued that oribatid mites have a particularly low metabolic rate, but this is not borne out by our analysis. In fact, earthworms are the lowest and springtails the highest. A good deal of biological variation of metabolic rate in soil invertebrates comes from the taxonomic identity of the species.

Our survey of metabolic rates only partly supports the GKGB framework. Metabolism varies with body size and temperature, but in earthworms and oribatids does not show clear trends in relation to ecology, at least not to the state of present analysis. In springtails, the group with the strongest morphological differentiation across the profile, the trend does exist on the average, but with large variation among species. Maybe resting metabolic rate is not the best indicator of metabolic adaptation. A property like *metabolic scope* (the difference between standard and active metabolism) would be a better candidate for soil profile-dependent adaptation. This is a question for further ecophysiological research.

6.2 Temperature responses

In the previous section we have adjusted metabolic rates to a reference temperature using the overall activation energy for the taxonomic group. In this section we take the issue one step further and ask the question: are there also differences between species or genotypes in the way soil invertebrates respond to temperature, allowing evolution of the temperature response as a whole?

To investigate the issue, we first consider some classical methods to describe the temperature responses of ectotherms. This applies to metabolic rate, but also to other biological rate processes such as growth, moulting and reproduction. For short, we refer to these processes jointly as "rate of development". Many different models have been proposed to describe how development rate of ectotherms may depend on temperature and four of them are highlighted here:

- *Linear*: Development is zero below a threshold temperature and increases linearly with increasing temperature. This model enjoys great popularity among entomologists. It is the basis behind "*degree-day summation*", a method in which the environmental temperature above a threshold is accumulated over time to predict the completion of a certain phase in the life of an insect, e.g., larval development. As long as the model holds, the temperature sum is a constant, characteristic for the completion of the specific developmental process. In a graph of development rate against temperature, the reciprocal of the slope is known as the "*thermal constant*", which equals the temperature sum for the completion of development (Figure 6.2). The degree-day method has also been applied in studies of soil invertebrates, e.g., to predict the phenology of oribatid mites in relation to environmental temperature schedules (Kaneko 1988). In a survey of temperature responses in nematodes, spiders and insects, Trudgill et al. (2005) found a negative correlation between the thermal constant and the threshold temperature: species with a high threshold temperature tend to have a small thermal constant (i.e., a steep response).

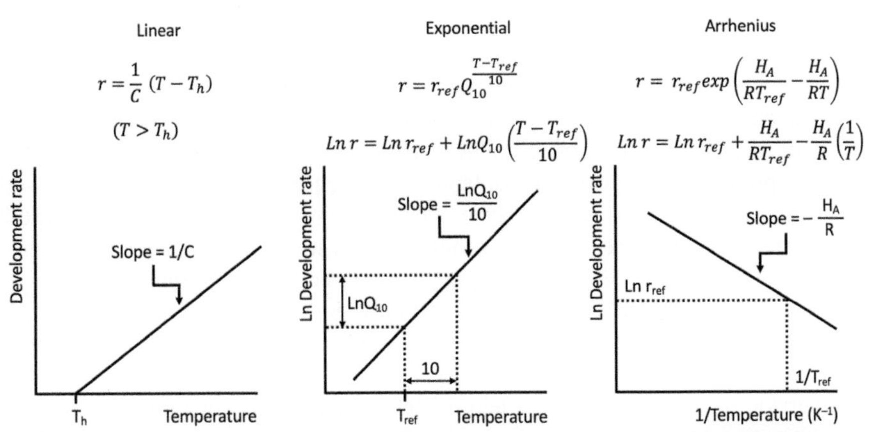

Figure 6.2. Illustration of the three main models for ectotherm temperature-dependent development. The formulas are written with explicit parameters: T_h = threshold temperature, C = thermal constant, Q_{10} = multiplicative increase of rate with a 10 degrees temperature rise, r_{ref} = development rate at reference temperature, T_{ref} = reference temperature, H_A = activation energy. Development rate in the exponential model is Ln-transformed so as to linearize the function. In the Arrhenius model this is achieved by plotting the Ln-transformed rate against the reciprocal of Kelvin temperature. C, Q_{10} and H_A may be considered measures of thermal responsiveness.

- *Exponential*: Development rate increases exponentially with temperature. This response originates from the theory of chemical reaction kinetics and is known as *Van 't Hoff's law*, named after Jacobus van 't Hoff, Dutch chemist and Nobel Prize winner. According to the equation, development rate increases with a fixed factor with equal steps on the temperature axis. For a step of 10°C this factor is called Q_{10}, by far the most common quantity to express temperature responsiveness. The Van't Hoff equation is valid only in the physiological metabolic range from 5 to 25 °C; it does not have a threshold temperature, nor an upper bound. Q_{10} is typically around 2, more often above than below, also for soil invertebrates. For instance, Zinkler (1966) found Q_{10} values for springtails and oribatid mites to vary between 1.9 and 2.7.
- *Arrhenius*: This model, derived by another Nobel prize winner, the Swedish physicist Svante Arrhenius, also derives from physical chemistry and can be seen as an extension of the Van 't Hoff equation. According to this model, development rate increases with an exponential factor determined by the ratio of thermal energy of the reactants relative to the activation energy of the reaction. The key parameter in the Arrhenius equation is *activation energy*, or *activation enthalpy*, H_A. The higher H_A, the steeper the temperature response. The model is usually visualized in a plot of the logarithm of reaction rate versus the reciprocal Kelvin temperature, which should give a straight line with slope $-H_A/R$ (Figure 6.2). The Arrhenius model was used in the analysis by Ehnes et al. (2011) that we discussed above, and also by Gillooly et al. (2001), Caruso et al. (2010), and many others. Activation energies for biological processes in ectotherms typically vary between 30 and 120 kJ/mol.
- *Sharpe/De Michele:* The most complete model for ectotherm development is due to Peter J.H. Sharpe at Texas University, and co-workers (Sharpe and De Michele 1977, Schoolfield et al. 1981). Their model departs from the Arrhenius equation but acknowledges the fact that both at low and high temperatures, reaction rates are suppressed by thermal inactivation. Characterizing these inactivation reactions by two more activation energies, a rather formidable equation results, with six parameters (not reproduced here), which nevertheless describes the complete physiological range of temperature responses in ectotherms very well (Wagner et al. 1984, Van der Have 2002). In Figure 6.3 the Sharpe-Schoolfield model is fitted to one of the most complete data sets on temperature dependence in soil invertebrates, egg development in the collembolan *Sminthurus viridis* (Davidson 1931). For most other species, however, measurements do not cover the complete temperature range, making it impossible to estimate all six parameters. For comparison of species, we therefore rely on the simplified version of the Sharpe-Schoolfield model, the Arrhenius equation with a single activation energy.

In Figure 6.1 we already saw that there are differences in activation energy between phylogenetic lineages. Such differences are also present among species within each lineage. We explore a data set on Collembola egg development to illustrate the argument. For other soil invertebrates a similar analysis is not possible due to dearth of data. In total 141 temperature responses of Collembola were collected,

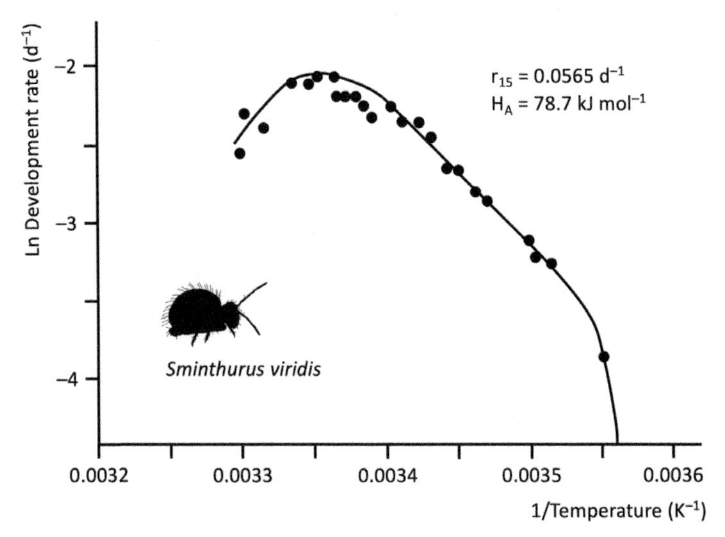

Figure 6.3. Arrhenius plot of the temperature response of egg development rate in *Sminthurus viridis*, using data from Davidson (1931). The continuous line is a least-squares fit of the Sharpe-Schoolfield equation (Sharpe and De Michele 1977, Schoolfield et al. 1981). The curve shows a linear segment in which the Arrhenius model holds, characterized by r_{15} and H_A, but bends downwards at the high and the low ends due to inactivation. After Van Straalen (1994).

mostly on egg development and moulting, but also on growth, reproduction and metabolism (Van Straalen 1994, Van Straalen and Van Diepen 1995). Interestingly, the different processes in the same animal appear to have a similar activation energy: if a species' moulting rate depends strongly on temperature, growth and reproduction depend on temperature equally strong. So, any one of the processes may be considered an indicator of temperature dependence in the species. A summary of the data for egg development (42 species) is given in Figure 6.4, where the species are grouped by Gisin's life-forms.

The differences between the ecological groups suggest a soil-profile-dependent trend: median activation energies are highest for epigeic species and lowest for euedaphic species. Interestingly, species living in caves (*troglomorphs*) have the lowest activation energy of all (Figure 6.4, right diagram). Since troglomorphism is a derived condition within the Collembola, we must assume that low activation energy evolved both in the soil-dwelling species and in cave-living lineages. In both cases this led to a shallow temperature response, approaching non-responsiveness. The data are a perfect illustration of Kennedy's argument in the GKGB hypothesis (see Chapter 1).

Like in nematodes and spiders (Trudgill et al. 2005) the Collembola data also show quite a strong correlation between development rate at 15°C and activation energy. Fast-developing species also respond strongly to temperature. This fits in an evolutionary scenario for the transition of hexapods from soil-living to surface-living. Not only their development speeded up, but also their temperature responsiveness. This scenario is confirmed by the fact that temperature responsiveness among foliar insects is generally larger than in the soil fauna. Berg et al. (2010) illustrated this

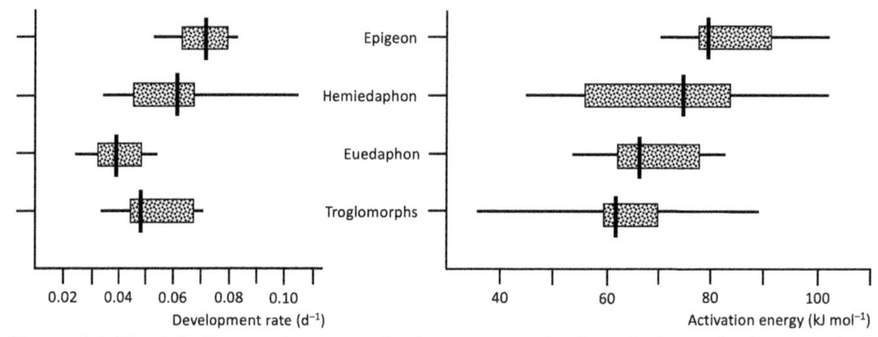

Figure 6.4. The left diagram shows egg development rates (reciprocal of egg development time), adjusted to 15°C, for 42 species of Collembola, grouped by Gisin's morphotypes. The right diagram shows activation energies for the same species. Both parameters were estimated by fitting the Arrhenius model to the original data (*cf.* Figure 6.2). For the morphotypes see Figure 1.2. For each group the median, interquartile range and full range of the data are given. From Van Straalen (1994).

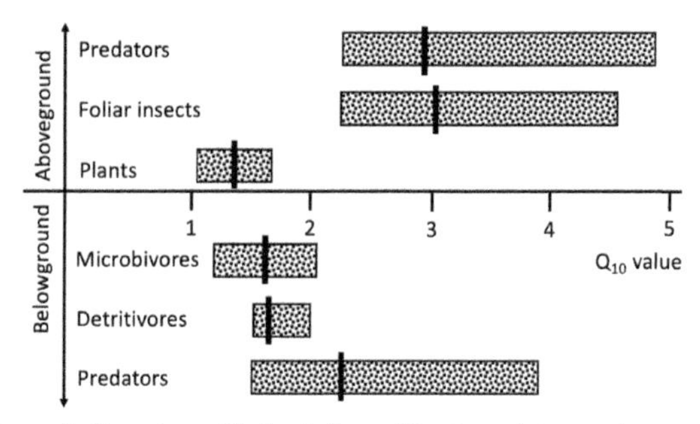

Figure 6.5. Overview of Q_{10}-values (derived from literature data on temperature responsiveness of metabolic rate and growth) for different functional groups of below-ground and above-ground communities. The vertical lines indicate means and the hatched boxes the range. Redrawn after Berg et al. (2010).

argument in a review of Q_{10}-values for above-ground and below-ground invertebrates (Figure 6.5). The authors noted that while climate change is raising environmental temperatures, a mismatch in aboveground-belowground interactions may occur, due to the fact that above-ground invertebrates respond differently to temperature than below-ground invertebrates.

Our analysis suggests that, at least in Collembola, temperature responsiveness is a trait that can evolve. Therefore, it is also expected to show genetic variation within a species. This was confirmed in a comparative study on juvenile growth which showed a more pronounced response to temperature in forest populations compared heath populations (Liefting and Ellers 2008). However, attempts to sort out the genetic mechanisms have been unsuccessful so far. In the model springtail for genetic research, *Orchesella cincta*, Driessen et al. (2007) attempted to estimate heritability of temperature responsiveness by exposing individual juvenile springtails subsequently to 12°C and 22°C. The difference in growth at these two temperatures was taken as a measure of the *thermal reaction norm* of the individual. However, nor artificial selection, nor parent-offspring regression could demonstrate a significant heritability (h^2) of this trait. Thermal responsiveness sometimes increased under selection for low values or decreased when selected for high values. The only consistent difference was a higher average thermal sensitivity in females compared to males. The conclusion of this work was that temperature responsiveness is determined by a complex non-additive and partly sex-linked genetic architecture.

The scientific literature is not unanimous on whether temperature responsiveness (measured by C, Q_{10} or H_A) can be considered a biometric (quantitative) trait. Under one view, phenotypic plasticity is defined as the covariance between genotype and environment and is considered a *bona fide* trait of which the heritability and response to selection can be studied (Scheiner and Lyman 1989, Scheiner 1993). In an alternative view, plasticity is not considered a single trait, but a by-product of selection in different environments (Via 1993, De Jong 1995). In this view there are no "*genes for plasticity*", separate from genes for trait means. The greatest problem, however, is the lack of insight into the genetic architecture of temperature responsiveness of any ectotherm, including soil invertebrates.

6.3 Cold hardiness

Frost is a critical issue for all invertebrates. According to many an entomological study, winter mortality is the most important factor contributing to year-to-year fluctuations of insect populations in temperate climates. Such studies have used *key factor analysis*, a statistical technique applied to time series of insect abundance data, introduced by the Canadian entomologist R.F. Morris (Morris 1959, Southwood 1967). Key factor analysis often showed that variation in winter mortality correlated highly with post-winter population abundance. Frost during winter is even more acute for soil invertebrates that live in the pore water, such as nematodes, tardigrades and earthworms: it is their very environment that freezes.

So, tolerance against low temperature and freezing, loosely called *cold-hardiness*, is a necessary requirement for any terrestrial invertebrate living at high latitude and high altitude. Significant parts of the land masses in the Holarctic experience frost every winter and in some of them (e.g., 60% of the land surface of Russia) permafrost is the rule. Since the earth's climate has seen many periods of severe cold, frost must also have been an issue in the evolutionary past, for those lineages that had undergone terrestrialization. In this section we ask the question: which mechanisms of cold hardiness have evolved in soil invertebrates and are there any differences between the ecological groups?

The cold-hardiness literature commonly makes a distinction between *freeze-avoidance* and *freeze-tolerance* (Wharton 1995, Holmstrup and Zachariassen 1996, Block 1996, Sømme 1999, 2000). Freeze-tolerant animals are those that can survive freezing and apply mechanisms to accommodate controlled freezing at relatively high (below-zero) temperatures. Freeze-avoiding animals aim to prevent freezing by all means, and rely upon supercooling. Among soil invertebrates, freeze-tolerance is found in nematodes, tardigrades and some species of earthworm (Wharton and Brown 1991, Sømme 1996, Wharton 2003, Hengherr et al. 2009, Berman et al. 2019). In freeze-tolerant species ice from the soil is allowed to spread across the body wall and this takes place at relatively high subzero temperatures (*ca.* –2 C°), a phenomenon called "*inoculative freezing*".

The distinction between freeze-tolerance and freeze-avoidance, although useful, is a simplification. The nematode *Panagrolaimus davidi* can undergo both complete freezing or freeze-avoidance, depending on the rate at which the temperature goes down (Wharton et al. 2003). In fact, even in so-called freeze-tolerant invertebrates the tissues themselves often do not tolerate ice formation. Intracellular freezing is rare, and lethal. Freeze-tolerance in many cases boils down to freeze-avoidance on the cellular level.

Cold hardiness can be measured by survival time under low temperature, but a particularly interesting variable is *supercooling point*. Pure water will usually not freeze exactly at 0°C, but at a temperature below zero, depending on conditions such as the presence of nucleators, water movement and the rate of cooling. This phenomenon is called supercooling. The strategy of a freeze-avoiding animal is therefore to push down the supercooling point as far as possible.

Supercooling points can be readily measured in small invertebrates by fixing them to a thermocouple, hung up in container with below-zero air temperature, while continuously monitoring the temperature of the animal. The temperature at which the animal freezes can be recognized by a small upwards bump in the decreasing temperature track, which is due to the heat of fusion becoming available upon freezing. Thanks to the ease of measurement, supercooling points have been reported on numerous occasions.

The Norwegian entomologist Lauritz Sømme had started to measure supercooling points in prostigmatid mites sampled at the nature reserve Vestfjella, Southern Norway (Sømme 1978). These animals exhibited considerable supercooling (down to –30°C). From Sømme's data and those of many others it appeared that there are appreciable differences between animal species, even in the same taxonomic category. Invertebrates collected from the Upper Kolyma Basin, in the Russian Far East, illustrate this variation, with supercooling points down to –45°C in some species of ant but only moderate supercooling in others (Berman and Leirikh 2018, Figure 6.6). That extreme supercooling has evolved more than once within the Formicidae illustrates that it is not a trait constrained by phylogeny but something that may evolve by adaptation in any taxon living under sufficiently cold conditions.

A similarly extreme supercooling capacity was noted by the British biologist William Block, who, working at the British Antarctic Survey, studied microarthropods at Signy Island in the Maritime Antarctic. In the springtail *Cryptopygus antarcticus* supercooling temperatures down to –30°C were no exception (Block et al. 1978). This also holds for nematodes; both free-living species and free-living stages of plant parasites can supercool to –30°C (Wharton et al. 1984).

In contrast to arthropods and nematodes, earthworms and enchytraeids are not known for extreme supercooling capacity. Even in one of the most cold-tolerant species, *Dendrobaena octaedra*, supercooling does not go down further than –12.2°C (Holmstrup 1994) and enchytraeids from Greenland reached –15°C (Slotsbo et al. 2008). Slugs are also poor supercoolers, surviving temperatures of only a few degrees below zero (Storey et al. 2007). The same holds for woodlice (Tanaka and Udagawa 1993). A moderate degree of supercooling is found in temperate Collembola, even during summertime; e.g., in forest floor springtails in the Netherlands, supercooling points varied from –15°C in winter to –5°C in summer (Van der Woude 1987).

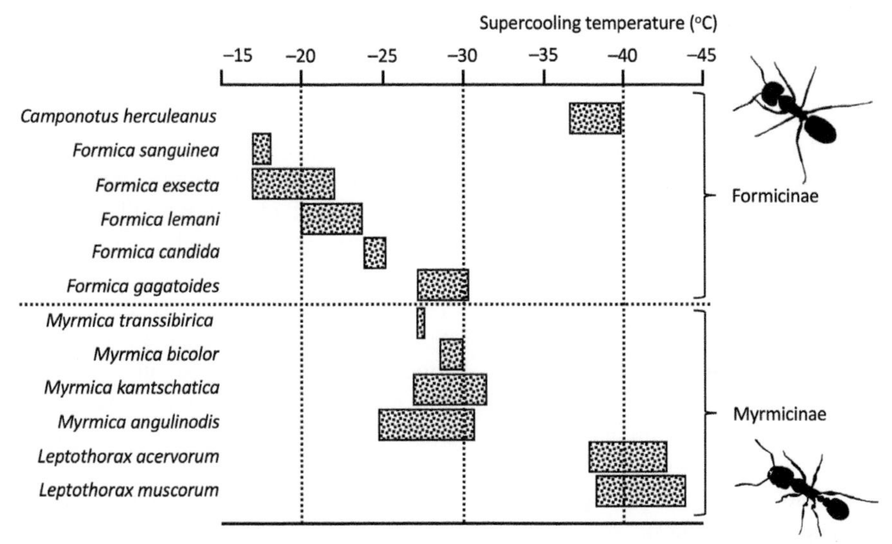

Figure 6.6. Variation in supercooling points measured for 12 species of ant, living in the Upper Kolyma Highlands, Russian Far East. Extreme supercooling is found in two subfamilies (Formicinae and Myrmicinae). From Berman and Leirikh (2018), with modifications.

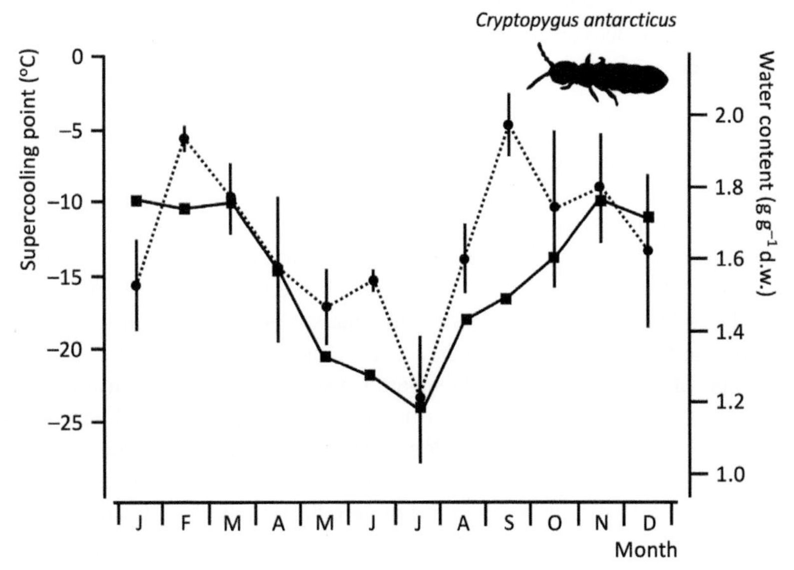

Figure 6.7. Changes in supercooling point (squares) and water content (circles) of springtails, *Cryptopygus antarcticus*, sampled at Signy Island, Maritime Antarctic. Water content data were averaged over 4 years and supercooling points are monthly averages over 8 years. Modified from Block (1996).

The capacity for supercooling is not a fixed property of the animal, it is highly inducible by environmental temperature. To illustrate this, Figure 6.7 shows the seasonal changes of supercooling point and water content in *Cryptopygus antarcticus* (Collembola, Isotomidae) during an annual cycle. The fact that supercooling temperatures are lower in winter implies that an actively regulated, temperature-dependent process modulates supercooling.

The data in Figure 6.7 also show that supercooling and water content are closely related: both go down during the cold Antarctic months of April to August. This is due to cold hardiness and drought tolerance relying on similar biochemical mechanisms. The dehydration power of low temperature (caused by the temperature dependence of water vapour pressure) could even be the primary effect of cold. We will see below that some mechanisms for cold hardiness actually rely on controlled dehydration.

In several of the soil arthropod studies, the distribution of supercooling points over the population is markedly bimodal. The population often consists of a high group and a low group (Block et al. 1978, Sømme 1978, Van der Woude 1987, Block 1996, Sjursen and Sinclair 2002). This bimodality is assumed to be due to variable gut content. It is assumed that particles in the gut act as ice nucleators, inducing the freezing. So, the first thing a freeze-susceptible animal should do is evacuate its gut. Indeed, it was shown that epigeic springtails (*Orchesella cincta*) sampled from tree bark, had empty guts during wintertime when their supercooling points were low (Van der Woude 1987).

What physiological mechanisms lie behind the induction of cold hardiness in soil invertebrates? Biochemical and molecular work has pointed out four principles:

- accumulation of glycerol and sugars
- cryoprotective dehydration
- induced synthesis of thermal hysteresis proteins
- modulation of phospholipid fatty acid composition.

Accumulation of glycerol or other "*polyols*" (compounds with two or more hydroxyl groups), such as glucose, trehalose, inositol, ethylene glycol, etc., is a most common biological mechanism, present in plants as well as animals. The action of these compounds relies on their highly hydrophilic nature affecting the likewise polar water molecules in such a way that it becomes more difficult for them to form ice crystals. In addition, the concentrations of solutes will increase osmolarity, thereby decreasing the freezing point of water, a phenomenon called *colligative action.* The mobilization of highly water-soluble sugars is usually from stored *glycogen,* which in many cases is seen to decrease under the development of cold hardiness.

Glucose accumulation is the dominant cryoprotecting mechanism in the enchytraeid worm *Enchytraeus albidus*. This was shown in experiments reported by Slotsbo et al. (2008), summarized in Figure 6.8. Already at –2°C, glucose is seen to be mobilized while glycogen concentrations decrease and this effect is enhanced at –14°C, at least for two populations from Greenland (Zackenberg and Nuuk) (Figure 6.8). Another population from Jena, Germany, was not as cold tolerant as the Greenland populations. The data suggest that temperature-induced glucose release from glycogen may be a mechanism under positive selection in Arctic populations (Fisker et al. 2014). Glucose is also needed for long-term anaerobic metabolism and so has a dual role during supercooling and freezing (Calderon et al. 2009).

In earthworms, the cocoons are considerably more cold-tolerant than the animals themselves. Earthworms do not rely on sustained supercooling but their cocoons

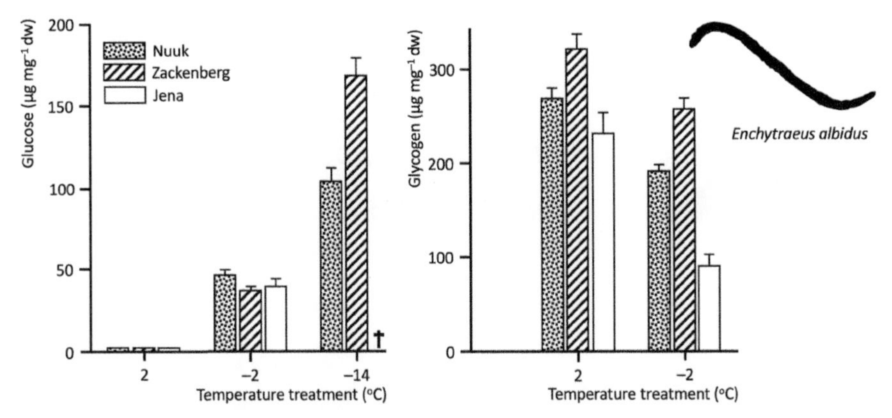

Figure 6.8. Glucose (left) and glycogen (right) concentrations in three populations of the potworm *Enchytraeus albidus* (Nuuk and Zackenberg from Greenland, Jena from Germany), experimentally subjected for 25 days to low temperatures (2, –2 and –14°C). Glucose was not measured in the German population exposed to –14°C due to heavy mortality. From Slotsbo et al. (2008), reproduced by permission from Elsevier.

can survive frozen soil easily. It was long thought that all earthworms are freeze-susceptible, but freeze-tolerance has been shown to exist in several cases, for example the Siberian species, *Eisenia nordenskioldi*, *Aporrectodea caliginosa* from Finland and even *Dendrobaena octaedra* from Denmark (Holmstrup and Zachariassen 1996, Holmstrup et al. 1999).

Cold hardiness in freeze-susceptible earthworms is achieved by a strategy called *cryoprotective dehydration*. This was discovered by the Danish physiologist Martin Holmstrup in the early 1990s (Holmstrup 1992, 1994, Holmstrup and Westh 1994, 1995), in experiments with cocoons of the northern parthenogenetic earthworm *Dendrobaena octaedra*. Cryoprotective dehydration is the release of osmotically active water from the tissues so as to make sure that the vapour pressure of the body fluids is always in equilibrium with the surrounding ice. The tissue itself does not freeze. The cold hardiness of earthworm cocoons relies on them being very resistant to dehydration: they can lose up to 80% of their normal water content (Petersen et al. 2008). The role of cryoprotectants is not to increase the osmolality of the body fluid, but to protect the cell membranes of the embryo (Holmstrup and Zachariassen 1996). A similar link between water content and cold hardiness was apparent in the Antarctic springtail (Figure 6.7) and cryoprotective dehydration was also identified in the Arctic collembolan, *Megaphorura arctica* (Holmstrup and Sømme 1998, Holmstrup 2018). This strategy may be common to all soil invertebrates with water-permeable cuticle.

A drawback of colligative cryoprotection, e.g., by polyol accumulation, is that it increases the viscosity of the body fluids. This may be less of a problem in plants, but in animals it will impede circulation and therefore has its limits. Many animals have an additional, non-colligative freeze-protection system due to *antifreeze proteins* (*AFPs*) also called *thermal hysteresis proteins* (*THPs*). Thermal hysteresis is the phenomenon that freezing and melting temperatures of a liquid are separate from each other. Antifreeze proteins augment this effect, they increase the hysteresis. It has

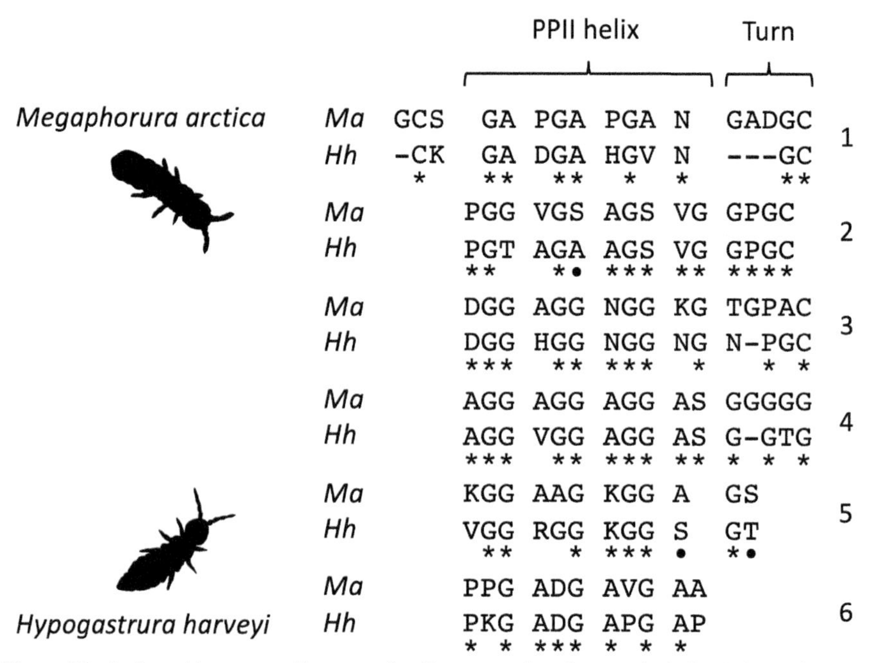

Figure 6.9. Amino acid sequence alignment of antifreeze proteins of two arctic Collembola. Amino acids are indicated by their single letter codes. The sequence reads from top left (N terminus) to down right (C terminus). Identical and similar residues are indicated by asterisks and dots, resp. The N-terminal signal sequence is not shown. The coding sequence contains six PPII helices (numbered 1–6) as well as linker sequences (turns). Helices 2, 4 and 6 take part in the outside ice-binding surface, helices 1, 3 and 5 are on the backside surface. Note the large number of glycines (G). Redrawn from Graham et al. (2020).

been suspected for a long time that these THPs, known from Antarctic fish, should also be present in soil invertebrates, and suggestions have been made repeatedly (Van der Woude 1987, Meier and Zettel 1997), but it was not until 2005 that the first AFP was isolated from a species of springtail active on snow, *Hypogastrura harveyi* (Graham and Davies 2005). Later such proteins were also identified in the Antarctic springtail *Gomphiocephalus hodgsoni* (Hawes et al. 2014) and the Arctic species *Megaphorura arctica* (Graham et al. 2020).

The action of antifreeze proteins is to bind to ice crystals and so inhibit their growth. This is achieved by an extraordinary high content of polar amino acids such as glycine or proline. The protein's secondary structure is characterized by several so-called *polyproline type II (PPII) helices*, domains with an extraordinary content of polar amino acids, often organized in arrays of three residues. AFPs usually come in a variety of isoforms, smaller and larger polypeptides, and often carry a signal sequence which targets them for excretion.

In Figure 6.9 an alignment is shown for the two antifreeze proteins characterized so far in Collembola, from *Hypogastrura harveyi* and *Megaphorura arctica* (Graham et al. 2020). The protein is predicted to form six PPII helices, as well as short "turn" sequences linking them. Of these six helices, three take part in the ice-binding outer surface. The many glycine residues make the protein very polar, a necessary property for binding to ice crystals. How the expression of AFP genes is regulated is not known, but induction by some kind of temperature-sensing system must be assumed.

Where collembolan AFPs come from is likewise unknown. They are very different from the antifreeze proteins of Antarctic fish and cod, which are different from each other as well. The similarity between the two springtails, however (belonging to two different families), is remarkable, given the expected very rapid evolution of these greatly adaptive proteins. Graham et al. (2020) estimated that the divergence could go back to the ice-ages of the Permian (300 Ma BP). Strong stabilizing selection would have conserved the coding sequence, while the genomic environment of the genes diverged over the years. A better view of the adaptive evolution of AFPs in soil invertebrates requires more sequence data and a molecular clock analysis.

The fourth mechanism contributing to cold hardiness is modulation of the fatty acid composition of cell membranes. Without adaptation, membranes tend to become more rigid at low temperature, which disturbs the many membrane-bound proteins such as ion channels, signal transducing receptors, import and export pumps, etc. All eukaryotic organisms therefore avail of a system to upregulate the fluidity of their cell membranes at low temperature. This is primarily done by modulating the proportion of unsaturated fatty acids using enzymes with desaturase activity, a process called *homoviscous adaptation.*

Van Dooremalen et al. (2011, 2013) measured the fatty acid composition in nine species of Collembola acclimated to different temperatures. As the *poly-unsaturated fatty acids* (*PUFAs*) are difficult to measure due to their instability during isolation and purification (Van Dooremalen et al. 2009) we recapitulate here only the data for fatty acids with one double bond (*mono-unsaturated fatty acids*, *MUFAs*). For epigeic and hemiedaphic species, these data demonstrate the expected increase of unsaturation at lower temperatures, from an average value of 21% at 20°C to 32% at 5°C, while the euedaphic species don't show a significant response at all (Table 6.4). The data are in perfect agreement with the trends in metabolic rate (Table 6.2) and egg development rate (Figure 6.4), which all indicate a greater temperature plasticity in surface-active springtails than in soil-dwelling species. This matches nicely our conclusion that epigeic Collembola are better equipped to evolve cold-hardiness than euedaphic species and this again fits with the GKGB hypothesis of Chapter 1.

In earthworms, polyunsaturated fatty acids likewise increased at 0°C compared to 20°C, but the response was similar for the freeze-tolerant *Eisenia nordenskioldi* and the freeze-avoiding species *Lumbricus rubellus* (Petersen and Holmstrup 2000). As both species are epigeic, this may be expected. It would be interesting to compare this to the temperature response of endogeic earthworms.

The four cold-hardiness strategies discussed here do not necessary exclude each other. Polyol accumulation, cryoprotective dehydration, antifreeze proteins and fatty acid modulation may be considered part of a physiological toolbox common to all animals. However, some evolutionary lineages rely on one mechanism more than others. It seems that cryoprotective dehydration is deployed mostly by species with water-soluble cuticle (nematodes, earthworms, enchytraeids and euedaphic springtails), while supercooling with cryoprotectants is most common among species living in open habitats, on the soil surface and in exposed above-ground habitats.

Table 6.4. Mono-unsaturated fatty acid content of nine species of springtail, experimentally acclimated to 5°C or 20°C. Data from Van Dooremalen et al. (2013).

	MUFA ± s.e.		
	5°C	**20°C**	**Delta**
Epigeic species			
Sminthurus viridis	18.8 ± 2.6	16.4 ± 2.3	2.40
Orchesella cincta	40.1 ± 0.6	24.2 ± 0.4	15.9*
Hemiedaphic species			
Tomocerus minor	38.3 ± 3.6	27.9 ± 2.8	10.4*
Isotoma riparia	28.2 ± 1.2	16.2 ± 0.8	12.0*
Desoria trispinata	33.8 ± 0.8	20.4 ± 1.6	13.4*
Euedaphic species			
Sinella tenebricosa	35.8 ± 0.5	39.2 ± 0.4	–3.40
Folsomia candida	50.0 ± 0.1	49.0 ±0.2	1.00
Protaphorura fimata	45.4 ± 0.1	48.7 ± 0.3	–3.30
Protaphorura subarmata	41.1 ± 0.6	42.5 ± 0.2	–1.40

MUFA = mono-unsaturated fatty acids, relative to total fatty acid content. Delta is the increase of MUFA when lowering temperature from 20°C to 5°C. An asterisk indicates a significant difference between the two temperature treatments.

Antifreeze proteins seem to be associated mainly with specialized *psychrophiles*, living on snow surfaces or under arctic conditions.

Another general conclusion is that cold-hardiness induction is generally stronger and better developed among surface active Collembola than among euedaphic ones. Epigeic Collembola resemble insects in this respect. For earthworms a similar trend could be true, but the picture is more complicated. Finnish populations of endogeic earthworm, e.g., *Aporrectodea caliginosa* appeared to be more freeze-tolerant than the epigeic *Lumbricus rubellus* sampled at the same place (Holmstrup and Overgaard 2007). In addition, adult worms can avoid frost by digging deeper, but cocoons cannot.

In all soil invertebrates, surface activity requires both cold-hardiness and drought tolerance, and partly relies on the same physiological basis, so a joint evolution of these properties is likely. The presence of drought resistance is sometimes seen as a "pre-adaptation" to cold-hardiness, but in fact they are complementary adaptations going hand in hand (Block 1996).

6.4 Drought resistance and anhydrobiosis

In addition to temperature, relative air humidity is another, maybe the main, driving force for adaptation of soil invertebrates. All surface-living soil invertebrates need to be resistant against desiccation to some extent. The most straightforward adaptation is to limit water loss across the external body surface. In experiments, this is measured by placing individual animals in non-saturated air on a sensitive electrobalance and recording the weight loss. The evaporation, expressed in mg of water per minute is related to the surface of the animal to obtain the "*transpiration rate*" (mg min^{-1} mm^{-2}).

This quantity is linearly related to the drying power of the air, measured from the vapour pressure of water, expressed, e.g., in mm Hg. The proportionality constant is called "*cuticular permeability*" (mg min^{-1} mm^{-2} $mmHg^{-1}$) and its reciprocal is called "*cuticular resistance*" (expressed in mmHg min mm^{-1}, noting that 1 mg of water is equivalent to 1 mm^3). This constant characterizes the species. The French ecophysiologist Guy Vannier likened it to the resistance in Ohm's law, defined as the quotient of voltage (here saturation deficit) to current (here evaporation flux) (Vannier 1974).

Using these definitions, species with different body shapes may be compared in a proper manner. For example, in woodlice, the shape of the body (flattened like in *Oniscus* or curved like in *Armadillidium*) significantly contributes to desiccation resistance, even at similar cuticular resistance (Broly et al. 2015). Body size has a non-linear effect: large species have a lower rate of water loss per unit of body weight (Vannier and Verhoef 1978, Csonka et al. 2018), due to the allometric relationship between surface and volume. Surface structures like hairs or scales, or patterns of cuticular lipid droplets may complicate matters further.

Terrestrial isopods have been favourite models for studying the evolution of drought resistance. The "water relations" of woodlice are among the textbook examples illustrating how the metabolic complex of respiration, transpiration and nitrogen excretion was adapted to match terrestrial conditions. The early publications of the British-born zoologist Eric B. Edney (1913–2000) are still a valuable read (Edney 1954) and his book "*Water Balance in Land Arthropods*" (Edney 1977) continued to be reprinted even up to 2017. Other classical publications are due to another British biologist, John Cloudsley-Thompson (1921–2013) (Cloudsley-Thompson 1962) and to the German-Israelian biologist Michael R. Warburg (1931–2014) (Warburg 1987, Hornung 2015).

These studies applied the classical approach of comparative zoology, placing the species of woodlice in a gradient from semi-aquatic littoral species, such as *Ligia*, via common grassland and woodland representatives, such as *Philoscia* and *Oniscus*, to drought-tolerant forms like *Armadillidium* and *Hemilepistus* (Edney 1968, Warburg 1968, Taylor and Carefoot 1993, Wright and Machin 1993, Habassi et al. 2020). However, we now know that such a linear comparison does not necessarily represent a correct evolutionary scenario. In Chapter 2 (see Figure 2.11) we learned that, according to modern phylogenetic reconstruction, *Ligia* evolved outside the Oniscidea proper. Isopod terrestrialization did not originate with *Ligia*, but occurred three times independently within the order of Isopoda.

In addition, the early physiological studies of isopod physiology often ignored the species-rich family of Trichoniscidae, woodlice with small body size and simple water relations. In the phylogenetic tree of Oniscidea, Trichoniscidae are positioned ancestral to the drought-tolerant families Porcellionidae and Armadillidiidae (Figure 2.11). So, rather than the linear sequence from *Ligia* to *Armadillidium*, a more likely evolutionary scenario is that drought-tolerant woodlice descended from a soil-living *Trichoniscus*-like isopod lineage with high water loss rate and small body size. Data on water loss rate published by Dias et al. (2013) agree with this reconstruction. Representatives of the Trichoniscidae, e.g., the genera *Trichoniscus*,

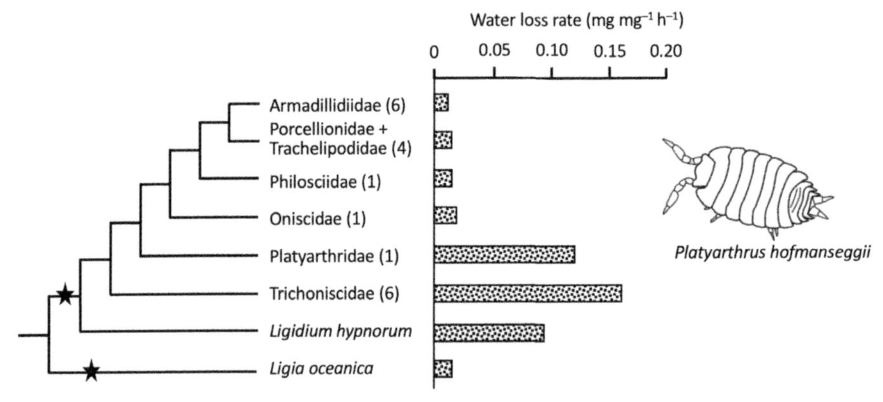

Figure 6.10. Water loss rate for several families of terrestrial woodlice, taken from Dias et al. (2013). Weight loss (mg per mg fresh weight per hour) was measured at 85% relative humidity for 22 species. The data were averaged per family; the number of species is given in parentheses. The phylogenetic tree is based on 18S rRNA sequences, as reported in Dias et al. (2013), except that *Ligia* was positioned separate from *Ligidium*, in agreement with Dimitriou et al. (2019). Porcellionidae is not monophyletic in Dias et al. (2013) and was joined with Trachelipodidae. Two assumed terrestrializations are indicated by asterisks; the third one is not visible in this phylogeny, but see Figure 2.11.

Haplophthalmus and *Hyloniscus*, have a much higher rate of water loss than *Oniscus* and *Porcellio* (Figure 6.10).

The relatively low rate of transpiration in *Ligia* documented by Dias et al. (2013) may relate to its osmoregulatory strategy in the intertidal environment. *Ligia oceanica* is known to regulate the osmotic value of its haemolymph at a high value, *hyperosmotic* to the medium even at 100% seawater (Todd 1963, Carefoot 1993). This creates a constant influx of water and depresses transpiration. As long as the animal can maintain its body fluids at an osmolarity high enough to attract water from unsaturated air it will not suffer from water shortage. The species *Platyarthrus hoffmanseggii*, phylogenetically intermediate between Trichoniscidae and Oniscidae, see Figure 6.10), could also be an exception, since it is *myrmecophilous* (living in ant nests) and is not expected to be selected for drought-resistance in these below-ground environments.

The other woodlice, starting with Oniscidae, evolved drought resistance to deal with low air humidity at the surface. This included changes in the cuticle, but the external surface still remained relatively water-permeable. Another crucial change took place in the respiratory and excretory systems. While the hygrophilic isopods respire mostly with the whole-body surface, there is a tendency in higher forms for respiration to be concentrated in the pleopods (Sutton 1972, Carefoot 1993). In Chapter 4 (Section 4.9) we saw that respiratory structures in arthropods are derived from the legs. Indeed, drought-tolerant isopods have evolved specialized respiratory chambers in their pleopods, called "lungs" or "*pseudotracheae*". These chambers, with their large internal surface but a small outside opening, allow the woodlouse to economize on water loss while maintaining efficiency of oxygen uptake.

Water loss in woodlice is also related to the system of nitrogen excretion. Woodlice are *ammonotelic*, i.e., they excrete ammonia as the end product of their nitrogen metabolism, which may be considered an evolutionary heritage from their aquatic past (Wieser 1972). Urine excreted by maxillary glands in the head is circulated externally across the body surface, during which ammonia is evaporated and the remaining water is recycled by the pleopods and the anus (Hoese 1981, 1982a,b, Carefoot 1993). In *Ligia* the channels run only ventrally, but in *Porcellio* the urine is also circulated through transversal channels in between the dorsal tergites. This peculiar arrangement to deal with the trade-off between water loss and nitrogen excretion is often considered unique to isopods, however, a similar system is found in other soil invertebrates such as springtails.

The kidneys (*labial nephridia*) of springtails are in the head, like in crustaceans, and discharge urine in a groove (*linea ventralis*) running ventrally in between the legs, over the thorax midline, up to the ventral tube on the second abdominal segment, where water may be resorbed (Eisenbeis 1974, Verhoef et al. 1979, Verhoef et al. 1983, Hopkin 1997). It is assumed that this system, like in isopods, allows ammonia excretion, while economizing on water loss. This interpretation is confirmed by observations on the morphology of the ventral groove: it is an open groove in drought-sensitive species, and is partly closed like a channel in drought-resistant springtails (Verhoef et al. 1983).

Surface-active woodlice and epigeic Collembola have solved the issue of water loss by morphological adaptations (less permeable cuticle, recycling of urine and specialized respiratory organs), but euedaphic Collembola and other soil-dwelling invertebrates with permeable cuticle still depend on completely saturated air. Among all life-history stages, the hatchlings are usually most sensitive and, under drought stress, reproduction is halted even before effects on growth are apparent (Wang et al. 2022). Even while animals are safe in saturated soil pores most of the time, they run a risk of drying out when these pores become under-saturated, e.g., in dry periods or close to plant roots. For example, at a haemolymph osmolarity of about 300 mM, a common value for springtails (Witteveen et al. 1987), net water loss will already occur at a relative air humidity of 99.4% (Bayley and Holmstrup 1999). These animals must survive short periods of water loss by physiological flexibility. How do they do that?

The mechanism was discovered by Bayley and Holmstrup (1999): euedaphic springtails survive temporary drought by accumulating sugars. In experiments with *Folsomia candida*, the authors showed that springtails initially dehydrate significantly when exposed to 98.2% RH, but they recovered within 4 days by raising the sugar content of their haemolymph to a value such that water could be attracted from the air (Figure 6.11). The mechanism is similar to induced cold-hardiness, discussed in Section 6.3.

The main sugars present in the haemolymph of *F. candida* turned out to be myoinositol and glucose. These two sugars alone constitute about 50% of the total osmotic value of the haemolymph under dry conditions. How the release of these sugars is induced is not known. In a whole-genome transcription profiling study, Timmermans et al. (2009) found a strong upregulation of carbohydrate metabolism in dehydrated springtails, but the crucial enzyme, *inositol-monophosphatase*, catalyzing the production of myoinositol from glucose, was hardly affected under

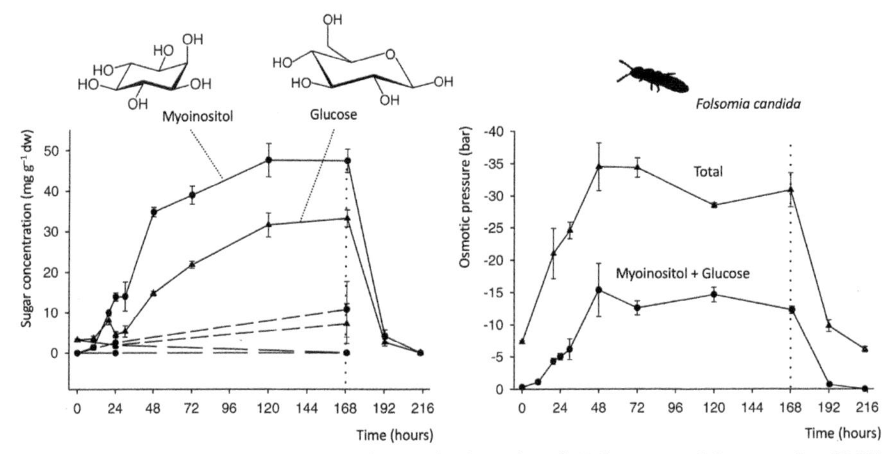

Figure 6.11. Left: increase of haemolymph sugar in the springtail *Folsomia candida* exposed to 98.2% relative humidity for 168 h, then returned to 100% RH. Animals were also exposed to 99.6% RH (dashed lines) and 100% RH (long dashes). Right: combined osmotic contribution of glucose and myoinositol (circles) and the total (measured) osmotic value (triangles). From Bayley and Holmstrup (1999). Reprinted with permission from AAAS. Structural formulas of the two main sugars are given above the graph.

desiccating conditions. How springtails upregulate their sugar metabolism remains unknown to date.

The most extreme form of drought tolerance found among soil invertebrates is *anhydrobiosis*. Our discussion on drought tolerance cannot be complete without highlighting this fascinating phenomenon. During anhydrobiosis an animal loses almost all its water (retaining sometimes less than 1% of the original amount), and continues to live in a quiescent state with practically zero metabolism. Although there is usually some mortality, it is common for 60–80% of the animals to revive after rehydration. The discovery of anhydrobiosis is usually attributed to the Dutch microscopist Antonie van Leeuwenhoek who described in 1702 how he could see "*animalcules*" in the water after he had rewetted completely dried-out sediment collected from a gutter (Wright et al. 1992, Wright 2001). The little animals were rotifers belonging to the class Bdelloidea.

We now know that anhydrobiosis occurs in several metazoan lineages including, in addition to Rotifera, Nematoda, Tardigrada, a few insects, brine shrimps and—often forgotten—Collembola (Wharton 2015). Anhydrobiosis in the aquatic environment is restricted to sediment-associated animals. In soil invertebrates anhydrobiosis is found mostly in pore water-dependent groups (nematodes and tardigrades). Collembola are the only truly terrestrial group showing anhydrobiosis. It is not found in earthworms, enchytraeids, slugs and snails, although some earthworms and gastropods show extensive drought-tolerance during aestivation, as we have seen in Section 3.7.

Anhydrobiosis is a form of *cryptobiosis*. Other types of cryptobiotic phenomena are *cryobiosis* (due to frost), *osmobiosis* (due to hyperosmolarity) and *anoxybiosis* (due to anaerobic conditions). Diapause responses and resting eggs are usually not considered cryptobiosis, although in some invertebrates, diapause eggs may undergo anhydrobiosis as well. The whole gamut of cryptobiotic and quiescence responses

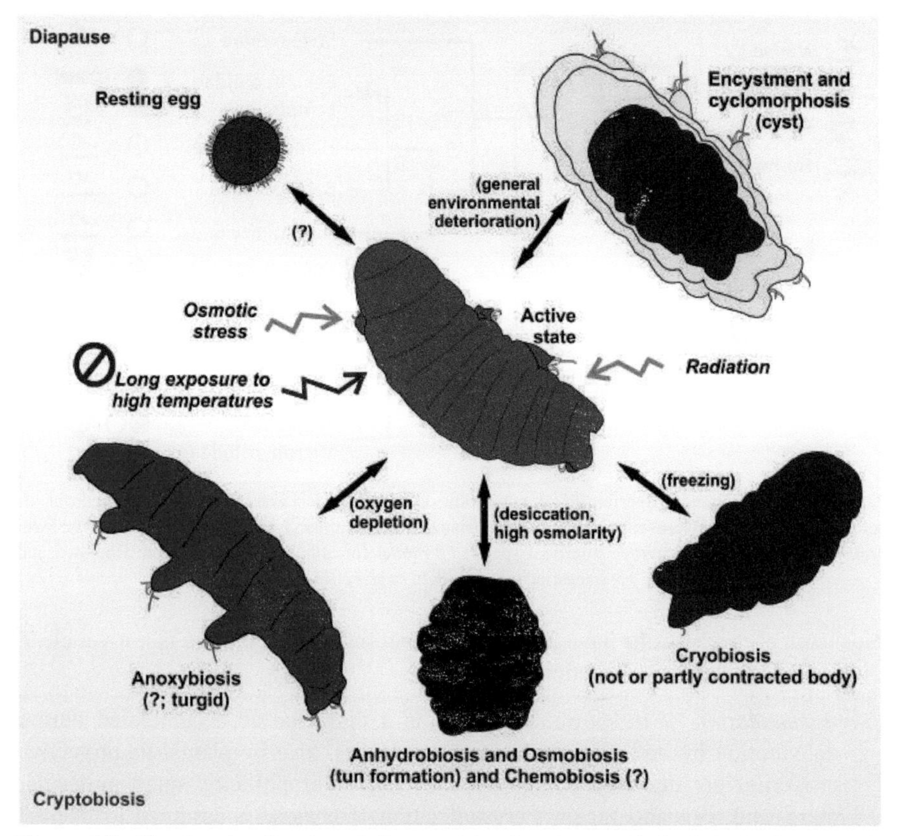

Figure 6.12. Showing the diversity of diapause (above) and cryptobiosis (below) phenomena in tardigrades. Each stress (frost, desiccation, hyperosmolarity, oxygen depletion) induces a different phenotype. There is no cryptobiotic stage that provides tolerance to chronically high temperatures. Reprinted from Møbjerg and Cardoso Neves (2021).

is seen in tardigrades (Figure 6.12). The phenotypes are different for each form of cryptobiosis.

Anhydrobiosis is characterized by a typical "*tun*" (barrel) stage and the transition to tun is called *tunnification*. When slowly dehydrating, the animal retracts its lobopods, shrivels along the segmental lines and loses more than 50% of its water, while extruding lipid droplets (Wright et al. 1992, Sømme 1996, Guidetti et al. 2011). Further dehydration to 0.5% may take place, during which the animal contracts completely and becomes hardly recognizable as an animal. Upon rehydration the tun phenotype reappears before the animal assumes its normal shape. Within tardigrades, anhydrobiosis is found in all lineages of Eutardigrada, the group that arose after terrestrialization of the phylum (Figure 6.13). It is not found in marine tardigrades (Heterotardigrada), except in the lineage of Echiniscoidea, which includes intertidal and limnoterrestrial species (*cf.* Figure 2.2).

The extreme stress resistance of the cryptobiotic stages of tardigrades has evoked a significant body of research due to its fundamental biological implications and possible medical applications, e.g., in the preservation of tissues. Several

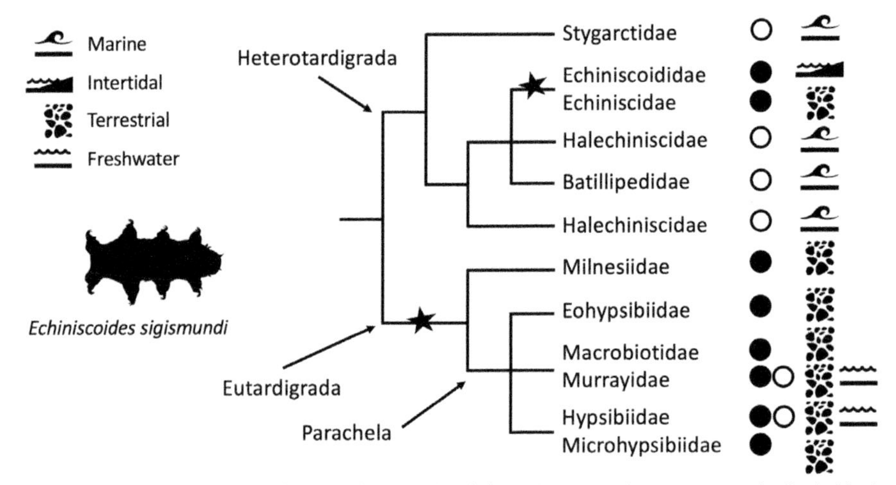

Figure 6.13. Phylogenetic distribution of anhydrobiosis in tardigrades. The occurrence of anhydrobiosis (closed circles) is closely connected to terrestrializations (asterisks) in the phylum. *Echiniscoides sigismundi* is a heterotardigrade intertidal species, a model for anhydrobiosis research. Redrawn and modified after Guidetti et al. (2011).

biochemical mechanisms have been studied, but the overall picture is not yet clear. Five groups of defensive reactions stand out:

- *Accumulation of trehalose.* Trehalose is a disaccharide accumulated during dehydration by both tardigrades and nematodes, also by plants. Its protective properties are due to a combination of high hydrophilicity, small molecular mass, and resistance against crystallization. Trehalose is assumed to interact with the phosphate groups on membrane phospholipids, keeping them apart and so stabilizing the membrane (Crowe et al. 1992, Gal et al. 2005, Boothby 2019).
- *LEA proteins. Late Embryogenesis Abundant* proteins (called after their dominance in a late stage of plant seed development, i.e., after abscission) are a diverse group of small hydrophilic peptides, which lack a secondary structure and are therefore also called "*intrinsically disordered proteins*" (Tunnacliffe and Wise 2007). They are expressed in all compartments of the cell, and contribute to the stabilization of larger proteins and membranes. They share certain amino acid motifs but given their sequence divergence it is doubtful whether they form a monophyletic group.
- *Tardigrade Disordered Proteins* (*TDPs*) form a group of intrinsically disordered proteins unique to tardigrades. Discovered only a few years ago, they are found in Parachela (a subgroup of Eutardigrada, see Figure 6.14) but not in other tardigrade lineages, such as the anhydrobiotic Milnesiiidae and Heterotardigrada (Hesgrove and Boothby 2020, Kamilari et al. 2019). Their function seems to be similar to trehalose: interacting with phosphates on membrane lipids. TDPs share some motifs with LEA-proteins, but are very different in other respects. Interestingly, all tardigrades have LEA protein genes, but these seem to be less functional in the lineages that rely on TDPs.

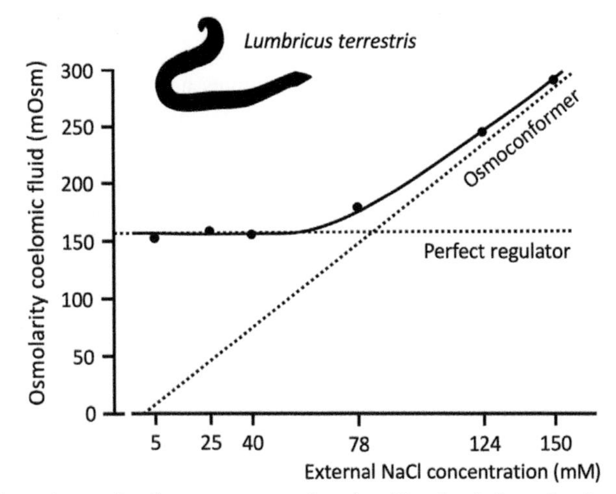

Figure 6.14. The earthworm *Lumbricus terrestris*, when placed in salt solutions of various concentrations, shows hyperosmotic regulation at low external salt concentrations but approaches osmoconformation at external salt levels above 50 mM. The solid line connects the data points (filled circles). Modified after Dietz and Alvarado (1970).

- *Heat shock proteins* (*Hsps*). This diverse group of proteins are well-known for their chaperone function supporting larger proteins and their role in signal transduction. They are highly upregulated under stress, and contribute to cellular homeostasis in almost all organisms. They are also seen to be upregulated during anhydrobiosis, although their role in the model species *Milnesium tardigradum* seems to be only minor (Reuner et al. 2010).
- *Antioxidant defence.* The transcriptome of anhydrobiotic invertebrates is also dominated by enzymes active in cellular redox homeostasis and defence against oxygen radicals, such as *superoxide dismutase*, *thioredoxin*, *glutathione S transferase*, *etc.* (Reardon et al. 2010, Mali et al. 2010, Wełnicz et al. (2011).

Not all soil invertebrates use these mechanisms to the same extent. Tardigrades generally do not rely strongly on trehalose. Marine tardigrades even lack a *trehalose-6-phosphate synthase* gene, which encodes the key enzyme to convert glucose into trehalose. All terrestrial tardigrades do have the enzyme, although not all of them accumulate trehalose during anhydrobiosis (Cesari et al. 2012). In contrast to tardigrades, anhydrobiotic nematodes do deploy sugar accumulation. Up to 30% of the body weight may consist of trehalose in the model species *Aphelenchus avenae* (Crowe et al. 1992). Dehydrated nematodes also accumulate inositol and glycerol (Womersley and Smith 1981). On the other hand, upregulation of heat shock proteins and enzymes combating oxygen radicals is seen in all lineages.

In 1968 the French zoologist Denis Poinsot discovered that the collembolan *Folsomides angularis* could also go into anhydrobiosis. In addition to this species, classified in the family Isotomidae, anhydrobiosis is documented for species in the genera *Brachystomella* (Brachystomellidae) and *Pseudachorutes* (Neanuridae) (Barra and Poinsot-Balaguer 1977, Belgnaoui and Barra 1989, Poinsot-Balaguer and Barra 1991, Arbea and Blasco-Zumeta 2001). The springtails attain a reddish colour, the body shows considerable longitudinal shortening, antennae and legs are retracted and folded against the body, all very much like the tunnification of tardigrades. In histological sections major changes are seen in the organization of the nucleus and nucleolus, suggesting DNA fragmentation and packaging in vesicles during anhydrobiosis, as documented for bdelloid rotifers (Gladyshev et al. 2008). DNA degradation and fragmentation are also observed in anhydrobiotic tardigrades (Rebecchi et al. 2009). The re-assembly of DNA during rehydration may predispose the organisms for horizontal gene transfer (see Section 6.11).

The scattered phylogenetic distribution of anhydrobiosis in the various soil invertebrate lineages suggests that it must be considered an adaptation connected to terrestrialization of the phylum. This seems to be more logical than assuming that it has arisen in different groups independently as a specialized adaptation (e.g., three times in different families of Collembola). The basic biochemical machinery that allows anhydrobiosis seems to be present in many terrestrial metazoans, especially Ecdysozoa, but only some lineages have exploited the toolbox to utter perfection (tardigrades) while it got lost in many others.

6.5 Osmoregulation

Osmoregulation, ion homeostasis, nitrogen excretion, cold-hardiness and drought resistance are all tightly connected and could be considered a single adaptive complex. Still it is useful to consider osmoregulation separately in the light of its evolutionary past. In comparative physiology the osmotic pressure of the body fluid in any terrestrial invertebrate is assumed to reflect its evolutionary ancestry (Little 1983). This suggestion is explored further here by correlating haemolymph osmotic values with the terrestrialization pathways that we derived in Chapter 2 from molecular phylogenies.

Like freshwater invertebrates, soil invertebrates maintain an internal osmotic value higher than their environment (*hyperosmotic regulation*). At low external salt concentrations, the internal osmotic value is more or less constant, but above a certain external value the animal cannot regulate anymore and becomes isosmotic with the environment; it becomes an *osmoconformer*. This classical pattern, described in many physiology textbooks (e.g., Schmidt-Nielsen 1997) also holds for the earthworm *Lumbricus terrestris* (Figure 6.14). The earthworm behaves much like a typical freshwater animal. A similar pattern is seen in the springtails *Isotoma viridis* and *Hypogastrura viatica* (Witteveen et al. 1987). However, the extent of regulation (the flatness of the curve in the low range) differs considerably between species. *L. terrestris* is much better regulator than the two springtails.

In addition, different groups of soil invertebrates regulate their haemolymph osmolarity at quite different levels. High osmotic values are observed in tardigrades, land hoppers, land crabs, woodlice, centipedes and spiders (Table 6.5). These are all groups that, according to our phylogenetic analysis in Chapter 2, colonized the land directly from the sea or through an intertidal intermediate. Relatively low osmotic values are observed in earthworms, snails and slugs, groups that colonized the land from freshwater. Millipedes and onychophorans also have low osmotic values, although millipedes are not assumed to have colonized the land through freshwater

Table 6.5. Osmotic values of body fluids (haemolymph or coelomic fluid) for different groups of soil invertebrates. Compiled from Todd (1963), Lindqvist (1970), Dietz and Alvarado (1970), Verhoef (1981), Little (1983), Carley et al. (1983), Riddle (1985), Witteveen et al. (1987), Verhoef and Prast (1989), Halberg et al. (2013) and Konopova et al. (2019). Putative terrestrialization scenarios are copied from Table 2.1.

Species	Taxonomic group	Osmotic value of body fluids (mOsm)	Putative evolutionary pathway to terrestrial life
Tardigrades			
Echiniscus testudo	Heterotardigrada, Echiniscoidea	507	Intertidal
Milnesium tardigradum	Eutardigrada, Milnesiidae	769	Direct
Richtersius coronifer	Eutardigrada, Richtersiidae	361	Direct
Macrobiotus cf. *hufelandi*	Eutardigrada, Macrobiotoidea	524	Direct
Earthworms			
Lumbricus terrestris	Clitellata, Crassiclitellata	145 – 165	Freshwater
Snails and slugs			
Helix pomatia	Panpulmonata, Stylommatophora	183	Freshwater
Deroceras reticulatum	Panpulmonata, Stylommatophora	345	Freshwater
Eutrochatella tankervillei	Cycloneritida, Helicinidae	67	Freshwater
Poteria lineata	Caenogastropoda, Cyclophoroidea	74	Freshwater
Pseudocyclotus laetus	Caenogastropoda, Truncatelloidea	103	Freshwater
Pomatias elegans	Caenogastropoda, Littorinoidea	254	Intertidal
Velvet worms			
Epiperipatus acacioi	Onychophora, Peripatidae	180	Direct
Peripatopsis sp.	Onychophora, Peripatopsidae	200	Direct
Landhoppers			
Talitrus	Amphipoda, Talitridae	400	Intertidal
Land crabs			
Austrothelphusa transversa	Decapoda, Gecarcinucidae	517	Freshwater
Cardisoma armatum	Decapoda, Gecarcinidae	744	Intertidal
Caenobita brevimanus	Decapoda, Paguroidea	800	Intertidal

Table 6.5 contd. ...

...Table 6.5 contd.

Species	Taxonomic group	Osmotic value of body fluids (mOsm)	Putative evolutionary pathway to terrestrial life
Woodlice			
Ligia oceanica	Isopoda, Ligiidae	1254	Intertidal
Porcellio scaber	Isopoda, Porcellionidae	620 – 700	Intertidal
Oniscus asellus	Isopoda, Oniscidae	590	Intertidal
Springtails			
Orchesella cincta	Collembola, Entomobryomorpha	350	Cavernicolous
Tomocerus minor	Collembola, Entomobryomorpha	320 – 350	Cavernicolous
Isotoma viridis	Collembola, Entomobryomorpha	292 – 404	Cavernicolous
Folsomia sexoculata	Collembola, Entomobryomorpha	394	Cavernicolous
Archisotoma pulchella	Collembola, Entomobryomorpha	510 – 911	Cavernicolous
Hypogastrura viatica	Collembola, Poduromorpha	448	Cavernicolous
Anurida maritima	Collembola, Poduromorpha	1127	Cavernicolous, (intertidal)
Jumping bristletails			
Petrobius maritimus	Archaeognatha, Machilidae	421	Cavernicolous, (intertidal)
Petrobius brevistylis	Archaeognatha, Machilidae	463	Cavernicolous, (intertidal)
Millipedes			
Iulus scandinavius	Diplopoda, Julida	236	Direct
Cylindroiulus londinensis	Diplopoda, Julida	214 – 232	Direct
Glomeris marginata	Diplopoda, Glomerida	158	Direct
Orthoporus texicolens	Diplopoda, Spirostreptida	175	Direct
Orthoporus ornatus	Diplopoda, Spirostreptida	204	Direct
Pachydesmus crassicutis	Diplopoda, Polydesmida	151 – 165	Direct

Table 6.5 contd. ...

...*Table 6.5 contd.*

Species	Taxonomic group	Osmotic value of body fluids (mOsm)	Putative evolutionary pathway to terrestrial life
Centipedes			
Lithobius forficatus	Chilopoda, Lithobiomorpha	350 – 370	Direct
Cormocephalus rubriceps	Chilopoda, Scolopendromorpha	379	Direct
Haplophilus subterraneus	Chilopoda, Geophilomorpha	438	Direct
Strigamia maritima	Chilopoda, Geophilomorpha	462	Direct
Hydroschendyla submaritima	Chilopoda, Geophilomorpha	498	Direct
Harvestmen			
Leiobunum vittatum	Arachnida, Opiliones	360	Direct
Scorpions			
Androctonus australis	Arachnida, Scorpiones	480	Direct
Spiders			
Tegenaria atrica	Arachnida, Araneae	400	Direct
Dugesiella hentzi	Arachnida, Araneae	360	Direct

and we don't have a terrestrialization scenario for onychophorans. Relatively high osmotic values are noted for springtails and other hexapods.

Overviewing the data, the relation between osmoregulation and the ancestral habitat of a group is less clear than often assumed (Little 1983). Instead, the data suggest a strong phylogenetic signal in the osmoregulatory capacities. The level of regulation seems to have been fixed during the evolutionary origin of the group and apparently did not change much since. This is, for instance, illustrated by a comparison of earthworms with tardigrades. These animals live in very comparable environments, maintaining a direct relationship with the same pore water, yet regulate the osmotic value of their body fluid at levels differing by a factor of 4 (Table 6.5). Another example is seen in woodlice, which have maintained a crustacean-like high osmotic value, while springtails, with a very similar habitat choice, regulate at about half these values. A remarkably high osmotic value is seen in the littoral isopod *Ligia oceanica* (even above sea water), which we have interpreted as a mechanism to limit evaporation (Section 6.4).

Osmoregulation is also subject to secondary evolutionary change. This is illustrated by the marine-littoral springtail species *Anurida maritima* and *Archisotoma pulchella*. Witteveen et al. (1987) demonstrated that *A. maritima* can be considered a truly salt water-adapted species. It does not regulate its internal osmotic value at all and behaves like an osmoconformer over a wide range of seawater concentrations.

Since *A. maritima* descends from a terrestrial lineage (*cf.* Figure 2.7) we must assume that its capacity for osmoregulation was lost when it evolved in the intertidal environment. The two littoral Collembola illustrate that the hexapod body plan does allow marine regression, but why so few species have chosen this pathway remains a mystery.

Osmoregulation and ionic regulation of soil invertebrates is concentrated in *nephridia* (kidneys), the gut (especially the *rectum*) and for some groups also in the skin. For example, earthworms achieve their hyperosmotic regulation by actively taking up chloride across the skin and producing hypo-osmotic urine from the nephridia (Little 1983). Insects, millipedes, arachnids and tardigrades are equipped with a set of tubules extending from the alimentary canal into the haemolymph, the so-called *Malpighian tubules*. Protura and some Diplura have a system of "*Malpighian papillae*", which are assumed to be homologous with the insect Malpighian tubules (Dallai and Burroni 1982), but these structures are lacking in Collembola.

Instead, Collembola have a unique organ involved in osmoregulation and ionic regulation, the ventral tube (*cf.* Figure 2.6). Physiological research with tiny microelectrodes has revealed the complex role of the ventral tube and its eversible vesicles for water balance, ionic balance and excretion (Konopova et al. 2019). The frontal part of the tube's vesicle is involved with uptake of water, Na^+ and Cl^-, as well as excretion of K^+ and H^+, while the posterior part is associated with uptake of H^+ and release of NH_4^+. These processes are summarized in Figure 6.15.

Interestingly, the ventral groove and the ventral tube of springtails serve exactly the same function as the external urine circulation of isopods ending in the uropodal complex. This seems like a remarkable case of convergent evolution. Alternatively, it emphasizes the relatively close relationship between Collembola and specific groups in the Crustacea, as genomic studies have confirmed (Faddeeva et al. 2015).

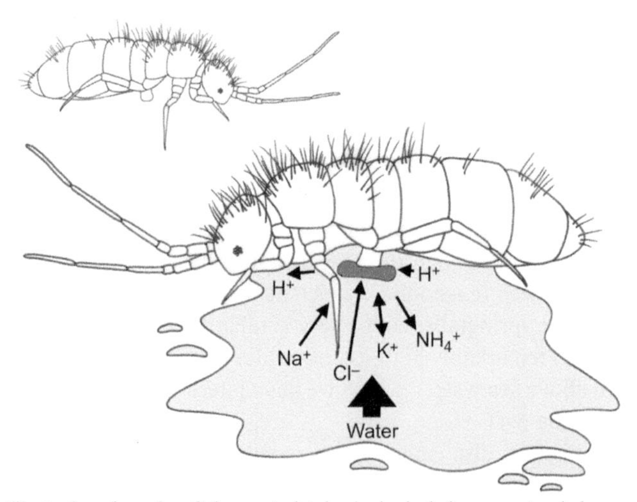

Figure 6.15. Illustrating the role of the ventral tube in ionic balance, water balance and ammonia excretion in Collembola. The investigated species is *Orchesella cincta*. Reprinted from Konopova et al. (2019), with permission from the Company of Biologists.

6.6 Surviving the heat

Drought is often caused by heat but heat is also a problem of its own. Many soil invertebrates avoid the heat by hiding away in the soil. In extreme environments a burrow may be necessary to escape scorching conditions at the surface (*cf.* Figure 1.6). However, some soil invertebrates such as spiders and beetles have adopted an active lifestyle at the surface, hunting for prey. These animals may profit from the high temperatures at the interface, but may also suffer from heat shocks. We have seen in Section 1.2 that surface temperatures may be extremely variable on a microscale. A small surface-active invertebrate that tolerates the average temperature must still beware of local hot spots above its thermal tolerance (*cf.* Figure 1.5). A heat shock may cause immobility, followed by rapid desiccation and death. How do these soil invertebrates avoid being caught by heat?

Many thermotolerant invertebrates survive heat shocks up to 40°C, or slightly above. They do that by deploying a wide range of biochemical measures that must prevent protein denaturation and membrane damage. Recent work on the wolf spider *Pardosa pseudoannulata*, a thermophilic species inhabiting rice fields, has shed some light on these mechanisms (Xiao et al. 2016). Transcription profiling showed that a heat shock causes a genome-wide change of gene expression. In total 466 genes were upregulated when spiders were exposed to 40°C for 1 h, compared to spiders kept at 25°C. In addition, 368 genes were downregulated by the heat shock. The biochemical processes to which these genes could be assigned indicated responses in various biochemical pathways (Table 6.6). In addition, nine genes encoding *heat shock proteins* were highly affected.

The transcription profile illustrates a characteristic response, seen in almost all ectotherms: high temperature treatment induces a rapid and massive induction of heat shock proteins, at the same time suppressing all other protein synthesis at the level of mRNA splicing. Since *Hsp* genes lack introns, their synthesis is not affected by inhibition of the spliceosome, which does affect almost all other proteins. It is also clear from transcriptomics that the heat shock profile is very different from the one

Table 6.6. Summary of genome-wide changes in gene expression in the spider *Pardosa pseudoannulata* when exposed for 1 h to 40°C, compared to spiders kept at 25°C. Percentage enrichment is the number of differential genes relative to the total number of genes in the genome classified under the pertinent pathway. Data from Xiao et al. (2016).

Biochemical pathway	Number of differentially expressed genes (40°C versus 25°C)	Percentage enrichment
Ribosome related	41	31.3%
Protein processing in endoplasmic reticulum	12	9.2%
Spliceosome	5	3.8%
Lysosome	4	3.1%
Oxidative phosphorylation	4	3.1%
Purine metabolism	4	3.1%
Arginine and proline metabolism	3	2.3%

seen under cold-acclimation. For example, antioxidant defence is one of the major mechanisms deployed under cold and drought (Section 6.3, 6.4), but is not seen under heat. From a physiological point of view, high and low temperatures are two quite different stress factors.

Among the various defence mechanisms against heat stress, heat shock proteins have received most attention by far. They are small to medium-sized cytoplasmic proteins classified in five categories, depending on their molecular mass (Feder and Hofmann 1999, Kregel 2002). They were discovered in 1962 in *Drosophila melanogaster* but now have been demonstrated throughout the tree of life, from bacteria to humans. Their phylogeny is difficult to reconstruct because they have been subject to many lineage-specific duplications and strong directional selection.

Heat shock proteins are especially known for their high inducibility and therefore are used as a model for the study of gene expression regulation. Their induction involves a cytoplasmic regulatory protein called *heat shock factor* (*HSF*), which under physiological conditions is kept inactivated in the cytoplasm by heat shock proteins. Under stress, *HSPs* are drawn towards denaturated proteins, relaxing the inactivation of *HSF*, which then escapes from the inhibition, translocates to the nucleus and acts as a transcriptional activator of *Hsp* genes. The subsequent surge of *de novo HSP* protein inhibits *HSF* again. It is a classic example of a biochemical feed-back loop. However, the regulatory network is actually much more complicated than can be discussed here. For details the reader is referred to Korsloot et al. (2004) and Van Straalen and Roelofs (2012).

Heat shock proteins are known to help folding peptides to a correct secondary structure directly after translation, and also to stabilize proteins damaged by denaturation. All *Hsps* respond to heat but many of them respond to other stress factors as well (cold, desiccation, toxic chemicals, osmotic stress). For example, the rapid induction of heat shock proteins allows short term survival under cold shock in the springtail *Folsomia candida* (Waagner et al. 2013). A particularly strong induction by heat is seen in the *Hsp70* family.

Heat shock protein genes have been identified in many soil invertebrates and are often studied as indicators of soil pollution, because of their inducibility by all kinds of stress factors. This is the reason that they are also called *stress proteins*. But presumably, their primary function is to protect the animal against adverse effects of heat, although this aspect of their function has received less attention than their role as general stress indicators.

In the wolf spider *P. pseudoannulata*, Sun et al. (2021) identified nine *Hsp* genes. Three of them were cloned and their expression analysed in response to heat treatment, using qPCR (Figure 6.16). This study nicely illustrates the strong inducibility of *Hsp70* by sublethal heat stress (expression increased by a factor of 150 compared to 20°C). The two other *Hsp* genes are not upregulated to the same extent (but still with a factor of 4 to 5). The strong inducibility of *Hsp70* in this thermophilic spider is assumed to contribute to its survival in hot environments, but in the absence of functional data, (e.g., using knock-out mutants) this is difficult to prove.

Hsp gene expression is typically of short duration. In the wolf spider, it was maximal after a 4 h heat shock, but returned to normal within 12 h, even if the stress continued (Sun et al. 2021). The actual tolerance to heat builds up more slowly and

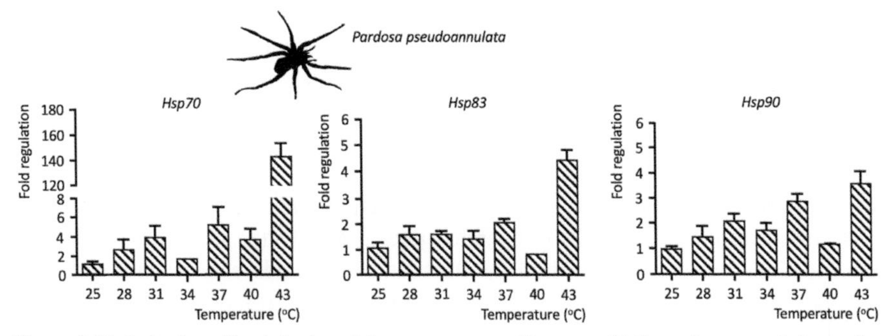

Figure 6.16. Induction of heat shock protein genes, measured by a quantitative polymerase chain reaction (qPCR) directed to three genes classified in the Hsp70, Hsp83 and Hsp90 families; mRNA abundance at six different temperature treatments (4 h) is quantified as fold regulation relative to 25°C. Reprinted, with modifications, from Sun et al. (2021), with permission from Elsevier.

this is correlated with the longer life-time of *HSP* protein (Bahrndorff et al. 2009). However, how *HSP* proteins contribute to thermotolerance on the level of the whole animal is still a mystery.

Assuming an adaptive value for the induction of *Hsps* one must assume that the proteins themselves or the rate of their induction are traits that can evolve under natural selection. Several studies have documented genetic variation in the protein-coding regions of *Hsps*, in the number of duplicates or, in some cases, the sequence of the *HSF* protein. However, such studies have mostly focused on heat tolerance of husbandry animals (Rong et al. 2019) and model species such as *Drosophila* (Frydenberg et al. 2003). There are hardly any studies on genetic variation of heat tolerance in soil invertebrates.

An exception is a study on the springtail *Orchesella cincta* (Bahrndorff et al. 2010). The level of *Hsp* induction in this species showed considerable variation between isofemale lines, suggesting genetic variation. Also, *HSP* protein expression and heat tolerance varied considerably, but the three variables were not correlated with each other, so how selection should work on this system is unclear.

It seems that genetic determination and heritability of heat tolerance is much more complicated than can be guessed from *Hsp* induction alone. Likewise, thermal reaction norms did not behave as a straightforward biometric character responding to directional selection (Driessen et al. 2007). Transcriptional regulation of gene expression typically involves 10 to 12 different proteins, which, added to promoter polymorphisms, represents an enormous potential for genetic variation (Wray et al. 2003). The evolution of regulatory responses is likewise complicated, due to the many possible interactions between the gene products. How heat tolerance has evolved in soil invertebrates remains largely unanswered to date.

6.7 Dealing with toxins

When invertebrates colonized the land and became soil invertebrates, many things had to change in their osmoregulation, temperature responses and metabolism, as we have seen above. In addition to these physiological adaptations, the terrestrialized invertebrates also had to adapt to a new diet. It is generally assumed that terrestrial animals, more than their aquatic ancestors, are exposed to a variety of toxins, such as feeding deterrents, plant secondary compounds and recalcitrant chemicals from litter decomposition. These chemicals are often referred to as *xenobiotic* (foreign to life), although in many cases they are produced by living organisms.

Animals in general have an extensive defence system against xenobiotics, loosely designated as *biotransformation*. The aim of this system is excretion of the toxin in some form or another. However, many xenobiotics are quite *lipophilic*, tend to accumulate and are not easily excreted due to low water solubility. Molecular modifications are usually required before such compounds can be removed from the body, as the main circulatory and excretory systems (blood, urine) are water-based. By introducing hydrophilic groups in the molecule (-OH, =O, -COOH) and by conjugating it to an endogenous compound with good water-solubility, excretion is usually accomplished.

The most problematic xenobiotics are those with a high *octanol-water partitioning coefficient* (K_{ow}) that are strongly lipophilic and hydrophobic. They tend to accumulate, in proportion to their $LogK_{ow}$, in tissues with a high lipid content such as the fat body of insects, and may cause tissue damage due to disturbance of membrane functions. Well-known low-molecular weight compounds are, e.g., aliphatic petroleum chemicals and chlorinated alkanes, which are mostly associated with groundwater. Nematodes, earthworms and potworms may be exposed to such compounds. Chemicals with higher $LogK_{ow}$, e.g., *polycyclic aromatic hydrocarbons* (*PAHs*), have a high affinity for dead organic matter and accumulate in the lower litter layers, the central place for activity of decomposer invertebrates. Also, pesticides applied directly to soil or as a seed treatment, may accumulate in the soil and cause problems to soil invertebrates. Finally, phenolic acids, derived from lignocellulose degradation, make up a large part of soil organic matter and several of them are toxic to animals.

When such chemicals enter soil invertebrates, through the gut or the skin, their biotransformation usually proceeds in three subsequent phases (Van Gestel et al. 2019, Figure 6.17):

1. Activation (usually oxidation) of the compound by an enzyme known as *cytochrome P450*, which acts in cooperation with *NADPH cytochrome P450 reductase* and other factors (phase I).
2. *Conjugation* of the activated product of phase I to an endogenous compound. A host of different enzymes is available for this task, depending on the compound, the tissue and the species. Slightly polar compounds may enter phase II directly, without being activated in phase I.
3. Excretion of the compound into circulation, urine, or other media, usually by means of membrane-spanning transporters belonging to the class of *ATP-binding cassette (ABC) transporters*. Hydrophilic compounds may pass on directly to phase III, without being activated or conjugated.

Biotransformation reactions are especially well developed in tissues with an uptake or a metabolic function. For soil invertebrates these metabolic tissues are the midgut of apterygotes and insect larvae, the fat body of insects, the hepatopancreas

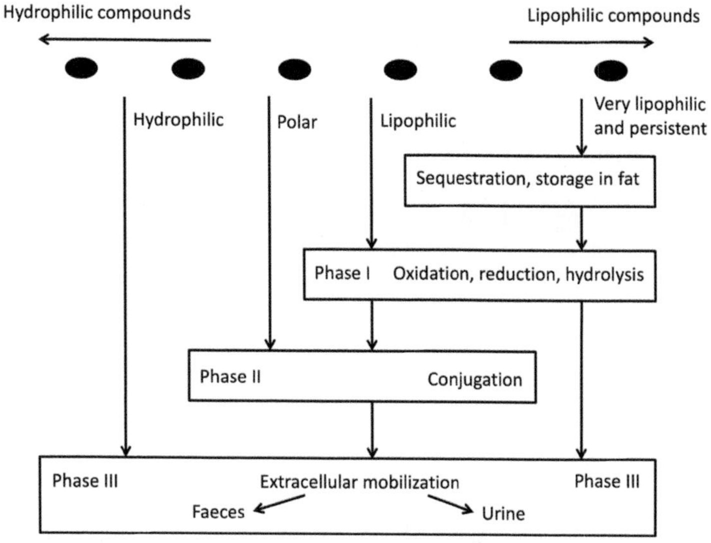

Figure 6.17. Schematic overview of the fate of xenobiotic compounds in animal cells. Depending on its lipophilicity, a compound is either stored for a long time, metabolized through phases 1 and II, or directly excreted. From Van Gestel et al. (2019).

of isopods, the midgut gland of snails and the chloragogen tissue of earthworms. With the low degree of organ differentiation in several invertebrates, the same tissue is often involved in uptake, biotransformation, as well as excretion, e.g., the gut epithelium of apterygotes.

The activity of biotransformation enzymes shows a lot of evolutionary divergence. On the average, it is higher in terrestrial animals compared to aquatic ones (Walker 1980, Stenersen et al. 1987, Siegfried and Young 1993). Among insects there is a remarkable correlation with the feeding habit: it is higher in polyphagous insects compared to oligophagous species (Brattsten 1979). Obviously, the diversity of chemical compounds in the diet or the environment has acted as a selective factor for enhanced biotransformation capacity.

Cytochrome P450 is a membrane-bound enzyme, associated with the *smooth endoplasmic reticulum* (sER). In biochemical experiments the sER of a cell is isolated as a set of small vesicles called microsomes. Therefore, cytochrome P450 protein and other membrane-bound enzymes are designated as *microsomal*. P450 is encoded by a gene called *CYP*, of which there are many copies in the genome, all slightly different from each other in terms of inducibility and substrate specificity. *Drosophila melanogaster* has 84 and *Caenorhabditis elegans* 74 *CYP* genes. In insects, there are four large clades, designated *CYP2*, *CYP3* and *CYP4*, as well as several mitochondrial *CYPs* (Feyereisen 2006). How these genes relate to each other, by descent (*orthology*) or by duplication within the same lineage (*paralogy*) is often difficult to tell.

Cytochrome P450 is a haem protein with an iron atom in the active centre. The reduced Fe^{2+} atom in the haem group binds molecular oxygen, and is oxidized to Fe^{3+} while splitting O_2; one O atom is introduced in the substrate, the other reacts with hydrogen to form water. Then the enzyme is reduced by accepting an electron from NADPH cytochrome P450 reductase. The overall reaction can be written as:

$$RH + O_2 + NADPH + H^+ \rightarrow ROH + H_2O + NADP^+$$

where R is an arbitrary substrate.

The reactions in phase II aim to connect the xenobiotic to an endogenous compound, so as to make it more water-soluble. Phase II reactions use the product of phase I, or, in case of less lipophilic compounds, address the xenobiotic directly. Examples are conjugation with sulphate by the enzyme *sulfotransferase*, with glutathione by *glutathione-S-transferase* (*GST*) and with glucuronic acid by *UDP-glucuronosyltransferase*. Some of these enzymes are located in the smooth endoplasmic reticulum, like P450, others are present in the cytoplasm. The phase II enzymes are also subject to extensive lineage-specific duplication and there are often tens of paralogs in any one genome, for instance the genome of *Drosophila melanogaster* has 37 GST genes alone.

A major class of proteins acting in phase III are *ATP-binding Cassette (ABC) transporters*. These are membrane-bound proteins that apply an ATP-dependent switch to drive foreign compounds out of the cell (Dermauw and Van Leeuwen 2014). Also in this group there are many genes, e.g., *Drosophila* has 56 ABC-transporter genes.

All three gene categories show indications of positive selection in the hexapod lineage. This was concluded in an evolutionary analysis of the transcriptome of two springtails, *Folsomia candida* and *Orchesella cincta*, in comparison to three crustaceans and three insects (Faddeeva et al. 2015). This confirms our suggestion above that adaptation to terrestrial life required an enhanced capacity to deal with xenobiotic chemicals.

Toxicological studies in vertebrates have shown that biotransformation enzymes are highly inducible by xenobiotic compounds, at the level of gene expression. A crucial regulatory pathway involves a cytoplasmic protein called *aryl hydrocarbon receptor* (*Ah receptor, AHR*). Xenobiotics that enter the cell and bind to this receptor are strong inducers of gene expression and cause enhanced *de novo* synthesis of biotransformation enzymes. The mechanism of induction has been extensively studied in rat liver, using dioxin as a model compound. We refer to textbooks of ecotoxicology for more information (Walker et al. 2001, Van Gestel et al. 2019).

Pioneering work on biotransformation enzymes in earthworms and other animals was done by the Norwegian physiologist Jørgen Herman Vogt Stenersen at the University of Oslo (see Stenersen 1992 for an overview). Stenersen and Øien (1981) determined the activity of glutathione-S-transferase in a variety of earthworm species (Table 6.7). The activity is classified here by Bouché's well known ecological categories. The data reveal a wide interspecific variation. One might intuitively expect that surface-active worms would require a higher biotransformation activity than endogeic worms, due to a more intense exposure to freshly fallen leaves that have not yet undergone leaching. However, this is not supported by the data in Table 6.7.

Table 6.7. Activity of glutathione-S-transferase (GST) towards the model substrate 2,4-dinitrobenzene in several species of earthworms, reported by Stenersen and Øien (1981).

Ecological category	Species	GST activity (nmole substrate transformed per s per mg of protein)
Surface living and litter feeding (épigés)	*Eisenia andrei*	12.5
	Dendrobaena veneta	3.2
	Lumbricus rubellus	13.9
Subsurface living, surface feeding (anéciques)	*Allobophora longa*	16.4
	Lumbricus terrestris	10.5
Soil living, subsurface feeding (endogés)	*Aporrectodea rosea*	35.9
	Aporrectodea caliginosa	7.5
	Allolobophora chlorotica	4.2
	Octolasion cyaneum	8.2

De Knecht et al. (2001) characterized biotransformation enzymes in the *hepatopancreas* of two isopods, *Porcellio scaber* and *Oniscus asellus*. These authors showed that aromatic hydroxylase, sulfotransferase, glutathione-S-transferase and UDP-glucosyltransferase were all present with high activity. However, the activities were not strongly enhanced by known inducers of biotransformation in vertebrates. This lack of inducibility was also found in earthworms (Stokke and Stenersen 1993). It seems that Ah-receptor-mediated regulation of xenobiotic metabolism is a vertebrate innovation and not present in invertebrates at all. The ancestral function of the AHR homolog might have been different, e.g., related to development (Hahn 2002).

Analytical-chemical studies in isopods have been focused on the model compound *pyrene*, a polycyclic aromatic hydrocarbon associated with incomplete burning and soot. The choice of this compound was motivated by the fact that it has a limited number of phase-I metabolites due to its symmetric molecular structure. In isopods pyrene is biotransformed to five different metabolites, including one phase I product, *1-hydroxypyrene*, and four phase-II metabolites, *pyrene-1-glucoside*, *pyrene-1-sulphate* and two uncharacterized compounds (Stroomberg et al. 1999, Figure 6.18). One of these unknowns was later identified as *pyrene-1-glycoside-malonate*, a novel conjugation product never seen in vertebrates. Springtails produce the same set of conjugates, except that the sulphate conjugate is not present, neither in *Folsomia candida* nor in *Orchesella cincta* (Stroomberg et al. 2004, Howsam and Van Straalen 2003). Pyrene-1-glucuronide, which is the most common metabolite in vertebrate metabolism, is never seen in soil invertebrates.

In earthworms the situation is again different. *Eisenia andrei*, when exposed to pyrene, makes metabolites only in small amounts; most of the parent compound is retrieved unchanged. Among three metabolites identified in low quantities, two were derived from 1-hydroxypyrene and one was an unknown other metabolite. In a study on the closely related species *Eisenia fetida*, Schmidt et al. (2017) found glucoside and sulphate conjugates to be the most dominant phase-II products. The sulphate conjugate was excreted in the medium. Therefore, I have labelled one of the

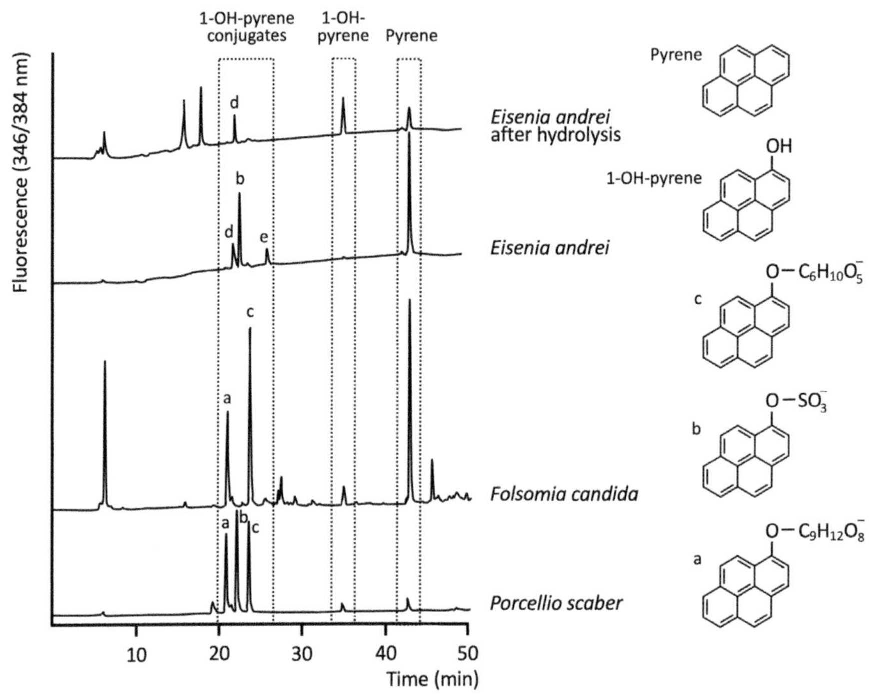

Figure 6.18. HPLC (high performance liquid chromatography) profiles, with fluorescence detection, for extracts of earthworms (*Eisenia andrei*), springtails (*Folsomia candida*) and woodlice (*Porcellio scaber*), exposed to the polycyclic aromatic hydrocarbon pyrene (shown in upper right corner). The major phase I metabolite is 1-hydroxypyrene. Five different phase II metabolites (conjugates) derived from 1-OH-pyrene are shown and three were identified as (a) pyrene-1-O-(6"-O-malonyl)glucoside, (b) pyrene-1-sulphate and (c) pyrene-1-glucoside. The identity of metabolites d and e remains unknown. Peak b in *E. andrei* is designated here as a sulphate conjugate, based on data by Schmidt et al. (2017) for *E. fetida*. The upper track shows the profile when the samples were treated with deconjugation enzymes: peaks b and e disappear and 1-OH-pyrene appears, proving that d and e are conjugates derived from 1-hydroxypyrene. Chromatographic traces were copied from Stroomberg et al. (2004) and mounted in the same frame.

earthworm metabolites not identified by Stroomberg et al. (2004) here as a sulphate (Figure 6.18).

This diversity of biotransformation pathways (isopods, springtails and earthworms) could relate to the presence or absence of specific enzymes, or to the variable availability of metabolic intermediates available for conjugation, e.g., related to the food. The presence of the malonyl-glycoside conjugate in isopods and springtails, and its absence in earthworms, points to the well-known phylogenetic relationship between hexapods and crustaceans (*cf.* Figure 2.14).

Given the great genetic diversity of biotransformation enzymes, with so many paralogs in a genome, evolution of biotransformation is expected to occur easily. However, this is only known in the context of pesticide resistance. For example, overexpression of *Cyp6g1*, due to insertion of a P element in its promoter, underlies resistance against DDT in *Drosophila* (Daborn et al. 2002). There are no such examples for pesticide resistance evolution in soil invertebrates. Whether they can evolve resistance to PAH by adaptations in the biotransformation system is likewise unknown.

Our knowledge of xenobiotic metabolism in soil invertebrates is still scanty. From what data we have we may conclude that it might be very different from what is known in vertebrates. The basic enzymatic machinery may be similar, but invertebrates have different phase II metabolites, some unique to specific lineages. Even the whole Ah-receptor mediated induction of cytochrome P450 seems to be lacking in invertebrates.

6.8 Adaptation to heavy metals

In addition to organic toxins, heavy metals constitute another challenge for soil invertebrates. Metals emitted to the environment by human activities (mining, metal smelting, waste disposal, traffic, consumer products) tend to end up in soils and sediments, the ultimate sinks. Due to their intrinsic persistence and accumulation in the lower litter layers, many soil invertebrates are exposed not to the average but to the highest concentrations. As all metals are toxic at high concentrations, dealing with excess metals is a crucial survival factor, the more so since uptake cannot be avoided. All organisms have dedicated uptake mechanisms for essential metals with physiological functions at low concentrations, but non-essential metals with similar chemical properties hitch-hike on these uptake mechanisms. The term xenobiotic is not well applicable to metals, but the famous saying of Paracelsus is: "*Poison is in everything. And no thing is without poison.*" It is the dose that makes a poison poisonous (Koeman 1996).

Metals that perform essential functions in biochemical processes, such as iron, manganese, copper, zinc, cobalt and molybdenum are expected to be regulated in the body. This also holds for soil invertebrates, although sometimes the regulatory

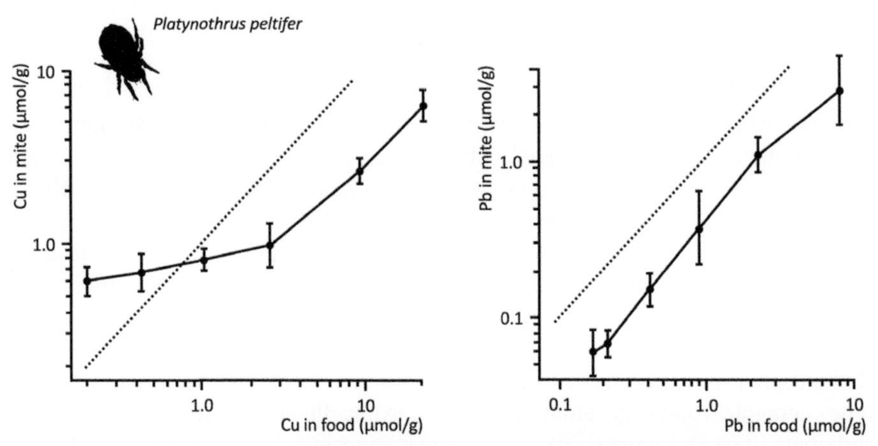

Figure 6.19. Comparison of whole-body concentrations of copper (left) and lead (right) in the oribatid mite *Platynothrus peltifer*, fed metal-contaminated food at various concentrations for 3 months. All concentrations are expressed per dry weight. Means are shown with standard deviations. The dotted line indicates equal concentrations in mite and food. From Denneman and Van Straalen (1991), redrawn and modified.

capacity is modest. In Figure 6.19 an example is given for copper in the oribatid mite *Platynothrus peltifer* as a function of the concentration in the food (Denneman and Van Straalen 1991). Copper is more or less regulated at a level between 0.6 and 0.9 μmol/g (75–100 μg/g), however, at higher concentrations in the food, the regulation falls down and copper increases in proportion to the concentration in the food, although at a lower level. The graph is very similar to the classical osmoregulatory response of earthworms (Figure 6.14). In contrast to copper, the non-essential metal lead is not regulated at all (Figure 6.19).

For essential metals the point at which regulation breaks down marks the beginning of toxicity. There are hardly any data in the dose range below that point, in which soil invertebrates have to deal with shortage of metals. There are some indications that metals such as manganese, that easily leach from surface soil under acid conditions, might be limiting for mites that have high body concentrations of this element (Van Straalen et al. 1988, Janssen and Hogervorst 1993). However, in general it is expected that the soil, due to its great binding capacity, is not a medium where shortage of essential metals is an important limiting factor.

Focussing on the toxic range, there are basically two ways in which animals can deal with metals: one is to sequester, bind or precipitate them in innoxious chemical forms, e.g., intracellular granules of carbonates and sulphides, phosphate salts in bone tissue or organic ligands in the cytoplasm. In this sequestered form metals are not toxic and can be stored until excretion. The other strategy is to excrete the metals as soon as possible. The two strategies do not exclude each other. For instance, to excrete excess of metals it may be necessary to bind them in a molecule which is then excreted, or to precipitate them in cells that are shed as a whole. All these strategies jointly are known as the *trafficking* scheme for metals inside tissues.

Onc characteristic site for temporary storage of metals is an intracellular granule. These granules (also called "*mineral concretions*" or "*spherocrystals*") have been described through electron microscopy in all soil invertebrates, especially in the midgut epithelium, but also in other digestive tissues such as the midgut glands of snails and spiders, the *hepatopancreas* of isopods and the *chloragogen tissue* of earthworms (Humbert 1974, Simkiss 1979, Brown 1982, Morgan 1984, Taylor and Simkiss 1984, Prosi and Dallinger 1988). In some cases, depending on the nutritional state of the animal, the epithelial cells are completely packed with them. Their mineral composition can be probed with *electron-induced X-ray emission*, which has revealed at least four different types (Hopkin et al. 1989, Hopkin 1989, Figure 6.20).

Different metals accumulate in granules of different types. The fate of a metal ion is determined by its chemical properties, in particular whether it is "*oxygen-seeking*" (class A, accumulating in type A granules with orthophosphate) or "*sulphur seeking*" (class B, accumulating in type B granules with sulphide precipitates) (Nieboer and Richardson 1980). Iron has an uptake and storage system of its own (*ferritin* and type C granules) and some animals also have extracellular granules made of calcium carbonate (type D, Figure 6.20).

The intracellular membrane-enclosed granules are residuals of the *lysosomal system*, which is particularly well developed in soil invertebrates. Lysosomal vesicles separate from the endoplasmic reticulum or from the plasma membrane and are used to digest food (taken up by *phagocytosis*) or to receive and digest cellular waste

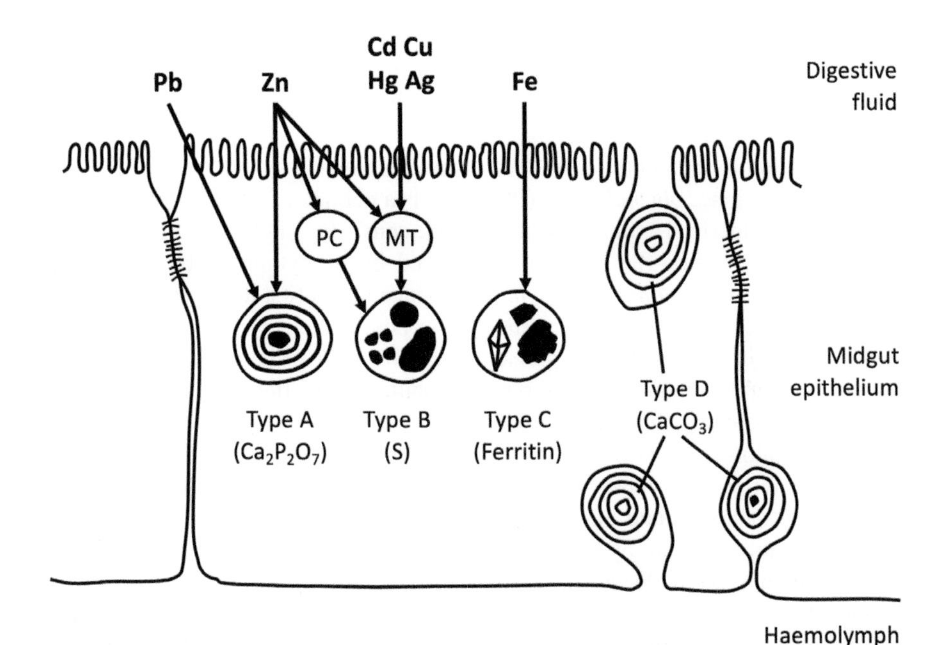

Figure 6.20. Schematic diagram of granules ("mineral concretions", "spherocrystals") in the midgut epithelium of soil invertebrates. The granules are classified based on their mineral composition. Different metals tend to accumulate in granules of different types. MT = metallothionein, PC = phytochelatin. Redrawn from Hopkin (1986), with modifications and additions.

products. Although the vesicles are bound for excretion by *exocytosis*, in some cases their prolonged residence in the cell acts as a store of metals. This seems to be the case for the type B granules in *S cells* ("small cells") of the isopod hepatopancreas, which are loaded with copper to the extent that sectioning the hepatopancreas scratches a histologist's microtome (S.P. Hopkin, p.c.). In springtails the granules are excreted not only by exocytosis, but also by renewal of the whole midgut epithelium, which takes place every moult (Van Straalen et al. 1987).

Metals may be deposited directly in mineral form through transporters in the lysosomal membrane, or they may be included in ligands first. One particularly important ligand is a metal-binding protein called *metallothionein* (*MT*). This term indicates a collection of low-molecular weight hydrophilic proteins, with an extraordinary high content of cysteine (up to 30% of amino acids). The cysteines are organized in metal-binding clusters, called *thiolate clusters*, where the binding is due to the sulphydryl group of cysteine. Usually six to seven bivalent metal ions are bound by nine to eleven cysteines and often there are two such clusters in a single metallothionein protein. It is assumed that metallothionein is taken up in lysosomal vesicles of the B type, or donates the metals to these vesicles (Figure 6.20).

Metallothioneins have been isolated and sequenced in several soil invertebrates including nematodes (Slice et al. 1990, Freedman et al. 1993), snails (Dallinger et al. 1997), springtails (Hensbergen et al. 1999, Nakamori et al. 2010) and earthworms (Stürzenbaum et al. 1998, 2001, Mustonen et al. 2014, Homa et al. 2016). The proteins of different animals share the characteristic metal-binding thiolate clusters but the sequences themselves are all very different from each other and are also very different from vertebrate metallothioneins. It is hardly possible to establish any phylogenetic relationship between the proteins; instead each animal phylum seems to have its own family (Binz and Kägi 1999).

In addition to metallothionein several soil invertebrates have a second system of metal-scavenging, *phytochelatin* (PC). This is a peptide of variable length which consists of polymerized *glutathione* (a tripeptide consisting of glutamic acid, cysteine and glycine). It is non-ribosomally synthesized in the cytoplasm by an enzyme called *phytochelatin synthase* (*PC synthase*). Since it was first discovered as a zinc-chelating molecule in plants it bears the prefix "phyto", however, PC is now known to occur in several invertebrates, including earthworms (Brulle et al. 2008). It may be assumed that PC delivers zinc to type B granules in invertebrates, like it delivers zinc to the vacuole in plants (Figure 6.20).

Metallothionein binds mostly copper and cadmium. In gastropods these functions are performed by two isoforms of MT, one mainly present in the midgut gland, highly inducible by cadmium and binding Cd, the other localized in the mantle, binding copper, but not inducible by Cu or Cd (Dallinger et al. 1997, 2000). Invertebrate metallothioneins normally do not bind much zinc; this function is performed by phytochelatin (Cobbett and Goldsbrough 2002). In vertebrates, however, MT is the main zinc-binding molecule and they lack the PC system. It is likely that the zinc-binding function of PC was taken over by metallothionein after loss of the PC synthase gene in the vertebrate lineage (Dehal et al. 2002).

The trafficking of metals through an invertebrate has been particularly well studied in the earthworm *Lumbricus rubellus* (Stürzenbaum et al. 2001, 2004). Using polyclonal antibodies raised against the cadmium-inducible *wMT-2* protein, Stürzenbaum et al. (2001) were able to track the transit of cadmium through the tissues (Figure 6.21). Cadmium taken up through the gut primarily accumulates in sulphur-rich vesicles of the *chloragogen* tissue, which surrounds the gut epithelium (Figure 6.21). Metallothionein was detected in vesicles in the proximal part of the *chloragocytes*, which confirms that metal-loaded MT is taken up by the lysosomal system. The vesicles, when loaded with metal are described as *chloragosomes* and resemble the B-type granules depicted in Figure 6.20 (Hopkin 1986). The chloragocytes regularly detach from the chloragogenous tissue and are found in the coelom, where they are known as *eleocytes*, a type of *coelomocyte* (see Section 6.9). In the coelom they are phagocyted by other *coelomocytes* called *amoebocytes*. These cells perform the same function as macrophages in vertebrate cellular immune defence. Finally, cadmium-MT is delivered to the kidney (*nephridium*), of which a pair is present in every segment. The nephridia of annelids are tubules forming a connection between the coelom and the external environment (see Figure 6.25). How exactly the nephridia excrete chloragosomes is not known.

The earthworm is the only soil invertebrate for which such a beautifully elaborated scheme of metal trafficking is available. For other species the scheme is different and involves fewer tissues. But what differs most between species is the residence time of the metal in the intestinal tissues. In some species the granules tend to stay in the tissue for a long time, even for the animal's whole life. This is especially valid for copper in isopods, which accumulates to an incredibly high level

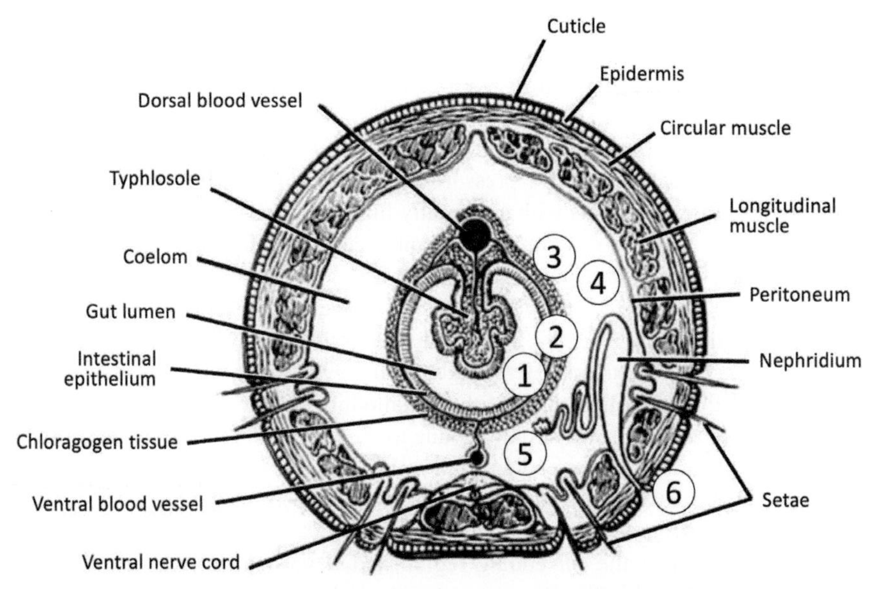

Figure 6.21. Cross section of an earthworm showing some of the internal organs and the trafficking of cadmium: (1) Uptake from the food by the intestinal epithelium, (2) transfer to the chloragogen tissue and accumulation in lysosomal vesicles at the proximal side of the chloragocytes, (3) excretion of chloragosomes into the coelom, (4) uptake by coelomocytes, (5) transfer to the nephridium, (6) excretion from the body. The reconstruction is based upon immunohistochemical analysis reported in Stürzenbaum et al. (2001, 2004).

in the hepatopancreas. The animal lives its life with an ever increasing "backpack" of safely stored metals. In springtails, however, the midgut epithelium is shed during every moult, removing about 80% of the metal burden each time (Van Straalen et al. 1987). Consequently, springtails have a high clearing rate and low levels of metals.

As the turnover times of metals vary considerably between lineages of soil invertebrates, so do the residues that are typically found in animals collected from contaminated ecosystems. Low concentrations are found in springtails and carabid beetles, intermediate levels in earthworms and high levels in isopods, pseudoscorpions and oribatids. These differences can be explained by the physiological trafficking strategy and the residence time of the metal-accumulating tissues or their granules. An exhaustive overview of metal kinetic parameters in soil invertebrates is provided by Ardestani et al. (2014).

A crucial aspect of metal trafficking, allowing protection against high doses, is that both MT and PC synthase are highly inducible. Especially the induction of MT by cadmium has developed into a model for gene regulation, like the heat shock response discussed above. A crucial regulatory protein in this pathway is *metal-responsive transcription factor* (*MTF*), which translocates to the nucleus upon entry of cadmium ions in the cell and activates the transcription of MT and other genes by binding to *metal responsive elements* (*MREs*) in gene promoters (Palmiter 1994). The regulatory cascade has been investigated in vertebrates in great detail, but whether invertebrates share the MTF-regulatory pathway is unclear. MTF is found in helicid snails (Höckner et al. 2009) and in *Drosophila* (Egli et al. 2003), but nematodes and earthworms seem to lack an MTF-encoding gene (Stürzenbaum et al. 2004), although they do have a Cd-inducible MT. So, these species might have a quite different regulatory pathway for MT induction.

As MT is the first line of defence against toxic metal ions, it may be expected that in contaminated environments, natural selection will enhance the efficiency by which MT is induced. Such a selective pressure is expected in industrially contaminated soils, mine tailings and soils with a naturally high metal content. Indeed, enhanced metal tolerance has been described in several soil invertebrates (Posthuma and Van Straalen 1993), although the mechanism is clear only in a few cases.

Following the "discovery" of metal tolerance in the springtail *Orchesella cincta* (Van Straalen et al. 1987, Posthuma et al. 1993) and elucidation of the MT protein and its DNA sequence (Hensbergen et al. 1999, 2000), Janssens et al. (2007) characterized the promoter region of the metallothionein gene. Promoters are often highly variable and may evolve quickly due to the fact that they are not constrained by the genetic code of open reading frames. It is assumed that, in general, 40% of promoter sequences are polymorphic (Wray et al. 2003). This was also the case in the *O. cincta Mt* promoter (*OcpMt*). Janssens et al. (2007) sequenced approximately 1600 bp upstream of the *Mt* coding region and described nine different *OcpMt* alleles, which differed in the arrangement and number of metal-responsive element (MRE) cassettes.

Figure 6.22a reproduces the genetic architecture of two of the alleles, the most common one (the wild type), designated A1 and an allele which appeared to be associated with metal tolerance, D2. Both alleles have six MREs but their arrangement in the proximal promoter is different. In addition to MREs there are several binding sites for *hydroxy-ecdysone receptor protein* (HERE). This suggests that MT expression might be synchronized with the moulting cycle, allowing cyclic excretion of metals by expelling the degenerated midgut epithelium at moult (Joosse and Buker 1979).

The inducibility of the various promoters was investigated using a luciferase construct driven by the *Orchesella Mt* promoters in a *Drosophila* cell line. The variation between the promoter alleles was considerable: two alleles were hardly responsive to cadmium; two others enhanced the expression of luciferase by a factor of 2000. Figure 6.22b shows the induction response as a function of the cadmium concentration in the medium, for the two selected alleles. The fact that both are responsive to Cd^{2+} suggests that the regulatory cascade of *Drosophila* cells (including *Drosophila MTF*) is acting upon the *Orchesella* promoters. The induction of promoter D2 is, however, considerably higher than the induction of A1 (Figure 6.22b).

The frequency of these alleles in field populations was screened in a survey of 23 *O. cincta* populations from Belgium, France, Germany and the Netherlands, with variable degrees of metal contamination (reference sites, industrially contaminated sites and mining sites). There was a positive correlation between the frequency of the high-expresser allele D2 and metal contamination of the soil. The frequency of the wild type allele A1 decreased with increasing contamination (Figure 6.22c). The data strongly suggest that mutations in the promoter that cause enhanced expression

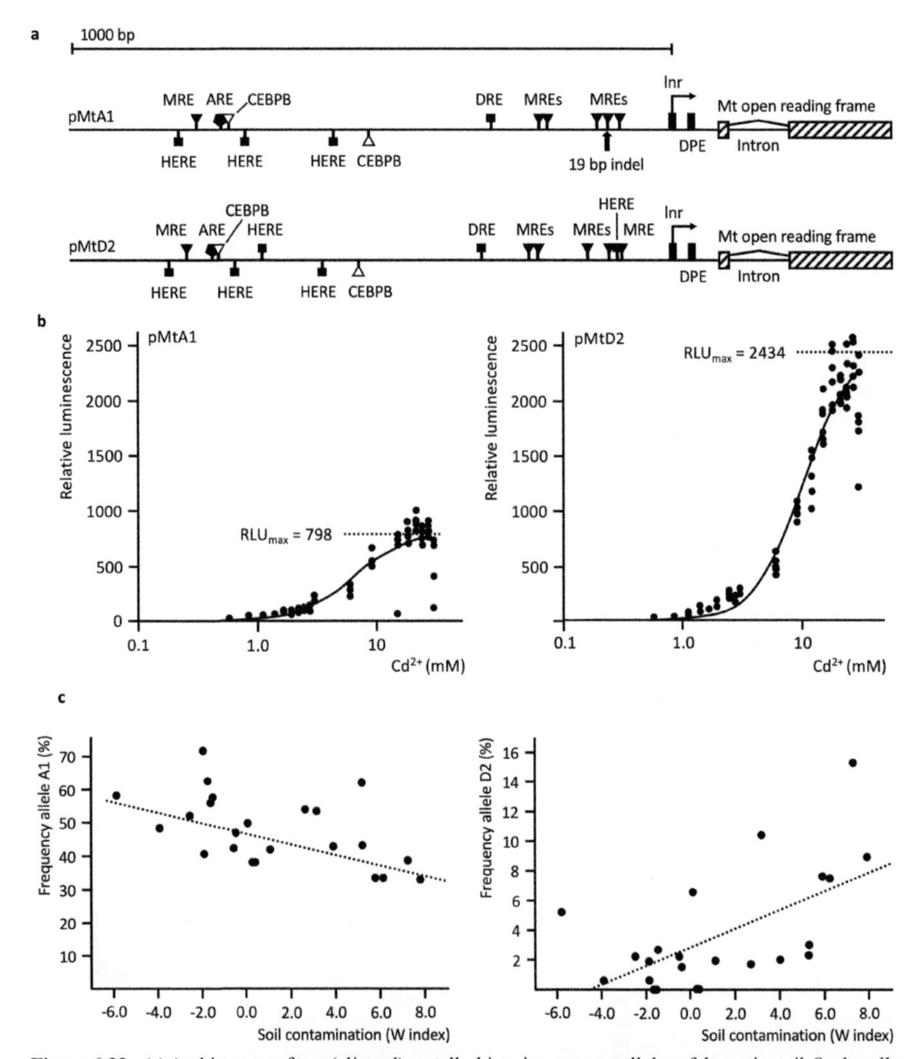

Figure 6.22. (a) Architecture of two (aligned) metallothionein promoter alleles of the springtail *Orchesella cincta* (out of 9 alleles identified), designated pMtA1 and pMtD2. MRE metal-responsive element, HERE 20-hydroxy-ecdysone-responsive element, ARE antioxidant-responsive element, CEBP CCAAT enhancer binding protein, DRE DNA replication-related element, Inr initiator element, DPE downstream promoter element. (b) Dose-response relationships for the action of cadmium on the two promoter alleles as a luciferase construct in a *Drosophila* S2 cell line. Responsiveness is expressed as luciferase units (RLU) relative to β-galactosidase. (c) Frequencies of pMtA1 and pMtD2 in 23 field populations (reference sites, metal mines and industrially contaminated woodlands in Belgium, France, Germany and the Netherlands), assessed using RFLPs of the PCR-amplified pMt locus (24 to 93 individuals per population). Site contamination is characterized by total concentrations of cadmium, lead, nickel, copper and zinc, joined in the W index defined as $W = \log \Pi_{i=1-n} (C_i/R_i)$, where C_i is the concentration of metal i, R_i is its reference concentration and n the number of metals considered (Widianarko et al. 2000). Sites with $W > 2$ are seriously contaminated, sites with $W \leq 0$ are clean. All data from Janssens et al. (2007, 2008), summarized and rearranged.

of metallothionein contribute to the evolution of metal tolerance in contaminated ecosystems.

A completely different tolerance mechanism is observed in the earthworm *Lumbricus rubellus*. Two genetically distinct lineages of this species were found to occur in a highly lead-polluted area, the Cwmystwyth mining site in Wales. Lineage A predominated in lead-contaminated calcareous "islands" of the mine soil, while lineage B was found in moderately polluted spots. Both lineages occurred in the clean microhabitats (Andre et al. 2010). Genotyping the worms by AFLPs and COII (*cf.* Table 3.9) revealed that the two lineages are genetically quite different from each other; the lead-tolerant worms could even be considered a separate (cryptic) species (Figure 6.23).

It is known that lead in biological tissues is often found associated with calcium and may enter the cell through calcium uptake mechanisms. In earthworms, Pb is sequestered in phosphate-rich granules of the chloragocytes, where it replaces Ca (Morgan and Morgan 1989). These type A granules (see Figure 6.20), are a sink for oxygen-seeking metals such as Pb and Ca. Although the granules are different from those in which cadmium accumulates, it is assumed that Pb leaves the body through the same pathway (delivery to coelomocytes and excretion through the nephridia, *cf.* Figure 6.21).

In accordance with the chemical similarity of Pb and Ca, it was found that when worms were exposed to Pb-contaminated soil, their transcriptome was highly

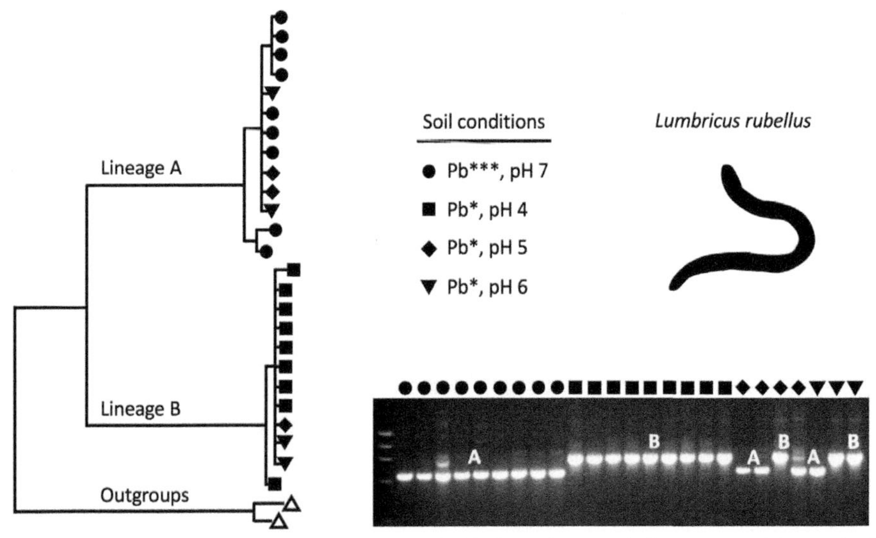

Figure 6.23. Left: genetic relationships among individual *Lumbricus rubellus*, recovered from a lead mine and genotyped by COII, with *L. castaneus* and *L. eiseni* as outgroups. The worms were sampled at different microhabitats in the Cwmystwyth valley in Wales. The phylogeny reveals two distinct lineages, A and B, with different microhabitat uses, specified in the legend. Right: Screening of individual worms for the expression of SERCA. RT-PCR amplification products were separated by electrophoresis on agarose gel. Two forward primers used in the PCR were designed so that each addressed one of the two isoforms of the gene. The microhabitats from which the worms were retrieved are given above the image. A and B indicate the two phylogenetic lineages. There is a very close match between COII genotype and SERCA isoform expression. Redrawn from Andre et al. (2010), with modifications.

enriched by genes with a function in Ca metabolism (Andre et al. 2010). Further research focused on a calcium transporter called *sarcoendoplasmic reticulum calcium transport ATPase* (*SERCA*). The gene encodes an ATP-driven ion pump, that transports calcium ions across intracellular endoplasmic membranes. It turned out that *L. rubellus* has two copies of this gene, which differ from each other in several amino acids. One of the isoforms was found to be expressed predominantly in the lead-tolerant worms, while the other was mainly expressed in lineage B (Figure 6.23).

The functional significance of this genetic differentiation is not yet clear. From studies on plants we know that some types of metal tolerance (especially zinc tolerance) are associated with alterations in metal transporters (Schat et al. 2000). So, the differential expression of SERCA genes in worms might point at a mechanism in which the trafficking of excess Pb is dealt with by switching among alternative calcium transporters. However, the SERCA protein is expressed in the endoplasmic reticulum, not in the cell membrane. In addition, in the mine soils, concentrations of Pb and Ca were highly correlated, allowing the possibility that altered Pb trafficking in worms is a consequence of high soil Ca, not an adaptation to Pb.

In fact, a genome-wide screen of genetic variation in British *L. rubellus* populations failed to resolve population structure between mine and control sites (Anderson et al. 2017). Earthworm populations seem to consist of multiple genetic lineages of which some tolerate metal-contaminated soils without the evolution of specific adaptations. Metal tolerance of a population is due to selection among existing lineages rather than evolution of novelty within an exposed population.

In summary, the studies reviewed above demonstrate that metal-contaminated soils may exert a selective pressure on soil invertebrates. In springtails, this conveys a fitness advantage to high-expresser alleles of the Cd-induced metallothionein promoter. The earthworm study suggests that differential expression of metal transporters could be a tolerance mechanism against Pb. Both studies are examples of evolution by pollution. It is also notable that both examples concern epigeic species. Whether metal tolerance is expected to evolve more rapidly among epigeic versus endogeic (euedaphic) soil invertebrates is difficult to say at the moment. More research is needed, combining mechanistic insights to toxicity, kinetics and ecology (Spurgeon et al. 2020).

6.9 Parasites, pathogens and symbionts

Soil invertebrates live in a microbial world. They often rely on microbes for food and they feed upon organic material pre-treated by microbes. However, some microbes are toxic and several are pathogenic. There is also a wide variety of microbial parasites attacking soil invertebrates. The regulatory and evolutionary effects emanating from this biotic community used to be largely ignored in soil ecology. This was partly due to the fact that traditional soil microbiology was limited to microbes that could be cultured in artificial media. With the advent of molecular techniques in the last decades, identification of environmental microbes and parasites became easier and this has changed the situation drastically.

Many soil invertebrates are host to parasitic larval stages of insects that complete their development by killing the host. For example, isopods may fall victim to Rhinophoridae, a family of flies completely specialized on woodlice. The female fly deposits eggs in places where isopods are expected and the first instar larva hatching from the egg penetrates an isopod through the soft connections between two segments. The larva grows in the haemolymph, consumes all internal organs and pupates in the cadaver. Wood et al. (2018) mention 18 species of isopod that are host to these flies, with 13 species of Rhinophoridae, making up 53 interactions.

A similar lifestyle, closer to predation than to parasitism, is seen in insects from the hymenopteran families Chalcididae, Diapriidae, Ichneumonidae, Sphecidae and Pompilidae. They attack a great variety of soil invertebrates such as harvestmen, spiders and insect larvae, usually by depositing their eggs upon or inside the living or paralyzed host with a specialized ovipositor. Unlike true parasites the larvae kill the host; therefore, they are called *parasitoids*. Saprophagous and mycophagous dipteran larvae as well as ants and carabid beetles are common hosts for parasitoid wasps. The community of soil and litter Hymenoptera (abbreviated "*SLH species*" by Ulrich 2004) often comprises as many species as their host community. For example, the wasp family Eucharitidae alone has around 500 species, all specialized on ants. SLH are an important but largely ignored component of the soil ecosystem (Ulrich 2004).

Some soil invertebrates, for example harvestmen, spiders and millipedes, suffer greatly from acarine ectoparasites. Most of these mites are classified in the large group Trombidiformes (*cf.* Figure 2.19), which includes the taxon Prostigmata. The genus *Leptus* (family Erythraeidae) is an extremely large group with many species specializing on harvestmen (Cokendolpher 1993). The mites on spiders mostly fall in another family, Trombidiidae (Durkin et al. 2021), while mites on millipedes are usually Mesostigmata and Astigmata (Farfan and Klompen 2012). Ectoparasitic mites damage the exoskeleton of the host and feed on the haemolymph. It is usually one of the nymphal stages (often the deutonymph) that is parasitic, while the adults are free-living. Some mites are commensalistic and only attach to the host exoskeleton to disperse (phoresy, see Section 3.9). The difference between phoresy and ectoparasitism is not always clear, as ectoparasites are also phoretic. The evolution from ectoparasitism to phoresy seems to have progressed further in millipedes than in harvestmen and spiders.

Table 6.8 gives an overview of soil invertebrate groups acting as hosts for parasites, parasitoids and pathogens (excluding pathogenic bacteria). For each group a rough estimate is given for the number of parasite species infecting that group. Overall, there are more host species than parasites, as many parasites infect several (related) hosts. A single species of parasitoid wasp may attack up to five different host species (Lachaud et al. 2012). In other cases, however, every host species has its own parasite, e.g., ants and entomopathogenic fungi often have a one-to-one relationship to each other. Other parasites are limited to a particular group of invertebrates; for instance, Trematoda (flukes) are found only in snails while Acanthocephala (thorny-headed worms) are found in a limited group of arthropods, including isopods, cockroaches and millipedes. On the other hand, parasitic nematodes infect almost all soil invertebrates, including isopods, beetles, harvestmen and snails (Table 6.8).

Table 6.8. Examples of host-parasite associations in soil invertebrates (not including pathogenic bacteria). For each group the number of species known as host and the number of known parasite species on that host group are given. References for this table are listed in Annex 8.5.

Number of species known as host	Number of species known as parasite	Example of host species	Parasite species on this host	Parasite group
Isopoda (woodlice)				
19	1	*Trichoniscus pusillus*	Invertebrate iridescent virus 31	Iridoviridae
1	1	*Armadillidium vulgare*	*Dispharynx nasuta*	Nematoda (roundworms)
6	1	*Armadillidium vulgare*	*Plagiorynchus cylindraceus*	Acanthocephala (thorny-headed worms)
18	13	*Porcellio scaber*	*Melanophora roralis*	Rhinophoridae (woodlouse flies)
Collembola (springtails)				
6	2	*Sminthurus viridis*	*Neozygites sminthuri*	Entomopathogenic fungi (Zygomycota)
Blattodea (cockroaches)				
1	1	*Periplaneta americana*	*Moniliformis moniliformis*	Acanthocephala (thorny-headed worms)
Dipteran larvae				
250	250	*Brachicoma devia*	*Trichopria inermis*	Parasitoid wasps (Hymenoptera)
Formicidae (ants)				
245	110	*Camponotus ligniperda*	*Ophiocordyceps unilateralis*	Entomopathogenic fungi (Ascomycota)
34	40	*Pachycondyla stigma*	*Kapala iridicolor*	Parasitoid wasps (Hymenoptera)
Passalidae (bess beetles)				
1	1	*Odontotaenius disjunctus*	*Chondronema passali*	Tylenchoidea (Nematoda)
Carabidae (ground beetles)				
86	29	*Nebria brevicollis*	*Gordonius violaceus*	Nematomorpha (hair worms)

Table 6.8 contd. ...

...Table 6.8 contd.

Number of species known as host	Number of species known as parasite	Example of host species	Parasite species on this host	Parasite group
Diplopoda (millipedes)				
1	1	*Cylindroiulus punctatus*	Invertebrate iridescent virus	Iridoviridae
1	1	*Archiulus moreleti*	*Gibbsia archiuli*	Apicomplexa (parasitic protists)
11	12	*Lusitanipus alternans*	*Diplopodomyces veneris*	Entomopathogenic fungi (Ascomycota)
6	6	*Parafontaria laminata*	*Rhigonema naylae*	Nematoda (roundworms)
3	3	*Cambala annulata*	*Gordionus lineatus*	Nematomorpha (hair worms)
2	1	*Floridobolus penneri*	*Macracanthorhynchus ingens*	Acanthocephala (thorny-headed worms)
76	40	*Ommatoiulus moreleti*	*Histiostoma feronianum*	Ectoparasitic mites (Astigmata)
Araneae (spiders)				
39	86	*Gongylidium rufipes*	*Ophiocordyceps verrucosa*	Araneopathogenic fungi (Ascomyota)
34	26	*Bolyphantes alticeps*	*Trombidium brevimanum*	Ectoparasitic mites (Trombidiformes)
Opiliones (harvestmen)				
29	10	*Phalangium opilio*	*Acanthorhynchus longispora*	Apicomplexa (parasitic protists)
6	4	*Leiobunum rotundum*	*Entomophaga batkoi*	Entomopathogenic fungi (Zygomycota)
1	1	*Opilio parietinus*	*Microsporidium weiseri*	Microsporidia (parasitic fungi)
9	8	*Phalangium opilio*	*Agamomermis phalangii*	Mermithidae (parasitic Nematoda)
18	25	*Mitopus morio*	*Leptus holmiae*	Ectoparasitic mites (Trombidiformes)
Lumbricidae (earthworms)				
1	1	*Lumbricus terrestris*	*Microcystis* sp.	Apicomplexa (parasitic protists)
Stylommatophora (land snails and slugs)				
1	1	*Patera binneyana*	*Pfeifferinella gugleri*	Apicomplexa (parasitic protists)
1	1	*Partula turgida*	*Steinhausia* spp.	Microsporidia (parasitic fungi)
30	8	*Deroceras reticulatum*	*Phasmarhabditis hermaphrodita*	Parasitic nematodes
21	2	*Aegista vulgivaga*	*Dicrocoelium chinensis*	Trematoda (parasitic flatworms, flukes)

Several parasites use soil invertebrates as an intermediate host; their definite host may be a mammal or a bird. For example, the nematode *Angiostrongylus vasorum*, found in slugs and snails, is a parasite of foxes and dogs (Majoros et al. 2010); the trematode *Brachylaima phaedusae* typically has three hosts: two of its life-stages are completed in different land snails, while a bird or a mammal is the definite host in which reproduction takes place (Waki et al. 2022). On the other hand, protist parasites like the unicellular apicomplexans, which are found in harvestmen, earthworms and snails (Table 6.8), are assumed to be taken up from the soil directly. They are often detected in the faeces of soil invertebrates, which explains their wide occurrence in soil surveys. Apicomplexa (mainly known for the human parasites *Plasmodium* and *Toxoplasma*) is the third most abundant protist group in soil (Fierer 2017, Del Campo et al. 2019, Figure 1.18).

The short overview in Table 6.8 is likely scraping only the surface of the wonderful diversity of parasites in soil invertebrates. Many pathogens require identification with molecular methods (e.g., the extremely diverse entomopathogenic fungi). Large-scale surveys, covering a significant part of the soil invertebrate community, are still to be done. In addition, Table 6.8 does not include pathogenic bacteria, for which an overview is difficult to get. Parasites, parasitoids and pathogens are important components of the soil food-web, with potentially great regulatory capacities.

Many parasites influence the behaviour of their hosts. A wide range of behavioural changes is shown by arthropods infected by entomopathogenic fungi (Roy et al. 2006). Among soil invertebrates, the ant tribus Camponotini is known for an enormous diversity of these fungi. An infected ant tends to climb to a high spot in the litter or mounts the vegetation and bites itself in the substrate before death. A long stalk of fungus then grows out of the head or the pronotum, to form a sporangium at the top. The behaviour imposed by the fungus obviously contributes to enhancing its dispersal.

In other cases, a parasite forces the host to seek a habitat where the definite host is expected or where reproduction must take place. This is done by *Gordonius*, a genus of nematomorph worm infecting beetles, crickets, grasshoppers and millipedes. The infected arthropod moves to the waterside and jumps in the water, triggering the worm to exit the host through the anus. Ground beetles infected by nematomorphs are best sought in wet habitats close to streams (Looney et al. 2012).

Another nice example of host behaviour manipulation is the acanthocephalan parasite *Plagiorynchus cylindraceus* infecting the isopod *Armadillidium vulgare* (Moore 1983). In field-collected pill bugs, prevalence of the parasite is low (0.15% of hosts infected, averaged over 4 years of observation). However, the prevalence of the parasite in isopods fed by parent starlings (*Sturnus vulgaris*) to their young appeared to be 12.7% (Figure 6.24). The latter frequency was measured by inducing nestlings to regurgitate the food after being fed by the parent. Obviously, starling parents do not feed at random with respect to infected isopods, they sample the infected ones nearly 100 times more often than expected. This interpretation was supported by behavioural experiments: *Armadillidium* tends to move less under cover, and more on light and dry substrates when infected by thorny-headed worms (Figure 6.24). What chemical signals from the parasite to the isopod's brain induce the woodlouse to abnormal, conspicuous behaviour is not known.

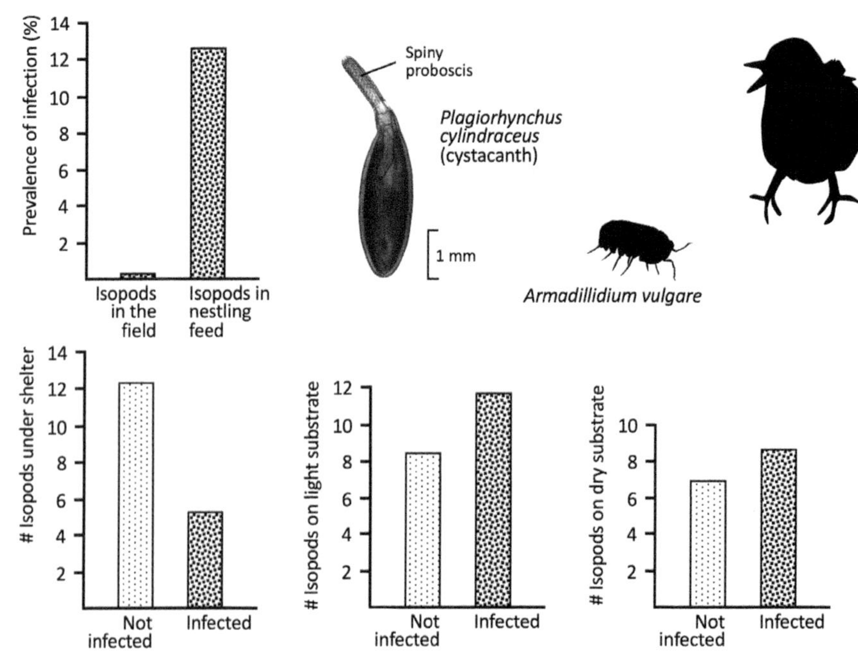

Figure 6.24. Prevalence of the parasite *Plagiorhynchus cylindraceus* (Acanthocephala) in *Armadillidium vulgare* is much greater in woodlice fed by starlings to their nestlings than in isopods collected from the field at random (top graph). This is corroborated by behavioural experiments showing a higher tendency to move outside a shelter (below left), on a white surface (below middle) or on a dry substrate (75% compared to 95% humidity) (below right). Data from Moore (1983). The acanthocephalan parasite shown is a cystacanth from the body cavity of *Trachelipus squamuliger*, reprinted from Dimitrova (2009) by permission of Springer Nature.

Soil invertebrates do not stand helpless against pathogens and parasites, nor are they all equally susceptible. For example, among three slug species, *Limax pseudoflavus* was not susceptible to infection by the nematode *Phasmarhabditis hermaphrodita*, while two other slugs, *Deroceras reticulatum* and *Milax gagates*, were readily infected and killed (Rae et al. 2008). Among Collembola, the euedaphic species *Folsomia candida* was not susceptible at all to the virulent entomopathogenic fungi *Beauveria bassiana* and *Metarhizium anisopliae*, which readily killed ants and termites (Broza et al. 2001). Up to now entomopathogenic fungi have been isolated only from epigeic Collembola (*Sminthurus viridis*, *Orchesella cincta*, *O. villosa* and *Deuterosminthurus sulphureus*), and one hemiedaphic species (*Isotoma anglica*), never from euedaphic species (Dromph et al. 2001). This suggests that fungivorous Collembola have evolved special mechanisms to counteract fungal pathogens, as an adaptation to life in soil.

All soil invertebrates are assumed to deploy their *innate immune system* to defend themselves against pathogens and parasites. The innate (non-adaptive) immune system of invertebrates is extensively studied only in model species like *Drosophila*, but the pathways involved are found in all invertebrates (Du Pasquier 2001, Hoffmann and Reichhart 2002). Among soil invertebrates, the immunity response of earthworms is the best investigated mechanism. The immune-active cells in the coelom (*coelomocytes*) are of three types (Morgan et al. 1999, Adamowicz 2005, Homa 2018):

- *Eleocytes*. These are cells detached from the *chloragogen* tissue surrounding the intestine. They are loaded with characteristic granules called *chloragosomes*. We met them before as vehicles for metal trafficking from the chloragogen tissue to the kidneys (Section 6.8).
- *Amoebocytes*. These cells (two subtypes are discerned) are the most common coelomocytes. They may form aggregates and *rosettes* around foreign cells or particles. Their main function is in cellular defence, i.e., to remove particles and bacteria by phagocytosis. This function is similar to the macrophages of vertebrates. They also contribute to the *encapsulation* response, i.e., the production of *melanin*, by which an impermeable shield is formed around a parasite.
- *Granulocytes*. These cells contain multiple electron-dense granules. They are assumed to be involved in humoral defence, i.e., the production of antibodies that are released upon intruders.

The terrestrialization of invertebrates must have boosted their immune system, to combat the great variety of new, soil-borne pathogens. How exactly this evolution took place and what immune components were expanded or suppressed is not known, due to lack of comparative immunological studies. The genome of *Folsomia candida* shows a significant expansion of stress-related gene families, in comparison to crustaceans (Faddeeva-Vakhrusheva et al. 2017), which suggests that upregulation of stress defence is indeed an aspect of terrestrialization.

Grading the various parasitic organisms listed in Table 6.8 by their effects on host fitness, the strongest fitness reduction is obviously caused by entomopathogenic fungi, as they must kill their host to complete the life-cycle. This is comparable to the action of parasitoid wasps and flies. True parasites are not expected to kill their host, although many have a fitness suppressive effect, e.g., due to inhibition of reproduction (trematodes in snails) or increased predation risk (acanthocephalans in isopods). However, there are also several cases in which the parasite has only a minor effect on host fitness. For example, the unicellular parasite *Monocystis* (Apicomplexa: Gregarinidae), which colonizes the seminal vesicles of earthworms, has only a weak effect on growth and no effect on reproduction, nor on foraging behaviour (Field et al. 2003, Field and Michiels 2005).

In addition to the many parasites and microbial pathogens plaguing soil invertebrates, a large community of microbes engages in commensalistic or mutualistic relationships. Such microbes are mostly concentrated in the gut, but in a few cases also in specific organs. We have already seen in Chapter 4 how many soil invertebrates are host to intracellular bacteria (*Wolbachia*, *Cardinium*, *Blattobacterium*, etc.). Extracellular symbionts outside the gut are less common; however, a quite remarkable case is the nephridial symbiotic community of earthworms.

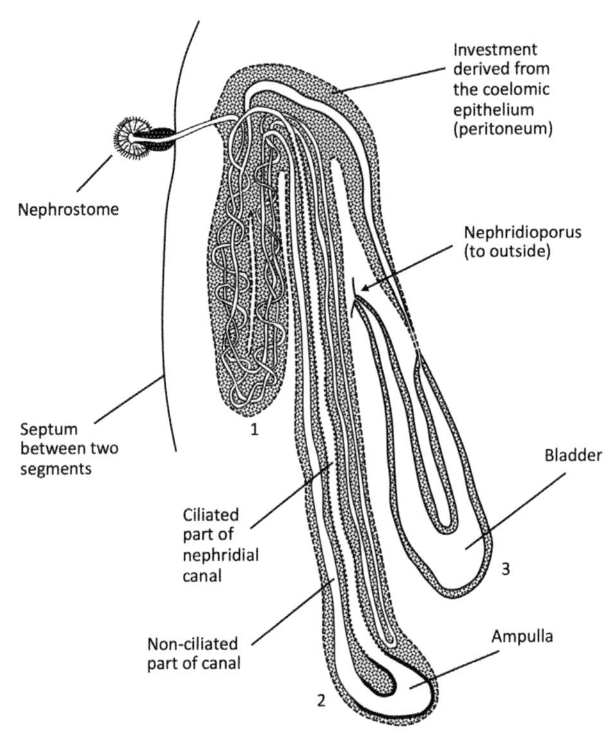

Figure 6.25. Diagram of the nephridium of *Lumbricus terrestris*. The kidney is a coelomoduct, opening in the coelom with a nephrostome, coiling through three lobes of the nephridium (1, 2, 3) and ending in a narrow external opening (nephridioporus). Redrawn from Knop (1926), originally by J. Maziarski 1905. Nephridial bacteria reside in the ampulla.

Already in 1926, the German microscopist Johannes Knop at the University of Greifswald, described the presence of bacteria in the nephridia of earthworms. The complicated structure of these kidneys, that we already met in Section 6.8, is reproduced in Figure 6.25. The nephridium is a *coelomoduct*, starting with a ciliated funnel in the coelom (*nephrostome*), giving way to a thin canal, which passes through the septum separating two segments, and then forms a three-lobed structure in the next segment. The canal loops three times through the first lobe and two times through the second lobe, to end in a sac-like structure, the functional bladder, in the third lobe. This ends in a tiny external opening, the *nephridioporus*. In the second lobe the canal widens in an *ampulla*. This is where the nephridial bacteria are located. Knop (1926) already noted that these bacteria do not enter the cells but reside in the lumen of the ampulla, fixed to their place with a gel-like substance.

The most common nephridial symbiont is *Verminephrobacter*, a bacterium known only from the nephridia of earthworms. Belonging to the common soil phylum Betaproteobacteria (*cf.* Figure 1.18), it is never found in the soil itself. It colonizes about 80% of lumbricid species and in some cases it is the only one. However, several earthworms have a mixed community including additional genera like *Nephrothrix*, *Flexibacter* (both Bacteriodetes), and an *Agromyces*-like bacterium from the Actinobacteria (Aira et al. 2018, Lund et al. 2018, Viana et al. 2018). At least 27 taxa from eight bacterial phyla are found scattered across the earthworm family Lumbricidae and some are also found in other earthworm families. These additional symbionts are more species-specific than *Verminephrobacter* and seem to have joined on a case-by-case basis.

Once symbionts, the bacteria do not colonize worms from the environment, they are vertically transmitted from parents to offspring during the production of cocoons. The bacteria are stored in high density inside the cocoon together with sperm and eggs. After fertilization, while the embryo develops, the symbionts colonize the nephridia as soon as they are formed. It is assumed they are transmitted only along the line of the cocoon-producing worms, very much like mitochondria are transmitted along the maternal line. However, in worms such as the compost worm *Eisenia fetida*, living in high densities with frequent matings, the bacteria could also be transferred biparentally or through "leaky" horizontal transmission (Viana et al. 2016).

The faithful vertical transfer is confirmed by a pattern of genetic distances between *Verminephrobacter* strains of different worm species (Møller et al. 2015). The phylogeny of symbionts closely parallels the phylogeny of their hosts (Figure 6.26). Hosts and symbionts seem to have co-diversified throughout their

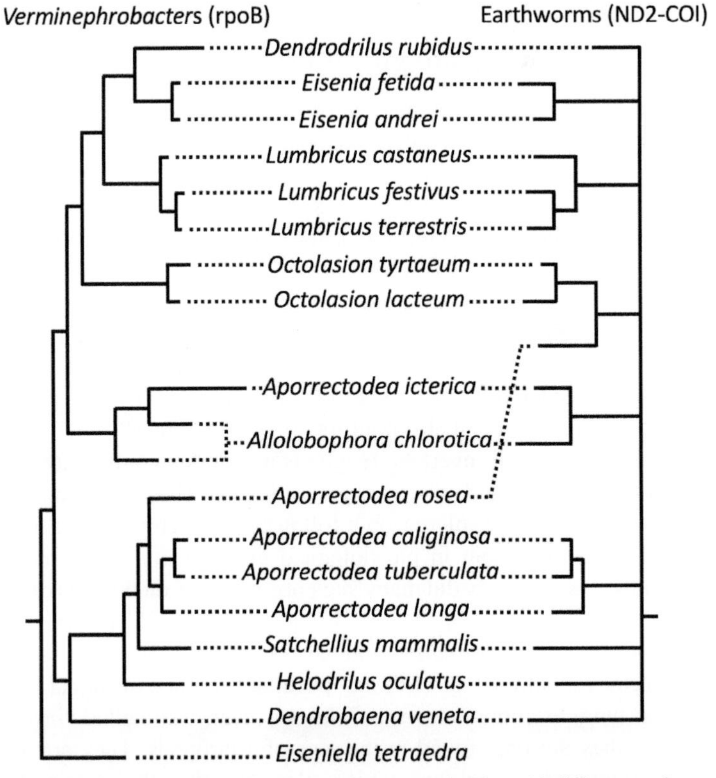

Figure 6.26. Molecular phylogeny of *Verminephrobacter* strains from 18 different earthworm species (left), aligned with the phylogeny of their hosts (right). The phylogeny of the bacteria is based upon DNA sequences of the RNA polymerase subunit B gene (rpoB); the earthworm phylogeny is based upon the mitochondrial gene NADH dehydrogenase joined with cytochrome c oxidase subunit I (ND2-COI). Redrawn from Møller et al. (2015).

evolutionary lifetime. Only *Aporrectodea rosea*, *A. ictera* and *Allobophora chlorotica* carry symbionts that are not monophyletic with each other. For the clade as a whole, the origin of the symbiosis is estimated to date back to the origin of the family Lumbricidae in the Cretaceous, around 100 Ma BP (Lund et al. 2010b). This does not hold for other nephridial symbionts, e.g., the phylogeny of *Flexibacter* reveals several host switches (Møller et al. 2015).

The presence of symbionts confers significant fitness benefits to the host, as worms without symbionts mature slower and produce fewer cocoons (Lund et al. 2010a, Viana et al. 2018). It is often assumed that the nephridial bacteria contribute to increased nitrogen retention. For example, the bacteria might produce proteases, which could degrade any protein remaining in the urine and ensure that its amino acids can still be resorbed. This interpretation is supported by the fact that the positive effects on reproduction and maturation are strongest when the worms are grown under nutrient-limiting conditions. However, the symbionts do not stimulate body-growth and no effects on nitrogen components of the cocoon could be found (Lund et al. 2010a). Worms in contaminated soils lacking the symbiont do not suffer from loss of fitness (Pass et al. 2015). Despite intensive research, the mutual benefits of this wonderful symbiosis have not yet been pinpointed.

6.10 The soil invertebrate gut microbiome

The gut of any animal is an environment of its own, with a very diverse community of microbes, usually comprising more cells than the host itself, with a gene complement far exceeding the gene complement of the host. For example, the human gut is estimated to contain 39 trillion cells (versus 30 trillion cells in the human body), with around 10,000 species and 8 million different genes (versus 21,000 genes in the human genome) (Sender et al. 2016). This complex microbial community influences many aspects of animal physiology, via direct nerve connections between entero-endocrine cells and the brain (Hoffman and Lumpkin 2018), or via hormones and immune signalling. Inclusion of the gut community in any assessment of an animal's functioning was considered "*a new imperative for the life sciences*" (McFall-Ngai et al. 2013). Does this statement also hold for soil invertebrates?

The microbiome of soil invertebrate guts is a complex community of bacteria, Archaea, protists, fungi and microscopic animals. These communities were traditionally studied by light microscopy, followed by attempts to culture specific isolates and characterize them in physiological experiments. The American 19th century naturalist Joseph Leidy did many such observations on the guts of millipedes and termites. In a beautiful teaching chart, he sketched the biocomplexity inside the gut of the millipede *Iulus marginatus* (Figure 6.27).

Leidy discovered, already in 1849, the peculiar presence of fungal filaments in the gut. These fungi are now known as *Trichomycetes*, also called "*arthropod gut fungi*" because they are only found in the guts of arthropods. They are a primitive group of Zygomycota, a phylum which also contains the entomopathogenic Entomophthorales. Their filamentous hyphae have a characteristic basal cell which attaches the filament firmly to the cuticle of the gut, usually at a specific position, most often the first part of the hindgut, which is lined with ectodermal cuticle (Nardi

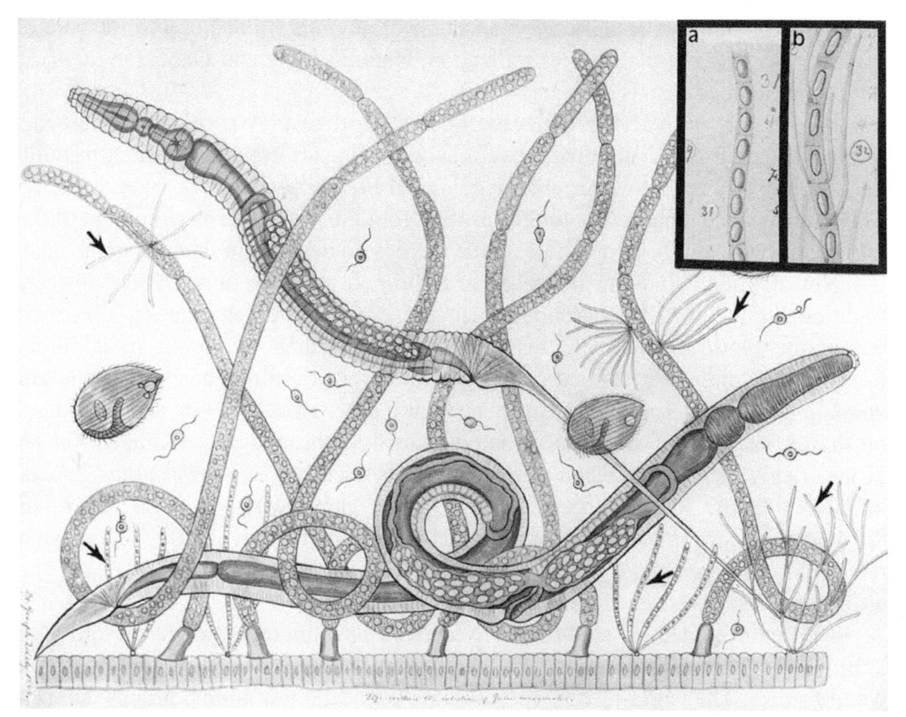

Figure 6.27. Joseph Leidy's teaching chart no. 60, showing a community of microscopic animals, protists, fungi and bacteria associated with the gut of the millipede *Iulus marginatus*. The drawing shows two gut-living nematodes, several filaments of trichomycete fungi attached to the epithelium and "bushes" of filamentous bacteria belonging to the group Lachnospiraceae (Firmicutes), attached to fungal hyphae and epithelial cells (arrows). These bacteria were described by Leidy (1849) as "*Arthromitus*". The large and small motile cells are parabasilid flagellates. The insets a and b show details from Leidy's termite chart. Reproduced from Margulis et al. (1998), courtesy of ANSP Archive Collection 532.

et al. 2016). The fungus sporulates in the gut, in synchrony with the moulting cycle of its host. As the cuticle of the hindgut is also moulted, the spores leave the body, but develop into a mycelium associated with the exuvium and sporulate again. These secondary spores may re-enter the host through the eating of faeces (Moss and Taylor 1996, Degawa 2009).

Within the insects, trichomycetes are found mainly in orders with aquatic larval stages, such as Ephemeroptera, Plecoptera and Trichoptera, as well as aquatic crustaceans. However, Collembola living in marshes and on the water surface (*Podura aquatica*, *Isotomurus palustris*) are also infected by trichomycetes. Truly terrestrial hosts include woodlice, passalid beetles and of course millipedes (Lichtwardt 1990, 2008, 2012). Whether these are all evolutionary independent colonizations cannot be decided at the moment. One possible scenario is that the ancestral fungi colonizing arthropod guts were aquatic (like the related parasitic group Chytridiomycota). The dispersal of spores from the trichomycete order Assellariales is still dependent on water (Degawa 2009), which explains why only water-associated Collembola are infected, not drought-tolerant terrestrial species. The colonization of millipedes and beetles must then be considered separate, later events. A possible relationship of trichomycete fungi with the terrestrialization of animals (in analogy to the role of mycorrhizal fungi in plant terrestrialization, Naranjo-Ortiz and Gabaldón 2019) is not to be excluded.

Whether trichomycetes produce enzymes aiding digestion is not known but seems likely given their association with several wood-eating arthropods (termites, passalid beetles, millipedes). However, the mycological literature considers the association as commensalistic, that is, the fungus profits from the protective environment of the gut and the availability of nutrients, while there is no benefit nor damage to the host.

Not all fungus-looking microbes in arthropod guts are fungi. Trichomycetes traditionally included two groups which are now considered protists, classified within the Opisthokonta but closer to Choanozoa than to Fungi. In addition, Leidy (1849) already described colonies of microbes forming "bushes" on the gut epithelium and on trichomycete hyphae (Figure 6.27, arrows). They are also found on the exoskeleton of parasitic nematodes inside arthropod guts (Kitagami et al. 2019). Leidy (1849) described these enigmatic microbes as "*Arthromitus*". After initial misidentification (Margulis et al. 1998) they are now known to be bacteria of the group Lachnospiraceae (phylum Firmicutes) (Thompson et al. 2012), somewhat related to the "*segmented filamentous bacteria*" in the gut of vertebrates and the gastrointestinal parasites of the genus *Clostridium*.

The main model for studies of invertebrate gut microbiomes is the order of termites (Isoptera). Termites include fungivores, humus feeders, detritivores and wood feeders. The wood-feeding termites are one of the few animal groups that can degrade *lignocellulose*, the complex of cellulose and hemicellulose encapsulated by lignin fibres. This digestive capacity has raised great interest among biotechnologists. Termite gut communities are explored for enzymes that may be used in a bioreactor to recycle valuable carbohydrates from agricultural waste (Brune 2014, Brune and Dietrich 2015).

The digestion of lignocellulose is symbiotic. Both host enzymes and enzymes from symbionts are involved. In lower termites, a crucial role is played by intestinal protists from the phyla Parabasalia and Preaxostyla: eukaryotic unicellular protists with multiple flagella, classified in the supergroup *Excavata*, a lineage with an unclear position at the base of the eukaryote tree of life (Burki et al. 2020). These intestinal flagellates are obligate symbionts of several invertebrates, with many evolutionary specializations, such as loss of mitochondria. Some parabasalids have extremely large cells and were described already by Leidy (1849) (see Figure 6.27). The flagellates ingest small wood particles by phagocytosis and digest them with potent cellulases and hemicellulases in intracellular vacuoles. In addition, flagellates of the genus *Trichonympha* carry a large population of intracellular symbiotic microbacteria classified in the phylum Elusimicrobia, which contribute to nitrogen provision (Zheng et al. 2015). So, the symbiosis involves three levels nested inside each other (termite, flagellate, endosymbiotic bacterium).

An overview of the digestive system of lower termites, with the various compartments and symbionts, is provided in Figure 6.28 (Brune 2014). The grinding and fragmenting action of the mandibles is a crucial first step, since this augments the surface for enzymatic action. The gut is similar to the one of cockroaches, but the tract is longer and subdivided in five compartments. An enteric valve between the

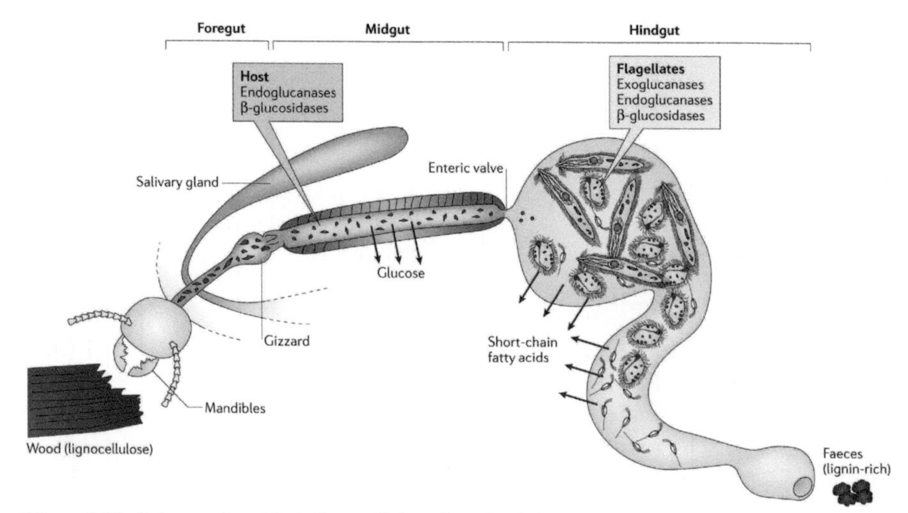

Figure 6.28. Scheme of symbiotic lignocellulose digestion in lower termite guts, showing the interplay of host enzymes (mainly in the midgut) and symbiont enzymes from flagellates (in the hind gut). Reprinted from Brune (2014), by permission from Springer Nature.

midgut and the hindgut prevents any reflux of particles from the voluminous hindgut sack, which has a large population of flagellates.

The acquisition of protists to support wood digestion was a major event in the evolution of the termites. Dated at about 130 Ma BP, it marked the divergence between the clades of Isoptera and Dictyoptera. Within the termites however, a second major evolutionary event took place, the loss of flagellates. This marks the transition from lower to higher termites. The youngest family, Termitidae, started to diversify while switching from wood to fungi, humus and litter. So, the changes in gut termite microbiomes reflect the major evolutionary events happening in the group (Dietrich et al. 2014).

In addition to microscopic studies, gut microbial communities, especially bacteria, have been studied intensively by DNA sequencing. One popular approach is to apply a PCR to gut DNA that amplifies a marker sequence universal to all bacteria, but containing a taxon-specific variable region. The 16S ribosomal RNA gene has proven to be extremely useful for this purpose. The microbial taxa identified after sequencing the PCR products, are called "*operational taxonomic units*" (*OTUs*). A taxonomic profile of the community is obtained by counting the number of clones or reads falling in each OTU.

An even more encompassing approach is to try and sequence not only rRNA genes, but all microbial DNA in the gut. This allows an overview of the complete gene complement of the microbial community (irrespective of the species where they came from). This approach, known as *metagenomics*, was introduced to soil ecology by the American microbiologist Jo Handelsman in 2002 (Handelsman et al. 2002). It is an extremely data-rich approach, requiring extensive bioinformatics to sort out the main trends and patterns (Van Straalen and Roelofs 2012).

For a number of soil invertebrates (isopods, springtails, cockroaches, termites, ants, earthworms) an overview of gut microbial communities has been obtained by DNA sequencing methods. Table 6.9 summarizes the taxonomic composition of these

Table 6.9. Taxonomic diversity by main phyla of microbial communities associated with soil invertebrates, as revealed by 16S rRNA profiling (% of clones or reads). Data from Bredon et al. (2018), Lapanje et al. (2010) (isopods); Agamennone et al. (2019), Xiang et al. (2019) (springtails); King et al. (2008), Do et al. (2014), Santana et al. (2015), Mikaelyan et al. (2015) (termites); Kautz et al. (2013) (ants); Knapp et al. (2010) (millipedes); Singh et al. (2015), Pass et al. (2015), Ma et al. (2017b), Liu et al. (2018), Sapkota et al. (2020), Thakur et al. (2021) (earthworms). For *F. candida*, *E. fetida* and *P. excavata* more than one report is given. For 17 termite species the data were pooled over feeding groups after Mikaelyan et al. (2015). For each microbial group the host species with highest abundance is highlighted in grey.

	Acidobacteria (%)	**Actino-bacteria (%)**	**Bacteriodetes (%)**	**Firmicutes (%)**	**Planctomycetes (%)**	**Proteobacteria (%)**	**Spirochaetes (%)**	**Others (%)***
Woodlice								
Armadillidium vulgare	-	1.7	4.6	8.2	0.1	80.6	0.5	4.3
Porcellio scaber	-	3.0	14.0	16.0	2.0	65.0	0.5	0.0
Springtails								
Folsomia candida 1	-	31.5	10.5	5.9	-	49.4	-	2.8
Folsomia candida 2[@]	2.6	1.3	16.1	3.4	1.3	70.3	-	5.1
Cockroaches								
Cryptocercus punctulatus	1.0	11.0	14.0	52.0	0.1	17.3	1.0	3.6
Parasphaeria boleiriana	0.5	8.0	16.0	53.0	0.5	19.5	1.5	1.0
Termites								
Coptotermes gestroi	-	0.5	11.6	22.5	1.1	17.8	17.4	29.1
Coptotermes curvignathus	-	6.0	65.0	15.0	0.1	7.3	2.0	4.6
Syntermes wheeleri	2.6	0.4	3.7	76.7	-	2.3	5.0	9.4
Three fungus-feeding termites	0.2	3.5	32.5	25.4	3.4	15.1	4.1	15.8
Three soil-feeding termites	0.4	3.9	15.1	55.8	0.7	7.8	8.2	8.1
Six humus-feeding termites	0.6	2.7	16.7	39.5	0.2	9.0	21.0	10.3
Five wood-feeding termites	1.2	1.5	7.6	11.3	0.1	2.9	65.4	10.0

Ants								
Cephalotes varians	-	5.0	0.8	-	-	61.5	-	32.7#
Millipedes								
Cylindroiulus fulviceps	-	-	26.1	18.0	-	35.3	1.6	19.0
Earthworms								
Eisenia fetida 1	-	18.6	1.2	27.1	-	48.6	-	4.5
Eisenia fetida 2	7.5	6.0	3.0	11.9	7.5	38.8	-	25.4
Eisenia fetida 3	-	7.5	8.6	3.0	1.0	65.2	-	14.7
Allolobophora chlorotica	10.1	19.6	3.8	8.9	-	17.0	-	40.6
Aporrectodea caliginosa	14.6	13.0	7.0	8.2	-	20.2	-	37.0
Aporrectodea tuberculata	15.1	29.1	2.5	0.6	-	16.5	-	36.2
Lumbricus herculeus	0.3	33.5	3.3	19.6	-	38.0	-	5.4
Lumbricus rubellus	3.2	28.0	5.9	5.3	-	52.3	-	5.3
Polypheretima elongata	-	34.1	8.5	8.5	-	35.7	-	13.1
Perionyx excavatus 1	-	36.4	3.1	15.5	-	31.8	-	13.1
Perionyx excavatus 2	12.0	1.3	1.3	9.3	4.0	44.0	1.3	26.6
Eudrilus eugeniae	-	13.0	0.8	1.5	-	77.9	-	6.8

*including unclassified groups and phyla not listed here, [@]pooled over four diet treatments

microbiomes, classified by bacterial phylum. A remarkably high diversity of gut microbes is seen in all soil invertebrates. It is no exception to find several hundreds of bacterial species in a single animal. The soil invertebrate microbial community is much more diverse than the specialized communities of higher insects (honeybees, fruit flies), and also more diverse than the aquatic and marine lineages from which soil invertebrates descend. It seems that the terrestrialization of invertebrates was accompanied by an increased diversity of gut microbes, which in later evolution of the lineage became more specialized and less diverse.

In most data sets there is considerable variability between strains or cultures of the same host species. Nevertheless, a certain fraction of the microbiome is constant across strains. For example, in a comparison of two *Folsomia candida* populations, 832 OTUs were found, of which 45% were present only in one strain and 17% only in the other. The OTUs shared across strains were the most abundant ones, making up 72% of the total number of reads in the one strain and 90% in the other (Agamennone et al. 2015). These core microbes may recolonize the gut from the pyloric region, after the gut epithelium is renewed and shed every moulting cycle (Tebbe et al. 2006). A similar population-differentiation of microbial gut communities is seen in the isopod *Armadillidium vulgare* (Dittmer et al. 2016).

The second main factor determining microbial community structure is the location in the gut. This is especially true for invertebrates with a clear compartmentalization of the digestive tract, such as termites (*cf.* Figure 6.28). In a survey of nine different higher termite species, the gut communities mostly clustered by compartment, independent of the phylogeny of the species (Mikaelyan et al. 2017). Acidity appeared to be a crucial determining factor. Another example is the association of certain bacteria with a specific intestinal compartment. For example, *Hepatoplasma crinochetorum*, a bacterium of the phylum Mollicutes, is found in the hepatopancreas of many isopods, and in lower numbers also in the gut (Leclercq et al. 2014, Dittmer et al. 2016).

The third factor affecting gut communities is dietary habit. Invertebrates with different feeding habits tend to have different gut communities. This is again obvious for termites, where the microbial communities cluster best by feeding guilds of the host (Mikaelyan et al. 2015). This is due to the fact that many microbes enter the host from the environment (horizontal transmission). However, an extensive review of microbial communities in 94 termite species (Bourguignon et al. 2018), demonstrated that co-speciation with the host, implying colony-to-offspring vertical transmission, is also an important factor. According to these authors, the distribution of microbes across termite species is best described by a mixed transmission model. Likewise, King et al. (2014) and Arora et al. (2022) emphasized the strong host-phylogenetic signal in termite gut bacterial communities, due to coevolution of species with dietary habit.

A final conclusion from the data in Table 6.9 is that the microbial composition of the invertebrate gut is very different from the soil. The two most abundant groups of bacteria in soil are Acidobacteria and Verrucomicrobia (Figure 1.18), but these phyla are hardly ever found in gut microbiomes (only, to a certain extent, in termites and earthworms). The gut microbiome often has a fair share of Actinobacteria (previously called "actinomycetes"), a lineage of Gram-positive anaerobic bacteria that form filaments resembling the mycelia of fungi. Also, Proteobacteria (mainly alphaproteobacteria) and Firmicutes are among the dominant representatives of invertebrate gut microbiomes. These three groups are, however, much rarer in the soil. The gut of soil invertebrates forms an—often anoxic—microzone in the soil allowing anaerobic processes such as denitrification, in an otherwise aerobic environment (Drake and Horn 2007, Zhu et al. 2021).

There are also patterns of microbiomes across the taxonomic groups, but these are difficult to interpret. Why would isopods and ants have a much higher abundance of Proteobacteria than other soil invertebrates? Why would Actinobacteria reach such a high densities in earthworms and springtails? And why are termites champions in Firmicutes and Spirochaetes? (Table 6.9). Most likely these patterns must be explained by physiological conditions of the gut, e.g., acidity and oxygenation. In addition, specific receptors in the gut epithelium may contribute to selectivity, leading to lineage-specific microbial communities, such as *Arthromitus* in termites and millipedes, *Hepatoplasma* in isopods and *Verminephrobacter* in earthworms.

In addition to 16S rRNA screening, which reveals the taxonomic diversity of the microbiome, an overview of functional diversity requires metagenomic sequencing. Of particular interest are genes encoding proteins active in carbohydrate metabolism since these are important contributors to the digestive function of the gut community. Enzymes and other proteins with catalytic or binding properties towards carbohydrates are jointly called *carbohydrate-active enzymes* or *CAZymes*. Data on DNA sequences for such proteins are collected in the CAZy database, which includes analytical tools for classification and characterization (http://www.cazy.org).

In an analysis of gut metagenomes of termites, isopods and springtails, Le Ngoc et al. (2021) identified hundreds of full-length CAZy genes, classified into 163 families and 6 functional classes. Termites, *Coptotermes gestroi*, had the largest number of CAZy genes (905), followed by springtails, *Folsomia candida* (648), and isopods, *Armadillidium vulgare* (627) (Figure 6.29). A total of 43 common CAZy families were shared between the three metagenomes. For example, several families of *glycoside hydrolases*, which hydrolyse the glucoside bonds between carbohydrates or between a carbohydrate and a non-carbohydrate moiety, are shared between the species. These enzymes are all needed for the degradation of common plant biomass components such as cellulose and hemicellulose.

On the other hand, some families appeared to be unique for one of the three species. Termites had 32 unique CAZy families, springtails 16 and isopods 11. Springtails had very few polysaccharide lyases in comparison to the other species, but the highest number of carbohydrate esterases (102 different proteins). Springtails lacked any pectinases, while isopods had multiple pectate lyases (10 proteins in total) and termites had only one. This could indicate that pectin, a major polysaccharide of plant cell walls, is not an important resource for springtails and termites, while for isopods it is.

A comprehensive functional interpretation of these patterns cannot yet be given. Metagenomics of soil invertebrates is still in an exploratory phase. But it may be expected, when more comparative data become available, that the systematic functional analysis of microbiomes will shed new light on the ecology and evolution

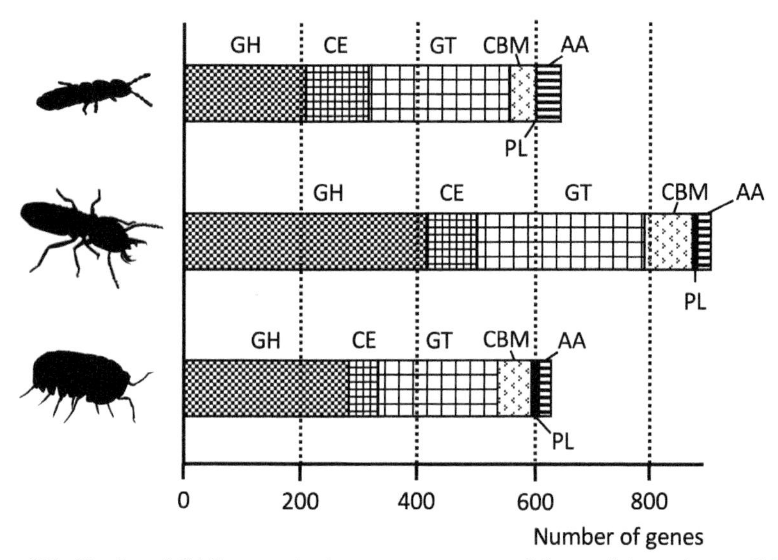

Figure 6.29. Number of CAZy genes in the gut metagenomes of three soil invertebrates: *Folsomia candida* (Collembola), *Coptotermes gestroi* (Isoptera) and *Armadillidium vulgare* (Isopoda). GH glycoside hydrolases, CE carbohydrate esterases, GT glycosyl transferases, CBM carbohydrate-binding molecules, PL polysaccharide lyases, AA auxiliary activities. From Le Ngoc et al. (2021).

of soil invertebrates. The "new imperative for the life sciences" (McFall-Ngai et al. 2013), applies to soil invertebrates maybe even more than to other animals.

6.11 Horizontal gene transfer

The intimate association between host and microbiome creates ample opportunities for horizontal gene transfer (*HGT*). This process, though quite common among prokaryotes (bacteria and Archaea), is rare for prokaryote-eukaryote transfers (Andersson 2005). When it happens, however, major evolutionary transitions may follow. For example, the evolution of tunicates as a separate lineage of the phylum Chordata, was accompanied by an HGT event involving enzymes for cellulose synthesis and degradation (Dehal et al. 2002, Matthysse et al. 2004). These enzymes, allowing synthesis of the tunic, are never found in other chordates. Did such rare events also happen in the evolution of soil invertebrates? Interestingly, among the eukaryotes subject to demonstrable prokaryote-eukaryote gene transfer, are three groups of soil invertebrates: nematodes, tardigrades and springtails.

Horizontally transferred genes are usually discovered in phylogenetic analysis and genome annotation. One spots a certain segment of a genome that clusters far outside the group of species to which the rest of the genome belongs. Boschetti et al. (2012) introduced the "*HGT index*", also called *h index*, which, for foreign genes in a metazoan genome, was defined as the difference between the highest non-metazoan and the highest metazoan *bitscore* (bitscore being a logarithmic measure of similarity between DNA sequences). A value of $h \geq 30$ was considered indicative of HGT.

However, phylogenetic evidence is rarely sufficient proof, due to contaminating bacterial DNA being assembled with the native genome. This is especially a problem when using short reads for genome assembly. More conclusive evidence may be obtained from genetic linkage with native genes, continuity on the same genomic contig, expression in host tissues and typical eukaryotic properties such as introns and signal peptides.

It is assumed that most HGT events are due to retroviruses, transposons and mobile elements (*transduction*). Many eukaryotic genomes are full of such genetic elements and HGT genes are often found to be associated with sequences of mobile elements or viruses. However, direct uptake of DNA through the diet may also play a role. Anyway, close proximity between donor and acceptor, i.e., transfer from symbionts, parasites and gut microbes into their hosts, is certainly a facilitating factor.

The common fate of HGT-genes is to become non-functional in the recipient genome. In most cases they are not expressed and develop into pseudogenes. This is often due to incompleteness of the sequence, lack of a promoter, or insertion in a non-active, heterochromatin, part of the genome. Even if the transferred segment has a proper open reading frame and happens to be inserted close to an active promoter, the gene may degrade over time due to lack of purifying selection, if the gene product does not add to the fitness of the host.

Whether HGT is a hallmark of animal genome evolution is a contested issue. Some authors (Crisp et al. 2015) have argued that HGT is an ongoing and important process, others have doubted the validity of this claim (Martin 2017, Salzberg 2017). However, there is now general agreement that HGT has been important in nematodes, tardigrades and springtails, as well as in several non-soil invertebrates such as bdelloid rotifers (Gladyshev et al. 2008). Tens to hundreds of foreign genes are found in the genomes of these animals, representing a few percent of the total number of transcribed genes.

It may not be a coincidence that groups with a relatively high rate of HGT include species that can undergo anhydrobiosis (*cf.* Section 6.4). We have seen above that anhydrobiosis comes with significant DNA degradation and fragmentation. In the reassembly process during rehydration, foreign DNA may be easily included in the genome. This would imply that HGT in nematodes, tardigrades and springtails dates back to the early terrestrial evolution of the group, when anhydrobiosis was more common than it is now.

In nematodes, horizontal gene transfer was discovered by Smant et al. (1998). These authors isolated a β-1,4-endoglucanase gene in two species of plant-parasitic cyst nematodes, *Globodera* and *Heterodera*. This was a remarkable finding since animals normally don't have cellulase genes. The foreign gene, most likely from a bacterium, was shown to encode an active enzyme, expressed in host tissues and contributing to the degradation of plant cell walls.

Many more HGT genes were identified later and their frequency seems to be particularly high in clade 12 of the nematode phylogeny (Tylenchida), which includes cyst nematodes and root-knot nematodes. In the genome of *Meloidogyne incognita* (southern root-knot nematode, a pest of cotton) 60 genes were identified belonging to six different protein families involved in plant cell-wall degradation (Haegeman et al. 2011, Mayer et al. 2011). Other functional categories involving HGT genes are stress defence and antioxidant defence. It is remarkable that most HGT genes in nematodes involve metabolic enzymes, and in particular many CAZys. Most likely,

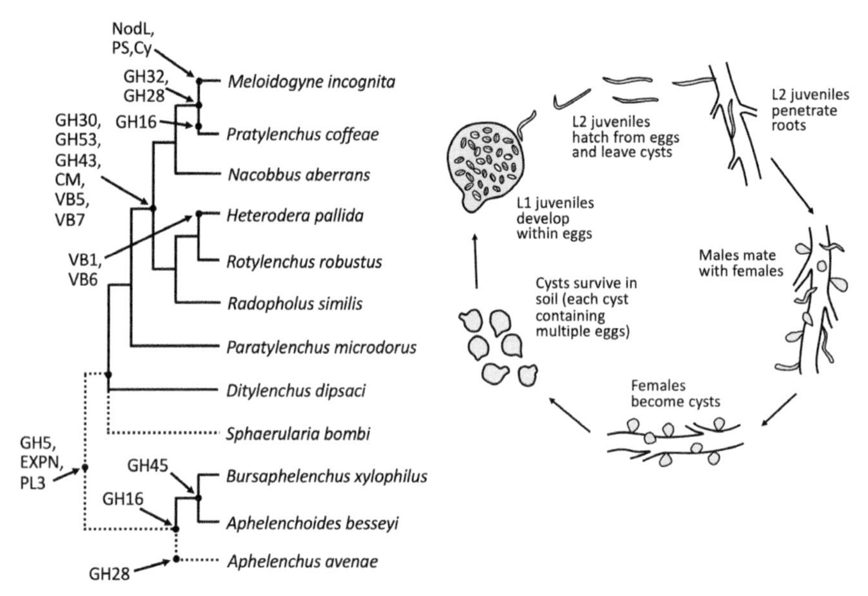

Figure 6.30. Left: Phylogeny of plant-parasitic nematodes in clades 12 and 10 of the phylum Nematoda. Lineages with solid lines are plant-parasitic; non-plant-related lineages are indicated by dashed lines. Assumed acquisitions of horizontally transferred genes are indicated by dots. GH glycoside hydrolase, PL polysaccharide lyase, EXPN expansin-like protein, VB gene associated with vitamin biosynthesis and salvage, CM chorismate mutase, PS polyglutamate synthase, Cy cyanate lyase, NodL nodulation factor. Redrawn from Haegeman et al. (2011), with modifications. Right: life-cycle of potato cyst nematode (*Globodera rostochiensis*), modified after Khan (2015). L1, L2 first and second larval stage.

these enzymes entered the genomes of nematodes while they were still bacterial feeders, since bacterivory is the ancestral mode of nutrition (see Section 2.1). The acquisition of cellulases must have been a major trigger for the evolution of plant parasitism, since all plant parasites must be able to penetrate the cellulose-based cell wall. Plant parasitism evolved at least three times in nematodes and there is a close phylogenetic correlation with HGT (Haegeman et al. 2011, Figure 6.30).

Tardigrades represent another case of ecologically relevant horizontal gene transfer. Boothby et al. (2015) reported a very high degree of HGT in the model species *Hypsibius dujardini*. No less than 6,663 genes (17.5% of the total) were identified as potentially derived from foreign, mainly bacterial, donors. However, this extremely high rate might have been due to co-assembly of contaminating bacterial DNA with tardigrade DNA. Later authors have reported much lower figures, e.g., Yoshida et al. (2017) argued that the highest-credibility set of HGT genes in *H. dujardini* should be no larger than 133, while a less-credible set contained 357 (1.8%). In another species, *Ramazottius varieornatus*, the number of HGT candidates was estimated as 1.6%. These rates are in the same order of magnitude as in other metazoans.

While tardigrades might not have a particularly high rate of HGT, some HGT genes appear to be related to their unique capacity for anhydrobiosis (*cf.* Section 6.4). In both *H. dujardini* and *R. varieornatus*, all catalase genes are of bacterial origin. Catalase is a key enzyme in oxidative stress defence, an important aspect of drought resistance. HGT is also involved in the accumulation of trehalose, a key protective molecule in drought resistance. As discussed in Section 6.4, Eutardigrada lack the enzyme needed for trehalose synthesis from glucose (*trehalose-6-phosphate synthase*, TPS). However, the terrestrial tardigrade *R. varieornatus* regained this enzyme through HGT. The species *H. dujardini*, which is restricted to freshwaters and streams, lacks a TPS and is poorly cryptobiotic compared to *R. varieornatus* (Yoshida et al. 2017). So, the evolution of extreme drought resistance in some tardigrade species was facilitated by the acquisition of foreign genes, underlining the importance of HGT as a rare but high-impact phenomenon.

The third soil invertebrate group for which horizontal gene transfer has been documented is the order of Collembola. In an assembly of the genome of *Orchesella cincta* (Entomobryidae), using strict criteria, including linkage to native genes on the same genomic scaffold, 253 genes were identified as true HGTs (Faddeeva-Vakhrusheva et al. 2016). These represented 1.2% of all genes (20,249). Interestingly, the donor organisms of these genes not only included bacteria (37.5%) but also fungi (30.4%) and protists (25.3%). The HGT set was enriched in functions related to carbohydrate metabolism (47 genes), proteolysis (43 genes) and oxidation-reduction reactions (34 genes).

A similar picture holds for the model species *Folsomia candida* (Isotomidae), but the extent of HGT in this species is larger (Faddeeva-Vakhrusheva et al. 2017). A total of 809 foreign genes were identified, which is 2.8% of all genes (28,734). Again, the donors were both bacteria (39.9%), fungi (32.8%) and protists (24.6%). Among fungi the arbuscular mycorrhizal fungus *Rhizophagus irregularis* (*Glomus intraradices*) was highly represented. As *F. candida* is a typical fungivorous species and prefers AMF (*cf.* Figure 1.17), this again underlines the importance of close contact as a factor enhancing the likelihood of HGT.

Like in *O. cincta*, most of the HGT genes in *F. candida* are CAZys, related to digestion of carbohydrates. However, among the HGT genes, springtails are unique in having genes from the *β-lactam biosynthesis* pathway. This pathway is responsible for the production of β-lactam antibiotics such as penicillins, cephalosporins and cephamycins. They occur in fungi of the genera *Penicillium* and *Aspergillus* and bacteria of the genus *Streptomyces*. The fungus *Penicillium* was the first organism in which antibiotic action was discovered by Alexander Flemming in 1928, but the origin of the pathway is most likely bacterial. The gene cluster was recruited (through an HGT event) by fungi about 370 Ma ago. In fungi the genes are regulated by eukaryotic promoters, very different from the ones in bacteria (Liras and Martin 2006).

The pathway includes two unique enzymes, *amino-adipoyl-cysteinyl-valine synthase* (*ACVS*) and *isopenicillin N synthase* (*IPNS*). ACVS is a non-ribosomal peptide synthase which catalyzes the fusion of valine, cysteine and amino-adipic acid to a tripeptide. Valine and cysteine are readily available amino acids, while amino-adipate, an intermediate in the synthesis of lysine, is produced in a special pathway (Suring et al. 2016). The second step is due to IPNS, an iron-containing haem protein, which introduces a β-lactam ring (a cyclic amide) in the molecule. All β-lactam compounds have this heterocyclic 4-ring. Several other reactions may follow, giving rise to a variety of chemical structures. A schematic overview of the pathways is given in Figure 6.31 (Brakhage et al. 2009).

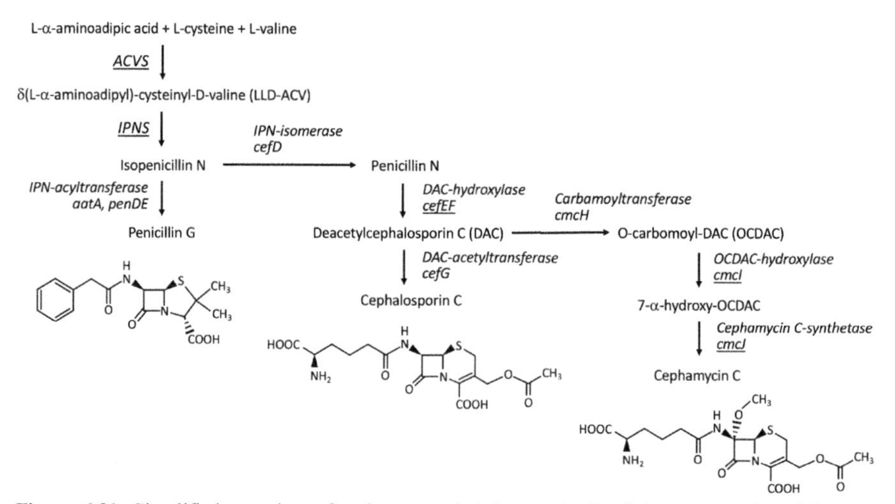

Figure 6.31. Simplified overview of pathways underlying synthesis of three categories of β-lactam antibiotics, adapted from Brakhage et al. (2009). Penicillins and cephalosporins are produced by fungi, cephamycins by bacteria. ACVS amino-adipoyl-cysteinyl-valine synthase, IPNS isopenicillin N synthase, cefDEFG members of cephamycin gene cluster, cmcHIJ, members of cephamycin C gene cluster. Note the characteristic 4-ring heterocyclic ring structure of all β-lactam compounds. Genes found in collembolan genomes are underlined (cf. Figure 6.32).

Roelofs et al. (2013) first discovered that *Folsomia candida* had an IPNS gene in its genome. A transcript of this gene was detected, because it was highly upregulated by oxidative stress induced by cadmium exposure. Later, several other β-lactam biosynthesis genes, including ACVS and two *cephamine C* genes, were found to be part of the cluster, plus an ABC transporter (Suring et al. 2017). *In situ* hybridization showed that the ACVS transcript was expressed in the gut epithelium and an ELISA assay demonstrated a chemical with a β-lactam ring in the gut. These observations strongly suggest that *F. candida* produces an antibiotic that is excreted in the gut lumen to manipulate the microbiome. The high tolerance of *Folsomia* against entomopathogenic fungi (see Section 6.10) could also be related to this capacity. The presence of *cmcI* and *cmcJ*, encoding enzymes in the (bacterial) pathway to cephamycine, suggests that the β-lactam compound is a cephamycine. However, *cefEF* and *cmcH* are missing from the cluster (Figures 6.31, 6.32). The exact nature of the active principle has not yet been elucidated (D. Roelofs, pers. comm.).

F. candida antibiotic genes cluster more closely with bacteria than with fungi, so the donor organism most likely was a bacterium. A major bacterial genus producing antibiotics is *Streptomyces*, which belongs to the Actinobacteria. This matches with the observation made above that Actinobacteria are one of the most abundant bacterial phyla in the microbiome of springtails (Table 6.9).

Can the possible production of antibiotics be considered an adaptation to life in soil? This question can be answered in the affirmative by looking at the distribution of β-lactam pathway genes in Collembola (Suring et al. 2017). An analysis of published animal genomes demonstrated that no other metazoan sequenced so far has β-lactam biosynthesis genes. They are also lacking in the sister groups of Collembola, Protura and Diplura. Within Collembola, however, the genes are widespread, detected in

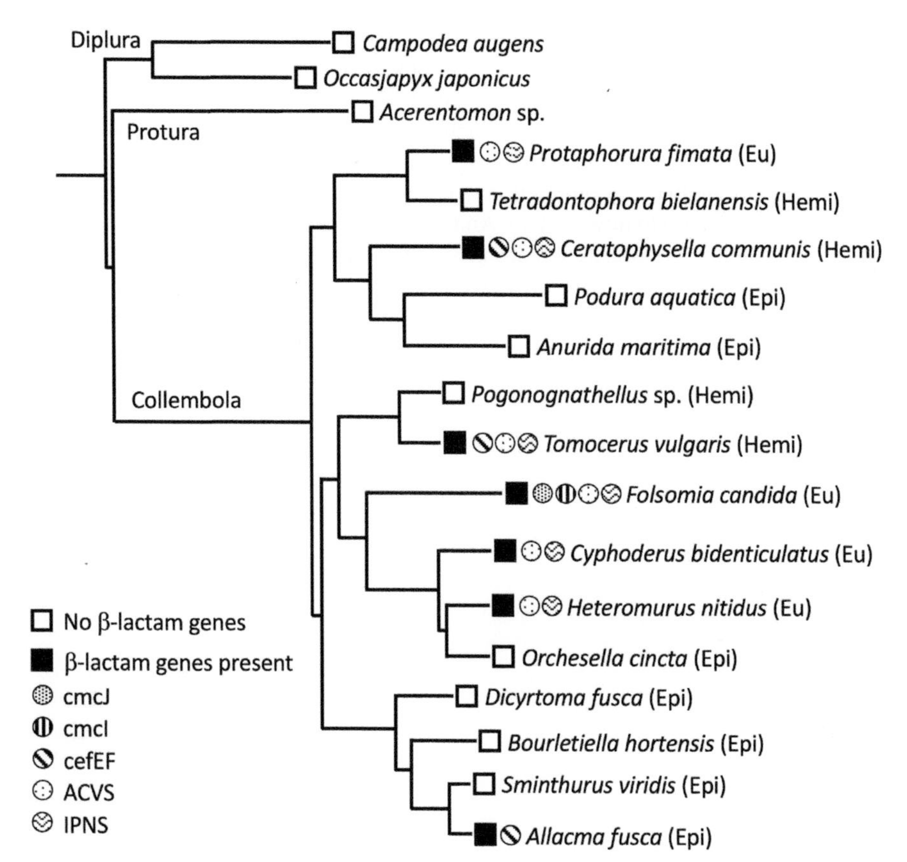

Figure 6.32. Transcriptome-based phylogenetic tree of fifteen species of Collembola, indicating the presence or absence of β-lactam biosynthesis genes, with species classified by life-forms according to Gisin. For gene codes see Figure 6.31. Three non-collembolan apterygotes served as outgroup (top of tree). Adapted from Suring et al. (2017).

60% of the species, though not all genes of the pathway are present in every species (Figure 6.32). ACVS and IPNS are always present jointly, so all positive species are expected to produce at least isopenicillin N, which has already an antibiotic action (*cf.* Figure 6.31).

The distribution of β-lactam biosynthesis genes over the collembolan species is scattered (Suring et al. 2017, Figure 6.32). This can be explained by assuming that the HGT event happened early in the evolution of Collembola, as part of their terrestrialization. The scattered distribution is explained by multiple losses. Relating the pattern of presence and absence to the collembolan life-forms reveals that all euedaphic species investigated have β-lactam biosynthesis genes, two of the four hemiedaphic species have them and one of the seven epigeic species. So, there is a strong suggestion in these data that the retention of antibiotic biosynthesis genes should be considered an adaptation to soil-living, while the genes were lost in the lineages adopting an epigeic life-style. This fits very well the idea formulated earlier that defence against pathogenic microbes is better developed in soil-living species than in the ones that live on the surface and in the vegetation.

Summarizing this short exploration of HGT events in soil invertebrates, we conclude that several lineages (at least nematodes, tardigrades and springtails) have foreign genes in their genome, but in low frequency (about 1% of the total number of genes, comparable to many other animals). The springtail *Folsomia candida* has the highest percentage up to now (2.8%). Among the genes that made their way into soil invertebrates many are related to carbohydrate metabolism and these have probably contributed to the expansion of digestive capacity. In addition, stress-related genes are prone to HGT. Horizontal gene transfer has contributed to some remarkable phenotypes such as anhydrobiosis in tardigrades and antibiotics production in Collembola.

CHAPTER 7

A Kaleidoscope of Adaptations

The Mollingunder Stones

For a while, leave your numbing comfort zones!
Inhale wet earth - fresh manure adds more zest.
No? Not a more open mind - still stressed?
Put an ear to the ground. Hear: saxophones.

Nematodes are doing their beastly best!
Larvae play the flute, centipedes rattle bones.
Snails ram the timpani, springtails vibraphones.
On cello spiders risk cardiac arrest ...

A slave choir chants hell notes from the ants nest.
Mites raise some horns, beetles four mellophones:
the high C is the ultimate test!

Worms blow their hearts out on trombones.
Feeling addressed - makes the presence manifest
of this band: "The Mollingunder Stones".

Jasper Aertsz

Chapter 7

A Kaleidoscope of Adaptations

Throughout this book I have used the hypotheses formulated by Ghilarov, Kennedy, Gisin and Bouché as guides to discuss the evolution of soil invertebrates (Chapter 1). In different wordings, these authors have all pointed out that the soil profile presents a strong selective gradient, running from humid, stable and spatially restricted conditions belowground, to dry, variable, sometimes extreme and spacious conditions on the surface. They also argued that soil invertebrates have adapted to these conditions in terms of morphology, reproductive mode, life-history patterns and physiology, and that convergent adaptive patterns have evolved in the different invertebrate lineages.

In general, the "GKGB hypothesis" was confirmed by the data collected in the various chapters of this book. The main trends in morphology, life-history, reproduction, behaviour and physiology are summarized in Figure 7.1. Soil invertebrates descended from marine ancestors, however, many groups lived in freshwater before they evolved further to colonize the soil. This is evident in several of the phylogenies presented in Chapter 2, where soil-living lineages are nested within freshwater stem groups. This situation, pointed out most explicitly by Ghilarov, holds particularly well for earthworms, enchytraeid worms, panpulmonate snails and caenogastropod snails. Other scenarios are applicable to other groups, such as the direct pathway (myriapods, chelicerates, tardigrades and onychophorans), as well as pathways involving an intertidal phase (amphipods, isopods, land crabs and ellobioid snails) or even cavernicolous stages (hexapods, flatworms).

It is important that comparative physiological studies of soil invertebrates take this evolutionary history into account. For example, according to phylogenetic reconstructions, terrestrial isopods such as *Porcellio* and *Armadillidium* did not evolve from a marine *Ligia*-like ancestor, as is often assumed, but from a later branch of *Trichoniscus*-like ancestors; the *Ligia* lineage was a separate terrestrialization. Similar complexities hold for other soil invertebrates. Some groups seem to have colonized the land only once (hexapods, myriapods, chelicerates, onychophorans), but many groups show a complicated pattern of multiple terrestrializations and marine regressions. In nematodes and to a lesser extent in snails, the number of terrestrializations is so large, that it has become difficult to disentangle them (the "habitat commuting" scenario).

In Chapter 2 I also pointed out that several terrestrializations came with significant changes in body shape and form. The hexapod body plan was something completely new, not like any of the crustacean ancestors. The same holds, to a lesser extent, for chelicerates and myriapods. These terrestrializations can be called "major", to contrast them with the terrestrializations that we see in isopods, annelid worms and snails, which I called "minor", not because they are less important, but because they did not change the body plan very much. In the major terrestrializations a developmental switch from free-swimming larvae to direct development was made. Many existing structures acquired new functions, the respiratory organs of arthropods, derived from leg buds, and the re-use of the snail mantle cavity as a lung being the great examples discussed in Chapter 4. These developmental switches deserve more study from a molecular-genetics perspective.

In summary, the analysis in this book has confirmed the Ghilarov part of the GKGB hypothesis, but at the same time things turned out to be more complicated. There are many more scenarios for terrestrialization than proposed by Ghilarov. In addition, we saw that in several lineages, not the deepest soil-living representatives are the most ancestral, but some intermediate life form, exemplified by the "hemiedaphon" in Gisin's classification of Collembola. It seems reasonable to assume that the first colonizers of the terrestrial environment were "half-way soil living" and gave rise to two lineages, one going deeper in the soil, the other evolving towards a surface-living life-style. Euedaphic (endogeic) groups have adaptations as specialized as the epigeic groups, as shown in Chapter 6. To complicate things further, in several groups of primarily above-ground living invertebrates, lineages have evolved that secondarily adapted to the soil habitat, in association with their feeding on plant roots.

The study of soil invertebrates has also pointed out several ecological facts that have not penetrated much in ecological theory. One striking example is the very shallow relationship between biomass turnover in relation to body size. Many, especially the smaller, soil invertebrates have a much lower population turnover than predicted from classical allometric relationships. This is obviously due to the fact that non-digging, small invertebrates live in the pores and crevices of the soil and so are subject to selective factors very different from surface-living species. The soil is a K-selective environment, much more than the surface, which is dominated by r selection. While in traditional r-K selection theory a large body size is seen as a K-selected trait, in fact it is small body size that concurs with K selection in soil, as discussed at length in Chapter 3.

These sometimes counterintuitive trends of life-history traits across the soil profile are seen in many independent groups of soil invertebrates (earthworms, oribatid mites, Collembola, centipedes), and represent one of the finest illustrations of the selective gradient of the soil, as laid out, in different wordings, by Kennedy and Bouché. I have placed particular emphasis upon the constraints imposed by determinate growth. The evolution of iteroparity is much easier for indeterminate growers, because at any age there is a residual reproductive value to be gained from growing larger. This simple developmental constraint favouring the evolution of iteroparity is very well illustrated by soil invertebrates, as shown by studies on spiders, snails, earthworms, springtails and isopods, discussed in Chapter 3.

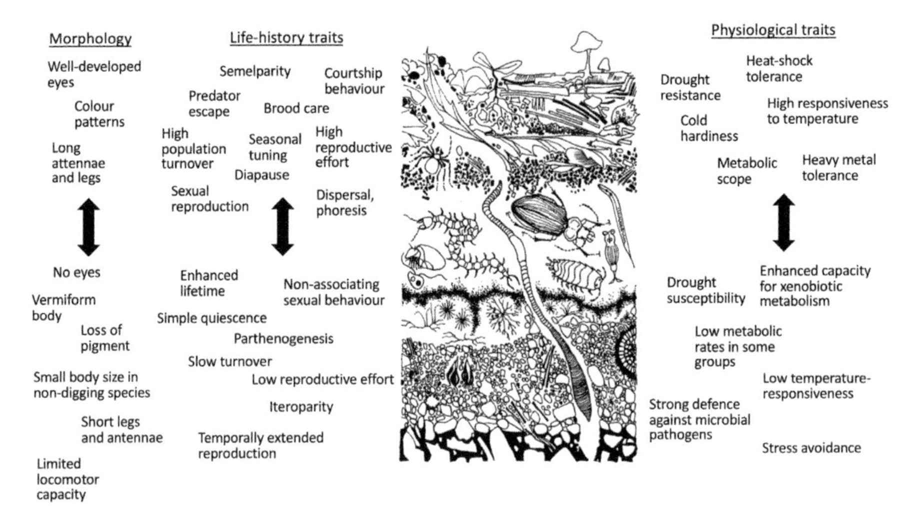

Figure 7.1. Simplified contrasts of morphology, life-history and physiological traits in species living at different positions of the soil profile, as discussed in various places in this book. The central image is from the title page of the proceedings volume of the 7th International Soil Zoology Colloquium, held in Syracuse, New York, edited by Daniel L. Dindal (1980).

The idea of "energy intensity" as a driving force in life-history evolution, a concept from the classical work of Kennedy, appeared to be particularly well applicable to patterns of temperature responses. It seems that in lineages exposed to stable and constant temperature (as in soil) the capacity to respond to temperature has flattened out. There is a trend for surface-living species to be more responsive to changing temperature, and this is reflected by a higher activation energy and higher Q_{10}. This trend is most clearly shown by Collembola, but is likely to be valid for many other soil invertebrates. However, the genetic architecture underlying temperature responsiveness is still unclear, and we don't have a model for the evolution of plasticity, that is, the temperature response as a whole.

Another fascinating aspect of soil invertebrate life-histories is that so many species are parthenogenetic. Parthenogenesis is very strongly associated with life in soil, most clearly shown by oribatid mites and Collembola. This has led soil ecologists to conclude that it is not the resource-richness of the environment that is promoting parthenogenesis, but the temporal stability and predictiveness of resources. This is in agreement with Graham Bell's "*tangled bank*" hypothesis for the evolution of sexuality. The study of parthenogenesis is, however, complicated by the fact that there are multiple proximal mechanisms that may cause it. In several lineages of arthropods, parthenogenesis is due to reproductive manipulators such as *Wolbachia*, but in parthenogenetic mites no *Wolbachia* can be found. Parthenogenesis of earthworms and isopods is associated with hybridization, leading to polyploid lineages. To understand the evolution of parthenogenesis, I call for a more mechanistic (cellular, molecular) approach to the problem.

In addition to parthenogenesis, several soil invertebrates are hermaphroditic, however, whether the soil as a selective environment has anything to do with it is doubtful. Simultaneous hermaphroditism is seen in several soil invertebrates including earthworms and snails. It seems to be mostly associated with low locomotor capacity (absence of legs). The phylogenetic distribution of hermaphroditism very much suggests that it should be considered a developmental constraint rather than an adaptation to something. The sequential hermaphroditism of nematodes seems to have an entirely different origin and most likely needs a different evolutionary explanation.

In terms of behaviour, soil invertebrates show a bewildering spectrum. Almost all behavioural studies on soil invertebrates deal with reproductive behaviours. This spectrum ranges from random deposition of spermatophores, seen in many Collembola, to the intricate courtship behaviours in pseudoscorpions and spiders. The fascinating complexity observed in some of these behaviours may be used to test predictions from sexual selection theory, the evolution of partner preference, the maximization of paternity, etc. In Chapter 5 of this book we have seen many examples.

Brood care and social behaviour are found in several lineages of soil invertebrates, but it is difficult to define what are the conditions favouring evolution of these behaviours. Most likely, there are different explanations applicable to different lineages. As discussed in Chapter 5, brood care is found in groups as different as ants, termites, earwigs, centipedes and spiders. A most fascinating diversity of parental care and subsocial behaviour is seen in spiders. Social behaviour of these animals can be considered a continuation of parental care. In centipedes brood care is found frequently in soil-living geophilomorphs, but within the evolution of the class, it is an ancestral condition, already present in chilopods before some of them adopted a soil-dwelling life-style. The well-known genetic condition of haplo-diploidy, the common explanation for brood care in ants and other hymenopterans, is not applicable to other soil invertebrates. So, the explanations for brood care are as different as the behaviours themselves.

On the physiological level, soil invertebrates show many beautiful examples of acclimation and adaptation. The regulation of coelomic osmotic pressure in earthworms is a classic example of hyper-osmoregulation, very similar to osmoregulation in freshwater invertebrates and fish. The lineages which have colonized the land from a freshwater environment (worms, snails) generally have lower osmotic values than lineages that went on land directly or through intertidal habitats (tardigrades, woodlice, land crabs), however, millipedes are an exception. It is often argued that the osmotic value under normal physiological conditions is indicative of the terrestrialization pathway in evolutionary history, but the relationship actually is not so clear. Osmoregulation may be a trait fixed in an early stage of land colonization, without much change in later terrestrial evolution. Loss of osmoregulation and return to osmoconformism is seen in species that secondarily adapted to intertidal conditions.

Adaptations with respect to respiration, metabolic rate, transpiration and nitrogen metabolism have been important in all soil invertebrates, and most soil invertebrates have found similar solutions. For example, in Chapter 6 I have pointed out the great functional similarity of the external water conducting systems of woodlice and springtails. Among the various physiological traits, metabolic rate seems the least variable and does not show consistent trends with the soil profile, in contrast to many suggestions in the literature. The only exceptions are seen in Collembola, where metabolic rate is lower among the euedaphic species compared to epigeic species. Maybe the experimental focus on resting metabolism obscures the importance of active metabolism and metabolic scope as traits evolving in the soil profile.

Cold hardiness research is another landmark of soil invertebrate studies. Several core insights of comparative physiology rely on cold hardiness research in earthworms and microarthropods. Measurements of body fluid osmotic pressure, supercooling points and sugar accumulation have provided important insights. In earthworms a new mechanism was discovered called cryoprotective dehydration. It shows how closely freeze-avoidance relies, not only on recruitment of cryoprotective molecules, but also on water relations and osmoregulation. Cold hardiness, like diapause, seems to be especially developed among surface-living species, as these obviously face the more extreme conditions.

Physiological studies used to be focused on animals above a certain body size, however, the application of molecular methods and highly sensitive analytical tools have removed this limitation. Using such methods, it is no problem at all to quantify heavy metal residues in individual mites, to isolate proteins from springtails and to measure enzyme activities in woodlouse hepatopancreas. In addition, DNA sequencing, transcriptome analysis and in situ hybridization are all applicable to tiny animals. These molecular methods have greatly advanced the field. For example, it has been possible to isolate and sequence antifreeze proteins in arctic springtails (Chapter 6). The thermal hysteresis that these proteins confer to the haemolymph allows the animals to survive frost and snow easily, a wonderful demonstration of how the study of extreme environments can enlighten our understanding of evolution.

Soil invertebrates also suffer from accumulation of toxicants in their environment. Toxicity testing using soil invertebrates is often applied to assess the risk of soil pollution. Soil invertebrates have developed quite an elaborate machinery for detoxification of organic substances, expanding the mechanisms of their freshwater and marine ancestors. In a few species, e.g., isopods and earthworms, these mechanisms have been studied in some detail. The data show that the biochemical detoxification pathways present in soil invertebrates may be very different from those in vertebrates. In particular, the regulation of biotransformation enzymes by the arylhydrocarbon-receptor system seems to be completely absent and the conjugation reactions are of a different kind.

Heavy metals are among the most threatening soil pollutants as they are retained by organic matter and clay and may accumulate in soil to concentrations far above those found elsewhere in the environment. Studies on earthworms and springtails have shown that heavy metals exert a selective pressure on invertebrate populations, favouring phenotypes with a genetically determined high resistance (Chapter 6). The study of metal tolerance in springtails, which involves overexpression of the metal-binding protein, metallothionein, is a nice illustration of the principle "evolution through pollution".

Finally, in Chapter 6 I have highlighted the soil invertebrate microbiome, including pathogens and parasites as factors determining ecology and evolution. This is a largely neglected field in my opinion. Chapter 6 shows how almost every soil invertebrate carries some kind of parasite (nematodes, acanthocephalans, hairworms) in its body. Some invertebrates, such as ants, harvestmen and spiders, suffer greatly from pathogenic fungi. On the other hand, the microbiome in the gut adds to the digestive capacity of the animal. Some groups, such as the lower termites, rely completely on flagellated protists to degrade lignocellulose. In addition, the microbiome is a source of horizontal gene transfer to the host and several examples are discussed in Chapter 6, illustrating the expansion of adaptive potential caused by such transfers.

Zooming out from the tiny invertebrates, we note that they live in environments that are degrading rapidly due to erosion, overuse, intensive farming, acidic rain, atmospheric nitrogen deposition, etc. This cannot be without effects on the soil microbial community and concomitantly, the many soil invertebrate communities relying on microbes. Nobody knows how many soil invertebrates and larval stages of above-ground insects die unseen from toxic bacteria or pathogenic fungi. As Jo Handelsman (2021) wrote in her impressive book "A World Without Soil": *the future of the earth beneath our feet has become precarious*. Although my book does not primarily address the many threats to soils, I hope, by illustrating the wonderful kaleidoscope of adaptations of soil invertebrates, to add to Jo's compelling appeal to save our soils.

Chapter 8

Annexes

8.1 Summary data and literature references for annual mean densities of soil invertebrate populations graphically presented in Figure 3.1

Studies were selected that reported annual mean densities derived from multiple sampling occasions in a year. In some cases, estimates were derived from graphs in the papers. The densities were all expressed in numbers per m^2, adjusted when expressed in other surface units, and transformed to ^{10}log. Data from pitfall traps were ignored. Data expressed per g of soil were recalculated per surface area when bulk density and sampling depth were specified. Each study was considered a single source for any one particular group of soil invertebrates. When a publication reported data on multiple sampling sites, the average log density was taken as one source. Data reported in books and review papers were also included, treated as a single source each. The median of all values (across *n* sources) is reported here, as well as the estimated interquartile range (Q_1-Q_3, comprising 50% of the observations).

Soil invertebrate group	Log Q1	Log Median	Log Q3	n	Sources
Nematodes	5.86	6.41	6.79	20	Blair et al. (1994), Boström and Sohlenius (1986), Chiba et al. (1976), Curry (1994), Ettema et al. (1998), Giller (1996), Hallas and Yeates (1972), Hodda and Wanless (1994), Hoste-Danyłow et al. (2013), Lavelle and Spain (2005), Paul and Clark (1989), Petersen and Luxton (1982), Phillipson et al. (1977), Schaefer and Schauermann (1990), Sohlenius (1980), Sohlenius and Boström (1986), Sohlenius and Boström (2001), Sohlenius (2002), Yeates (1981), Yeates (1991)
Tardigrades	3.62	4.34	4.87	7	Hallas and Yeates (1972), Harada and Ito (2006), Ito (1999), Ito and Abe (2001), Nelson et al. (2018), Petersen and Luxton (1982), Schaefer and Schauerman (1990)
Dipteran larvae	1.93	2.89	3.26	12	Curry (1994), Frouz (1999), Frouz et al. (2015), Heynen (1988), Ito et al. (2002), Kopeszki (1993), Lavelle and Spain (2005), Loranger et al. (1998), Paul and Clark (1989), Petersen and Luxton (1982), Schaefer and Schauermann (1990), Tamura et al. (1969)
Ants	2.33	3.02	3.82	13	Curry (1994), Frith and Frith (1990), Giller (1996), Ito et al. (2002), Jensen (1978a), Lavelle and Spain (2005), Loranger et al. (1998), Meehan et al. (2006), Nielsen (1974), Nielsen et al. (1976), Osler and Beattie (1999), Petersen and Luxton (1982), Tamura et al. (1969)
Ground bugs	1.60	1.80	2.05	5	Curry (1994), Ito et al. (2002), Loranger et al. (1998), Osler and Beattie (1999), Solhøy (1972)
Termites	1.59	2.23	2.75	7	Chiba et al. (1975), Curry (1994), Ito et al. (2002), Lavelle and Spain (2005), Loranger et al. (1998), Osler and Beattie (1999), Petersen and Luxton (1982)

contd. ...

...contd.

Soil invertebrate group	Log Q1	Log Median	Log Q3	n	Sources
Diplurans	1.77	2.24	2.53	8	Axelsson et al. (1984), Chiba et al. (1975), Curry (1994), Ito et al. (2002), Loranger et al. (1998), Osler and Beattie (1999), Rajagopal and Ganesha Bhat (1995), Schaefer and Schauermann (1990)
Proturans	2.55	3.04	3.49	11	Axelsson et al. (1984), Chiba et al. (1975), Curry (1994), Kopeszki (1993), Leinaas (1978), Loranger et al. (1998), Osler and Beattie (1999), Price (1973), Rajagopal and Ganesha Bhat (1995), Petersen and Luxton (1982), Schaefer and Schauermann (1990)
Springtails	3.84	4.30	4.65	55	Axelsson et al. (1984), Badejo and Van Straalen (1993), Bandyopadhyaya et al. (2002), Blair et al. (1994), Chagnon et al. (2000), Chernova and Kuznetsova (2000), Chiba et al. (1975), Curry (1994), Davis and Murphy (1961), Davis (1963), Dhillon and Gibson (1962), Ehrnsberger et al. (1997), Fjellberg (1975), Giller (1996), Hale (1966), Hasegawa (2002), Hoste-Danyłow et al. (2013), Hutson and Veitch (1983), Ito et al. (2002), Juceviča and Melecis (2002), Kopeszki (1993), Kopeszki and Meyer (1994), Kuznetsova (1994), Kuznetsova and Krest'yaninova (1998), Leinaas (1978a), Leinaas (1978b), Loranger et al. (1998), MacLean et al. (1977), Meehan et al. (2006), Milne (1962), Mitchell (1977), Niijima (1975), Noordam and Van der Vaart-de Vlieger (1943), Osler and Beattie (1999), Paul and Clark (1989), Petersen (1980), Petersen and Luxton (1982), Petersen and Gjelstrup (1987), Petersen (2000), Pflug and Wolters (2002), Pomeroy (1977), Poole (1961), Price (1973), Purvis and Curry (1980), Rajagopal and Ganesha Bhat (1995), Schaefer and Schauermann (1990), Sgardelis et al. (1999), Shaw (2003), Solhøy (1972), Takeda (1995a), Tamura (1967), Teuben and Smidt (1992), Usher (1970), Van Straalen (1989), Wood (1967)

Woodlice	1.70	2.41	2.78	18	Chiba et al. (1975), Curry (1994), Frith and Frith (1990), Hoste-Danyłow et al. (2013), Ito et al. (2002), Lavelle and Spain (2005), Loranger et al. (1998), Ma et al. (1991a), Osler and Beattie (1999), Paul and Clark (1989), Paoletti and Hassall (1999), Petersen and Luxton (1982), Phillipson (1983), Schaefer and Schauermann (1990), Scheu and Poser (1996), Soma and Saitô (1979), Sunderland et al. (1976), Sutton (1968)
Landhoppers	1.06	1.76	2.79	4	Friend and Richardson (1986), Frith and Frith (1990), Ito et al. (2002), Osler and Beattie (1999)
Millipedes	1.17	1.74	2.29	16	Blower and Miller (1977), Chiba et al. (1975), Giller (1996), Frith and Frith (1990), Hopkin and Read (1992), Hoste-Danyłow et al. (2013), Ito et al. (2002), Kime and Wauthy (1984), Lavelle and Spain (2005), Loranger et al. (1998), Meyer (1985), Osler and Beattie (1999), Petersen and Luxton (1982), Phillipson and Meyer (1984), Schaefer and Schauermann (1990), Scheu and Poser (1996), Tamura et al. (1969)
Pauropods	2.26	2.69	3.74	8	Axelsson et al. (1984), Chiba et al. (1975), Lagerlöf and Scheller (1989), Loranger et al. (1998), Osler and Beattie (1999), Petersen and Luxton (1982), Price (1973), Rajagopal and Ganesha Bhat (1995)
Symphylans	1.42	2.11	2.92	8	Axelsson et al. (1984), Chiba et al. (1975), Ito et al. (2002), Lagerlöf and Scheller (1989), Loranger et al. (1998), Osler and Beattie (1999), Petersen and Luxton (1982), Rajagopal and Ganesha Bhat (1995)
Centipedes	0.68	1.49	2.11	12	Chiba et al. (1975), Frith and Frith (1990), Hoste-Danyłow et al. (2013), Ito et al. (2002), Loranger et al. (1998), Meehan et al. (2006), Osler and Beattie (1999), Petersen and Luxton (1982), Schaefer and Schauermann (1990), Shaw (2003), Soma and Saitô (1979), Tamura et al. (1969)

contd. ...

contd. ...

Soil invertebrate group	Log Q1	Log Median	Log Q3	n	Sources
Spiders	1.27	1.76	2.50	18	Chiba et al. (1975), Curry (1994), Dumpert and Platen (1985), Frith and Frith (1990), Giller (1996), Hoste-Danyłow et al. (2013), Ito et al. (2002), Lavelle and Spain (2005), Loranger et al. (1998), Meehan et al. (2006), Osler and Beattie (1999), Paul and Clark (1989), Pętal et al. (1971), Petersen and Luxton (1982), Schaefer and Schauermann (1990), Solhøy (1972), Soma and Saitô (1979), Tamura et al. (1969)
Pseudoscorpions	1.20	2.06	2.55	11	Aguiar *et al*. (2006), Chiba et al. (1975), Gabbutt (1967), Goddard (1976), Ito et al. (2002), Loranger et al. (1998), Meyer et al. (1985), Osler and Beattie (1999), Rajagopal and Ganesha Bhat (1995), Schaefer and Schauermann (1990), Yamamoto et al. (2001)
Oribatid mites	3.79	4.18	4.69	35	Blair et al. (1994), Cancela da Fonseca et al. (1995), Chiba et al. (1975), Curry (1994), Davis (1963), Emmanuel et al. (1985), Grishina et al. (1995), Hutson (1980), Hutson and Veitch (1983), Ito et al. (2002), Kaneko (1985), Kaneko (1995), Khalil et al. (1999), Khalil et al. (2016), Lasebikan (1974), Luxton (1981b), Meehan et al. (2006), Mitchell (1977), Osler and Beattie (1999), Petersen and Luxton (1982), Petersen and Gjelstrup (1987), Price (1973), Rajagopal and Ganesha Bhat (1995), Schaefer and Schauermann (1990), Sgardelis et al. (1993), Stamou and Sgardelis (1989), Takeda (1995b), Teuben and Smidt (1992), Thomas (1979), Wallwork (1983), Wood (1967), Zaitsev (1997), Zaitsev et al. (2002), Zaitsev and Wolters (2006)

Mesostigmata	3.23	3.61	4.00	22	Axelsson et al. (1984), Blair et al. (1994), Chiba et al. (1975), Curry (1994), Davis (1963), Emmanuel et al. (1985), Grishina et al. (1995). Hoste-Danyłow et al. (2013), Hutson (1980), Hutson and Veitch (1983), Ito et al. (2002), Osler and Beattie (1999), Petersen and Gjelstrup (1987). Petersen and Luxton (1982), Price (1973), Rajagopal and Ganesha Bhat (1995), Schaefer and Schauermann (1990), Sgardelis et al. (1999), Takeda (1995b), Wallwork (1983). Wood (1967)
Harvestmen	0.66	0.89	1.22	5	Chiba et al. (1975), Ito et al. (2002). Osler and Beattie (1999), Schaefer and Schauermann (1990), Solhøy (1972)
Land snails	0.72	2.08	2.81	8	Fog (1979), Frith and Frith (1990), Ito et al. (2002), Hoste-Danyłow et al. (2013), Mason (1970), Paul and Clark (1989), Phillipson and Abel (1983), Schaefer and Schauermann (1990)
Enchytraeids	3.92	4.33	4.62	14	Abrahamsen (1969), Blair et al. (1994), Chiba et al. (1976), Curry (1994), Didden (1993), Hoste-Danyłow et al. (2013), Huhta (1984), Ito et al. (2002), Lavelle and Spain (2005), MacLean et al. (1977). Nurminen (1967), Paul and Clark (1989). Petersen and Luxton (1982), Schaefer and Schauermann (1990), Swift et al. (1979)
Earthworms	1.48	1.90	2.32	32	Abrahamsen (1972), Axelsson et al. (1984), Baker (1999), Blair et al. (1994), Curry (1994), Curry et al. (1995), Doube and Schmidt (1997), Ernst and Emmerling (2009), Frith and Frith (1990), Giller (1996), Hoste-Danyłow et al. (2013), Ito et al. (2002), Johnston (2019), Jones et al. (2001a), Lavelle (1977), Lavelle and Spain (2005), Nordström and Rundgren (1974), Nowak (1975), Paoletti (1999), Paul and Clark (1989), Petersen and Luxton (1982), Ponge et al. (1999), Redmond et al. (2014), Römbke (1985), Salmon (2001), Schaefer and Schauermann (1990), Scheu and Poser (1996), Shakir and Dindal (1997), Swift et al. (1979), Tamura et al. (1969), Tiwari and Mishra (1995), Zorn et al. (2005)

8.2 Summary of demographic theory showing how production is calculated from population structure classified by body weight

The rate of production can be calculated directly from field data on population structure with animals classified by body weight. It measures the mass leaving the population per time unit through mortality (Pm). For any stationary population, biomass mortality is equal to the sum of reproduction and growth rate of all individuals jointly (Pr + Pg). The table below provides a formal proof of this equality. For details see Van Straalen (1985b).

Argument	Formulas	Variables
Consider a stationary population with age structure	$n(x)$	n = number of individuals per age unit x = age
Integration over all ages gives the total population size	$N = \int_0^{x_m} n(x)dx$	N = total population size x_m = maximal age
A growth curve specifies body size as a function of age	w = G(x)	w = individual mass G(x) = growth function
The biomass of the population is obtained by multiplying the numbers at each age by the weight at that age and integrating over all ages	$B = \int_0^{x_m} N(x)\, G(x)dx$	B = biomass ("standing crop") in mass units
The individual rate of growth is a function of body size	$g(x) = \dfrac{dG(x)}{dx}$	g(x) = growth rate (mass per time)
Total production due to growth is obtained by multiplying each age class by its growth rate and integrating over all ages	$P_g = \int_0^{x_m} N(x)\, g(x)dx$	P_g = production due to body growth (mass per time unit)
Total production due to reproduction is obtained by multiplying the rate of entry by the mass of the recruits (animals of age 0, eggs or hatchlings)	$P_r = n(0)\, w_0$	n(0) = number of recruits entering the population per time unit (total reproduction) w_0 = G(0) = body mass of recruits

Total production by the population is the sum of reproduction and growth	$P = P_g + P_r$	P = production (mass per time unit)
Total production is also equal to biomass leaving the population by mortality	$P_m = -\int_0^{x_m} \frac{dN(x)}{dx} G(x)dx$	P_m = biomass leaving the population through mortality (mass per time unit) dN/dx = mortality at age x (numbers dying per age unit)
These two measures of production are mathematically identical	$P_g + P_r = P_m$	This can be demonstrated by applying integration by parts to the equation for P_m and noting that $N(x_m) = 0$
The ratio of production to biomass is the biomass turnover rate	$\frac{P}{B} = \theta_m$	θ_m = biomass turnover rate (per time unit)
The inverse of biomass turnover is the average residence time of biomass in the population	$e_m = \frac{1}{\theta_m}$	e_m = average residence time of biomass
Likewise. the recruitment rate divided by the population size is the turnover of individuals	$\frac{n(0)}{N} = \theta_n$	θ_n = turnover rate of individuals (per time unit)
The inverse of individual turnover is the average residence time of individuals in the population	$e_m = \frac{1}{\theta_m}$	e_n = average lifetime. life expectancy at recruitment
If mortality rate is constant. turnover of individuals equals turnover of biomass	$\theta_m = \theta_n$ $e_m = e_n$	This can be shown by stating $-dN(x)/dx = \theta_n N(x)$ in the equation for P_m

8.3 Data and literature references on P/B ratios of soil invertebrates in relation to body size, graphically presented in Figure 3.3

Data on biomass turnover (yr^{-1}) of soil invertebrate populations were collected from the literature and used as reported or calculated from data on production, expressed in $mg/m^2/yr$ and biomass (mg/m^2). Production was measured in several different ways. either from the sum of growth and reproduction or from mortality. The principle is explained in Annexure 8.2.

Species	Adult body mass (mg)	P/B (yr^{-1})	Source		Species	Adult body mass (mg)	P/B (yr^{-1})	Source
Collembola					**Isopods**			
Orchesella cincta	0.5097	9.40	Van Straalen (1989)		*Trichoniscus pusillus*	13.6	3.50	Phillipson (1983)
Tomocerus minor	0.8461	6.80	Van Straalen (1989)		*Armadillidium vulgare*	50.0	2.03	Saito (1969)
Tetracanthella sylvatica	0.0565	6.10	Takeda (1976)		*Ligidium japonicum*	12.0	2.09	Saito (1969)
Onychiurus tricampatus	0.0180	6.30	Hale (1980)		*Porcellio scaber*	40.0	1.79	Saito (1969)
Onychiurus tricampatus	0.0180	4.60	Hale (1980)		*Ligia dilatata*	180.0	3.75	Koop and Field (1981)
Onychiurus procampatus	0.0631	4.80	Healey (1967)		*Ligia oceanica*	200.0	2.73	Willows (1987)
Isotoma trispinata	0.0278	20.0	Tanaka (1970)		*Burmoniscus ocellatus*	29.1	2.60	Lam et al. (1991)

Onychiurus sp.	0.0278	10.6	Tanaka (1970		*Tracheoniscus rathkei*	35.0	3.00	White (1968)
Lepidocyrtus lignorum	0.0990	8.76	Leinaas and Bleken (1983)		**Ants**			
Folsomia octoculata	0.0278	13.14	Takeda (1984)		*Pogonomyrmex badius*	6.60	1.23	Jensen (1978a)
Oribatid mites					*Myrmica rubra*	2.00	1.27	Jensen (1978a)
Steganacarus spinosus	0.0195	7.50	Luxton (1975, 1981c)		*Lasius niger*	1.93	1.43	Jensen (1978a)
Steganacarus magnus	0.3110	2.80	Luxton (1975, 1981c)		*Pogonomyrmex occidentalis*	10.0	1.29	Jensen (1978a)
Steganacarus magnus	0.2704	0.80	Thomas (1979)		*Lasius alienus*	1.53	4.53	Jensen (1978a)
Damaeus claviceps	0.1760	4.10	Luxton (1975, 1981c)		*Formica pratensis*	10.8	2.59	Jensen (1978b)
Galumna lanceata	0.0450	4.80	Luxton (1975, 1981c)		*Lasius flavus*	2.87	2.90	Jensen (1978b)
Gustavia microcephala	0.0455	6.00	Luxton (1975, 1981c)		**Earthworms**			
Nothrus silvestris	0.0487	4.90	Luxton (1975, 1981c)		*Drawida wilsii*	90.05	4.99	Srivastava et al. (2013)
Nothrus palustris	0.1792	4.10	Luxton (1975, 1981c)		*Lumbricus terrestris*	768.9	0.43	Lakhani and Satchell (1970)

contd. ...

...contd.

Species	Adult body mass (mg)	P/B (yr^{-1})	Source		Species	Adult body mass (mg)	P/B (yr^{-1})	Source
Nothrus palustris	0.0565	2.97	Thomas (1979)		*Millsonia anomala*	1935	2.05	Lavelle (1977)
Xenillus stegeocranus	0.1480	4.40	Luxton (1975, 1981c)		*Allolobophora chlorotica*	144.5	3.21	Curry et al. (1995)
Chamobates cuspidatus	0.0088	2.20	Luxton (1975, 1981c)		*Allolobophora caliginosa*	260.0	1.10	Nowak (1975)
Tectocepheus velatus	0.0047	2.60	Luxton (1975, 1981c)		*Octochaetona surensis*	557.3	0.47	Senepati and Dash (1981)
Achiptera coleoptrata	0.0245	3.70	Luxton (1975, 1981c)		*Lampito mauritii*	105.0	4.63	Dash and Patra (1977)
Ceratozetes gracilis	0.0245	3.00	Luxton (1975, 1981c)		**Millipedes**			
Hypochthonius rufulus	0.0207	4.30	Luxton (1975, 1981c)		*Leptoiulus saltuvagus*	16.89	0.73	Meyer (1985)
Carabodes labyrinthicus	0.0415	10.2	Luxton (1975. 1981c)		*Ophyiulus pilosus*	20.0	1.77	Blower and Miller (1974)
Eulohmannia ribagai	0.0200	4.40	Luxton (1975, 1981c)		*Triacontazona caroli*	1.30	0.86	Meyer (1979)
Ceratozetes kananaskis	0.0081	0.17	Mitchell (1979)		*Orobainosoma fonticulorum*	2.00	0.89	Meyer (1979)

Platynothrus peltifer	0.0598	2.97	Thomas (1979)		**Nematodes (total)**			
Platynothrus peltifer	0.0598	0.71	Vera (1993)		Meadow. Poland	0.0003	5.00	Sohlenius (1980)
Hermannia gibba	0.1895	2.29	Thomas (1979)		Pine forest. Sweden	0.0003	8.74	Sohlenius (1980)
Parachiptera punctata	0.0335	2.71	Thomas (1979)		Beech Forest. England	0.0002	5.16	Phillipson et al. (1977)
Mesostigmatid mites								
Olodiscus minimus	0.0120	7.00	Athias-Binche (1985)					
Neodiscopoma catalonica	0.0240	5.30	Athias-Binche (1985)					
Urodinychus carinatus	0.0770	9.50	Athias-Binche (1985)					
Armaturopoda cariacea	0.0135	5.20	Athias-Binche (1985)					
Trachytes lamda	0.0190	15.0	Athias-Binche (1985)					
Polyaspinus quadrangularis	0.0200	3.50	Athias-Binche (1985)					

8.4 Normalizing measured metabolic rates to reference body weight and temperature

Using the metabolic model proposed by Gillooly et al. (2001) measurements of metabolic rates made for animals of different body sizes at different temperatures can be adjusted to standard temperature and standard body size, to allow comparisons between species within an evolutionary lineage. The Gillooly model combines the Boltzmann factor with Kleiber's allometric scaling and is reparameterized here to include the three biologically meaningful parameters: metabolic rate at reference temperature, body size scaling parameter and activation energy. Numerical values for activation energies and allometric scaling coefficients are taken from Ehnes et al. (2011) for each group separately.

It is assumed that metabolism scales allometrically with body size and that temperature has a size-independent multiplicative effect, described by thc Boltzmann factor.	$M = aW^b \exp\left(-\frac{E}{kT}\right)$	M = metabolic rate (J h^{-1}) a = normalisation factor (J h^{-1}) W = body weight (mg) b = allometric scaling exponent E = activation energy (eV) k = Boltzmann's constant = 8.617333×10^{-5} (eV K^{-1}) T = temperature (K)
Expressing activation energy in Joules per mole changes the equation to	$H_A = \frac{R}{k}E$ $M = aW^b \exp\left(-\frac{H_A}{RT}\right)$	H_A = activation energy (J mol^{-1}) R = universal gas constant = 8.314 (J K^{-1} mol^{-1}) H_A = 96,480 E
Transforming to natural logarithms	$lnM = lna + b\ lnW - \frac{H_A}{R}\left(\frac{1}{T}\right)$	This equation defines a linear regression of lnM on lnW and 1/T, with intercept ln a, and two slope parameters, b and $-H_A/R$. The three parameters may be estimated by multiple linear regression. The intercept determines the reference metabolic rate, M_{ref}

Define reference metabolic rate M_{ref} as the metabolic rate arbitrarily chosen reference body-weight and reference temperature	$M_{ref} = aW^{b}_{ref}\,exp\left(-\frac{H_A}{RT_{ref}}\right)$	W_{ref} = reference body weight (mg) T_{ref} = reference temperature (K)
Evaluating M/M_{ref} and reworking the equation	$M_{ref} = M\left(\frac{W_{ref}}{W}\right)^{b} \exp\left(\frac{H_A\left(1-\frac{T}{T_{ref}}\right)}{RT}\right)$	M = measured metabolic rate ($J\ h^{-1}$) M_{ref} = metabolic rate at body weight W_{ref} and temperature T_{ref} ($J\ h^{-1}$)
Example: assume metabolic rate was measured as 1.40 $J\ h^{-1}$ at 19 °C for an earthworm of 875 mg. M_{ref} is calculated using the allometric scaling coefficient and activation energy for earthworms (Ehnes et al. 2011)	$M_{ref} = 1.4\left(\frac{1000}{875}\right)^{0.8} \exp\left(\frac{42768\left(1-\frac{292}{288}\right)}{8.314x292}\right)$ $= 1.22\ J\ h^{-1}$	M = 1.40 $J\ h^{-1}$ W = 875 mg W_{ref} = 1000 mg b = 0.8006928 T = 273 + 19 = 292 K T_{ref} = 273 + 15 = 288 K E = 0.4432832 eV H_A = 42,768 $J\ mol^{-1}$ M_{ref} = metabolic rate at 1000 mg body weight and 15°C temperature = 1.22 $J\ h^{-1}$

8.5 References for host-parasite associations in Table 6.8

Papers were retrieved from the literature on the occurrence of parasites in soil invertebrates in which host and parasite were identified to species; examples of parasite host systems are given in Table 6.8.

Host group	References
Isopods	Moore (1983), Sassaman and Garthwaite (1984), Moore and Lasswell (1986), Wijnhoven and Berg (1999), Dimitrova (2009), Wood et al. (2018)
Springtails	Steenberg et al. (1996), Keller and Steenberg (1997), Dromph et al. (2001)
Cockroaches	Moore (1984)
Dipteran larvae	Nixon (1980), Ulrich (2004)
Ants	Andersen et al. (2012), Espadaler and Santamaria (2012), Lachaud et al. (2012), Silva and Feitosa (2017), Silva et al. (2019), Silva and Casadei-Ferreira (2020), antwiki.org
Beetles	Davis and Prouty (2019), Poinar et al. (2004), Looney et al. (2012), Ernst et al. (2016)
Millipedes	Hopkin and Read (1992), Schmidt-Rhaesa et al. (2009), Farfan and Klompen (2012), Santamaria et al. (2014), Chiu et al. (2020), Nagae et al. (2021)
Spiders	Fain and Jocqué (1996), Mąkol and Felska (2011), Mąkol et al. (2017), Shresta et al. (2019), Durkin et al. (2021)
Harvestmen	Cokendolpher (1993)
Earthworms	Field et al. (2004), Field and Michiels (2005, 2006)
Snails and slugs	Rae et al. (2008), Majoros et al. (2010), Ivanova et al. (2019), McAllister and Hnida (2021), Waki et al. (2021, 2022), O'Brien and Pellet (2022)

References

Abe, M. and Kuroda, R. (2019). The development of CRISPR for a mollusc establishes the form in *Lsdia1* as the long-sought gene for snail dextral/sinistral coiling. *Development*, 146: dev175976.

Abrahamsen, G. (1969). Sampling design in studies of population densities in Enchytraeidae (Oligochaeta). *Oikos*, 20: 54–66.

Abrahamsen, G. (1972). Ecological study of Lumbricidae (Oligochaeta) in Norwegian coniferous forest soils. *Pedobiologia*, 12: 267–281.

Adamo, S.A. and Chase, J.M. (1990). The "love dart" of the snail *Helix aspersa* injects a pheromone that decreases courtship duration. *Journal of Experimental Zoology*, 255: 80–87.

Adamo, S.A. and Hoy, R.R. (1994). Mating behaviour of the field cricket *Gryllus bimaculatus* and its dependence on social and environmental cues. *Animal Behaviour*, 47: 857–868.

Adamowicz, A. (2005). Morphology and ultrastructure of the earthworm *Dendrobaena veneta* (Lumbricidae) coelomocytes. *Tissue and Cell*, 37: 125–133.

Agamennone, V., Jakupovic, D., Weedon, J.T., Suring, W.J., Van Straalen, N.M., Roelofs, D. et al. (2015). The microbiome of *Folsomia candida:* an assessment of bacterial diversity in a *Wolbachia*-containing animal. *FEMS Microbiology Ecology*, 91: fiv123.

Agamennone, V., Le Ngoc, G., Van Straalen, N.M., Brouwer, A. and Roelofs, D. (2019). Antimicrobial activity and carbohydrate metabolism in the bacterial metagenome of the soil-living invertebrate *Folsomia candida*. *Scientific Reports*, 9: 7308.

Agnarsson, I., Avilés, L., Coddington, J.A. and Maddison, W.P. (2006). Sociality in theridiid spiders: repeated origins of an evolutionary dead-end. *Evolution*, 60: 2342–2351.

Aguiar, N.O., Gualberto, T.L. and Franklin, E. (2006). A medium-spatial scale distribution pattern of Pseudoscorpionida (Arachnida) in a gradient of topography (altitude and inclination), soil factors, and litter in a Central Amazonia forest reserve, Brazil. *Brazilian Journal of Biology*, 66: 791–802.

Ahmed, M., Roberts, N.G., Adediran, F., Smythe, A.B., Kocot, K.M. and Holovachov, O. (2022). Phylogenomic analysis of the phylum Nematoda: conflicts and congruences with morphology, 18S rRNA, and mitogenomes. *Frontiers in Ecology and Evolution*, 9: 769565.

Ahn, S.-J., Martin, R., Rao, S. and Choi, M.-Y. (2017). Neuropeptides predicted from the transcriptome analysis of the gray garden slug *Deroceras reticulatum*. *Peptides*, 93: 51–65.

Ahtiainen, J.J., Alatalo, R.V., Mappes, J. and Vertainen, L. (2003). Fluctuating asymmetry and sexual performance in the drumming wolf spider *Hygrolycosa rubrofasciata*. *Annales Zoologici Fennici*, 40: 281–292.

Aira, M., Pérez-Losada, M. and Domínguez, J. (2018). Diversity, structure and sources of bacterial communities in earthworm cocoons. *Scientific Reports*, 8: 6632.

Al-Assiuty, A.I.M., Bayoumi, B.M., Khalil, M.A. and Van Straalen, N.M. (1993). Egg number and abundance of ten Egyptian oribatid mite species (Acari: Cryptostigmata) in relation to habitat quality. *European Journal of Soil Biology*, 29: 59–65.

Albo, M.J., Bilde, T. and Uhl, G. (2013). Sperm storage mediated by cryptic female choice for nuptial gifts. *Proceedings of the Royal Society B*, 280: 20131735.

Allee, W.C. (1926). Studies in animal aggregations: causes and effects of bunching in land isopods. *Journal of Experimental Zoology*, 45: 255–277.

Allen, K.R. (1951). The Horokiwi stream. A study of a trout population. *New Zealand Marine Department Fisheries Bulletin*, 10: 1–231.

Allwood, J., Gleeson, D., Mayer, G., Daniels, S., Beggs, J.R. and Buckley, T.R. (2010). Support for vicariant origins of the New Zealand Onychophora. *Journal of Biogeography*, 37: 669–681.

Álvarez-Presas, M., Baguña, J. and Riutort, M. (2008). Molecular phylogeny of land and freshwater planarians (Tricladida, Platyhelminthes): From freshwater to land and back. *Molecular Phylogenetics and Evolution*, 47: 555–568.

Andersen, D.C. (1987). Below-ground herbivory in natural communities: a review emphasizing fossorial animals. *The Quarterly Review of Biology*, 62: 261–286.

Andersen, S.B., Ferrari, M., Evans, H.C., Elliot, S.L., Boomsma, J.J. and Hughes, D.P. (2012). Disease dynamics in a specialized parasite of ant societies. *PLoS One*, 7: e36352.

Anderson, J.M. (1975). The enigma of soil animal species diversity. pp. 51–58. *In*: Vanek, J. (ed.). *Progress in Soil Zoology*. Junk B.V., The Hague.

Anderson, C., Cunha, L., Sechi, P., Kille, P. and Spurgeon, D. (2017). Genetic variation in populations of the earthworm, *Lumbricus rubellus*, across contaminated sites. *BMC Genetics*, 18: 97.

Anderson, F.E., Williams, B.W., Horn, K.M., Erséus, C., Halanych, K.M., Santos, S.R. et al. (2017). Phylogenomic analyses of Crassiclitellata support major Northern and Southern Hemisphere clades and a Pangaean origin for earthworms. *BMC Evolutionary Biology*, 17: 123.

Andersson, J.O. (2005). Lateral gene transfer in eukaryotes. *Cellular and Molecular Life Sciences*, 62: 1182–1197.

Andre, J., King, R.A., Stürzenbaum, S.R., Kille, P., Hodson, M.E. and Morgan, A.J. (2010). Molecular genetic differentiation in earthworms inhabiting a heterogeneous Pb-polluted landscape. *Environmental Pollution*, 158: 883–890.

Anthes, N., David, P., Auld, J.R., Hoffer, J.N.A., Jarne, P., Koene, J.M. et al. (2010). Bateman gradients in hermaphrodites: an extended approach to quantify sexual selection. *The American Naturalist*, 176: 249–263.

Angelini, D.R. and Kaufman, T.C. (2005). Comparative developmental genetics and the evolution of arthropod body plans. *Annual Review of Genetics*, 39: 95–119.

Angle, J.C., Morin, T.H., Solden, L.M., Narrowe, A.B., Smith, G.J., Borton, M.A. et al. (2017). Methanogenesis in oxygenated soils is a substantial fraction of wetland methane emissions. *Nature Communications*, 8: 1567.

Anthes, N. (2010). Mate choice and reproductive conflict in simultaneous hermaphrodites. pp. 329–357. *In*: Kappeler, P. (ed.). *Animal Behaviour: Evolution and Mechanisms*. Springer, Heidelberg.

Antoł, A. and Czarnoleski, M. (2018). Size dependence of offspring production in isopods: a synthesis. *ZooKeys*, 801: 337–357.

Araújo, J.P.M. and Hughes, D.P. (2016). Diversity of entomopathogenic fungi: which groups conquered the insect body? *Advances in Genetics*, 94: 1–39.

Arbea, J.I. and Blasco-Zumeta, J. (2001). Ecología de los Colémbolos (Hexapoda, Collembola) en Los Monegros (Zaragoza, España). *Boletin de las Sociedad Entomológica Aragonesa*, 28: 35–48.

Ardestani, M.M., Van Straalen, N.M. and Van Gestel, C.A.M. (2014). Uptake and elimination kinetics of metals in soil invertebrates: A review. *Environmental Pollution*, 193: 277–295.

Arenas-Mena, C. (2010). Indirect development, transdifferentiation and the macroregulatory evolution of metazoans. *Philosophical Transactions of the Royal Society B*, 365: 653–669.

Arora, J., Kinjo, Y., Šobotník, J., Buček, A., Clitheroe, C., Stiblik, P. et al. (2022). The functional evolution of termite gut microbiota. *Microbiome*, 10: 78.

Athias-Binche, F. (1985). Analyses démographiques des populations d' Uropodides (Arachnides: Anactinotriches) de la hêtraie de la Massane, France. *Pedobiologia*, 28: 225–253.

Audet, P. (2012). AM symbiosis and other plant-soil interactions in relation to environmental stress. pp. 233–264. *In*: Ahmad, P. and Prasad, M.N.V. (eds.). *Environmental Adaptations and Stress Tolerance of Plants in the Era of Global Change*. Springer, New York.

Averof, M. and Cohen, S.M. (1997). Evolutionary origin of insect wings from ancestral gills. *Nature*, 385: 627–630.

Ávila-Jiménez, M.L. and Coulson, S.J. (2011). A Holarctic biogeographical analysis of the Collembola (Arthropoda, Hexapoda) unravels recent post-glacial colonization patterns. *Insects*, 2: 273–296.

Avilés, L., Maddison, W.P. and Agnarsson, I. (2006). A new independently derived social spider with explosive profileration and female size dimorphism. *Biotropica*, 38: 743–753.

Axelsson, B., Lohm, U. and Persson, T. (1984). Enchytraeids, lumbricids and soil arthropods in a northern deciduous woodland- a quantitative study. *Holarctic Ecology*, 7: 91–103.

Baars, M.A. (1979a). Catches in pitfall traps in relation to mean densities of carabid beetles. *Oecologia*, 41: 25–46.

Baars, M.A. (1979b). Patterns of movement of radioactive beetles. *Oecologia*, 44: 125–140.

Badejo, M.A. and Van Straalen, N.M. (1993). Seasonal abundance of springtails in two contrasting environments. *Biotropica*, 25: 222–228.

Baermann, G. (1917). Eine einfache Methode zur Auffindung von Ankylostomum–(Nematoden)–Larven in Erdproben. *Geneeskundig Tijdschrift voor Nederlandsch Indië*, 57: 131–137.

Baguña, J. (2012). The planarian neoblast: the rambling history of its origins and some current black boxes. *International Journal of Developmental Biology*, 56: 19–37.

Bahram, M., Hildebrand, F., Forslund, S.K., Anderson, J.L., Soudzilovskaia, N.A., Bodegom, P.M. et al. (2018). Structure and function of the global topsoil microbiome. *Nature*, 560: 233–237.

Bahrndorff, S., Mariën, J., Loeschcke, V. and Ellers, J. (2009). Dynamics of heat-induced thermal stress resistance and Hsp70 expression in the springtail, *Orchesella cincta*. *Functional Ecology*, 23: 233–239.

Bahrndorff, S., Mariën, J., Loeschcke, V. and Ellers, J. (2010). Genetic variation in heat resistance and HSP70 expression in inbred isofemale lines of the springtail *Orchesella cincta*. *Climate Research*, 43: 41–47.

Baker, G.H. (1999). Spatial and temporal patterns in the abundance and biomass of earthworm populations in pastures in southern Australia. *Pedobiologia*, 43: 487–496.

Baldi, C., Cho, S. and Ellis, R.E. (2009). Mutations in two independent pathways are sufficient to create hermaphroditic nematodes. *Science*, 326: 1002–1005.

Ballesteros, J.A. and Sharma, P.P. (2019). A critical appraisal of the placement of Xiphosura (Chelicerata) with account of known sources of phylogenetic error. *Systematic Biology*, 68: 869–917.

Bandyopadhyaya, I., Choudhuri, D.K. and Ponge, J.-F. (2002). Effects of some physical factors and agricultural practices on Collembola in a multiple cropping programme in West Bengal (India). *European Journal of Soil Biology*, 38: 111–117.

Banse, K. and Mosher, S. (1980). Adult body mass and annual production/biomass relationships of field populations. *Ecological Monographs*, 50: 355–379.

Bardgett, R.D. (2005). *The Biology of Soil. A Community and Ecosystem Approach*. Oxford University Press, Oxford.

Barnett, K. and Johnson, S.N. (2013). Living in the soil matrix: abiotic factors affecting root herbivores. *Advances in Insect Physiology*, 45: 1–52.

Barot, S., Blouin, M., Fontaine, S., Jouquet, P., Lata, J.C. and Mathieu, J. (2007). A tale of four stories: soil ecology, theory, evolution and the publication system. *PLoS One*, 2: e1248.

Barra, J.A. and Poinsot-Balaguer, N. (1977). Modifications ultrastructurales accompagnant l'anhydrobiose chez un Collembole: *Folsomides variabilis*. *Revue d'Écologie et de Biologie du Sol*, 14: 189–197.

Bartels, P.J., Apodaca, J.J., Mora, C. and Nelson, D.R. (2016). A global biodiversity estimate of a poorly known taxon: phylum Tardigrada. *Zoological Journal of the Linnean Society*, 178: 730–736.

Bateman, A.J. (1948). Intra-sexual selection in *Drosophila*. *Heredity*, 105: 349–508.

Bauer, T. (1981). Prey capture and structure of the visual space of an insect that hunts by sight on the litter layer (*Notiophilus biguttatus* F., Carabidae, Coleoptera). *Behavioral Ecology and Sociobiology*, 8: 91–97.

Bayley, M. and Holmstrup, M. (1999). Water vapor absorption in arthropods by accumulation of myoinositol and glucose. *Science*, 285: 1909–1911.

Bayley, M., Overgaard, J., Sødergaard Høj, A., Malmendal, A., Holmstrup, M. and Wang, T. (2010). Metabolic changes during estivation in the common earthworm *Aporrectodea caliginosa*. *Physiology and Biochemical Zoology*, 83: 541–550.

Beese, K., Armbruster, G.F.J., Beier, K. and Baur, B. (2008). Evolution of the sperm-storage organs in the carrefour of stylommatophoran gastropods. *Journal of Zoological Systematics and Evolutionary Research*, 47: 49–60.

Belgnaoui, S. and Barra, J.-A. (1989). Water loss and survival in the anhydrobiotic Collembola *Folsomides angularis*. *Revue d'Ecologie et de Biologie du Sol*, 26: 123–132.

Bell, A.W. (1959). *Enchytraeus fragmentosus*, a new species of naturally fragmenting oligochaete worm. *Science*, 129: 1278.

Bell, G. (1976). On breeding more than once. *The American Naturalist*, 110: 57–77.

Bell, G. (1982). *The Masterpiece of Nature. The Evolution and Genetics of Sexuality*. University of California Press, Berkeley.

Belozerov, V.N. (2008). Diapause and quiescence as two main kinds of dormancy and their significance in life cycles of mites and ticks (Chelicerata: Arachnida: Acari). Part 1. Acariformes. *Acarina*, 16: 79–130.

Bely, A.E. and Sikes, J.M. (2010). Acoel and platyhelminth models for stem-cell research. *Journal of Biology*, 9: 14.

Bengtsson, G., Ohlsson, L. and Rundgren, S. (1985a). Influence of fungi on growth and survival of *Onychiurus armatus* (Collembola) in a metal polluted soil. *Oecologia*, 68: 63–68.

Bengtsson, G., Gunnarsson, T. and Rundgren, S. (1985b). Influence of metals on reproduction, mortality and population growth in *Onychiurus armatus* (Collembola). *Journal of Applied Ecology*, 22: 967–978.

Bengtsson, G., Erlandsson, A. and Rundgren, S. (1988). Fungal odour attracts soil Collembola. *Soil Biology and Biochemistry*, 20: 25–30.

Bengtsson, G., Hedlund, K. and Rundgren, S. (1991). Selective odor perception in the soil Collembola *Onychiurus armatus*. *Journal of Chemical Ecology*, 17: 2113–2125.

Bengtsson, J. (1994). Temporal predictability in forest soil communities. *Journal of Animal Ecology*, 63: 653–665.

Benítez-Álvarez, L., Leal-Zanchet, A.M., Oceguera-Figueroa, A., Lopes Ferreira, R., De Medieros Bento, D., Braccini, J. et al. (2020). Phylogeny and biogeography of the Cavernicola (Platyhelminthes: Tricladida): Relicts of an epigean group sheltering in caves? *Molecular Phylogenetics and Evolution*, 145: 106709.

Benke, M., Reise, H., Montagne-Wajer, K. and Koene, J.M. (2010). Cutaneous application of an accessory-gland secretion after sperm exchange in a terrestrial slug (Mollusca; Pulmonata). *Zoology*, 113: 118–124.

Berlese, A. (1905). Apparecchio per raccogliere presto ed in gran numero piccoli artropodi. *Redia*, 2: 85–89.

Berg, M.P., Soesbergen, M., Tempelman, D. and Wijnhoven, H. (2008). *Verspreidingsatlas Nederlandse landpissebedden, duizendpoten en miljoenpoten (Isopoda, Chilopoda, Diplopoda)*. European Invertebrate Survey - Nederland, Leiden and Vrije Universiteit - Afdeling Dierecologie, Amsterdam.

Berg, M.P., Kiers, E.T., Driessen, G., Van der Heijden, M., Kooi, B.W., Kuenen, F. et al. (2010). Adapt or disperse: understanding species persistence in a changing world. *Global Change Biology*, 16: 587–598.

Berman, D.I. and Leirikh, A.N. (2018). Cold hardiness of mass soil invertebrate animals of northeastern Asia: 1. Cold hardiness and the mechanisms of its maintenance. *Biology Bulletin*, 45: 669–679.

Berman, D.I., Bulakhova, N.A., Meshcheryakova, E.N. and Shekhovtsov, S.V. (2019). Cold resistance and the distribution of genetic lineages of the earthworm *Eisenia nordenskioldi* (Oligochaeta, Lumbricidae). *Biology Bulletin*, 46: 430–437.

Bernadou, A., Kramer, B.H. and Korb, J. (2021). Major evolutionary transitions in social insects, the importance of worker sterility and life history trade-offs. *Frontiers in Ecology and Evolution*, 9: 732907.

Bert, W., Karssen, G. and Helder, J. (2011). Phylogeny and evolution of nematodes. pp. 45–59. *In*: Jones, J., Gheysen, G. and Fenoll, C (eds.).*Genomics and Molecular Genetics of Plant-Nematode Interactions*. Springer Science+Business Media B.V., Dordrecht.

Bertolani, R. (2001). Evolution of the reproductive mechanisms in tardigrades – a review. *Zoologische Anzeiger*, 240: 247–252.

Betsch-Pinot, M.-C. (1977). Les parades sexuelles primitives chez les Collemboles Symphypléones. *Revue d'Écologie et de Biologie du Sol*, 14: 15–19.

Binz, P.A. and Kägi, J.H.R. (1999). Metallothionein: Molecular evolution and classification. pp. 7–13. *In*: Klaassen, C.D. (ed.). *Metallothionein IV*. Birkhäuser Verlag, Basel.

Birch, L.C. (1948). The intrinsic rate of natural increase of an insect population. *Journal of Animal Ecology*, 17: 15–26.

Birkemoe, T. and Leinaas, H.P. (1999). Reproductive biology of the arctic collembolan *Hypogastrura tullbergi*. *Ecography*, 22: 31–39.

Birkemoe, T., Jacobsen, R.M., Sverdrup-Thygeson, A. and Biedermann, P.H.W. (2018). Insect-fungus interactions in dead wood systems. pp. 377–427. *In*: Ulyshen, M.D. (ed.). *Saproxylic Insects, Zoological Monographs 1*. Springer Nature Basel.

Birkhofer, K., Henschel, J.R. and Scheu, S. (2006). Spatial pattern analysis in a territorial spider: evidence for multi-scale effects. *Ecography*, 29: 641–648.

Birkhofer, K., Henschel, J. and Lubin, Y. (2012). Effects of extreme climatic events on small-scale spatial patterns: a 20–year study of the distribution of a desert spider. *Oecologia*, 170: 651–657.

Bitsch, C. and Bitsch, J. (2000). The phylogenetic interrelationships of the higher taxa of apterygote insects. *Zoologica Scripta*, 29: 131–156.

Blair, J.M., Parmelee, R.V. and Wyman, R.L. (1994). A comparison of the forest floor invertebrate communities of four forest types in the northeastern U.S. *Pedobiologia*, 38: 146–160.

Blancquaert, J.-P. and Mertens, J. (1977). Mating behaviour in *Sphaeridia pumilis* (Collembola). *Pedobiologia*, 17: 343–349.

Blancquaert, J.-P. and Mertens, J. (1979). Postembryonal development in *Megalothorax minimus* (Willem, 1900), (Collembola). *Revue d'Écologie et de Biologie du Sol*, 16: 125–130.

Blancquaert, J.-P., Coessens, R. and Mertens, J. (1981a). Life history of some Symphypleona (Collembola) under experimental conditions I- Embryonal development and diapause. *Revue d'Écologie et de Biologie du Sol*, 18: 115–126.

Blancquaert, J.-P., Coessens, R. and Mertens, J. (1981b). Life history of some Symphypleona (Collembola) under experimental conditions II.- Postembryonal development and reproduction. *Revue d'Écologie et de Biologie du Sol*, 18: 373–390.

Blaxter, M.L., De Ley, P., Garey, J.R., Liu, L.X., Scheldeman, P., Vierstraete, A. et al. (1998). A molecular evolutionary framework for the phylum Nematoda. *Nature*, 392: 71–75.

Block, W., Young, S.R., Conradi-Larsen, E.M. and Sømme, L. (1978). Cold tolerance of two Antarctic terrestrial arthropods. *Experientia*, 34: 1166–1167.

Block, W. (1996). Cold or drought—the lesser of two evils for terrestrial arthropods? *European Journal of Entomology*, 93: 325–339.

Blossey, B. and Hunt-Joshi, T.R. (2003). Belowground herbivory by insects: Influence on plants and aboveground herbivores. *Annual Review of Entomology*, 48: 521–547.

Blossfeld, S. and Gansert, D. (2007). A novel non-invasive optical method for quantitative visualization of pH dynamics in the rhizosphere of plants. *Plant, Cell and Environment*, 30: 176–186.

Blouin, M., Hodson, M.E., Delgado, E.A., Baker, G., Brussaard, L., Butt, K.R. et al. (2013). A review of earthworm impact on soil function and ecosystem services. *European Journal of Soil Science*, 64: 161–182.

Blower, J.G. and Miller, P.F. (1974). The life-cycle and ecology of *Ophyiulus pilosus* (Newport) in Britain. *Symposia of the Zoological Society of London*, 32: 503–525.

Blower, J.G. and Miller, P.F. (1977). The life-history of the julid millipede *Cylindroiulus nitidus* in a Derbyshire wood. *Journal of Zoology*, 183: 339–351.

Boag, B., Palmer, L.F., Neilson, R. and Chambers, S.J. (1994). Distribution and prevalence of the predatory planarian *Artioposthia triangulata* (Dendy) (Tricladida: Terricola) in Scotland. *Annals of Applied Biology*, 124: 165–171.

Boetzl, F.A., Ries, E., Schneider, G. and Krauss, J. (2018). It's a matter of design—how pitfall trap design affects trap samples and possible predictions. *Peer J*, 6: e5078.

Bogovic, P. and Strle, F. (2015). Tick-borne encephalitis: A review of epidemiology, clinical characteristics, and management. *World Journal of Clinical Cases*, 3: 430–441.

Bongers, T. and Ferris, H. (1999). Nematode community structure as a bioindicator in environmental monitoring. *Trends in Ecology and Evolution*, 14: 224–228.

Bonkowski, M., Villenave, C. and Griffiths, B. (2009). Rhizosphere fauna: the functional and structural diversity of intimate interactions of soil fauna with plant roots. *Plant and Soil*, 321: 213–233.

Bonte, D. and Maelfait, J.-P. (2001). Life history, habitat use and dispersal of a dune wolf spider (*Pardosa monticola* (Clerck, 1757) Lycosidae, Araneae) in the Flemish coastal dunes (Belgium). *Belgian Journal of Zoology*, 131: 145–157.

Bonte, D., Lens, L. and Maelfait, J.-P. (2004). Lack of homeward orientation and increased mobility result in high emigration rates from low-quality fragments in a dune wolf spider. *Journal of Animal Ecology*, 73: 643–650.

Boothby, T.C., Tenlen, J.R., Smith, F.W., Wang, J.R., Patanella, K.A., Nishimura, E.O. et al. (2015). Evidence for extensive horizontal gene transfer from the draft genome of a tardigrade. *Proceedings of the National Academy of Sciences of the United States of America*, 112: 15976–15981.

Boothby, T.C. (2019). Mechanisms and evolution of resistance to environmental extremes in animals. *EvoDevo*, 10: 30.

Boschetti, C., Carr, A., Crisp, A., Eyres, I., Wang-Koh, Y., Lubzens, E. et al. (2012). Biochemical diversification through foreign gene expression in bdelloid rotifers. *PLoS Genetics*, 8: e1003035.

Boström, S. and Sohlenius, B. (1986). Short-term dynamics of nematode communities in arable soil. Influence of a perennial and an annual cropping system. *Pedobiologia*, 29: 345–357.

Bouché, M.B. (1977). Strategies lombriciennes. pp. 122–132. *In*: Lohm, U. and Persson, T. (eds.). *Soil Organisms as Components of Ecosystems*. Ecological Bulletins, Stockholm.

Bourguignon, T., Lo, N., Dietrich, C., Šobotník, J., Sidek, S., Roisin, Y. et al. (2018). Rampant host switching shaped the termite gut microbiome. *Current Biology*, 28: 649–654.

Bowden, J., Haines, I.H. and Mercer, D. (1976). Climbing Collembola. *Pedobiologia*, 16: 298–312.

Bowden, J.J. and Buddle, C.M. (2012). Life history of tundra-dwelling wolf spiders (Araneae: Lycosidae) from the Yukon Territory, Canada. *Canadian Journal of Zoology*, 90: 714–721.

Boxshall, G.A. (2004). The evolution of arthropod limbs. *Biological Reviews of the Cambridge Philosophical Society*, 79: 253–300.

Boyden, A. (1950). Is parthenogenesis sexual or asexual reproduction? *Nature*, 166: 820.

Brakhage, A.A., Thön, M., Spröte, P., Scharf, D.H., Al-Abdallah, Q., Wolke, S.M. et al. (2009). Aspects on evolution of fungal β-lactam biosynthesis gene clusters and recruitment of *trans*-acting factors. *Phytochemistry*, 70: 1801–1811.

Brandt, M. and Mahsberg, D. (2002). Bugs with a backpack: the function of nymphal camouflage in the West African assassin bugs *Paredocla* and *Acanthaspis* spp. *Animal Behaviour*, 63: 277–284.

Brattsten, L.B. (1979). Ecological significance of mixed-function oxidations. *Drug Metabolism Review*, 10: 35–58.

Bredon, M., Dittmer, J., Noël, C., Moumen, B. and Bouchon, D. (2018). Lignocellulose degradation at the holobiont level: teamwork in a keystone soil invertebrate. *Microbiome*, 6: 162.

Brena, C. and Akam, M. (2013). An analysis of segmentation dynamics throughout embryogenesis in the centipede *Strigamia maritima*. *BMC Biology*, 11: 112.

Brenneis, G., Bogomolova, E.V., Arango, C.P. and Krapp, F. (2017). From egg to "no-body": an overview and revision of developmental pathways in the ancient arthropod lineage Pycnogonida. *Frontiers in Zoology*, 14: 6.

Bretfeld, G. (1963). Zur Anatomie und Embryologie der Rumpfmuskulatur und der abdominalen Anhänge der Collembolen. *Zoologische Jahrbücher Anatomie*, 80: 309–384.

Bretfeld, G. (1970). Grundzuge des Paarungsverhaltens europäischer Bourletiellini (Collembola, Sminthuridae) und daraus abgeleitete taxonomisch- nomenklatorische Folgerungen. *Zeitschrift fur Zoologische Systematik und Evolutionsforschung*, 8: 259–273.

Bretfeld, G. (1971). Das Paarungsverhalten europäischer Bourletiellini (Sminthuridae). *Revue d'Écologie et de Biologie du Sol*, 8: 145–153.

Bringmark, E. (1989). Spatial variation in soil pH of beech forests in relation to buffering properties and soil depths. *Oikos*, 54: 165–177.

Brockett, B.F.T. and Hassall, M. (2005). The existence of an Allee effect in populations of *Porcellio scaber* (Isopoda: Oniscidea). *European Journal of Soil Biology*, 41: 123–127.

Broly, P., Mullier, R., Deneubourg, J.-L. and Devigne, C. (2012). Aggregation in woodlice: social interaction and density effects. *ZooKeys*, 176: 133–144.

Broly, P., Deville, P. and Maillet, S. (2013a). The origin of terrestrial isopods (Crustacea: Isopoda: Oniscidea). *Evolutionary Ecology*, 27: 461–476.

Broly, P., Deneubourg, J.-L. and Devigne, C. (2013b). Benefits of aggregation in woodlice: a factor in the terrestrialization procsess? *Insectes Sociaux*, 60: 419–435.

Broly, P., Devigne, L., Deneubourg, J.-L. and Devigne, C. (2014). Effects of group size on aggregation against desiccation in woodlice (Isopoda: Oniscidea). *Physiological Entomology*, 39: 165–171.

Broly, P., Devigne, C. and Deneubourg, J.-L. (2015). Body shape in terrestrial isopods: a morphological mechanism to resist desiccation? *Journal of Morphology*, 276: 1283–1289.

Broly, P., Mullier, R., Devigne, C. and Deneubourg, J.-L. (2016). Evidence of self-organization in a gregarious land-dwelling crustacean (Isopoda: Oniscidea). *Animal Cognition*, 19: 181–192.

Brouwers, N.C. and Newton, A.C. (2009). Movement rates of woodland invertebrates: a systematic review of empirical evidence. *Insect Conservation and Diversity*, 2: 10–22.

Brown, B.E. (1982). The form and function of metal containing "granules" in invertebrate tissues. *Biological Reviews*, 57: 621–667.

Broza, M., Pereira, R.M. and Stimac, J.L. (2001). The nonsusceptibility of soil Collembola to insect pathogens and their potential as scavengers of microbial pesticides. *Pedobiologia*, 45: 523–534.

Brulle, F., Cocquerelle, C., Wamalah, A.N., Morgan, A.J., Kille, P., Leprêtre, A. et al. (2008). cDNA cloning and expression analysis of *Eisenia fetida* (Annelida: Oligochaeta) phytochelatin synthase under cadmium exposure. *Ecotoxicology and Environmental Safety*, 71: 47–55.

Brune, A. (2014). Symbiotic digestion of lignocellulose in termite guts. *Nature Reviews Microbiology*, 12: 168–180.

Brune, A. and Dietrich, C. (2015). The gut microbiota of termites: digesting the diversity in the light of ecology and evolution. *Annual Review of Microbiology*, 69: 145–166.

Brunsting, A.M.H. and Heessen, H.J.L. (1984). Density regulation in the carabid beetle *Pterostichus oblongopunctatus*. *Journal of Animal Ecology*, 53: 751–760.

Brussaard, L., Behan-Pelletier, V.M., Bignell, D.E., Brown, V.K., Didden, W., Folgarait, P. et al. (1997). Biodiversity and ecosystem functioning in soil. *Ambio*, 26: 563–570.

Brussaard, L. (1998). Soil fauna guilds, functional groups and ecosystem processes. *Applied Soil Ecology*, 9: 123–135.

Burger, M. (2007). Sperm dumping in a haplogyne spider. *Journal of Zoology*, 273: 74–81.

Burki, F., Roger, A.J., Brown, M.W. and Simpson, A.G.B. (2020). The new tree of eukaryotes. *Trends in Ecology and Evolution*, 35: 43–55.

Butt, K., Frederickson, J. and Morris, R.M. (1994). Effect of earthworm density on the growth and reproduction of *Lumbricus terrestris* L. (Oligochaeta: Lumbricidae) in culture. *Pedobiologia*, 38: 254–261.

Calderon, S., Holmstrup, M., Westh, P. and Overgaard, J. (2009). Dual roles of glucose in the freeze-tolerant earthworm *Dendrobaena octaedra*: cryoprotection and fuel for metabolism. *The Journal of Experimental Biology*, 212: 959–966.

Cameron, E.K., Bayne, E.M. and Coltman, D.W. (2008). Genetic structure of invasive earthworms *Dendrobaena octaedra* in the boreal forests of Alberta: insights into introduction mechanisms. *Molecular Ecology*, 17: 1189–1197.

Cancela da Fonseca, J.P., Boudjema, G., Sarkar, S. and Julien, J.M. (1995). Can statistical analyses improve our knowledge about relationships between environmental factors and soil microarthropod communities. pp. 42–51. *In*: Edwards, C.A., Abe, T. and Striganova, B.R. (eds.). *Structure and Function of Soil Communities*. Kyoto University Press, Kyoto.

Cannavacciuolo, M., Bellido, A., Cluzeau, D., Gascuel, C. and Trehen, P. (1998). A geostatistical approach to the study of earthworm distribution in grassland. *Applied Soil Ecology*, 9: 345–349.

Capinera, J.L. and Leppla, N.C. (2001). Shortwinged mole cricket, *Scapteriscus abbreviatus* Scudder; southern mole cricket, *Scapteriscus borelli* Giglio-Tos; and tawny mole cricket, *Scapteriscus vicinus* Scudder (Insecta: Orthoptera: Gryllotalpidae). *Institute of Food and Agricultural Sciences*, University of Florida, Gainesville.

Carapelli, A., Frati, F., Nardi, F., Dallai, R. and Simon, C. (2000). Molecular phylogeny of the apterygotan insects based on nuclear and mitochondrial genes. *Pedobiologia*, 44: 361–373.

Carapelli, A., Nardi, F., Dallai, R. and Frati, F. (2006). A review of molecular data for the phylogeny of basal hexapods. *Pedobiologia*, 50: 191–204.

Carapelli, A., Liò, P., Nardi, F., Van der Wath, E. and Frati, F. (2007). Phylogenetic analysis of mitochondrial protein coding genes confirms the reciprocal paraphyly of Hexapoda and Crustacea. *BMC Evolutionary Biology*, 7 (Suppl 2), S8.

Carefoot, T.H. (1993). Physiology of terrestrial isopods. *Comparative Biochemistry and Physiology*, 106A: 413–429.

Carley, W.W., Caracciolo, E.A. and Mason, R.T. (1983). Cell and coelomic fluid volume regulation in the earthworm *Lumbricus terrestris*. *Comparative Biochemistry and Physiology*, 74A: 569–575.

Carpenter, A. (1988). The biology of *Campodea staphylinus* (Campodeidae: Diplura) in a grassland soil. *Pedobiologia*, 32: 31–38.

Carroll, S.B., Grenier, J.K. and Weatherbee, S.D. (2005). *From DNA to Diversity*. Blackwell Publishing, Malden.

Carter, M.J., Lardies, M.A., Nespolo, R.F. and Bozinovic, F. (2004). Heritability of progeny size in a terrestrial isopod: transgenerational environmental effects on a life history trait. *Heredity*, 93: 455–459.

Caruso, T., Garlaschelli, D., Bargagli, R. and Convey, P. (2010). Testing metabolic scaling theory using intraspecific allometries in Antarctic microarthropods. *Oikos*, 119: 935–945.

Cesari, M., Altiero, T. and Rebecchi, L. (2012). Identification of the trehalose-6–phosphate synthase (tps) gene in desiccation tolerant and intolerant tardigrades. *Italian Journal of Zoology*, 79: 530–540.

Chagnon, M., Paré, D. and Hébert, C. (2000). Relationships between soil chemistry, microbial biomass, and the collembolan fauna of southern Québec sugar maple stands. *Ecoscience*, 7: 301–316.

Chapman, J.W., Reynolds, D.R., Smith, A.D., Riley, J.R., Telfer, M.G. and Woiwod, I.P. (2005). Mass aerial migration in the carabid beetle *Notiophilus biguttatus*. *Ecological Entomology*, 30: 264–272.

Charlesworth, B. (1980). *Evolution in Age-structured Populations*. Cambridge University Press, Cambridge.

Charnov, E.L., Maynard Smith, J. and Bull, J.J. (1976). Why be an hermaphrodite? *Nature*, 263: 125–126.

Chase, R. and Blanchard, K.C. (2006). The snail's love dart delivers mucus to increase paternity. *Proceedings of the Royal Society B*, 273: 1471–1475.

Chen, B. and Wise, D.H. (1999). Bottom-up limitation of predaceous arthropods in a detritus-based terrestrial food-web. *Ecology*, 80: 761–772.

Chen, W.-Y., Li, W.-H., Ju, Y.-R., Liao, C.-M. and Liao, V.H.-C. (2017). Life cycle toxicity assessment of earthworms exposed to cadmium-contaminated soils. *Ecotoxicology*, 26: 360–369.

Chernova, N.M. and Kuznetsova, N.A. (2000). Collembolan community organization and its temporal predictability. *Pedobiologia*, 44: 451–466.

Chernova, N.M., Potapov, M.B., Savenkova, Y.Y. and Bokova, A.I. (2010). Ecological significance of parthenogenesis in Collembola. *Entomological Reviews*, 90: 23–38.

Chiara, V., Portugal, P.R. and Jeanson, R. (2019). Social intolerance is a consequence, not a cause, of dispersal in spiders. *PLoS Biology*, 17: e3000319.

Chiba, S., Takuya, A., Aoki, J., Imadaté, G., Ishikawa, K., Kondoh, M. et al. (1975). Studies on the productivity of soil animals in Pasoh Forest Reserve, West Malaysia. I. Seasonal change in the density of soil mesofauna: Acari, Collembola and others. *Science Reports of the Hirosaki University*, 22: 87–124.

Chiba, S., Abe, T., Kondoh, M., Shiba, M. and Watanabe, H. (1976). Studies on the productivity of soil animals in Pasoh Forest Reserve, West Malaysia. II. Seasonal change in the density of Nematoda and Enchytraeidae. *Science Reports of the Hirosaki University*, 28: 74–78.

Chiu, M.-C., Huang, C.-G., Wu, W.-J., Lin, Z.-H., Chen, H.-W. and Shiao, S.-F. (2020). A new millipede-parasitizing horsehair worm, *Gordius chianshanus* sp. nov., at medium altitudes in Taiwan (Nematomorpha, Gordiida). *ZooKeys*, 941: 25–48.

Choi, W.I. and Ryoo, M.I. (2003). A matrix model for predicting seasonal fluctuations in field populations of *Paronychiurus kimi* (Collembola: Onychiuridae). *Ecological Modelling*, 162: 259–265.

Choi, W.I., Moorhead, D.L., Neher, D.A. and Ryoo, M.I. (2006). A modelling study of soil temperature and moisture effects on population dynamics of *Paronychiurus kimi* (Collembola: Onychiuridae). *Biology and Fertility of Soils*, 43: 69–75.

Chown, S.L., Marais, E., TerBlanche, J.S., Klok, C.J., Lighton, J.R.B. and Blackburn, T.M. (2007). Scaling of insect metabolic rate is inconsistent with the nutrient supply network model. *Functional Ecology*, 21: 282–290.

Christensen, B. and O'Connor, F.B. (1958). Pseudofertilization in the genus *Lumbricillus* (Enchytraeidae). *Nature*, 181: 1085–1086.

Christensen, B. (1959). Asexual reproduction in the Enchytraeidae. *Nature*, 184: 1159–1160.

Christensen, B. (1961). Studies on the cyto-taxonomy and reproduction in the Enchytraeidae with notes on parthenogenesis and polyploidy in the animal kingdom. *Hereditas*, 47: 387–450.

Christensen, B., Hvilsom, M.M. and Pedersen, B.V. (1992). Genetic variation in coexisting sexual diploid and parthenogenetic triploid forms of *Fredericia galba* (Enchytraeidae, Oligochaeta) in a heterogenous environment. *Hereditas*, 117: 153–162.

Christensen, B., Pedersen, B.V. and Hvilsom, M.M. (2002). Persisting clone pool differences in sexual/asexual *Buchholzia appendiculata* (Enchytraeidae, Oligochaeta) as revealed by genetic markers. *Pedobiologia*, 46: 90–99.

Christensen, B. and Glenner, H. (2010). Molecular phylogeny of Enchytraeidae (Oligochaeta) indicates separate invasions of the terrestrial environment. *Journal of Zoological Systematics and Evolutionary Research*, 48: 208–212.

Christensen, O.M. and Mather, J.G. (1995). Colonisation by the land planarian *Artioposthia triangulata* and impact on lumbricid earthworms at a horticultural site. *Pedobiologia*, 39: 144–154.

Christiansen, K. and Bellinger, P. (1995). The biogeography of Collembola. *Polskie Pismo Entomologiczne*, 64: 279–294.

Christias, C., Couvaraki, C., Georgopoulos, S.G., Macris, B. and Vomvoyanni, V. (1975). Protein content and amino acid composition of certain fungi evaluated for microbial protein production. *Applied Microbiology*, 29: 250–254.

Christoffersen, M.L. (2012). Phylogeny of basal descendants of cocoon-forming annelids (Clitellata). *Turkish Journal of Zoology*, 36: 95–119.

Ciarelli, S., Van Straalen, N.M., Klap, V.A. and Van Wezel, A.P. (1999). Effects of sediment bioturbation by the estuarine amphipod *Corophium volutator* on fluoranthene resuspension and transfer into the mussel (*Mytilus edulis*). *Environmental Toxicology and Chemistry*, 18: 318–328.

Cicconardi, F., Nardi, F., Emerson, B.C., Frati, F. and Fanciulli, P.P. (2010). Deep phylogenetic divisions and long-term persistence of forest invertebrates (Hexapoda: Collembola) in the North-Western Mediterranean basin. *Molecular Ecology*, 19: 386–400.

Cicconardi, F., Fanciulli, P.P. and Emerson, B.C. (2013). Collembola, the biological species concept and the underestimation of global species richness. *Molecular Ecology*, 22: 5382–5396.

Clarke, R.D. and Grant, P.R. (1968). An experimental study of the role of spiders as predators in a forest litter community. Part I. *Ecology*, 49: 1152–1154.

Cloudsley-Thompson, J.L. (1962). Microclimates and the distribution of terrestrial arthropods. *Annual Review of Entomology*, 7: 199–222.

Cobbett, C. and Goldsbrough, P. (2002). Phytochelatins and metallothioneins: roles in heavy metal detoxification and homeostasis. *Annual Review of Plant Biology*, 53: 159–182.

Coddington, J.A. and Levi, H.W. (1991). Systematics and evolution of spiders (Araneae). *Annual Review of Ecology and Systematics*, 22: 565–592.

Cokendolpher, J.C. (1993). Pathogens and parasites of Opiliones (Arthropoda: Arachnida). *Journal of Arachnology*, 21: 120–146.

Coleman, D.C., Callaham Jr., M.H. and Crossley Jr. D.A. (2017). *Fundamentals of Soil Ecology, 3rd Edition*. Academic Press, London.

Collado, R., Has-Cordes, E. and Schmelz, R.M. (2012). Microtaxonomy of fragmenting *Enchytraeus* species using molecular markers, with a comment on species complexes in enchytraeids. *Turkish Journal of Zoology*, 36: 85–94.

Collins, G.E., Hogg, I.D., Convey, P., Barnes, A.D. and McDonald, I.R. (2019). Spatial and temporal scales matter when assessing the species and genetic diversity of springtails (Collembola) in Antarctica. *Frontiers in Ecology and Evolution*, 7: 76.

Collins, G.E., Hogg, I.D., Convey, P., Sancho, L.G., Cowan, D.A., Lyons, W.B. et al. (2020). Genetic diversity of soil invertebrates corroborates timing estimates for past collapses of the West Antarctic Ice Sheet. *Proceedings of the National Academy of Sciences of the United States of America*, 117: 22293–22302.

Convey, P. (1994). Growth and survival strategy of the Antarctic mite *Alaskozetes antarcticus*. *Ecography*, 17: 97–107.

Convey, P. (1996). Overwintering strategies of terrestrial invertebrates in Antarctica—the significance of flexibility in extremely seasonal environments. *European Journal of Entomology*, 93: 489–505.

Cook, C.E., Yue, Q. and Akam, M. (2005). Mitochondrial genomes suggest that hexapods and crustaceans are mutually paraphyletic. *Proceedings of the Royal Society of London, Series B*, 272: 1295–1304.

Cook Hill, R., Egydio de Carvalho, C., Salogiannis, J., Schlager, B., Pilgrim, D. and Haag, E.S. (2006). Genetic flexibility in the convergent evolution of hermaphroditism in *Caenorhabditis elegans*. *Developmental Cell*, 10: 531–538.

Copley, J. (2000). Ecology goes underground. *Nature*, 406: 452–454.

Cordaux, R., Michel-Salzat, A. and Bouchon, D. (2001). *Wolbachia* infection in crustaceans: novel hosts and potential routes for horizontal transmission. *Journal of Evolutionary Biology*, 14: 237–243.

Cordaux, R., Michel-Salzat, A., Frelon-Raimond, M., Rigaud, T. and Bouchon, D. (2004). Evidence for a new feminizing *Wolbachia* strain in the isopod *Armadillidium vulgare*: evolutionary implications. *Heredity*, 93: 78–84.

Costa, D., Timmermans, M.J.T.N., Sousa, J.P., Ribeiro, R., Roelofs, D. and Van Straalen, N.M. (2013). Genetic structure of soil invertebrate populations: Collembolans, earthworms and isopods. *Applied Soil Ecology*, 68: 61–66.

Cottarelli, V., Bruno, M.C., Spena, M.T. and Grasso, R. (2012). Studies on subterranean copepods from Italy, with descriptions of two new epikrastic species from a cave in Sicily. *Zoological Studies*, 51: 556–582.

Crisp, A., Boschetti, C., Perry, M., Tunnacliffe, A. and Micklem, G. (2015). Expression of multiple horizontally acquired genes is a hallmark of both vertebrate and invertebrate genomes. *Genome Biology*, 16: 50.

Croll, R.P. (2009). Developing nervous systems in molluscs: navigating the twists and turns of a complex life cycle. *Brain Behaviour and Evolution*, 74: 164–176.

Crommentuijn, T., Brils, J. and Van Straalen, N.M. (1993). Influence of cadmium on life-history characteristics of *Folsomia candida* (Willem). *Ecotoxicology and Environmental Safety*, 26: 216–227.

Crommentuijn, T., Stäb, J.A., Doornekamp, A., Estoppey, O. and Van Gestel, C.A.M. (1995). Comparative ecotoxicity of cadmium, chlorpyrifos and triphenyltin hydroxide for four clones of the parthenogenetic collembolan *Folsomia candida* in an artificial soil. *Functional Ecology*, 9: 734–742.

Crowe, J.H., Hoekstra, F.A. and Crowe, L.M. (1992). Anhydrobiosis. *Annual Review of Physiology*, 54: 579–599.

Crowther, T.W., Maynard, D.S., Leff, J.W., Oldfield, E.E., McCulley, R.L., Fierer, N. et al. (2014). Predicting the responsiveness of soil biodiversity to deforestation: a cross-biome study. *Global Change Biology*, 20: 2983–2994.

Csonka, D., Halasy, K., Buczkó, K. and Hornung, E. (2018). Mophological traits—desiccation resistance—habitat characteristis: a possible key for distribution in woodlice (Isopoda, Oniscidae). *ZooKeys*, 801: 481–499.

Culver, D.C. (2017). Kenneth A. Christiansen (1924–2017). *Subterranean Biology*, 24: 53–61.

Curry, J.P. (1994). *Grassland Invertebrates*. Chapman and Hall, Inc., London.

Curry, J.P., Byrne, D. and Boyle, K.E. (1995). The earthworm population of a winter wheat cereal field and its effects on soil and nitrogen turnover. *Biology and Fertility of Soils*, 19: 166–172.

Czarnetzki, A.B. and Tebbe, C.C. (2004). Detection and phylogenetic analysis of *Wolbachia* in Collembola. *Environmental Microbiology*, 6: 35–44.

Czechowski, P., Sands, C.J., Adams, B.J., D'Haese, C.A., Gibson, J.A.E., McInnes, S.J. et al. (2012). Antarctic Tardigrada: a first step in understanding molecular operational taxonomic units (MOTUs) and biogeography of cryptic meiofauna. *Invertebrate Systematics*, 26: 526–538.

Dabert, M., Witalinski, W., Kazmierski, A., Olszanowski, Z. and Dabert, J. (2010). Molecular phylogeny of acariform mites (Acari, Arachnida): Strong conflict between phylogenetic signal and long-branch attraction artifacts. *Molecular Phylogenetics and Evolution*, 56: 222–241.

Daborn, P.J., Yen, J.L., Bogwitz, M.R., Le Goff, G., Feil, E., Jeffers, S. et al. (2002). A single P450 allele associated with insecticide resistance in *Drosophila*. *Science*, 297: 2253–2256.

Dallai, R. and Burroni, D. (1982). Fine structure of the pyloric region and malpighian papillae of Diplura. *Memorie della Società entomologica italiana*, 60: 125–135.

Dallai, R., Sbordoni, V., Cobolli Sbordoni, M. and De Matthaeis, E. (1983). Chromosome and enzyme diversity in some species of Neanuridae (Collembola). *Pedobiologia*, 25: 301–311.

Dallai, R., Fanciulli, P.P. and Frati, F. (2000). Aberrant spermatogenesis and the peculiar mechanism of sex determination in symphypleonan Collembola. *The Journal of Heredity*, 91: 351–358.

Dallai, R., Fanciulli, P.P., Carapelli, A. and Frati, F. (2001). Aberrant spermatogenesis and sex determination in Bourletiellidae (Hexapoda, Collembola), and their evolutionary significance. *Zoomorphology*, 120: 237–245.

Dallinger, R., Berger, B., Hunziker, P. and Kägi, J.H.R. (1997). Metallothionein in snail Cd and Cu metabolism. *Nature*, 388: 237–238.

Dallinger, R., Berger, B., Gruber, C., Hunziker, P. and Stürzenbaum, S. (2000). Metallothioneins in terrestrial invertebrates: structural aspects, biological significance and implications for their use as biomarkers. *Cellular and Molecular Biology*, 46: 331–346.

Damen, W.G.M., Saridaki, T. and Averof, M. (2002). Diverse adaptations of an ancestral gill: a common evolutionary origin for wings, breathing organs, and spinnerets. *Current Biology*, 12: 1711–1716.

Damuth, J. (1981). Population density and body size in mammals. *Nature*, 290: 699–700.

Darwin, C. (1859). *The Origin of Species*. John Murray, London.

Darwin, C. (1871). *The Descent of Man, and Selection in Relation to Sex.* John Murray, London.

Darwin, C. (1881). *The Formation of Vegetable Mound through the Action of Worms*. Cambridge University Press, Cambridge.

Dash, M.C. and Patra, U.C. (1977). Density, biomass and energy budget of a tropical earthworm population from a grassland site in Orissa, India. *Revue d'Ecologie et de Biologie du Sol*, 14: 461–471.

David, J.-F., Célérier, M.-L. and Geoffroy, J.-J. (1999). Periods of dormancy and cohort-splitting in the millipede *Polydesmus angustus* (Diplopoda: Polydesmidae). *European Journal of Entomology*, 96: 111–116.

David, J.-F., Geoffroy, J.-J. and Célérier, M.-L. (2003). First evidence for photoperiodic regulation of the life cycle in a millipede species, *Polydesmus angustus* (Diplopoda: Polydesmidae). *Journal of Zoology*, 260: 111–116.

David, J.-F. (2009). Female reproductive patterns in the millipede *Polydesmus angustus* (Diplopoda: Polydesmidae) and their significance for cohort-splitting. *European Journal of Entomology*, 106: 211–216.

David, J.-F. and Geoffroy, J.-J. (2011). Cohort-splitting in the millipede *Polydesmus angustus* (Diplopoda: Polydesmidae): No evidence for maternal effects on life-cycle duration. *European Journal of Entomology*, 108: 371–376.

Davidson, J. (1931). The influence of temperature on the incubation period of the eggs of *Sminthurus viridis* L. (Collembola). *Australian Journal of Experimental Biology and Medical Sciences*, 8: 143–152.

Davis, B.N.K. and Murphy, P.W. (1961). An analysis of the Acarina and Collembola fauna of land reclaimed from opencast iron-stone mining. *University of Nottingham School of Agriculture Report*, 3–7.

Davis, B.N.K. (1963). A study of micro-arthropod communities in mineral soils near Corby, Northants. *Journal of Animal Ecology*, 32: 49–71.

Davis, A.K. and Prouty, C. (2019). The sicker the better: nematode-infested passalus beetles provide enhanced ecosystem services. *Biology Letters*, 15: 20180842.

De Boer, P.A.C.M., Jansen, R.F., Koene, J.M. and Ter Maat, A. (1997). Nervous control of male sexual drive in the hermaphroditic snail *Lymnaea stagnalis*. *The Journal of Experimental Biology*, 200: 941–951.

De Deyn, G.B., Raaijmakers, C.E., Zoomer, H.R., Berg, M.P., De Ruiter, P.C., Verhoef, H.A. et al. (2003). Soil invertebrate fauna enhances grassland succession and diversity. *Nature*, 422: 711–713.

Degawa, Y. (2009). Secondary spore formation in *Orchesellaria mauguioi* (Asellariales, Trichomycetes) and its taxonomic and ecological implications. *Myoscience*, 50: 247–252.

Dehal, P., Satou, Y., Campbell, R.K., Chapman, J., Degnan, B., De Tomaso, A. et al. (2002). The draft genome of *Ciona intestinalis*: insights into the chordate and vertebrate origins. *Science*, 298: 2157–2167.

De Jong, G. (1995). Phenotypic plasticity as a product of selection in a variable environment. *The American Naturalist*, 145: 493–512.

De Knecht, J.A., Stroomberg, G.J., Tump, C., Helms, M., Verweij, R.A., Commandeur, J. et al. (2001). Characterization of enzymes involved in biotransformation of polycyclic aromatic hydrocarbons in terrestrial isopods. *Environmental Toxicology and Chemistry*, 20: 1457–1464.

Del Campo, J., Heger, T.J., Rodríguez-Martínez, R., Worden, A.Z., Richards, T.A., Massana, R. et al. (2019). Assessing the diversity and distribution of apicomplexans in host and free-living environments using high-throughput amplicon data and a phylogenetically informed reference framework. *Frontiers in Microbiology*, 10: 2373.

Delgado-Baquerizo, M., Oliverio, A.M., Brewer, T.E., Benavent-González, A., Eldridge, D.J., Bardgett, R.D. et al. (2018). A global atlas of the dominant bacteria found in soil. *Science*, 359: 320–325.

De Lima e Silva, C., Brennan, N., Brouwer, J.M., Commandeur, D., Verweij, R.A. and Van Gestel, C.A.M. (2017). Comparative toxicity of imidacloprid and thiacloprid to different species of soil invertebrates. *Ecotoxicology*, 26: 555–564.

De Lima e Silva, C., Van Haren, C., Mainardi, G., De Rooij, W., Ligtelijn, M., Van Straalen, N.M. et al. (2021). Bringing ecology into ecotoxicology: Life-cycle toxicity of two neonicotinoids to four species of springtails in LUFA 2.2 natural soil. *Chemosphere*, 263: 128245.

De Meeûs, T., Prugnolle, F. and Agnew, P. (2007). Asexual reproduction: Genetics and evolutionary aspects. *Cellular and Molecular Life Sciences*, 64: 1355–1372.

Den Boer, P.J. (1968). Spreading of risk and stabilization of animal numbers. *Acta Biotheoretica*, 18: 165–194.

Den Boer, P.J. (1970). Stabilization of animal numbers and the heterogeneity of the environment: The problem of the persistence of sparse populations. pp. 77–97. *In:* Den Boer, P.J. and Gradwell, G.R. (eds.). *Dynamics of Populations*. Pudoc, Wageningen.

Den Boer, P.J. (1981). On the survival of populations in a heterogeneous and variable environment. *Oecologia*, 50: 39–53.

Denneman, C.A.J. and Van Straalen, N.M. (1991). The toxicity of lead and copper in reproduction tests using the oribatid mite *Platynothrus peltifer*. *Pedobiologia*, 35: 305–311.

Denver, D.R., Clark, K.A. and Raboin, M.J. (2011). Reproductive mode evolution in nematodes: Insights from molecular phylogenies and recently discovered species. *Molecular Phylogenetics and Evolution*, 61: 584–592.

Dermauw, W. and Van Leeuwen, T. (2014). The ABC gene family in arthropods: Comparative genomics and role in insecticide transport and resistance. *Insect Biochemistry and Molecular Biology*, 45: 89–110.

De Ruiter, P.C., Ouborg, N.J. and Ernsting, G. (1988). Density dependent mortality in the springtail species *Orchesella cincta* due to predation by the carabid beetle *Notiophilus biguttatus*. *Entomologia experimentalis et applicata*, 48: 25–30.

De Ruiter, P.C., Neutel, A.-M. and Moore, J.C. (1994). Modelling food webs and nutrient cycling in agro-ecosystems. *Trends in Ecology and Evolution*, 9: 378–383.

De Ruiter, P.C., Neutel, A.-M. and Moore, J.C. (1995). Energetics, patterns of interaction strength, and stability in real ecosystems. *Science*, 269: 1257–1260.

D'Haese, C.A. (2002). Were the first springtails semi-aquatic? A phylogenetic approach by means of 28S rDNA and optimization alignment. *Proceedings of the Royal Society of London, Series B*, 269: 1143–1151.

Dhillon, B.S. and Gibson, N.H.E. (1962). A study of the Acarina and Collembola of agricultural soils. I. Numbers and distribution in undisturbed grassland. *Pedobiologia*, 1: 189–209.

Dias, A.T.C., Krab, E.J., Mariën, J., Zimmer, M., Cornelissen, J.H.C., Ellers, J. et al. (2013). Traits underpinning desiccation resistance explain distribution patterns of terrestrial isopods. *Oecologia*, 172: 667–677.

Dias, B.C., Da Silva Souza, E., Hara, M.R. and Willemart, R.H. (2014). Intense leg tapping behavior by the harvestman *Mischonyx cuspidatus* (Gonyleptidae): an undescribed defensive behavior in Opiliones? *The Journal of Arachnology*, 42: 123–125.

Díaz Cosín, D.J., Ruiz, M.P., Ramajo, M. and Gutierrez, M. (2006). Is the aestivation of the earthworm *Hormogaster elisae* a paradiapause? *Invertebrate Biology*, 125: 250–255.

Díaz Cosín, D.J., Novo, M. and Fernández, R. (2011). Reproduction of earthworms: Sexual selection and parthenogenesis. pp. 69–86. *In:* Karaca, A. (ed.). *Biology of Earthworms, Soil Biology 24*. Springer Verlag, Berlin-Heidelberg.

Didden, W.A.M. (1993). Ecology of terrestrial Enchytraeidae. *Pedobiologia*, 37: 2–29.

Diekkrüger, B. and Röske, H. (1995). Modelling the dynamics of *Isotoma notabilis* (Collembola) on sites of different agricultural usage. *Pedobiologia*, 39: 58–73.

Dietrich, C., Köhler, T. and Brune, A. (2014). The cockroach origin of the termite gut microbiota: patterns in bacterial community structure reflect major evolutionary events. *Applied and Environmental Microbiology*, 80: 2261–2269.

Dietz, T.H. and Alvarado, R.H. (1970). Osmotic and ionic regulation in *Lumbricus terrestris* L. *Biological Bulletin*, 138: 247–261.

Dimitriou, A.C., Taiti, S. and Sfenthourakis, S. (2019). Genetic evidence against monophyly of Oniscidea implies a need to revise scenarios for the origin of terrestrial isopods. *Scientific Reports*, 9: 18508.

Dimitrova, Z.M. (2009). Occurrence of cystacanths of *Plagiorhynchus cylindraceus* (Acanthocephala) in the terrestrial isopods *Trachelipus squamuliger* and *Armadillidium vulgare* (Oniscidea) in Bulgaria. *Acta Parasitologica*, 54: 53–56.

Dindal, D.L. (ed.) (1980). Soil Biology as Related to Land Use Practices. *Proceedings of the VIIth International Colloquium of Soil Zoology*. US-EPA Office of Pesticides and Toxic Substances, Washington.

Dittmer, J., Lesobre, J., Moumen, B. and Bouchon, D. (2016). Host origin and tissue microhabitat shaping the microbiota of the terrestrial isopod *Armadillidium vulgare*. *FEMS Microbiology Ecology*, 92: fiw063.

Do, T.H., Nguyen, T.T., Nguyen, T.N., Le, Q.G., Nguyen, C., Kimura, K. et al. (2014). Mining biomass-degrading enzymes through Illumina-based *de novo* sequencing and metagenomic analysis of free-living bacteria in the gut of the lower termite *Coptotermes gestroi* in Vietnam. *Journal of Bioscience and Bioengineering*, 118: 665–671.

Domes, K., Althammer, M., Norton, R.A., Scheu, S. and Maraun, M. (2007a). The phylogenetic relationship between Astigmata and Oribatida (Acari) as indicated by molecular markers. *Experimental and Applied Acarology*, 42: 159–171.

Domes, K., Norton, R.A., Maraun, M. and Scheu, S. (2007b). Reevolution of sexuality breaks Dollo's law. *Proceedings of the National Academy of Sciences of the United States of America*, 104: 7139–7144.

Domínguez, J. and Edwards, C.A. (1997). Effects of stocking rate and moisture content on the growth and maturation of *Eisenia andrei* (Oligochaeta) in pig manure. *Soil Biology and Biochemistry*, 29: 743–747.

Domínguez, J., Velando, A., Aira, M. and Monroy, F. (2003). Uniparental reproduction of *Eisenia fetida* and *E. andrei* (Oligochaeta: Lumbricidae). *Pedobiologia*, 47: 530–534.

Domínguez, J. and Velando, A. (2013). Sexual selection in earthworms: Mate choice, sperm competition, differential allocation and partner manipulation. *Applied Soil Ecology*, 69: 21–27.

Domínguez, J., Aira, M., Breinholt, J.W., Stojanovich, M., James, S.W. and Pérez-Losada, M. (2015). Underground evolution: New roots for the old tree of lumbricid earthworms. *Molecular Phylogenetics and Evolution*, 83: 7–19.

Doube, B.M. and Schmidt, O. (1997). Can the abundance or activity of soil macrofauna be used to indicate the biological health of soils? pp. 265–295. *In:* Pankhurst, C.E., Doube, B.M. and Gupta, V.V.S.R. (eds.). *Biological Indicators of Soil Health*. CAB International, Wallingford.

Dózsa-Farkas, K. (1995). *Enchytraeus dudichi* sp. n., a new fragmenting *Enchytraeus* species from Iran. *Opuscula Zoologica Budapest*, 27-28: 41–44.

Drake, H.L. and Horn, M.A. (2007). As the worm turns: the earthworm gut as a transient habitat for soil microbiomes. *Annual Review of Microbiology*, 61: 169–189.

Driessen, G., Ellers, J. and Van Straalen, N.M. (2007). Variation, selection and heritability of thermal reaction norms for juvenile growth in *Orchesella cincta*. *European Journal of Entomology*, 104: 39–46.

Dromph, K.M., Eilenberg, J. and Esbjerg, P. (2001). Natural occurrence of entomophthoralean fungi pathogenic to collembolans. *Journal of Invertebrate Pathology*, 78: 226–231.

Duhamel, M., Pel, R., Ooms, A., Bücking, H., Jansa, J., Ellers, J. et al. (2013). Do fungivores trigger the transfer of protective metabolites from host plants to arbuscular mycorrhizal hyphae? *Ecology*, 94: 2019–2029.

Dumke, M. (2016). Extended maternal care and offspring interactions in the subsocial Australian crab spider, *Xysticus bimaculatus*. *Australian Journal of Zoology*, 64: 344–352.

Dumont, H.J. (1995). The evolution of groundwater Cladocera. *Hydrobiologia*, 307: 69–74.

Dumont, H.J. and Negrea, S. (1996). A conspectus of the Cladocera of the subterranean waters of the world. *Hydrobiologia*, 325: 1–30.

Dumpert, K. and Platen, R. (1985). Zur Biologie eines Buchenwaldbodens 4. Die Spinnenfauna. *Carolinea*, 42.

Dunger, W., Schulz, H.-J. and Zimdars, B. (2002). Colonization behaviour of Collembola under different conditions of dispersal. *Pedobiologia*, 46: 316–327.

Dunlop, J.A. and Alberti, G. (2007). The affinities of mites and ticks: a review. *Journal of Zoological Systematics and Evolutionary Research*, 46: 1–18.

Du Pasquier, L. (2001). The immune system of invertebrates and vertebrates. *Comparative Biochemistry and Physiology Part B*, 129: 1–15.

Dupont, L., Grésille, Y., Richard, B., Decaëns, T. and Mathieu, J. (2015). Dispersal constraints and fine-scale spatial genetic structure in two earthworm species. *Biological Journal of the Linnean Society*, 114: 335–347.

Dupont, L., Torres-Leguizamon, M., René-Corail, P. and Mathieu, J. (2017). Landscape features impact connectivity between soil populations: a comparative study of gene flow in earthworms. *Molecular Ecology*, 26: 3128–3140.

Dupont, L., Pauwels, M., Dume, C., Deschins, V., Audusseau, H., Gigon, A. et al. (2019). Genetic variation of the epigeic earthworm *Lumbricus castaneus* populations in urban soils of the Paris region (France) revealed using eight newly developed microsatellite markers. *Applied Soil Ecology*, 135: 33–37.

Dupont-Nivet, M., Guiller, A. and Bonnet, J.C. (1997). Genetic and environmental variability of adult size in some stocks of the edible snail, *Helix aspersa*. *Journal of Zoology (London)*, 241: 757–765.

Dupont-Nivet, M., Mallard, J., Bonnet, J.C. and Blanc, J.M. (1998). Quantitative genetics of reproductive traits in the edible snail *Helix aspersa* Müller. *Journal of Experimental Zoology*, 281: 220–227.

Durkin, E.S., Cassidy, S.T., Gilbert, R., Richardson, E.A., Roth, A.M. and Shablin, S. (2021). Parasites of spiders: Their impact on host behavior and ecology. *Journal of Arachnology*, 49: 281–298.

Duron, O., Hurst, G.D.D., Hornett, E.A., Josling, J.A. and Engelstädter, J. (2008). High incidence of the maternally inherited bacterium *Cardinium* in spiders. *Molecular Ecology*, 17: 1427–1437.

Eberhard, W.G. (2004). Why study spider sex: special traits of spiders facilitate studies of sperm competition and cryptic female choice. *The Journal of Arachnology*, 32: 545–556.

Edgecombe, G.D. and Giribet, G. (2007). Evolutionary biology of centipedes (Myriapoda: Chilopoda). *Annual Review of Entomology*, 52: 151–170.

Edgecombe, G.D. (2010). Arthropod phylogeny: an overview from the perspectives of morphology, molecular data and the fossil record. *Arthropod Structure and Development*, 39: 74–87.

Edgecombe, G.D., Bonato, L. and Giribet, G. (2010). Brooding in *Mecistocephalus togensis* (Geophilomorpha: Placodesmata) and the evolution of parental care in centipedes (Chilopoda). *International Journal of Myriapodology*, 3: 139–144.

Edgecombe, G.D., Strullu-Derrien, C., Góral, T., Hetherington, A.J., Thompson, C. and Koch, M. (2020). Aquatic stem group myriapods close a gap between molecular divergence dates and the terrestrial fossil record. *Proceedings of the National Academy of Sciences of the United States of America*, 117: 8966–8972.

Edney, E.B. (1954). Woodlice and the land habitat. *Biological Reviews of the Cambridge Philosophical Society*, 29: 185–219.

Edney, E.B. (1968). Transition from water to land in isopod crustaceans. *American Zoologist*, 8: 309–326.

Edney, E.B. (1977). *Water Balance in Land Arthropods*. Springer-Verlag, Berlin.

Edwards, C.A. and Lofty, J.R. (1972). *Biology of Earthworms*. Chapman and Hall Ltd, London.

Edwards, C.A. and Thompson, A.R. (1973). Pesticides and the soil fauna. *Residue Reviews*, 45: 1–79.

Edwards, C.A. (2004). *Earthworm Ecology. Second Edition*. St. Lucie Press, Boca Raton.

Egli, D., Selvaraj, A., Yepiskoposyan, H., Zhang, B., Hafen, E., Georgiev, O. et al. (2003). Knockout of "metal-responsive transcription factor" MTF-1 in *Drosophila* by homologous recombination reveals its central role in heavy metal homeostasis. *The EMBO Journal*, 22: 100–108.

Ehnes, R.B., Rall, B.C. and Brose, U. (2011). Phylogenetic grouping, curvature and metabolic scaling in terrestrial invertebrates. *Ecology Letters*, 14: 993–1000.

Ehrnsberger, R., Sterzynska, M. and Szeptycki, A. (1997). Apterygota of the North Sea marsh—community structure and vertical distribution. *Pedobiologia*, 41: 123–130.

Eichinger, E., Bruckner, A. and Stemmer, M. (2007). Earthworm expulsion by formalin has severe and lasting side effects on soil biota and plants. *Ecotoxicology and Environmental Safety*, 67: 260–266.

Eijsackers, H. (2011). Earthworms as colonizers of natural and cultivated soil environments. *Applied Soil Ecology*, 50: 1–13.

Eisenbeis, G. (1974). Licht- und elektronenmikroskopische Untersuchungen zur Ultrastruktur des Transportepithels am Ventraltubus arthropleoner Collembolen (Insecta). *Cytobiologie*, 9: 180–202.

Ellis, R.E. and Lin, S.-Y. (2014). The evolutionary origins and consequences of self-fertility in nematodes. *F100Prime Reports*, 6: 62.

Emlen, D.J., Marangelo, J., Ball, B. and Cunningham, C.W. (2005). Diversity in the weapons of sexual selection: horn evolution in the beetle genus *Onthophagus* (Coleoptera: Scarabaeidae). *Evolution*, 59: 1060–1084.

Emlen, D.J. and Philips, T.K. (2006). Phylogenetic evidence for an association between tunneling behaviour and the evolution of horns in dung beetles (Coleoptera: Scarabaeidae: Scarabaeinae). *Coleopterists Society Monograph*, 5: 47–56.

Emmanuel, N., Curry, J.P. and Evans, G.O. (1985). The soil Acari of barley plots with different cultural treatments. *Experimental and Applied Acarology*, 1: 101–113.

Engel, M.S. and Grimaldi, D.A. (2004). New light shed on the oldest insect. *Nature*, 427: 627–630.

Engel, J., Hertzog, L., Tiede, J., Wagg, C., Ebeling, A., Briesen, H. et al. (2017). Pitfall trap sampling bias depends on body mass, temperature, and trap number: insights from an individual-based model. *Ecosphere*, 8: e01790.

Engelmann, M.D. (1968). The role of soil arthropods in community energetics. *American Zoologist*, 8: 61–69.

Enghoff, H. (1984). Phylogeny of millipedes—a cladistic analysis. *Zeitschrift für zoologische Systematik und Evolutionsforschung*, 22: 8–26.

Eppley, S.M. and Jesson, L.K. (2008). Moving to mate: the evolution of separate and combined sexes in multicellular organisms. *Journal of Evolutionary Biology*, 21: 727–736.

Erdmann, W. and Kaczmarek, Ł. (2017). Tardigrades in space research—past and future. *Origins of Life and Evolution of Biospheres*, 47: 545–553.

Ernst, G. and Emmerling, C. (2009). Impact of five tillage systems on soil organic carbon and the density, biomass, and community composition of earthworms after a ten year period. *European Journal of Soil Biology*, 45: 247–251.

Ernst, C.M., Hanelt, B. and Buddle, C.M. (2016). Parasitism of ground beetles (Coleoptera: Carabidae) by a species of hairworm (Nematomorpha: Gordiida) in Arctic Canada. *Journal of Parasitology*, 102: 327–335.

Ernsting, G. and Jansen, J.W. (1978). Interspecific and intraspecific selection by the predator *Notiophilus biguttatus* F. (Carabidae) concerning two collembolan prey species. *Oecologia*, 33: 173–183.

Ernsting, G. and Mulder, A.J. (1981). Components of predatory behaviour underlying density-dependent prey-size selection by *Notiophilus biguttatus* F. (Carabidae, Coleoptera). *Oecologia*, 51: 169–174.

Ernsting, G. and Fokkema, D.S. (1983). Antennal damage and regeneration in springtails (Collembola) in relation to predation. *Netherlands Journal of Zoology*, 33: 476–484.

Ernsting, G. and Isaaks, A. (2002). Gamete production and sexual dimorphism in an insect (*Orchesella cincta*) with indeterminate growth. *Ecological Entomology*, 27: 145–151.

Erséus, C. and Källersjö, M. (2004). 18S rDNA phylogeny of Clitellata (Annelida). *Zoologica Scripta*, 33: 187–196.

Erséus, C. (2005). Phylogeny of oligochaetous Clitellata. *Hydrobiologia*, 535/536: 357–372.

Erséus, C., Rota, E., Matamoros, L. and De Wit, P. (2010). Molecular phylogeny of Enchytraeidae (Annelida, Clitellata). *Molecular Phylogenetics and Evolution*, 57: 849–858.

Erséus, C., Williams, B.W., Horn, K.M., Halanych, K.M., Santos, S.R., James, S.W. et al. (2020). Phylogenomic analyses reveal a Palaeozoic radiation and support a freshwater origin for clitellate annelids. *Zoologica Scripta*, 49: 614–640.

Erwin, D.H., Laflamme, M., Tweedt, S.M., Sperling, E.A., Pisani, D. and Peterson, K.J. (2011). The Cambrian conundrum: early divergence and later ecological success in the early history of animals. *Science*, 334: 1091–1097.

Espalader, X. and Santamaria, S. (2012). Ecto- and endoparasitic fungi on ants from the Holarctic region. *Psyche*, 2012: 168478.

Ettema, C.H., Coleman, D.C., Vellidis, G., Lowrance, R. and Rathburn, S.L. (1998). Spatiotemporal distributions of bacterivorous nematodes and soil resources in a restored riparian wetland. *Ecology*, 79: 2721–2734.

Ettema, C. and Wardle, D.A. (2002). Spatial soil ecology. *Trends in Ecology and Evolution*, 17: 177–183.

Evans, K. and Stone, A.R. (1977). A review of the distribution and biology of the potato cyst-nematodes *Globodera rostochiensis* and *G. pallida*. *International Journal of Pest Management*, 23: 178–189.

Faber, J.H. (1991). Functional classification of soil fauna: a new approach. *Oikos*, 62: 110–117.

Faddeeva, A., Studer, R.A., Kraaijeveld, K., Ylstra, B., Mariën, J., Op den Camp, H.J.M. et al. (2015). Collembolan transcriptomes highlight molecular evolution of hexapods and provide clues to the adaptation to terrestrial life. *PLoS One*, 10: e0130600.

Faddeeva-Vakhrusheva, A., Derks, M.F.L., Yahya Anvar, S., Agamennone, V., Suring, W., Smit, S. et al. (2016). Gene family evolution reflects adaptation to soil environmental stressors in the genome of the collembolan *Orchesella cincta*. *Genome Biology and Evolution*, 8: 2106–2117.

Faddeeva-Vakhrusheva, A., Kraaijeveld, K., Derks, M.F.L., Yahya Anvar, S., Agamennone, V., Suring, W. et al. (2017). Coping with living in the soil: the genome of the parthenogenetic springtail *Folsomia candida*. *BMC Genomics*, 18: 493.

Fain, A. and Jocqué, R. (1996). A new larva of the genus *Leptus* latreille, 1796 (Acari: Erythraeidae) parasitic on a spider from rwanda. *International Journal of Acarology*, 22: 101–108.

Fallon, A.M. (2021). Growth and maintenance of *Wolbachia* in insect cell lines. *Insects*, 12: 706.

Fanciulli, P.P., Frati, F., Dallai, R. and Rusek, J. (1991). High genetic divergence among populations of *Tetradontophora bielanensis* (Insecta, Collembola) in Europe. *Revue d'Écologie et de Biologie du Sol*, 28: 165–173.

Fanciulli, P.P., Gaju, M., Bach, C. and Frati, F. (1997). Genetic differentiation and detection of cryptic species in the genus *Lepismachilis* (Insecta: Microcoryphia) from the Western Mediterranean region. *Biological Journal of the Linnean Society*, 62: 533–551.

Fanciulli, P.P., Melegari, D., Carapelli, A., Frati, F. and Dallai, R. (2000). Population structure, gene flow and evolutionary relationships in four species of the genera *Tomocerus* and *Pogonognathellus* (Collembola, Tomoceridae). *Biological Journal of the Linnean Society*, 70: 221–238.

Fanciulli, P.P., Summa, D., Dallai, R. and Frati, F. (2001). High levels of genetic variability and population differentiation in *Gressittacantha terranova* (Collembola, Hexapoda) from Victoria Land, Antarctica. *Antarctic Science*, 13: 246–254.

Farfan, M.A. and Klompen, H. (2012). Phoretic mite associates of millipedes (Diplopoda, Julidae) in the northern Atlantic region (North America, Europe). *International Journal of Myriapodology*, 7: 62–91.

Faurby, S., Jørgensen, A., Kristensen, R.M. and Funch, P. (2012). Distribution and speciation in marine intertidal tardigrades: testing the roles of climatic and geographical isolation. *Journal of Biogeography*, 39: 1596–1607.

Feder, M.E. and Hofmann, G.E. (1999). Heat shock proteins, molecular chaperones and the stress response: evolutionary and ecological physiology. *Annual Review of Physiology*, 61: 243–282.

Feller, C. (1997). The concept of humus in the past three centuries. *Advances in GeoEcology*, 29: 15–46.

Feng, H., Guo, J., Wang, W., Song, X. and Yu, S. (2019). Soil depth determines the composition and diversity of bacterial and archaeal communities in a poplar plantation. *Forests*, 10: 550.

Fernández, R., Almodóvar, A., Novo, M., Simancas, B. and Díaz Cosín, D.J. (2012). Adding complexity to the complex: New insights into the phylogeny, diversification and origin of parthenogenesis in the *Aporrectodea caliginosa* species complex. *Molecular Phylogenetics and Evolution*, 64: 368–379.

Fernández, R., Edgecombe, G.D. and Giribet, G. (2017). Phylogenomics illuminates the backbone of the Myriapoda Tree of Life and reconciles morphological and molecular phylogenies. *Scientific Reports*, 8: 83.

Feyereisen, R. (2006). Evolution of insect P450. *Biochemical Society Transactions*, 34: 1252–1255.

Field, S.G., Schirp, H.J. and Michiels, N.K. (2003). The influence of *Monocystis* sp. infection on growth and mating behaviour of the earthworm *Lumbricus terrestris*. *Canadian Journal of Zoology*, 81: 1161–1167.

Field, S.G., Kurtz, J., Cooper, E.L. and Michiels, N.K. (2004). Evaluation of an innate immune reaction to parasites in earthworms. *Journal of Invertebrate Pathology*, 86: 45–49.

Field, S.G. and Michiels, N.K. (2005). Parasitism and growth in the earthworm *Lumbricus terrestris*: fitness costs of the gregarine parasite *Monocystis* sp. *Parasitology*, 130: 397–403.

Field, S.G. and Michiels, N.K. (2006). Does the acephaline gregarine *Monocystis* sp. modify the surface behaviour of its earthworm host *Lumbricus terrestris*? *Soil Biology and Biochemistry*, 38: 1334–1339.

Fierer, N., Grandy, A.S., Six, J. and Paul, E.A. (2009). Searching for unifying principles in soil ecology. *Soil Biology and Biochemistry*, 41: 2249–2256.

Fierer, N. (2017). Embracing the unknown: disentangling the complexities of the soil microbiome. *Nature Reviews Microbiology*, 15: 579–589.

Fierer, N. (2019). Earthworms' place on Earth. *Science*, 366: 425–426.

Fischer, B.M., Schatz, H. and Maraun, M. (2010). Community structure, trophic position and reproductive mode of soil and bark-living oribatid mites in an alpine grassland ecosystem. *Experimental and Applied Acarology*, 52: 221–237.

Fisker, K.V., Overgaard, J., Sørensen, J.G., Slotsbo, S. and Holmstrup, M. (2014). Roles of carbohydrate reserves for local adaptation to low temperatures in the freeze tolerant oligochaete *Enchytraeus albidus*. *Journal of Comparative Physiology B*, 184: 167–177.

Fjellberg, A. (1975). Organization and dynamics of Collembola populations on Hardangervidda. pp. 73–79. *In:* Wielgolaski, F.E. (ed.). *Fennoscandian Tundra Ecosystems. Part 2 Animals and Systems Analysis. Ecological Studies 17*. Springer Verlag, Berlin.

Fjellberg, A. (1978). Generic switch-over in *Isotoma nivea* Schäffer, 1896. A new case of cyclomorphosis in Collembola (Isotomidae). *Norwegian Journal of Entomology*, 25: 221–222.

Fog, K. (1979). Studies on decomposing wooden stumps. III. Different relations among some gastropod species and species groups to the stump microflora, weather changes and pH. *Pedobiologia*, 19: 200–212.

Fortin, M.-J. and Dale, M. (2005). *Spatial Analysis. A Guide for Ecologists*. Cambridge University Press, Cambridge.

Fountain, M.T. and Hopkin, S.P. (2005). *Folsomia candida* (Collembola): a "standard" soil arthropod. *Annual Review of Entomology*, 50: 201–222.

Franklin, I.R. and Allendorf, F.W. (2014). The 50/500 rule is still valid - Reply to Frankham et al. *Biological Conservation*, 176: 284–285.

Frati, F., Spinsanti, G. and Dallai, R. (2001). Genetic variation of mtCOII gene sequences in the collembolan *Isotoma klovstadi* from Victoria Land, Antarctica: evidence for population differentiation. *Polar Biology*, 24: 934–940.

Frati, F., Negri, I., Fanciulli, P.P., Pellecchia, M., De Paola, V., Scali, V. et al. (2004). High levels of genetic differentiation between *Wolbachia*-infected and non-infected populations of *Folsomia candida* (Collembola, Isotomidae). *Pedobiologia*, 48: 461–468.

Freedman, J.H., Slice, L.W., Dixon, D., Fire, A. and Rubin, C.S. (1993). The novel metallothionein genes of *Caenorhabditis elegans*. Structural organization and inducible, cell-specific expression. *The Journal of Biological Chemistry*, 268: 2554–2564.

Freeman, J.A. (1952). Occurrence of Collembola in the air. *Proceedings of the Royal Entomological Society of London, Series A*, 27: 28.

Friend, J.A. and Richardson, A.M.M. (1986). Biology of terrestrial amphipods. *Annual Review of Entomology*, 31: 25–48.

Frith, D. and Frith, C. (1990). Seasonality of litter invertebrate populations in an Australian upland tropical rain forest. *Biotropica*, 22: 181–190.

Frouz, J. (1999). Use of soil dwelling Diptera (Insecta, Diptera) as bioindicators: a review of ecological requirements and response to disturbance. *Agriculture, Ecosystems and Environment*, 74: 167–186.

Frouz, J., Jedlička, P., Šimáčková, H. and Lhotákova, Z. (2015). The life cycle, population dynamics, and contribution to litter decomposition of *Penthetria holosericea* (Diptera: Bibionidae) in an alder forest. *European Journal of Soil Biology*, 71: 21–27.

Frydenberg, J., Hoffmann, A.A. and Loeschke, V. (2003). DNA sequence variation and latitudinal associations in *hsp23*: *hsp26* and *hsp27* from natural populations of *Drosophila melanogaster*. *Molecular Ecology*, 12: 2025–2032.

Fussey, G.D. (1984). The distribution of the two forms of the woodlouse *Trichoniscus pusillus* Brandt (Isopoda: Oniscoidea) in the British Isles: a reassessment of geographic parthenogenesis. *Biological Journal of the Linnean Society*, 23: 309–321.

Gabbutt, P.D. (1967). Quantitative sampling of the pseudoscorpion *Chthonius ischnocheles* from beech litter. *Journal of Zoology*, 151: 469–478.

Gabbutt, P.D. (1969). Life-histories of some British pseudoscorpions inhabiting leaf litter. pp. 229–235. *In:* Sheals, J.G. (ed.). *The Soil Ecosystem*. The Systematics Association, London.

Gal, T.Z., Glazer, I. and Koltai, H. (2005). Stressed worms: Responding in the post-genomics era. *Molecular and Biochemical Parasitology*, 143: 1–5.

Galli, L., Capurro, M., Colasanto, E., Molyneux, T., Murray, A., Torti, C. et al. (2020). A synopsis of the ecology of Protura (Arthropoda: Hexapoda). *Revue Suisse de Zoologie*, 126: 155–164.

García, L.F., Franco, V., Robledo-Ospina, L.E., Viera, C., Lacava, M. and Willemart, R.H. (2016). The predation strategy of the recluse spider *Loxosceles rufipes* (Lucas, 1834) against four prey species. *Journal of Insect Behavior*, 29: 515–526.

Garey, J.R., McInnes, S.J. and Nichols, P.B. (2008). Global diversity of tardigrades (Tardigrada) in freshwater. *Hydrobiologia*, 595: 101–106.

Garwood, R.J., Edgecombe, G.D., Charbonnier, S., Chabard, D., Sotty, D. and Giribet, G. (2016). Carboniferous Onychophora from Montceau-les-Mines, France, and onychophoran terrestrialization. *Invertebrate Biology*, 135: 179–190.

Geiger, R. (1961). *Das Klima der bodennahen Luftschicht. Ein Lehrbuch der Mikroklimatologie*. Friedr. Vieweg and Sohn, Braunschweig.

Geiger, R., Aron, R.H. and Todhunter, P. (2009). *The Climate Near the Ground. Seventh Edition.* Rowman and Littlefield Publishers, Washington D.C.

Gérard, G. and Berthet, P. (1966). A statistical study of microdistribution of Oribatei (Acari). Part II: The transformation of data. *Oikos*, 17: 142–149.

Gerisch, B., Tharyan, R.G., Mak, J., Denzel, S.I., Popkes-van Oepen, T., Henn, N. et al. (2020). HLH-30/TFEB is a master regulator of reproductive quiescence. *Developmental Cell*, 53: 316–329.

Gerth, M., Gansauge, M.-T., Weigert, A. and Bleidorn, C. (2014). Phylogenomic analyses uncover origin and spread of the *Wolbachia* pandemic. *Nature Communications*, 5: 5117.

Gervascio, T., Czekanski-Moir, J., Rundell, J. and Webster, N.B. (2017). Dating the terrestrial invasion of Cyclophoroidea (Mollusca: Gastropoda) using a fossilized birth-death model. Poster presented at the Evolution Meeting 2017 in Portland, OR, USA.

Ghilarov, M.S. (1949). *Peculiarities of Soil as an Environment and its Role in the Insect Evolution*. USSR Academy of Sciences Publishers, Moscow.

Ghilarov, M.S. (1958). L'importance du sol dans l'origine et l'evolution des insectes. pp. 443–451. *In:* Becker, E.C. (ed.). *Proceedings of the Tenth International Congress of Entomology* 17–25 August 1956 Mortimer, Montreal.

Ghiselin, M.T. (1969). The evolution of hermaphroditism among animals. *Quarterly Review of Biology*, 44: 189–208.

Gibert, J. (2001). Basic attributes of groundwater ecosystems. pp. 39–52. *In:* Griebler, C., Danielopol, D.L., Gibert, J., Nachtnebel, H.P. and Notenboom, J. (eds.). *Groundwater Ecology. A Tool for Management of Water Resources*. Office for Official Publications of the European Communities Luxembourg.

Gibson, J.S. and Uetz, G.W. (2008). Seismic communication and mate choice in wolf spiders: components of male seismic signals and mating success. *Animal Behaviour*, 75: 1253–1262.

Giller, P.S. (1996). The diversity of soil communities, the "poor man's tropical rainforest". *Biodiversity and Conservation*, 5: 135–168.

Gillooly, J.F., Brown, J.H., West, G.B., Savage, V.M. and Charnov, E.L. (2001). Effects of size and temperature on metabolic rate. *Science*, 293: 2248–2251.

Giribet, G., Edgecombe, G.D., Carpenter, J.M., D'Haese, C.A. and Wheeler, W.C. (2004). Is Ellipura monophyletic? A combined analysis of basal hexapod relationships with emphasis on the origin of insects. *Organisms Diversity and Evolution*, 4: 319–340.

Giribet, G. and Sharma, P.P. (2015). Evolutionary biology of harvestmen (Arachnida, Opiliones). *Annual Review of Entomology*, 60: 157–175.

Giribet, G. and Edgecombe, G.D. (2019). The phylogeny and evolutionary history of arthropods. *Current Biology*, 29: R592–R602.

Gisin, H. (1943). Ökologie und lebensgemeinschaften der Collembolen im Schweizerischen Exkursionsgebiet Basels. *Revue Suisse de Zoologie*, 50: 131–224.
Gisin, H. (1960). *Collembolenfauna Europas*. Musée d'Histoire Naturelle, Genève.
Gist, C.S., Crossley, D.A.J. and Merchant, V.A. (1974). An analysis of life tables for *Sinella curviseta* (Collembola). *Environmental Entomology*, 3: 840–844.
Gladyshev, E.A., Meselson, M. and Arhipova, I.R. (2008). Massive horizontal gene transfer in bdelloid rotifers. *Science*, 320: 1210–1213.
Glenner, H., Thomsen, P.F., Hebsgaard, M.B., Sørensen, M.V. and Willerslev, E. (2006). The origin of insects. *Science*, 314: 1883–1884.
Glime, J.M. (2017). Tardigrade densities and richness. pp. 5-5-2 - 5-5-21. *In:* Glime, J.M. (ed.). *Bryophyte Ecology, Volume 2: Bryological Interaction*. Michigan Technological University.
Goddard, S.J. (1976). Population dynamics, distribution patterns and life cycles of *Neobisium muscorum* and *Chthonius orthodactylus*. *Journal of Zoology*, 178: 295–304.
Goldbach, B. (2000). The evolutionary changes in the reproductive biology of the Archaeognatha (Insecta). *Zoologische Anzeiger*, 239: 215–229.
Goldstein, D.B. and Schlötterer, C. (1999). *Microsatellites. Evolution and Applications*. Oxford University Press, Oxford.
Golovatch, S.I. and Kime, R.D. (2009). Millipede (Diplopoda) distributions: a review. *Soil Organisms*, 81: 565–597.
Gols, R., Ernsting, G. and Van Straalen, N.M. (2004). Paternity analysis in a hexapod with indirect sperm transfer. *Journal of Insect Behavior*, 17: 317–328.
Gordon, S.D. and Uetz, G.W. (2011). Multimodal communication in wolf spiders on different substrates: evidence for behavioural plasticity. *Animal Behaviour*, 81: 367–375.
Goto, H.E. (1960). Facultative parthenogenesis in Collembola (Insecta). *Nature*, 188: 958–959.
Gould, S.J. (1977). *Ontogeny and Phylogeny*. The Belknap Press of Harvard University Press, Cambridge.
Gould, S.J. and Vrba, E.S. (1982). Exaptation – a missing term in the science of form. *Paleobiology*, 8: 4–15.
Graham, L.A. and Davies, P.L. (2005). Glycine-rich antifreeze proteins from snow fleas. *Science*, 310: 461.
Graham, L.A., Boddington, M.E., Holmstrup, M. and Davies, P.L. (2020). Antifreeze protein complements cryoprotective dehydration in the freeze-avoiding springtail *Megaphorura arctica*. *Scientific Reports*, 10: 3047.
Grandcolas, P., Guilbert, E., Robillard, T., D'Haese, C.A., Murienne, J. and Legendre, F. (2004). Mapping characters on a tree with or without the outgroups. *Cladistics*, 20: 579–582.
Green, C.D. (1964). The life history and fecundity of *Folsomia candida* (Willem) var. *distincta* (Bagnall) (Collembola: Isotomidae). *Proceedings of the Royal Entomological Society of London, Series A*, 39.
Greenslade, P. and Whalley, P.E.S. (1986). The systematic position of *Rhyniella praecursor* Hirst and Maurik (Collembola), the earliest known hexapod. pp. 319–323. *In:* Dallai, R. (ed.). *2nd International Seminar on Apterygota*. University of Siena, Siena.
Grégoire-Wibo, C. (1974). Bioécologie de *Folsomia quadrioculata* (Insecta, Collembola). *Pedobiologia*, 14: 199–207.
Grégoire-Wibo, C. and Snider, R.M. (1977). The intrinsic rate of natural increase, its interest to ecology and its application to various species of Collembola. pp. 442–448. *In:* Lohm, U. and Persson, T. (eds.). *Soil Organisms as Components of Ecosystems*. Ecological Bulletins, Stockholm.
Grégoire-Wibo, C. and Snider, R.M. (1983). Temperature-related mechanisms of population persistence in *Folsomia candida* and *Protaphorura armata* (Insecta: Collembola). *Pedobiologia*, 25: 413–418.
Grimnes, K.A. and Snider, R.M. (1981). An analysis of egg production in four strains of *Folsomia candida* (Collembola). *Pedobiologia*, 22: 224–231.
Grishina, L.G., Nikolskij, V.V. and Wasylik, A. (1995). Communities of Acarina in the soils of potato field of Western Siberia. *Polish Ecological Studies*, 21: 293–309.
Grober, M.S. and Rodgers, E.W. (2007). The evolution of hermaphroditism. *Journal of Theoretical Biology*, 251: 190–192.
Guerra, C.A., Bardgett, R.D., Caon, L., Crowther, T.W., Delgado-Baquerizo, M., Montanarella, L. et al. (2021). Tracking, targeting, and conserving soil biodiversity. *Science*, 371: 239–241.

Guidetti, R., Altiero, T. and Rebecchi, L. (2011). On dormancy strategies in tardigrades. *Journal of Insect Physiology*, 57: 567–576.
Guil, N. and Giribet, G. (2012). A comprehensive molecular phylogeny of tardigrades – adding genes and taxa to a poorly resolved phylum-level phylogeny. *Cladistics*, 28: 21–49.
Guil, N., Jørgensen, A. and Kristensen, R. (2019). An upgraded comprehensive multilocus phylogeny of the Tardigrada tree of life. *Zoologica Scripta*, 48: 120–137.
Haag, C.R. and Ebert, D. (2004). A new hypothesis to explain geographic parthenogenesis. *Annales Zoologici Fennici*, 41: 53s9–544.
Haag, E.S., Fitch, D.H.A. and Delattre, M. (2018). From "the worm" to "the worms" and back again: the evolutionary developmental biology of nematodes. *Genetics*, 210: 397–433.
Habassi, A., Khemaissa, H. and Nasri-Ammar, K. (2020). Ecophysiological responses of the desert isopod *Hemilepistus reaumurii* to the combined effects of thermoperiod and photoperiod. *Biologia*, 75: 2251–2260.
Haegeman, A., Jones, J.T. and Danchin, E.G.J. (2011). Horizontal gene transfer in nematodes: a catalyst for plant parasitism? *Molecular Plant-Microbe Interactions*, 24: 879–887.
Hagstrum, D.W. and Hagstrum, W.R. (1970). A simple device for producing fluctuating temperatures, with an evaluation of the ecological significance of fluctuating temperatures. *Annals of the Entomological Society of America*, 63: 1385–1389.
Hahn, M.E. (2002). Aryl hydrocarbon receptors: diversity and evolution. *Chemico-Biological Interactions*, 141: 131–160.
Halberg, K.A., Larsen, K.W., Jørgensen, A., Ramløv, H. and Møbjerg, N. (2013). Inorganic ion composition in Tardigrada: cryptobionts contain a large fraction of unidentified organic solutes. *The Journal of Experimental Biology*, 216: 1235–1243.
Hale, W.G. (1965). Observations on the breeding biology of Collembola. *Pedobiologia*, 5: 146–152, 161–177.
Hale, W.G. (1966). A population study of moorland Collembola. *Pedobiologia*, 6: 65–99.
Hale, W.G. (1980). Production and energy flow in two species of *Onychiurus* (Collembola, Insecta Apterygota). *Pedobiologia*, 20: 274–287.
Hallas, T.E. and Yeates, G.W. (1972). Tardigrada of the soil and litter of a Danish beech forest. *Pedobiologia*, 12: 287–301.
Hamilton, A.L. (1969). On estimating annual production. *Limnology and Oceanography*, 14: 771–782.
Handelsman, J., Liles, M., Mann, D., Riesenfeld, C. and Goodman, R.M. (2002). Cloning the metagenome: culture-independent access to the diversity and functions of the uncultivated microbial world. *Methods in Microbiology*, 33: 241–255.
Handelsman, J. (2021). *A World Without Soil. The Past, Present, and Precarious Future of the Earth Beneath our Feet*. Yale University Press, New Haven.
Hanski, I. and Gilpin, M. (1991). Metapopulation dynamics: brief history and conceptual domain. *Biological Journal of the Linnean Society*, 42: 3–16.
Hanski, I., Woiwod, I. and Perry, J. (1993). Density dependence, population persistence, and largely futile arguments. *Oecologia*, 95: 595–598.
Harada, H. and Ito, M.T. (2006). Soil-inhabiting tardigrade communities in forests of Central Japan. *Hydrobiologia*, 558: 119–127.
Hartl, D.L. and Clark, A.G. (1997). *Principles of Population Genetics. Third Edition*. Sinauer Associates, Inc., Sunderland.
Harvey, M.S. (1992). The phylogeny and classification of the Pseudoscorpionida (Chelicerata: Arachnida). *Invertebrate Taxonomy*, 6: 1373–1435.
Harzsch, S., Vilpoux, K., Blackburn, D.C., Platchetzki, D., Brown, N.L., Melzer, R. et al. (2006). Evolution of arthropod visual systems: development of the eyes and central visual pathways in the horseshoe crab *Limulus polyphemus* Linnaeus, 1758 (Chelicerata, Xiphosura). *Developmental Dynamics*, 235: 2641–2655.
Hasegawa, M. (2002). The response of collembolan community to the amount and composition of organic matter of a forest floor. *Pedobiologia*, 46: 353–364.
Hassall, M. and Rushton, P. (1985). The adaptive significance of coprophagous behaviour in the terrestrial isopod *Porcellio scaber*. *Pedobiologia*, 28: 169–175.

Hassall, M. and Dangerfield, J.M. (1990). Density-dependent processes in the population dynamics of *Armadillidium vulgare* (Isopoda: Oniscidae). *Journal of Animal Ecology*, 59: 941–958.

Hawes, T.C., Worland, M.R., Convey, P. and Bale, J.S. (2007). Aerial dispersal of springtails on the Antarctic Peninsula: implications for local distribution and demography. *Antarctic Science*, 19: 3–10.

Hawes, T.C., Marshall, C.J. and Wharton, D.A. (2014). A 9 kDa antifreeze protein from the Antarctic springtail, *Gomphiocephalus hodgsoni*. *Cryobiology*, 69: 181–183.

Hayes, K.A., Burks, R.L., Castro-Vazquez, A., Darby, P.C., Heras, H., Martín, P.R. et al. (2015). Insights from an integrated view of the biology of apple snails (Caenogastropoda: Ampullariidae). *Malacologia*, 58: 245–302.

Healey, I.N. (1967). The energy flow through a population of soil Collembola. pp. 695–708. *In:* Petrusewicz, K. (ed.). *Secondary Productivity of Terrestrial Ecosystems*. Warszawa- Krakow.

Hedlund, K., Ek, H., Gunnarsson, T. and Svegborn, C. (1990). Mate choice and male competition in *Orchesella cincta* (Collembola). *Experientia*, 46: 524–526.

Hedlund, K., Bengtsson, G. and Rundgren, S. (1995). Fungal odour discrimination in two sympatric species of fungivorous collembolans. *Functional Ecology*, 9: 869–875.

Heethoff, M., Domes, K., Laumann, M., Maraun, M., Norton, R.A. and Scheu, S. (2007). High genetic divergences indicate ancient separation of parthenogenetic lineages of the oribatid mite *Platynothrus peltifer*. *Journal of Evolutionary Biology*, 20: 392–402.

Heethoff, M., Norton, R.A., Scheu, S. and Maraun, M. (2009). Parthenogenesis in oribatid mites (Acari, Oribatida): Evolution without sex. pp. 241–257. *In*: Schön, I., Martens, K. and Van Dijk, P. (eds.). *Lost Sex. The Evolutionary Biology of Parthenogenesis.* Springer Science+Business Media B.V., Dordrecht.

Helmus, M.R., Mahler, D.L. and Losos, J.B. (2014). Island biogeography of the Anthropocene. *Nature*, 513: 543–546.

Hendriks, A.J. (1999). Allometric scaling of rate, age and density parameters in ecological models. *Oikos*, 86: 293–310.

Hendriks, A.J. (2007). The power of size: A meta-analysis reveals consistency of allometric regressions. *Ecological Modelling*, 205: 196–208.

Hengherr, S., Worland, M.R., Reuner, A., Brümmer, F. and Schill, R.O. (2009). Freeze tolerance, supercooling points and ice formation: comparative studies on the subzero temperature survival of limno-terrestrial tardigrades. *The Journal of Experimental Biology*, 212: 802–807.

Hensbergen, P.J., Donker, M.H., Van Velzen, M.J.M., Roelofs, D., Van der Schors, R.C., Hunziker, P.E. et al. (1999). Primary structure of a cadmium-induced metallothionein from the insect *Orchesella cincta* (Collembola). *European Journal of Biochemistry*, 259: 197–203.

Hensbergen, P.J., Van Velzen, M.J.M., Adi Nugroho, R., Donker, M.H. and Van Straalen, N.M. (2000). Metallothionein-bound cadmium in the gut of the insect *Orchesella cincta* (Collembola) in relation to dietary cadmium exposure. *Comparative Biochemistry and Physiology Part C*, 125: 17–24.

Hesgrove, C. and Boothby, T.C. (2020). The biology of tardigrade disordered proteins in extreme stress tolerance. *Cell Communication and Signaling*, 18: 178.

Heynen, C. (1988). Zur Biologie eines Buchenwaldbodens. 11. Die Dipterenlarven. *Carolinea*, 46: 115–130.

Hill, R.W., Wyse, G.A. and Anderson, M. (2008). *Animal Physiology. Second Edition*. Sinauer Associates, Sunderland.

Höckner, M., Stefanon, K., Schuler, D., Fantur, R., De Vaufleury, A. and Dallinger, R. (2009). Coping with cadmium exposure in various ways: the two helicid snails *Helix pomatia* and *Cantarus aspersus* share the Metal Transcription Factor-2: but differ in promoter organization and transcription of their Cd-metallothionein genes. *Journal of Experimental Zoology*, 311A, 776–787.

Hodda, M. and Wanless, F.R. (1994). Nematodes from an English chalk grassland: population ecology. *Pedobiologia*, 38: 530–545.

Hoese, B. (1981). Morphologie und Funktion des Wasserleitungssystems der terrestrischen Isopoden (Crustacea, Isopoda, Oniscoidea). *Zoomorphology*, 98: 135–167.

Hoese, B. (1982a). Der Ligia-Typ des Wasserleitungssystems bei terrestrischen Isopoden und seine Entwicklung in der familie Ligiidae (Crustacea, Isopoda, Oniscoidea). *Zoologische Jahrbücher Abteilung für Systematik*, 108: 225–261.

Hoese, B. (1982b). Morphologie und Evolution der Lungen bei den terrestrischen Isopode (Crustacea, Isopoda, Oniscoidea). *Zoologische Jahrbücher Anatomie*, 107: 396–422.

Hoffman, B.U. and Lumpkin, E.A. (2018). A gut feeling. *Science*, 361: 1203–1204.

Hoffmann, J.A. and Reichhart, J.-M. (2002). *Drosophila* innate immunity: an evolutionary perspective. *Nature Immunology*, 3: 121–126.

Hogg, I.D. and Hebert, P.D.N. (2004). Biological identification of springtails (Hexapoda: Collembola) from the Canadian arctic, using mitochondrial DNA barcodes. *Canadian Journal of Zoology*, 82: 749–754.

Holighaus, G. and Rohlfs, M. (2019). Volatile and non-volatile fungal oxylipins in fungus-invertebrate interactions. *Fungal Ecology*, 38: 28–36.

Holmstrup, M. (1992). Cold hardiness strategy in cocoons of the lumbricid earthworm *Dendrobaena octaedra* (Savigny). *Comparative Biochemistry and Physiology*, 102A: 49–52.

Holmstrup, M. (1994). Physiology of cold hardiness in cocoons of five earthworm taxa (Lumbricidae: Oligochaeta). *Journal of Comparative Physiology B*, 164: 222–228.

Holmstrup, M. and Westh, P. (1994). Dehydration of earthworm cocoons exposed to cold: a novel hardiness mechanism. *Journal of Comparative Physiology B*, 164: 312–315.

Holmstrup, M. and Westh, P. (1995). Effects of dehydration on water relations and survival of lumbricid earthworm egg capsules. *Journal of Comparative Physiology B*, 165: 377–383.

Holmstrup, M. and Zachariassen, K.E. (1996). Physiology of cold hardiness in earthworms. *Comparative Biochemistry and Physiology*, 115A: 91–101.

Holmstrup, M. and Sømme, L. (1998). Dehydration and cold hardiness in the Arctic collembolan *Onychiurus arcticus* Tullberg 1876. *Journal of Comparative Physiology B*, 168: 197–203.

Holmstrup, M., Costanzo, J.P. and Lee Jr., R.E. (1999). Cryoprotective and osmotic responses to cold acclimation and freezing in freeze-tolerant and freeze-intolerant earthworms. *Journal of Comparative Physiology B*, 169: 207–214.

Holmstrup, M. (2001). Sensitivity of life history parameters in the earthworm *Aporrectodea caliginosa* to small changes in soil water potential. *Soil Biology and Biochemistry*, 33: 1217–1223.

Holmstrup, M. and Overgaard, J. (2007). Freeze tolerance in *Aporrectodea caliginosa* and other earthworms from Finland. *Cryobiology*, 55: 80–86.

Holmstrup, M. (2018). The springtail *Megaphorura arctica* survives extremely high osmolality of body fluids during drought. *Journal of Comparative Physiology B*, 188: 939–945.

Holsinger, J.R. (1993). Biodiversity of subterranean amphipod crustaceans: global patterns and zoogeographic implications. *Journal of Natural History*, 27: 821–835.

Holterman, M., Van der Wurff, A.W.G., Van den Elsen, S., Van Megen, H., Bongers, T., Holovachov, O. et al. (2006). Phylum-wide analysis of SSU rDNA reveals deep phylogenetic relationships among nematodes and accelerated evolution toward crown clades. *Molecular Biology and Evolution*, 23: 1792–1800.

Holterman, M., Holovachov, O., Van den Elsen, S., Van Megen, H., Bongers, T., Bakker, J. et al. (2008). Small subunit ribosomal DNA-based phylogeny of basal Chromadoria (Nematoda) suggests that transitions from marine to terrestrial habitats (and *vice versa*) require relatively simple adaptations. *Molecular Phylogenetics and Evolution*, 48: 758–763.

Holterman, M., Schratzberger, M. and Helder, J. (2019). Nematodes as evolutionary commuters between marine, freshwater and terrestrial habitats. *Biological Journal of the Linnean Society*, 128: 756–767.

Holyak, M. and Lawton, J.H. (1993). Comment arising from a paper by Wolda and Dennis: using and interpreting the results of tests for density dependence. *Oecologia*, 95: 592–594.

Holznagel, W.E., Colgan, D.J. and Lydeard, C. (2010). Pulmonate phylogeny based on 28S rRNA gene sequences: A framework for discussing habitat transitions and character transformation. *Molecular Phylogenetics and Evolution*, 57: 1017–1025.

Homa, J., Stürzenbaum, S. and Kolaczkowska, E. (2016). Metallothionein 2 and heat shock protein 72 protect *Allolobophora chlorotica* from cadmium but not nickel or copper exposure: body malformation and coelomocyte functioning. *Archives of Environmental Contamination and Toxicology*, 71: 267–277.

Homa, J. (2018). Earthworm coelomocyte extracellular traps: structural and functional similarities with neutrophil NETs. *Cell and Tissue Research*, 371: 407–414.

Honeker, L.K., Graves, K.R., Tfally, M.M., Krechmer, J.E. and Meredith, L.K. (2021). The volatilome: a vital piece of the complete soil metabolome. *Frontiers in Environmental Science*, 9: 649905.

Hopkin, S.P. (1986). Ecophysiological strategies of terrestrial arthropods for surviving heavy metal pollution. pp. 263–266. *In:* Velthuis, H.H.W. (ed.). *Proc. 3rd. European Congress of Entomology*. NEV, Amsterdam.

Hopkin, S.P., Hames, C.A.C. and Dray, A. (1989). X-ray microanalytical mapping of the intracellular distribution of pollutant metals. *Microscopy and Analysis*, November 1989: 23–27.

Hopkin, S.P. (1989). *Ecophysiology of Metals in Terrestrial Invertebrates*. Elsevier Applied Science, London.

Hopkin, S.P. and Read, H.J. (1992). *The Biology of Millipedes*. Oxford University Press.

Hopkin, S.P. (1997). *Biology of the Springtails (Insecta: Collembola)*. Oxford University Press, Oxford.

Hornung, E. (2015). In commemoration of Prof. M.R. Warburg and of his contribution to terrestrial isopod biology (31 May 1931: Berlin – 9 February 2014: Haifa). *ZooKeys*, 515: 1–11.

Höss, S., Menzel, R., Gessler, F., Nguyen, H.T., Jehle, J.A. and Traunspurger, W. (2013). Effects of insecticidal crystal proteins (Cry proteins) produced by genetically modified maize (Bt maize) on the nematode *Caenorhabditis elegans*. *Environmental Pollution*, 178: 147–151.

Hoste-Danyłow, A., Olieva-Makulec, K., Olejniczak, I., Hajdamowicz, I., Stańska, M., Marczak, D. et al. (2013). The shape of the intraspecific metabolic-rate–body-size relationship affects interspecific biomass and abundance distributions of soil animals within a forest ecosystem. *Annales Zoologici Fennici*, 289–302.

Howsam, M.J. and Van Straalen, N.M. (2003). Pyrene metabolism in the springtail *Orchesella cincta* L. (Collembola, Entomobryidae). *Environmental Toxicology and Chemistry*, 22: 1481–1486.

Hu, Y., Linz, D.M. and Moczek, A.P. (2019). Beetle horns evolved from wing serial homologs. *Science*, 366: 1004–1007.

Huang, J., Xu, Q., Sun, Z.J., Tang, G.L. and Su, Z.Y. (2007). Identifying earthworms through DNA barcodes. *Pedobiologia*, 51: 301–309.

Huettel, R.N. and Golden, A.M. (1991). Nathan Augustus Cobb: the father of nematology in the United States. *Annual Review of Phytopathology*, 29: 15–26.

Hug, L.A., Baker, B.J., Anantharaman, K., Brown, C.T., Probst, A.J., Castelle, C.J. et al. (2016). A new view of the tree of life. *Nature Microbiology*, 1: 16048.

Huhta, V. (1984). Response of *Cognettia sphagnetorum* (Enchytraeidae) to manipulation of pH and nutrient status in coniferous forest soil. *Pedobiologia*, 27: 245–260.

Humbert, W. (1974). Localisation, structure et genèse des concrétions minérales dans le mésentéron des Collemboles Tomoceridae (Insecta, Collembola). *Zeitschrift für Morphologie der Tiere*, 78: 93–109.

Hutson, B.R. (1980). Colonization of industrial reclamation sites by Acari, Collembola and other invertebrates. *Journal of Applied Ecology*, 17: 255–275.

Hutson, B.R. and Veitch, L.G. (1983). Mean annual population density of Collembola and Acari in the soil and litter of three indigenous South Australian forests. *Australian Journal of Ecology*, 8: 113–126.

Hynes, H.B.N. and Coleman, M.J. (1968). A simple method of assessing the annual production of stream benthos. *Limnology and Oceanography*, 13: 569–573.

Inomata, K., Kobari, F., Yoshida-Noro, C., Myohara, M. and Tochinai, S. (2000). Possible neural control of asexually reproductive fragmentation in *Enchytraeus japonensis* (Oligochaeta, Enchytraeidae). *Invertebrate Reproduction and Development*, 37: 35–42.

Ito, M. (1999). Ecological distribution, abundance and habitat preference of terrestrial tardigrades in various forests on the Northern slop of Mt. Fuji, Central Japan. *Zoologische Anzeiger*, 238: 225–234.

Ito, M.T. and Abe, W. (2001). Micro-distribution of soil inhabiting tardigrades (Tardigrada) in a sub-alpine coniferous forest of Japan. *Zoologische Anzeiger*, 240: 403–407.

Ito, M., Hasegawa, M., Iwamoto, K. and Kitayama, K. (2002). Patterns of soil macrofauna in relation to elevation and geology on the slope of Mount Kinabalu, Sabah, Malaysia. *Sabah Parks Nature Journal*, 5: 153–163.

Itoh, R., Hisamatsu, M., Matsunaga, M. and Hishida, F. (1995). Growth and reproduction of a collembolan species, *Folsomia candida* (Willem), under experimental conditions. *Journal of the College of Arts and Sciences, Showa University*, 26: 23–30.

Ivanova, E., Clausi, M., Sparacio, I. and Spiridonov, S. (2019). Preliminary data on the parasite survey of terrestrial gastropods of Sicily. *Russian Journal of Nematology*, 27: 37–45.

Jacob, F. (1977). Evolution and tinkering. *Science*, 196: 1161–1166.

Jaenike, J. and Selander, R.K. (1979). Evolution and ecology of parthenogenesis in earthworms. *American Zoologist*, 19: 729–737.

Jakšová, P., L'uptáčik, P. and Miklisová, D. (2019). Distribution of Oribatida (Acari) along a depth gradient in forested scree slopes. *Subterranean Biology*, 31: 29–48.

James, S. (1998). Earthworms and earth history. pp. 3–14. *In:* Edwards, C.A. (ed.) *Earthworm Ecology*. St. Lucie Press, Boca Raton.

James, S.W. and Davidson, S.K. (2012). Molecular phylogeny of earthworms (Annelida: Crassiclitellata) based on 28S, 18S and 16S gene sequences. *Invertebrate Systematics*, 26: 213–229.

James, S.W., Csuzdi, C., Chang, C.-H., Aspe, N.M., Jiménez, J.J., Feijoo, A. et al. (2021). Comment on "Global distribution of earthworm diversity". *Science*, 10.1126/science.abe4629.

Jamieson, B.G.M., Tillier, S., Tillier, A., Justine, J.-L., Ling, E., James, S. et al. (2002). Phylogeny of the Megascolecidae and Crassiclitellata (Annelida, Oligochaeta): combined versus partitioned analysis using nuclear (28S) and mitochondrial (12S, 16S) rDNA. *Zoosystema*, 24: 707–734.

Janicke, T., Häderer, I.K., Lajeunesse, M.J. and Anthes, N. (2016). Darwinian sex roles confirmed across the animal kingdom. *Science Advances*, 2: e1500983.

Janssen, G.M. and Joosse, E.N.G. (1987). Reproduction and growth in Collembola under laboratory conditions. *Pedobiologia*, 30: 1–8.

Janssen, G.M., De Jong, G., Joosse, E.N.G. and Scharloo, W. (1988). A negative maternal effect in springtails. *Evolution*, 42: 828–834.

Janssen, M.P.M. and Hogervorst, R.F. (1993). Metal accumulation in soil arthropods in relation to micro-nutrients. *Environmental Pollution*, 79: 181–189.

Janssens, T.K.S., Mariën, J., Cenijn, P., Legler, J., Van Straalen, N.M. and Roelofs, D. (2007). Recombinational micro-evolution of functionally different metallothionein promoter alleles from *Orchesella cincta*. *BMC Evolutionary Biology*, 7: 88.

Janssens, T.K.S., Del Rio Lopez, R., Mariën, J., Timmermans, M.J.T.N., Montagne-Wajer, M., Van Straalen, N.M. et al. (2008). Comparative population analysis of metallothionein promoter alleles suggests stress-induced microevolution in the field. *Environmental Science and Technology*, 42: 3873–3878.

Jarne, P. and Auld, J.R. (2006). Animals mix it up too: the distribution of self-fertilization among hermaphroditic animals. *Evolution*, 60: 1816–1824.

Jarvis, N.J. (2020). A review of non-equilibrium water flow and solute transport in soil macropores: principles, controlling factors and consequences for water quality. *European Journal of Soil Science*, 71: 279–302.

Jensen, T.F. (1978a). Annual production and respiration in ant populations. *Oikos*, 31: 207–213.

Jensen, T.F. (1978b). An energy budget for a field population of *Formica pratense* Retz. (Hymenoptera: Formicidae). *Natura Jutlandica*, 20: 203–226.

Jeratthitikul, E., Beantaowong, U. and Panha, S. (2017). DNA barcoding of the Thai species of terrestrial earthworms in the genera *Amynthas* and *Metaphire* (Haplotaxida: Megascolecidae). *European Journal of Soil Biology*, 81: 39–47.

Jiménez, J.J., Brown, G.G., Decaëns, T., Feijoo, A. and Lavelle, P. (2000). Differences in the timing of diapause and patterns of aestivation in tropical earthworms. *Pedobiologia*, 44: 677–694.

Jiménez, J.J., Filser, J. and The KEYSOM Team. (2020). *Soil Fauna: Key to Soil Organic Matter Dynamics and Modelling*. COST, Brussels.

Johnson, C., Krsek, M., Wellington, E.M.H., Stott, A.W., Cole, L., Bardgett, R.D. et al. (2005). Soil invertebrates disrupt carbon flow through fungal networks. *Science*, 309: 1047.

Johnston, A.S.A. (2019). Land management modulates the environmental controls on global earthworm communities. *Global Ecology and Biogeography*, 28: 1787–1795.

Jones, C.G., Lawton, J.H. and Shachak, M. (1994). Organisms as ecosystem engineers. *Oikos*, 69: 373–386.

Jones, H.D., Santoro, G., Boag, B. and Neilson, R. (2001a). The diversity of earthworms in 200 Scottish fields and the possible effect of New Zealand flatworms (*Arthurdendyus triangulatus*) on earthworm populations. *Annals of Applied Biology*, 139: 75–92.

Jones, H.D., Green, J., Harrison, K. and Palin, D.W. (2001b). Further monthly records (1994 to 2000) of size and abundance in a population of the "Australian" flatworm, *Australoplana sanguinea alba* in the U.K. *Belgian Journal of Zoology*, 131: 217–220.

Joosse, E.N.G. (1970). The formation and biological significance of aggregations in the distribution of Collembola. *Netherlands Journal of Zoology*, 20: 299–314.

Joosse, E.N.G. and Veltkamp, E. (1970). Some aspects of growth, moulting and reproduction in five species of surface dwelling Collembola. *Netherlands Journal of Zoology*, 20: 315–328.

Joosse, E.N.G. (1971). Ecological aspects of aggregation in Collembola. *Revue d'Écologie et de Biologie du Sol*, 8: 91–97.

Joosse, E.N.G. and Verhoef, H.A. (1974). On the aggregational habits of surface dwelling Collembola. *Pedobiologia*, 14: 245–249.

Joosse, E.N.G. and Koelman, T.A.C.M. (1979). Evidence for the presence of aggregation pheromones in *Onychiurus armatus*, a pest insect in sugar beet. *Entomologia experimentalis et applicata*, 26: 197–201.

Joosse, E.N.G. and Buker, J.B. (1979). Uptake and excretion of lead by litter-dwelling Collembola. *Environmental Pollution*, 18: 235–240.

Joosse, E.N.G. and Verhoef, H.A. (1987). Developments in ecophysiological research on soil invertebrates. *Advances in Ecological Research*, 16: 175–248.

Jucevičа, E. and Melecis, V. (2002). Long-term dynamics of Collembola in a pine forest ecosystem. *Pedobiologia*, 46: 365–372.

Judson, P.P. and Normark, B.B. (1996). Ancient asexual scandals. *Trends in Ecology and Evolution*, 11: 41–46.

Kamilari, M., Jørgensen, A., Schiøtt, M. and Møbjerg, N. (2019). Comparative transcriptomics suggest unique molecular adaptations within tardigrade lineages. *BMC Genomics*, 20: 607.

Kammenga, J.E., Busschers, M., Van Straalen, N.M., Jepson, P.C. and Bakker, J. (1996). Stress induced fitness reduction is not determined by the most sensitive life-cycle trait. *Functional Ecology*, 10: 106–111.

Kammenga, J. and Laskowski, R. (Eds.). (2000). *Demography in Ecotoxicology*. John Wiley and Sons Ltd., Chichester.

Kampfraath, A.A., Klasson, L., Anvar, S.Y., Vossen, R.H.A.M., Roelofs, D., Kraaijeveld, K. et al. (2019). Genome expansion of an obligate parthenogenesis-associated *Wolbachia* poses an exception to the symbiont reduction model. *BMC Genomics*, 20: 106.

Kampichler, C. and Geissen, V. (2005). Temporal predictability of soil microarthropod communities in temperate forests. *Pedobiologia*, 49: 41–50.

Kaneko, N. (1985). A comparison of oribatid mite communities in two different soil types in a cool temperate forest in Japan. *Pedobiologia*, 28: 255–264.

Kaneko, N. (1988). Life history of *Oppiella nova* (Oudemans) (Oribatei) in cool temperate forest soils in Japan. *Acarologia*, 24: 215–221.

Kaneko, N. (1989). Life histories of four oribatid mite species in a mull type soil in a cool temperate forest in Japan. *Pedobiologia*, 33: 117–126.

Kaneko, N. (1995). Community organization of oribatid mites in various forest soils. pp. 21–33. *In:* Edwards, C.A., Abe, T. and Striganova, B.R. (eds.). *Structure and Function of Soil Communities*. Kyoto University Press, Kyoto.

Kaneko, N., McLean, M.A. and Parkinson, D. (1995). Grazing preference of *Onychiurus subtenuis* (Collembola) and *Oppiella nova* (Oribatei) for fungal species inoculated on pine needles. *Pedobiologia*, 39: 538–546.

Karaca, A. (2011). *Biology of Earthworms*. Springer Verlag, Berlin.

Karasawa, S. and Hijii, N. (2008). Vertical stratification of oribatid (Acari: Oribatida) communities in relation to their morphological and life-history traits and tree structures in a subtropical forest in Japan. *Ecological Research*, 23: 57–69.

Karmegam, N. and Daniel, T. (2009). Growth, reproductive biology and life cycle of the vermicomposting earthworm, *Perionyx ceylanensis* Mich. (Oligochaeta: Megascolecidae). *Bioresource Technology*, 100: 4790–4796.

Katz, A.D. (2020). Inferring evolutionary timescales without independent timing information: an assessment of "universal" insect rates to calibrate a Collembola (Hexapoda) molecular clock. *Genes*, 11: 1172.

Kaufmann, O. (1932). Einige Bemerkungen über den Einfluss von Temperaturschwankungen auf die Entwicklungsdauer und Streuung bei Insekten und seine graphische Darstellung durch Kettenlinie und Hyperbel. *Zeitung für Morphologie und Ökologie der Tiere*, 25: 353–361.

Kautenburger, R. (2006). Genetic structure among earthworms (*Lumbricus terrestris* L.) from different sampling sites in western Germany based on random amplified polymorphic DNA. *Pedobiologia*, 50: 257–266.

Kautz, S., Rubin, B.E.R., Russell, J.A. and Moreau, C.S. (2013). Surveying the microbiome of ants: comparing 454 pyrosequencing with traditional methods to uncover bacterial diversity. *Applied and Environmental Microbiology*, 79: 525–534.

Kaygorodova, I.A. and Sherbakov, D.Y. (2006). Molecular study of the systematic position of Baikalian oligochaetes in Clitellata. *Russian Journal of Genetics*, 42: 1390–1397.

Keller, S. and Steenberg, T. (1997). *Neozygites sminthuri* sp. nov. (Zygomycetes, Entomophthorales), a pathogen of the springtail *Sminthurus viridis* L. (Collembola, Sminthuridae). *Sydowia*, 49: 21–24.

Keller, L. (2009). Adaptation and the genetics of social behaviour. *Philosophical Transactions of the Royal Society B*, 364: 3209–3216.

Kennedy, C.H. (1928). Evolutionary level in relation to geographic, seasonal and diurnal distribution of insects. *Ecology*, 9: 367–379.

Keshavarz Jamshidian, M., Verweij, R.A., Van Gestel, C.A.M. and Van Straalen, N.M. (2017). Toxicokinetics and time-variable toxicity of cadmium in *Oppia nitens* Koch (Acari: Oribatida). *Environmental Toxicology and Chemistry*, 36: 408–413.

Khadjeh, S., Turetzek, N., Pechmann, M., Schwager, E.E., Wimmer, E.A., Damen, W.G.M. et al. (2012). Divergent role of the Hox gene *Antennapedia* in spiders is reponsible for the convergent evolution of abdominal limb repression. *Proceedings of the National Academy of Sciences of the United States of America*, 109: 4921–4926.

Khalil, M.A., Abdel-Lateif, H.M. and Al-Assiuty, A.-N.I. (1999). Changes in oribatid faunal structure associated with land conversion from annual crop into orchard. *Pedobiologia*, 43: 85–96.

Khalil, M.A., Al-Assiuty, A.-N.I. and Van Straalen, N.M. (2011). Egg number varies with population density: a study of three oribatid mite species in orchards in Egypt. *Acarologia*, 51: 251–258.

Khalil, M.A., Al-Assiuty, A.-N.I., Van Straalen, N. and Al-Assiuty, B.A. (2016). Changes in soil oribatid communities associated with conversion from conventional to organic agriculture. *Experimental and Applied Acarology*, 68: 183–196.

Khan, M.R. (2015). Nematode diseases of crops in India. pp. 183–224. *In:* Awasthi, L.P. (ed.). *Recent Advances in the Diagnosis and Management of Plant Diseases*. Springer India, New Delhi.

Kiers, E.T., Rousseau, R.A., West, S.A. and Denison, R.F. (2003). Host sanctions and the legume-rhizobium mutualism. *Nature*, 425: 78–81.

Kiers, E.T. and Van der Heijden, M.G.A. (2006). Mutualistic stability in the arbuscular mycorrhizal symbiosis: exploring hypotheses of evolutionary cooperation. *Ecology*, 87: 1627–1636.

Kiers, E.T. and Denison, R.F. (2008). Sanctions, cooperation, and the stability of plant-rhizosphere mutualisms. *Annual Review of Ecology and Systematics*, 19: 215–236.

Kiers, E.T., Duhamel, M., Beesetty, Y., Mensah, J.A., Franken, O., Verbruggen, E. et al. (2011). Reciprocal rewards stabilize cooperation in the mycorrhizal symbiosis. *Science*, 333: 880–882.

Kim, M. and Or, D. (2019). Microscale pH variations during drying of soils and desert biocrusts affect HONO and NH_3 emissions. *Nature Communications*, 10: 3944.

Kime, R.D. and Wauthy, G. (1984). Aspects of relationships between millipedes, soil texture and temperature in deciduous forests. *Pedobiologia*, 26: 387–402.

Kime, R.D. and Golovatch, S.I. (2000). Trends in the ecological strategies and evolution of millipedes (Diplopoda). *Biological Journal of the Linnean Society*, 69: 333–349.

Kimura, K., Shibuya, K. and Chiba, S. (2013). The mucus of a land snail love-dart suppresses subsequent matings in darted individuals. *Animal Behaviour*, 85: 631–635.

Kimura, K., Shibuya, K. and Chiba, S. (2016). Effect of injection of love-dart mucus on physical vigour in land snails: can remating suppression be explained by physical damage? *Ethology Ecology and Evolution*, 28: 284–294.

King, R.A., Tibble, A.L. and Symondson, W.O.C. (2008). Opening a can of worms: unprecedented sympatric cryptic diversity within British lumbricid earthworms. *Molecular Ecology*, 17: 4684–4698.

King, J.H.P., Mahadi, N.M., Bong, C.F.J., Ong, K.H. and Hassan, O. (2014). Bacterial microbiome of *Coptotermes curvignathus* (Isoptera: Rhinotermitidae) reflects the coevolution of species and dietary pattern. *Insect Science*, 21: 584–596.

Kiontke, K.C., Félix, sM.-A., Ailion, M., Rockman, M.V., Braendle, C., Pénigault, J.-P. et al. (2011). A phylogeny and molecular barcodes for *Caenorhabditis*, with numerous new species from rotting fruits. *BMC Evolutionary Biology*, 11: 339.

Kitagami, Y., Kanzaki, N. and Matsuda, Y. (2019). First report of filamentous bacteria associated with *Rhigonema* sp. (Nematoda: Rhigonematidae) dwelling in the hindgut of *Riukiaria* sp. (Diplopoda: Xystodesmidae). *Helminthologia*, 56: 219–228.

Klarica, J., Kloss-Brandstätter, A., Traugot, M. and Juen, A. (2012). Comparing four mitochondrial genes in earthworms – Implications for identification, phylogenetics, and discovery of cryptic species. *Soil Biology and Biochemistry*, 45: 23–30.

Kleiber, M. (1961). *The Fire of Life. An Introduction to Animal Energetics*. John Wiley and Sons Inc., New York.

Klekowski, R.Z. and Duncan, A. (1975). Physiological approach to ecological energetics. pp. 15–64. *In:* Grodzinski, W., Klekowski, R.Z. and Duncan, A. (eds.). *Methods for Ecological Bioenergetics*. Blackwell, Oxford.

Klimov, P.B., OConnor, B.M., Chetverikov, P.E., Bolton, S.J., Pepato, A.R., Mortazavi, A.L. et al. (2018). Comprehensive phylogeny of acariform mites (Acariformes) provides insights on the origin of the four-legged mites (Eriophyoidea), a long branch. *Molecular Phylogenetics and Evolution*, 119: 105–117.

Klok, C., De Roos, A.M., Marinissen, J.C.Y., Baveco, H.M. and Ma, W.-C. (1997). Assessing the effects of abiotic environmental stress on population growth in *Lumbricus rubellus* (Lumbricidae, Oligochaeta). *Soil Biology and Biochemistry*, 29: 287–293.

Knapp, B.A., Seeber, J., Rief, A., Meyer, E. and Insam, H. (2010). Bacterial community composition of the gut microbiota of *Cylindroiulus fulviceps* (Diplopoda) as revealed by molecular fingerprinting and cloning. *Folia Microbiologia*, 55: 489–496.

Knop, J. (1926). Bakterien und Bakteroiden bei Oligochäten. *Zeitschrift für Morphologie und Ökologie der Tiere*, 6: 588–624.

Koch, M. (1997). Monophyly and phylogenetic position of the Diplura (Hexapoda). *Pedobiologia*, 41: 9–12.

Kocot, K.M., Halanych, K.M. and Krug, P.J. (2013). Phylogenomics supports Panpulmonata: Opisthobranch paraphyly and key evolutionary steps in a major radiation of gastropod molluscs. *Molecular Phylogenetics and Evolution*, 69: 764–771.

Koehler, H.H. (1999). Predatory mites (Gamasina, Mesostigmata). *Agriculture, Ecosystems and Environment*, 74: 395–410.

Koeman, J.H. (1996). Toxicology, history and scope of the field. pp. 3–14. *In:* Niesink, R.J.M., De Vries, J. and Hollinger, M.A. (eds.). *Toxicology. Principles and Applications*. CRC Press, Boca Raton.

Koene, J.M. and Chase, J.M. (1998). The love dart of *Helix aspersa* Müller is not a gift of calcium. *Journal of Molluscan Studies*, 64: 75–80.

Koene, J.M., Sundermann, G. and Michiels, N.K. (2002). On the function of body piercing during copulation in earthworms. *Invertebrate Reproduction and Development*, 41: 35–40.

Koene, J.M., Pfőrtner, T. and Michiels, N.K. (2005). Piercing the partner's skin influences sperm uptake in the earthworm *Lumbricus terrestris*. *Behavioral Ecology and Sociobiology*, 59: 243–249.

Koene, J.M. and Schulenburg, H. (2005). Shooting darts: co-evolution and counter-adaptation in hermaphroditic snails. *BMC Evolutionary Biology*, 5: 25.

Koene, J.M., Liew, T.-S., Montagne-Wajer, K. and Schilthuizen, M. (2013). A syringe-like love dart injects male accessory gland products in a tropical hermaphrodite. *PLoS One*, 8: e69968.

Köhler, F. and Mayer, G.B. (2016). Aphally in the stylommatophoran land snail *Phaedusa* (Clausiliidae: Phaedusinae) in Timor and its systematic implications. *Molluscan Research*, 36: 239–246.

Kölliker, M. (2007). Benefits and costs of earwig (*Forficula forficularia*) family life. *Behavioral Ecology and Sociobiology*, 61: 1489–1497.

König, S., Mehlich, A.M., Büllesbach, J. and Michiels, N. (2006). Allohormones in *Lumbricus terrestris*? Mass spectrometry of the setal gland product indicates possible role of ubiquitin. *Invertebrate Reproduction and Development*, 49: 103–111.

Kononova, M.M. (1961). *Soil Organic Matter. Its Nature, its Role in Soil Formation and in Soil Fertility*. Pergamon Press, Oxford.

Konopova, B. and Akam, M. (2014). The *Hox* genes *Ultrabithorax* and *abdominal-A* specify three different types of abdominal appendage in the springtail *Orchesella cincta* (Collembola). *EvoDevo*, 5: 2.

Konopova, B., Kolosov, D. and O'Donnell, M.J. (2019). Water and ion transport across the eversible vesicles in the collophore of the springtail *Orchesella cincta*. *Journal of Experimental Biology*, 222: jeb.2006691.

Kooistra, M.J. and Brussaard, L. (1995). A micromorphological approach to the study of soil structure-soil biota interactions. pp. 55–69. *In:* Edwards, C.A., Abe, T. and Striganova, B.R. (eds.). *Structure and Function of Soil Communities*. Kyoto University Press, Kyoto.

Koop, K. and Field, J.G. (1981). Energy transformation by the supralittoral isopod *Ligia dilatata* Brandt. *Journal of Experimental Marine Biology and Ecology*, 53: 221–233.

Kopeszki, H. (1993). Auswirkungen von Düngergaben auf die Mesofauna, insbesondere Collembolenfauna, verschiedener Waldstandorte in Böhmerwald. *Zoologische Anzeiger*, 3/4: 83–98.

Kopeszki, H. and Meyer, E. (1994). Artenzusammenstellung und Abundanz von Collembolen in Waldböden Voralbergs (Österreich). *Berichte des Naturwissenschaftlich-Medizinischen Vereins in Innsbruck*, 81: 151–166.

Korsloot, A., Van Gestel, C.A.M. and Van Straalen, N.M. (2004). *Environmental Stress and Cellular Response in Arthropods*. CRC Press, Boca Raton.

Kozłowski, J. (1992). Optimal allocation of resources to growth and reproduction: Implications for age and size at maturity. *Trends in Ecology and Evolution*, 7: 15–19.

Kozłowski, J. (1996). Optimal allocation of resources explains interspecific life-history patterns in animals with indeterminate growth. *Proceedings of the Royal Society of London, Series B*, 263: 559–566.

Kozlowski, M.W. and Aoxiang, S. (2006). Ritual behaviors associated with spermatophore transfer in *Deuterosminthurus bicinctus* (Collembola: Bourletiellidae). *Journal of Ethology*, 24: 103–109.

Kramer, J. and Meunier, J. (2016). Maternal condition determines offspring behavior towards family members in the European earwig. *Behavioral Ecology*, 27: 494–500.

Krebs, C.J. (1972). *Ecology. The Experimental Analysis of Distribution and Abundance*. Harper and Row, New York.

Kregel, K.C. (2002). Heat shock proteins: modifying factors in physiological stress responses and acquired thermotolerance. *Journal of Applied Physiology*, 92: 2177–2196.

Kristensen, R.M. and Hallas, T.E. (1980). The tidal genus *Echiniscoides* and its variability, with erection of Echiniscoididae fam.n. (Tardigrada). *Zoologica Scripta*, 9: 113–127.

Kuiters, A.T. and Denneman, C.A.J. (1987). Water-soluble phenolic substances in soils under several coniferous and deciduous tree species. *Soil Biology and Biochemistry*, 19: 765–769.

Kuo, D.-H. (2017). The polychaete-to-clitellate transition: An EvoDevo perspective. *Developmental Biology*, 427: 230–240.

Kurek, P., Nowakowski, K., Rutkowski, T., Ważna, A., Cichocki, J., Zacharyasiewicz, M. et al. (2020). Underground diversity: Uropodina mites (Acari: Mesostigmata) from European badger (*Meles meles*) nests. *Experimental and Applied Acarology*, 82: 503–513.

Kurita, Y. and Wada, H. (2011). Evidence that gastropod torsion is driven by asymmetric cell profileration activated by TGF-β signalling. *Biology Letters*, 7: 759–762.

Kurup, N.C. and Prabhoo, N.R. (1977). Facultative parthenogenesis in *Cyphoderus javanus* (Collembola: Insecta). *Current Science*, 46: 168.

Kutschera, U. and Elliott, J.M. (2010). Charles Darwin's observations on the behaviour of earthworms and the evolutionary history of the giant endemic species from Germany, *Lumbricus badensis* (Oligochaeta: Lumbricidae). *Applied and Environmental Soil Science*, 2: 823047.

Kuznetsova, N. (1994). Collembola guild structure as an indicator of tree plantation conditions in urban areas. *Memorabilia Zoologica*, 49: 197–205.

Kuznetsova, N.A. and Krest'yaninova, A.I. (1998). Dynamics of springtail communities (Collembola) in hydrological series of pine forests in southern taiga. *Entomological Review*, 78: 969–981.

Kvist, S. and Siddall, M.E. (2013). Phylogenomics of Annelida revisited: a cladistic approach using genome-wide expressed sequence tag data mining and examining the effects of missing data. *Cladistics*, 29: 435–448.

Lachaud, J.-P., Cerdan, P. and Pérez-Lachaud, G. (2012). Poneromorph ants associated with parasitoid wasps of the genus *Kapala* Cameron (Hymenoptera: Eucharitidae) in French Guiana. *Psyche*, 2012: 393486.

Lagerlöf, J. and Scheller, U. (1989). Abundance and activity of Pauropoda and Symphyla (Myriapoda) in four cropping systems. *Pedobiologia*, 33: 315–321.

Lakhani, K.H. and Satchell, J.E. (1970). Production by *Lumbricus terrestris*. *Journal of Animal Ecology*, 39: 473–492.

Lam, P.K.S., Dudgeon, D. and Ma, H.H.T. (1991). Ecological energetics of populations of four sympatric isopods in a Hongkong forest. *Journal of Tropical Ecology*, 7: 475–490.

Lapanje, A., Zrimec, A., Drobne, D. and Rupnik, M. (2010). Long-term Hg pollution-induced structural shifts of bacterial community in the terrestrial isopod (*Porcellio scaber*) gut. *Environmental Pollution*, 158: 3186–3193.

Larink, O. (1997). Apomorphic and plesiomorphic characteristics in Archaeognatha, Monura and Zygentoma. *Pedobiologia*, 41: 3–8.

Larsen, B.B., Miller, E.C., Rhodes, M.K. and Wiens, J.J. (2017). Inordinate fondness multiplied and redistributed: the number of species on earth and the new pie of life. *The Quarterly Review of Biology*, 92: 229–265.

Lasebikan, B.A. (1974). Preliminary communication on microarthropods from a tropical rain forest in Nigeria. *Pedobiologia*, 14: 402–411.

Laskowski, R. (1997). Estimating fitness costs of pollution in iteroparous invertebrates. pp. 305–319. *In:* Van Straalen, N.M. and Løkke, H. (eds.) *Ecological Risk Assessment of Contaminants in Soil*. Chapman and Hall, London.

Lauber, C.L., Hamady, M., Knight, R. and Fierer, N. (2009). Pyrosequencing-based assessment of soil pH as a predictor of soil bacterial community structure at the continental scale. *Applied and Environmental Microbiology*, 75: 5111–5120.

Laudien, M. (1973). Effect of temperature on processes of growth and development. pp. 355–377. *In:* Precht, H., Christophersen, J. and Larcher, W. (eds.). *Temperature and Life*. Springer Verlag, New York.

Laumann, M., Norton, R.A., Weigmann, G., Scheu, S., Maraun, M. and Heethoff, M. (2007). Speciation in the parthenogenetic oribatid mite genus *Tectocepheus* (Acari Oribatida) as indicated by molecular phylogeny. *Pedobiologia*, 51: 111–122.

Lavelle, P. (1977). Bilan énergétique des populations naturelles du ver de terre geophage *Millsonia anomala* (Oligochetes-Acanthodrillidae) dans la savanne de Lamto (Cote d'Ivoire). *Geo-Eco-Trop*, 1: 149–157.

Lavelle, P. (1981). Stratégies de reproduction chez les vers de terre. *Acta Oecologia. Oecologia Generalis*, 2: 117–133.

Lavelle, P. (1983). The structure of earthworm communities. pp. 449–466. *In:* Satchell, J.E. (ed.). *Earthworm Ecology. From Darwin to Vermiculture*. Chapman and Hall, London.

Lavelle, P., Bignell, D.E., Lepage, M., Wolters, V., Roger, P., Ineson, P. et al. (1997). Soil function in a changing world: the role of invertebrate ecosystem engineers. *European Journal of Soil Biology*, 33: 159–193.

Lavelle, P. and Spain, A.V. (2005). *Soil Ecology*. Springer, Dordrecht.

Lavigne, D.M. (1982). Similarity in energy budgets of animal populations. *Journal of Animal Ecology*, 51: 195–206.

Lavy, D. and Verhoef, H.A. (1996). Effects of food quality on growth and body composition of the collembolan *Orchesella cincta*. *Physiological Entomology*, 21: 64–70.

Lavy, D., Van Rijn, M.J., Zoomer, H.R. and Verhoef, H.A. (2001). Dietary effects on growth, reproduction, body composition and stress resistance in the terrestrial isopods *Oniscus asellus* and *Porcellio scaber*. *Physiological Entomology*, 26: 18–25.

Leclercq, S., Dittmer, J., Bouchon, D. and Cordeaux, R. (2014). Phylogenomics of "*Candidatus* Hepatoplasma crinochetorum", a lineage of Mollicutes associated with noninsect arthropods. *Genome Biology and Evolution*, 62: 407–415.

Lee, Q. and Widden, P. (1996). *Folsomia candida*, a "fungivorous" collembolan, feeds preferentially on nematodes rather than soil fungi. *Soil Biology and Biochemistry*, 28: 689–690.
Lehtonen, J., Jennions, M.D. and Kokko, H. (2012). The many costs of sex. *Trends in Ecology and Evolution*, 27: 172–178.
Lei, J., Guo, X., Zeng, Y., Zhou, J., Gao, Q. and Yang, Y. (2021). Temporal changes in global soil respiration since 1987. *Nature Communications*, 12: 403.
Leidy, J. (1849). Dr. Leidy offered the following observations. *Proceedings of the Academy of Natural Sciences of Philadelphia*, 4: 224–233.
Leinaas, H.P. (1978). Seasonal variation in sampling efficiency of Collembola and Protura. *Oikos*, 31: 307–312.
Leinaas, H.P. (1981). Cyclomorphosis in *Hypogastrura lapponica* (Axelson, 1902) (= *H. frigida* [Axelson, 1905] syn. nov.) (Collembola, Poduridae). Morphological adaptations and selection for winter dispersal. *Zeitschrift für Zoologische Systematik und Evolutionsforschung*, 19: 278–285.
Leinaas, H.P. (1983). Synchronized moulting controlled by communication in group-living Collembola. *Science*, 219: 193–195.
Leinaas, H.P. and Bleken, E. (1983). Egg diapause and demographic strategy in *Lepidocyrtus lignorum* Fabricius (Collembola; Entomobryidae). *Oecologia*, 58: 194–199.
Le Ngoc, G. (2021). *Isolation and Characterization of Novel Enzymatic Activities from Gut Metagenomes to Support Lignocellulose Breakdown.* PhD Thesis, Vrije Universiteit Amsterdam.
Leonard, J.L., Pearse, J.S. and Harper, A.B. (2002). Comparative reproductive biology of *Ariolimax californicus* and *A. dolichophallus* (Gastropoda: Stylommatophora). *Invertebrate Reproduction and Development*, 41: 83–93.
Leonard, J.L. (2006). Sexual selection: lessons from hermaphroditic mating systems. *Integrative and Comparative Biology*, 46: 349–367.
Lesna, I. and Sabelis, M.W. (1999). Diet-dependent female choice for males with "good genes" in a soil predatory mite. *Nature*, 401: 581–584.
Lesna, I., Conijn, C., Sabelis, M. and Van Straalen, N.M. (2000). Biological control of the bulb mite, *Rhizoglyphus robini*, by the predatory mite, *Hypoaspis aculeifer*, on lilies: Predator-prey dynamics in the soil, under greenhouse and field conditions. *Biocontrol Science and Technology*, 10: 179–193.
Lesoway, M.P. and Henry, J.Q. (2021). Twisted shells, spiral cells, and asymmetries: evo-devo lessons learned from gastropods. pp. 749–766. *In: Evolutionary Developmental Biology*. A Reference Guide. Springer Nature Switzerland Cham.
Lever, J. (1979). On torsion in gastropods. pp. 5–23. *In:* Van der Spoel, S., Van Bruggen, A.C. and Lever, J. (eds.). *Pathways in Malacology*. Bohn, Scheltema and Holkema, Utrecht.
Lewis, J.G.E. (1965). The food and reproductive cycles of the centipedes *Lithobius variegatus* and *Lithobius forficatus* in a Yorkshire woodland. *Proceedings of the Zoological Society of London*, 144: 269–283.
Lewis, S.C., Dyal, L.A., Hilburn, C.F., Weitz, S., Liau, W.-S., LaMunyon, C.W. et al. (2009). Molecular evolution in *Panagrolaimus* nematodes: origins of parthenogenesis, hermaphroditism and the Antarctic species *P. davidi*. *BMC Evolutionary Biology*, 9: 15.
Lichtwardt, R.W. (1990). Fungi associated with passalid beetles and their mites. *Mycologia*, 91: 694–702.
Lichtwardt, R.W. (2012). Trichomycete gut fungi from tropical regions of the world. *Biodiversity and Conservation*, 21: 2397–2402.
Lichtwardt, R.W. (2008). Trichomycetes and the arthropod gut. pp. 3–19. *In:* Esser, K. (ed.). *The Mycota. A Comprehensive Treatise on Fungi as Experimental Systems for Basic and Applied Research. Volme VI: Human and Animal Relationships, 2nd Edition, ed. by A.A. Brakhage and P.F. Zipfel.* Springer Verlag, Berlin.
Liefting, M. and Ellers, J. (2008). Habitat-specific differences in thermal plasticity in natural populations of a soil arthropod. *Biological Journal of the Linnean Society*, 94: 265–271.
Lighton, J.R.B. (2008). *Measuring Metabolic Rates. A Manual for Scientists.* Oxford University Press, Oxford.
Lindenmann, W. (1950). Untersuchungen zur postembryonalen Entwicklung schweizerischer Orchesellen. *Revue Suisse de Zoologie*, 57: 353–428.
Linderstrøm-Lang, K. (1937). Principle of the cartesian diver applied to gasometric technique. *Nature*, 140: 108.

Lindqvist, O.V. (1970). The blood osmotic pressure of the terrestrial isopods *Porcellio scaber* Latr. and *Oniscus asellus* L., with reference to the effect of temperature and body size. *Comparative Biochemistry and Physiology*, 37: 503–510.
Link, A., Triebskorn, R. and Köhler, H.-R. (2019). TFG-β signalling is involved in torsion and shell positioning in the giant ramshorn snail *Marisa cornuarietis* (Gastropoda: Ampullariidae). *Journal of Molluscan Studies*, 85: 1–10.
Liras, P. and Martin, J.F. (2006). Gene clusters for beta-lactam antibiotics and control of their expression: why have clusters evolved, and from where did they originate? *International Microbiology*, 9: 9–19.
Little, C. (1983). *The Colonisation of Land. Origins and Adaptations of Terrestrial Animals.* Cambridge University Press, Cambridge.
Litwin, A., Nowak, M. and Różalska, S. (2020). Entomopathogenic fungi: unconventional applications. *Reviews in Environmental Science and Biotechnology*, 19: 23–42.
Liu, J. and Wu, D. (2017). Chemical attraction of conspecifics in *Folsomia candida* (Collembola). *Journal of Insect Behavior*, 30: 331–341.
Liu, D., Lian, B., Wu, C. and Guo, P. (2018). A comparative study of gut microbiota profiles of earthworms fed in three different substrates. *Symbiosis*, 74: 21–29.
Lloyd, M. (1967). Mean crowding. *Journal of Animal Ecology*, 36: 1–30.
Lodi, M. and Koene, J.M. (2016a). The love-darts of land snails: integrating physiology, morphology and behaviour. *Journal of Molluscan Studies*, 82: 1–10.
Lodi, M. and Koene, J.M. (2016b). On the effect specificity of accessory gland products transferred by the love-dart of land snails. *BMC Evolutionary Biology*, 16: 104.
Looney, C., Hanelt, B. and Zack, R.S. (2012). New records of nematomorph parasites (Nematomorpha: Gordiida) of ground beetles (Coleoptera: Carabidae) and camel crickets (Orthoptera: Raphidophoridae) in Washington State. *Journal of Parasitology*, 98: 554–559.
Lopez, P., Casane, D. and Philippe, H. (2002). Heterotachy, and important process of protein evolution. *Molecular Biology and Evolution*, 19: 1–7.
Loranger, G., Ponge, J.-F., Blanchart, E. and Lavelle, P. (1998). Impact of earthworms on the diversity of microarthropods in a vertisol (Martinique). *Biology and Fertility of Soils*, 27: 21–26.
Lowry, J. and Myers, A.A. (2013). A phylogeny and classification of the Amphipoda with the establishment of the new order Ingolfiellida (Crustacea: Pericardia). *Zootaxa*, 3610: 1–80.
Lozano-Fernandez, J., Carton, R., Tanner, A.R., Puttick, M.N., Blaxter, M., Vinther, J. et al. (2016). A molecular palaeobiological exploration of arthropod terrestrialization. *Philosophical Transactions of the Royal Society B*, 371: 20150133.
Lozano-Fernandez, J., Giacomelli, M., Fleming, J.F., Chen, A., Vinther, J., Thomsen, P.F. et al. (2019). Pancrustacean evolution illuminated by taxon-rich genomic-scale data sets with an expanded remipede sampling. *Genome Biology and Evolution*, 11: 2055–2070.
Lozano-Fernandez, J., Tanner, A.R., Puttick, M.N., Vinther, J., Edgecombe, G.D. and Pisani, D. (2020). A Cambrian-Ordovician terrestrialization of arachnids. *Frontiers in Genetics*, 11: 182.
Luan, Y.-X., Mallat, J.M., Xie, R.-D., Yang, Y.-M. and Yin, W.-Y. (2005). The phylogenetic position of three basal-hexapod groups (Protura, Diplura and Collembola) based on ribosomal RNA gene sequences. *Molecular Biology and Evolution*, 22: 1597–1592.
Lubin, Y.D. and Henschel, J.R. (1990). Foraging at the thermal limit: burrowing spiders (*Seothyra*, Eresidae) in the Namib desert dunes. *Oecologia*, 84: 461–467.
Lund, M.B., Holmstrup, M., Lomstein, B.A., Damgaard, C. and Schramm, A. (2010a). Beneficial effects of *Verminephrobacter* nephridial symbionts on the fitness of the earthworm *Aporrectodea tuberculata. Applied and Environmental Microbiology*, 76: 4738–4743.
Lund, M.B., Davidson, S.K., Holmstrup, M., James, S., Kjeldsen, K.U., Stahl, D.A. et al. (2010b). Diversity and host specificity of the *Verminephrobacter*–earthworm symbiosis. *Environmental Microbiology*, 12: 2142–2151.
Lund, M.B., Mogensen, M.F., Marshall, I.P.G., Albertsen, M., Viana, F. and Schramm, A. (2018). Genomic insights into the *Agromyces*-like symbiont of earthworms and its distribution among host species. *FEMS Microbiology Ecology*, 94: fiy068.
Luxton, M. (1975). Studies on the oribatid mites of a Danish beech wood soil. II. Biomass, calorimetry, and respirometry. *Pedobiologia*, 15: 161–200.

Luxton, M. (1981a). Studies on the oribatid mites of a Danish beech wood soil. IV. Developmental biology. *Pedobiologia*, 21: 312–340.

Luxton, M. (1981b). Studies on the oribatid mites of a Danish beech wood soil VI. Seasonal population changes. *Pedobiologia*, 21: 387–409.

Luxton, M. (1981c). Studies on the oribatid mites of a Danish beech wood soil. VII. Energy budgets. *Pedobiologia*, 22: 77–111.

Ma, H.H.T., Dudgeon, D. and Lam, P.K.S. (1991a). Seasonal changes in populations of three sympatric isopods in a Hongkong forest. *Journal of Zoology (London)*, 224: 347–365.

Ma, H.H.T., Lam, P.K.S. and Dudgeon, D. (1991b). Inter- and intraspecific variation in the life-histories of three sympatric isopods in a Hongkong forest. *Journal of Zoology (London)*, 224: 677–687.

Ma, Y., Chen, W.J., Li, Z.-H., Zhang, F., Gao, Y. and Luan, Y.-X. (2017a). Revisiting the phylogeny of *Wolbachia* in Collembola. *Ecology and Evolution*, 7: 2009–2017.

Ma, L., Xie, Y., Han, Z., Giesy, J.P. and Zhang, X. (2017b). Responses of earthworms and microbial communities in their guts to Triclosan. *Chemosphere*, 168: 1194–1202.

Ma, S., He, F., Tian, D., Zou, D., Yan, Z., Yang, Y. et al. (2018). Variations and determinants of carbon content in plants: a global synthesis. *Biogeosciences*, 15: 693–702.

MacArthur, R.H. (1962). Some generalized theorems of natural selection. *Proceedings of the National Academy of Sciences of the United States of America*, 48: 1893–1897.

MacArthur, R.H. and Wilson, E.O. (1967). *The Theory of Island Biogeography*. Princeton University Press, Princeton.

MacArthur, R. (1968). Selection for life tables in periodic environments. *The American Naturalist*, 102: 381–383.

Maclagan, D.S. (1932). An ecological study of the "lucerne flea" (*Smynthurus viridis*, Linn.) - I. *Bulletin of Entomological Research*, 23: 101–145.

MacLean, S.F., Douce, G.K., Morgan, E.A. and Skeel, M.A. (1977). Community organization in the soil invertebrates of Alaskan arctic tundra. pp. 90–101. *In:* Lohm, U. and Persson, T. (eds.). *Soil Organisms as Components of Ecosystems*. Ecological Bulletins, Stockholm.

Madec, L., Guiller, A., Coutellec-Vreto, M.-A. and Desbuquois, C. (1998). Size-fecundity relationships in the land snail *Helix aspersa*: preliminary results on a form outside the norm. *Invertebrate Reproduction and Development*, 34: 83–90.

Majoros, G., Fukár, O. and Farkas, R. (2010). Autochthonous infection of dogs and slugs with *Angiostrongylus vasorum* in Hungary. *Veterinary Parasitology*, 174: 351–354.

Mąkol, J. and Felska, M. (2011). New records of spiders (Araneae) as hosts of terrestrial Parasitengona mites (Actinotrichida: Prostigmata). *The Journal of Arachnology*, 39: 352–354.

Mąkol, J., Felska, M. and Król, Z. (2017). New genus and species of microtrombidiid mite (Actinotrichida: Trombidioidea, Microtrombidiidae) parasitizing spiders (Araneae: Araneidae) in Costa Rica. *Acarologia*, 57: 517–527.

Mali, B., Grohme, M.A., Förster, F., Dandekar, T., Schnölzer, M., Reuter, D. et al. (2010). Transcriptome survey of the anhydrobiotic tardigrade *Milnesium tardigradum* in comparison with *Hypsibius dujardini* and *Richtersius coronifer*. *BMC Genomics*, 11: 168.

Mansfield, J.H. (2013). cis-regulatory change associated with snake body-plan evolution. *Proceedings of the National Academy of Sciences of the United States of America*, 110: 10473–10474.

Maraun, M., Heethoff, M., Schneider, K., Scheu, S., Weigmann, G., Cianciolo, J. et al. (2004). Molecular phylogeny of oribatid mites (Oribatida, Acari): evidence for multiple radiations of parthenogenetic lineages. *Experimental and Applied Acarology*, 33: 183–201.

Maraun, M., Fronczek, S., Marian, F., Sandmann, D. and Scheu, S. (2013). More sex at higher altitudes: Changes in the frequency of parthenogenesis in oribatid mites in tropical montane forests. *Pedobiologia*, 56: 185–190.

Marc, P., Canard, A. and Ysnel, F. (1999). Spiders (Araneae) useful for pest limitation and bioindication. *Agriculture, Ecosystems and Environment*, 74: 229–273.

Marcadé, I., Souty-Grosset, C., Bouchon, D., Rigaud, T. and Raimond, R. (1999). Mitochondrial DNA variability and *Wolbachia* infection in two sibling woodlice species. *Heredity*, 83: 71–78.

Marek, P.E., Buzatto, B.A., Shear, W.A., Means, J.C., Black, D.G., Harvey, M.S. et al. (2021). The first true millipede–1306 legs. *Scientific Reports*, 11: 23126.

Margulis, L., Jorgensen, J.Z., Dolan, S., Kolchinsky, R., Rainey, F.A. and Lo, S.-C. (1998). The *Arthromitus* stage of *Bacillus cereus*: Intestinal symbionts of animals. *Proceedings of the National Academy of Sciences of the United States of America*, 95: 1236–1241.
Marinissen, J.C.Y. and Van den Bosch, F. (1992). Colonization of new habitats by earthworms. *Oecologia*, 91: 371–376.
Marley, N.J., McInnes, S.J. and Sands, C.J. (2011). Phylum Tardigrada: A re-evaluation of the Parachela. *Zootaxa*, 2819: 51–64.
Marotta, R., Ferraguti, M., Erséus, C. and Gustavsson, L.M. (2008). Combined-data phylogenetics and character evolution of Clitellata (Annelida) using 18S rDNA and morphology. *Zoological Journal of the Linnean Society*, 154: 1–26.
Martin, W.F. (2017). Too much eukaryote LGT. *Bioessays*, 39: 1700115.
Martínez-Ansemil, E., Creuzé des Châtelliers, M., Martin, P. and Sambugar, B. (2012). The Parvidrilidae—a diversified groundwater family: description of six new species from southern Europe, and clues for its phylogenetic position within Clitellata (Annelida). *Zoological Journal of the Linnean Society*, 166: 530–558.
Martinsson, S. and Erséus, C. (2014). Cryptic diversity in the well-studied terrestrial worm *Cognettia sphagnetorum* (Clitellata: Enchytraeidae). *Pedobiologia*, 57: 27–35.
Martinsson, S. and Erséus, C. (2021). Cryptic Clitellata: molecular species delimitation of clitellate worms (Annelida): an overview. *Diversity*, 13: 36.
Mason, C.F. (1970). Snail populations, beech litter production, and the role of snails in litter decomposition. *Oecologia*, 5: 215–239.
Mason, K.S., Kwapich, C.L. and Tschinkel, W.R. (2015). Respiration, worker body size, tempo and activity in whole colonies of ants. *Physiological Entomology*, 40: 149–165.
Massoud, Z. (1976). Essai de synthèse sur la phylogénie des Collemboles. *Revue d'Écologie et de Biologie du Sol*, 13: 241–252.
Mathieu, J., Caro, G. and Dupont, L. (2018). Methods for studying earthworm dispersal. *Applied Soil Ecology*, 123: 339–344.
Matthysse, A.G., Deschet, K., Williams, M., Marry, M., White, A.R. and Smith, W.C. (2004). A functional cellulose synthase from ascidian epidermis. *Proceedings of the National Academy of Sciences of the United States of America*, 101: 986–991.
Mayer, W.E., Herrmann, M. and Sommer, R.J. (2007). Phylogeny of the nematode genus *Pristionchus* and implications for biodiversity, biogeography and the evolution of hermaphroditism. *BMC Evolutionary Biology*, 7: 104.
Mayer, W.E., Schuster, L.N., Bartelmes, G., Dieterich, C. and Sommer, R.J. (2011). Horizontal gene transfer of microbial cellulases into nematode genomes is associated with functional assimilation and gene turnover. *BMC Evolutionary Biology*, 11: 13.
Maynard Smith, J. (1986). Contemplating life without sex. *Nature*, 324: 300–301.
McAllister, C.T. and Hnida, J.A. (2021). *Pfeifferinella gugleri* (Apicomplexa: Pfeifferinellidae) in half-lidded ovals, *Patera binneyana* (Gastropoda: Stylommatophora: Polygyridae), from Oklahoma, U.S.A. *Comparative Parasitology*, 88: 1–3.
McElwee, J.J., Schuster, E., Blanc, E., Thomas, J.H. and Gems, D. (2004). Shared transcriptional signature in *Caenorhabditis elegans* dauer larvae and long-lived *daf-2* mutants implicates detoxification system in longevity assurance. *The Journal of Biological Chemistry*, 279: 44533–44543.
McFall-Ngai, M., Hadfield, M.G., Bosch, T.C.G., Carey, H.V., Domazet-Loso, T., Douglas, A.E. et al. (2013). Animals in a bacterial world, a new imperative for the life sciences. *Proceedings of the National Academy of Sciences of the United States of America*, 110: 3229–3236.
McGaughran, A., Stevens, M.I., Hogg, I.D. and Carapelli, A. (2011). Extreme glacial legacies: a synthesis of the Antarctic springtail phylogeographic record. *Insects*, 2: 62–82.
Meehan, T.D. (2006). Mass and temperature dependence of metabolic rate in litter and soil invertebrates. *Physiology and Biochemical Zoology*, 79: 878–884.
Meehan, T.D., Drumm, P.K., Farrar, R.S., Oral, K., Lanier, K.E., Pennington, E.A. et al. (2006). Energetic equivalence in a soil arthropod community from an aspen-conifer forest. *Pedobiologia*, 50: 307–312.
Meier, P. and Zettel, J. (1997). Cold hardiness in *Entomobrya nivalis* (Collembola, Entomobryidae): annual cycle of polyols and antifreeze proteins, and antifreeze triggering by temperature and photoperiod. *Journal of Comparative Physiology B*, 167: 297–304.

Mertens, J. and Bourgoignie, R. (1977). Aggregation pheromone in *Hypogastrura viatica* (Collembola). *Behavioral Ecology and Sociobiology*, 2: 41–48.

Meyer, E. (1979). Life-cycles and ecology of high alpine Nematophora. pp. 295–306. *In:* Camatini, M. (ed.). *Myriapod Biology*. Academic Press London.

Meyer, E. (1985). Distribution, activity, life-history and standing crop of Julidae (Diplopoda, Myriapoda) in the Central High Alps (Tyrol, Austria). *Holarctic Ecology*, 8: 141–150.

Meyer, E., Wäger, H. and Thaler, K. (1985). Struktur und jahreszeitliche Dynamik von *Neobisium*-Populationen in zwei Höhenstufen in Nordtirol (Österreich) (Arachnida: Pseudoscorpiones). *Revue d'Écologie et de Biologie du Sol*, 22: 221–232.

Michel-Salzat, A., Cordaux, R. and Bouchon, D. (2001). *Wolbachia* diversity in the *Porcellionides pruinosus* complex of species (Crustacea: Oniscidea): evidence for host-dependent patterns of infection. *Heredity*, 87: 428–434.

Michiels, N.K., Holmer, A. and Vorndran, I.C. (2001). Precopulatory assessment in relation to body size in the earthworm *Lumbricus terrestris*: avoidance of dangerous liaisons? *Behavioural Ecology*, 12: 612–618.

Michiels, N.K. and Koene, J.M. (2006). Sexual selection favors harmful mating in hermaphrodites more than in gonochorists. *Integrative and Comparative Biology*, 46: 473–480.

Mikaelyan, A., Dietrich, C., Köhler, T., Poulsen, M., Sillam-Dussès, D. and Brune, A. (2015). Diet is the primary determinant of bacterial community structure in the guts of higher termites. *Molecular Ecology*, 24: 5284–5295.

Mikaelyan, A., Meuser, K. and Brune, A. (2017). Microenvironmental heterogeneity of gut compartments drives bacterial community structure in wood- and humus-feeding higher termites. *FEMS Microbiology Ecology*, 93, fiw210.

Milne, S. (1962). Phenology of a natural population of soil Collembola. *Pedobiologia*, 2: 41–52.

Misof, B., Lin, S., Meusemann, K., Peters, R.S., Donath, A., Mayer, C. et al. (2014). Phylogenomics resolves the timing and pattern of insect evolution. *Science*, 346: 763–767.

Mitchell, M.J. (1977). Population dynamics of oribatid mites (Acari, Cryptostigmata) in an aspen woodland soil. *Pedobiologia*, 17: 305–319.

Mitchell, M.J. (1979). Energetics of oribatid mites (Acari: Cryptostigmata) in an aspen Woodland soil. *Pedobiologia*, 19: 89–98.

Mitić, B.M., Antić, D.Ž., Ilić, B.S., Makarov, S.E., Lučić, L.R. and Ćurčić, B.P.M. (2012). Parental care in *Cryptos hortensis* (Donovan) (Chilopoda: Scolopendromorpha) from Serbia, The Balkan Peninsula. *Archives of Biological Sciences*, 64: 1117–1121.

Møbjerg, N. and Dahl, C. (1996). Studies on the morphology and ultrastructure of the Malpighian tubules of *Halobiotus crispae* Kristensen, 1982 (Eutardigrada). *Zoological Journal of the Linnean Society*, 116: 85–99.

Møbjerg, N. and Cardoso Neves, R. (2021). New insights into survival strategies of tardigrades. *Comparative Biochemistry and Physiology, Part A*, 254: 110890.

Mocquard, J.P., Juchault, P. and Souty-Grosset, C. (1989). The role of environmental factors (temperature and photoperiod) in the reproduction of the terrestrial isopod *Armadillidium vulgare* (Latreille, 1804). *Monitore Zoologico Italiano, Monografia*, 4: 455–475.

Moczek, A.P. and Emlen, D.J. (1999). Proximate determination of male horn dimorphism in the beetle *Onthophagus taurus* (Coleoptera: Scarabaeidae). *Journal of Evolutionary Biology*, 12: 27–37.

Møller, P., Lund, M.B. and Schramm, A. (2015). Evolution of the tripartite symbiosis between earthworms, *Verminephrobacter* and *Flexibacter*-like bacteria. *Frontiers in Microbiology*, 6: 529.

Monge-Najera, J. (1995). Phylogeny, biogeography and reproductive trends in the Onychophora. *Zoological Journal of the Linnean Society*, 114: 21–60.

Monroy, F., Aira, M., Velando, A. and Domínguez, J. (2005). Size-assortative mating in the earthworm *Eisenia fetida* (Oligochaeta, Lumbricidae). *Journal of Ethology*, 23: 69–70.

Montgomery, Jr., T.H. (1903). Studies on the habits of spiders, particularly those of the mating period. *Proceedings of the Academy of Natural Sciences of Philadelphia*, 55: 59–149.

Moore, J. (1983). Responses of an avian predator and its isopod prey to an acanthocephalan parasite. *Ecology*, 64: 1000–1015.

Moore, J. (1984). Parasites that change the behavior of their host. *Scientific American*, 250: 108–115.

Moore, J. and Lasswell, J. (1986). Altered behavior in isopods (*Armadillidium vulgare*) infected with the nematode *Dispharynx nasuta*. *Journal of Parasitology*, 72: 186–189.

Morgan, A.J. (1984). The localization of heavy metals in the tissues of terrestrial invertebrates by electron microprobe X-ray analysis. *Scanning Electron Microscopy*, 4: 1847–1865.

Morgan, E.J. and Morgan, A.J. (1989). The effect of lead incorporation on the elemental composition of earthworm (Annelida, Oligochaeta) chloragosome granules. *Histochemistry*, 92: 237–241.

Morgan, A.J., Stürzenbaum, S.R. and Kille, P. (1999). A short overview of molecular biomarker strategies with particular regard to recent developments in earthworms. *Pedobiologia*, 43: 574–584.

Morisita, M. (1962). Iδ-index, a measure of dispersion of individuals. *Researches on Population Ecology*, 4: 1–7.

Morris, R.F. (1959). Single-factor analysis in population dynamics. *Ecology*, 40: 580–588.

Moss, S.T. and Taylor, J. (1996). Mycobionts in the guts of millipedes—the Eccrinales. *Mycologist*, 10: 121–124.

Mougenot, F., Combe, M. and Jeanson, R. (2012). Ontogenesis and dynamics of aggregation in a solitary spider. *Animal Behaviour*, 84: 391–398.

Mulder, C., Cohen, J.E., Setälä, H., Bloem, J. and Breure, A.M. (2005). Bacterial traits, organism mass, and numerical abundance in the detrital soil food web of Dutch agricultural grasslands. *Ecology Letters*, 8: 80–90.

Mulder, C. (2006). Driving forces from soil invertebrates to ecosystem functioning: the allometric perspective. *Naturwissenschaften*, 93: 467–479.

Murienne, J., Daniels, S.R., Buckley, T.R., Mayer, G. and Giribet, G. (2014). A living fossil tale of Pangaean biogeography. *Proceedings of the Royal Society B*, 281: 20132648.

Mustonen, M., Haimi, J., Väisänen, A. and Knott, K.E. (2014). Metallothionein gene expression differs in earthworm populations with different exposure history. *Ecotoxicology*, 23: 1732–1743.

Nagae, S., Sato, K., Tanabe, T. and Hasegawa, K. (2021). Symbiosis of the millipede parasitic nematodes Rhigonematoidea and Thelastomatoidea with evolutionary different origins. *BMC Evolutionary Biology*, 21: 120.

Nakamori, T., Fujimori, A., Kinoshita, K., Ban-nai, T., Kubota, Y. and Yoshida, S. (2010). mRNA expression of a cadmium-responsive gene is a sensitive biomarker of cadmium exposure in the soil collembolan *Folsomia candida*. *Environmental Pollution*, 158: 1689–1695.

Nannipieri, P. (2020). Soil is still an unknown biological system. *Applied Sciences*, 10: 3717.

Naranjo-Ortiz, M.A. and Gabaldón, T. (2019). Fungal evolution: major ecological adaptations and evolutionary transitions. *Biological Reviews*, 94: 1443–1476.

Nardi, F., Spinsanti, G., Boore, J.L., Carapelli, A., Dallai, R. and Frati, F. (2003). Hexapod origins: monophyletic or paraphyletic? *Science*, 299: 1887–1889.

Nardi, J.B., Bee, C.M. and Taylor, S.J. (2016). Compartmentalization of microbial communities that inhabit the hindguts of millipedes. *Arthropod Structure and Development*, 45: 462–474.

Nasri-Ammar, K., Souty-Grosset, C. and Mocquard, J.P. (2001). Time measurement in the photoperiodic induction of sexual rest in the terrestrial Isopod *Armadillidium vulgare* (Latreille). *Comptes Rendues d'Academie des Sciences de Paris, Sciences de la vie*, 324: 701–707.

Nathan, D., Getz, W.M., Revilla, E., Holyoak, M., Saltz, D. and Smouse, P.E. (2008). A movement ecology paradigm for unifying organismal movement research. *Proceedings of the National Academy of Sciences of the United States of America*, 105: 19052–19059.

Nelson, D. (2002). Current status of the Tardigrada: evolution and ecology. *Integrative and Comparative Biology*, 42: 652–659.

Nelson, D.R., Bartels, P.J. and Guil, N. (2018). Tardigrade ecology. pp. 163–210. *In:* Schill, R.O. (ed.). *Water Bears: The Biology of Tardigrades. Zoological Monographs 2*. Springer Nature, Cham

Newmark, P.A. and Sánchez Alvarado, A. (2002). Not your father's planarian: a classical model enters the era of functional genomics. *Nature Reviews Genetics*, 3: 210–219.

Nichols, P.B., Nelson, D.R. and Garey, J.R. (2006). A family level analysis of tardigrade phylogeny. *Hydrobiologia*, 558: 53–60.

Nichols, E., Spector, S., Louzada, J., Larsen, T., A'mezquita, S., Favila, M.E. et al. (2008). Ecological functions and ecosystem services provided by Scarabaeinae dung beetles. *Biological Conservation*, 141: 1461–1474.

Nickerson, J.C., Snyder, D.E. and Oliver, C.C. (1979). Acoustical burrows constructed by mole crickets. *Annals of the Entomological Society of America*, 72: 438–440.

Nicklas, N.L. and Hoffmann, R.J. (1981). Apomictic parthenogenesis in a hermaphroditic terrestrial slug, *Deroceras laeve* (Müller). *Biological Bulletin*, 160: 123–135.

Nieboer, E. and Richardson, D.H.S. (1980). The replacement of the nondescript term "heavy metals" by a biologically and chemically significant classification of metal ions. *Environmental Pollution Series B*, 1: 3–26.

Nielsen, M.G. (1974). Number and biomass of worker ants in a sandy heath area in Denmark. *Natura Jutlandica*, 17: 91–95.

Nielsen, M.G., Skyberg, N. and Winther, L. (1976). Studies on *Lasius flavus* F. (Hymenoptera, Formicidae): I. Population density, biomass and distribution of nests. *Entomologiske Meddelelser*, 44: 65–75.

Nielsen, C. (1995). *Animal Evolution. Interrelationships of the Living Phyla.* Oxford University Press, Oxford.

Nielsen, C. (2013). Life cycle evolution: was the eumetazoan ancestor a holopelagic, planktotrophic gastraea? *BMC Evolutionary Biology*, 13: 171.

Nielsen, C. (2018). Origin of the trochophora larva. *Royal Society Open Science*, 5: 180042.

Niijima, K. (1973). Experimental studies on the life history, fecundity and growth of *Sinella curviseta* (Apterygota, Collembola). *Pedobiologia*, 13: 186–204.

Niijima, K. (1975). Seasonal changes in collembolan populations in a warm temperate forest of Japan II. Population dynamics of the dominant species. *Pedobiologia*, 15: 40–52.

Niijima, K., Nii, M. and Yoshimura, J. (2021). Eight-year periodical outbreaks of the train millipede. *Royal Society Open Science*, 8: 201399.

Nijhout, H.F. (2019). The multistep morphing of beetle horns. *Science*, 366: 946–947.

Nilsson, E. and Bengtsson, G. (2004). Endogenous free fatty acids repel and attract Collembola. *Journal of Chemical Ecology*, 30: 1431–1443.

Niu, G., Johnson, R.M. and Berenbaum, M.R. (2011). Toxicity of mycotoxins to honeybees and its amelioration by propolis. *Apidologie*, 42: 79–87.

Nixon, G.E.J. (1980). Diapriidae (Diapriinae) Hymenoptera, Proctotrupoidea. *In:* Fitton, M.G. (ed.). *Handbook for the Identification of British Insects. Volume VIII, Part 3.* Royal Entomological Society of London, London.

Noordam, D.J. and Van der Vaart-de Vlieger, S.H. (1943). Een onderzoek naar samenstelling en beteekenis van de fauna van eikenstrooisel. *Boschbouw Tijdschrift*, 16: 470–492.

Nordström, S. and Rundgren, S. (1974). Environmental factors and lumbricid associations in Southern Sweden. *Pedobiologia*, 14: 1–27.

Norton, R.A. and Palmer, S.C. (1991). The distribution, mechanisms and evolutionary significance of parthenogenesis in oribatid mites. pp. 107–136. *In:* Schuster, R. and Murphy, P.W. (eds.). *The Acari. Reproduction, Development and Life-history Strategies.* Chapman and Hall, London.

Notenboom, J. (1991). Marine regression and the evolution of groundwater dwelling amphipods (Crustacea). *Journal of Biogeography*, 18: 437–454.

Notenboom, J., Cruys, K., Hoekstra, J. and Van Beelen, P. (1992). Effect of ambient oxygen concentration upon the acute toxicity of chlorophenols and heavy metals to the groundwater copepod *Parastenocaris germanica* (Crustacea). *Ecotoxicology and Environmental Safety*, 24: 131–143.

Novo, M., Almodóvar, A., Fernández, R.M., Gutiérrez, M. and Díaz Cosín, D.J. (2010). Mate choice of an endogeic earthworm revealed by microsatellite markers. *Pedobiologia*, 53: 375–379.

Nowak, E. (1975). Population density of earthworms and some elements of their production in several grassland environments. *Ekologia Polska*, 23: 495–491.

Nowak, M.A., Tarnita, C.E. and Wilson, E.O. (2010). The evolution of eusociality. *Nature*, 466: 1057–1062.

Nowakowska, A., Caputa, M. and Rogalska, J. (2010). Natural aestivation and antioxidant defence in *Helix pomatia*: effect of acclimation to various external conditions. *Journal of Molluscan Studies*, 76: 354–359.

Nurminen, M. (1967). Ecology of enchytraeids (Oligochaeta) in Finnish coniferous forest soil. *Annales Zoologici Fennici*, 4: 147–157.

Nuutinen, V. and Butt, K.R. (1997a). Pre-mating behaviour of the earthworm *Lumbricus terrestris*. *Soil Biology and Biochemistry*, 29: 307–308.

Nuutinen, V. and Butt, K.R. (1997b). The mating behaviour of the earthworm *Lumbricus terrestris* (Oligochaeta: Lumbricidae). *Journal of Zoology*, 242: 783–798.

O'Brien, M.F. and Pellet, S. (2022). Diseases of Gastropoda. *Frontiers in Immunology*, 12: 802920.

Ojala, R. and Huhta, V. (2001). Dispersal of microarthropods in forest soil. *Pedobiologia*, 45: 443–450.

Oliveira, I.d.S., Bai, M., Jahn, H., Gross, V., Martin, C., Hammel, J.U. et al. (2016). Earliest onychophoran in amber reveals Gondwanan migration patterns. *Current Biology*, 26: 2594–2601.

Oostenbrink, M. (1954). Een doelmatige methode voor het toetsen van aaltjesbestrijdingsmiddelen in grond met *Hoplolaimus uniformis* als proefdier. *Mededelingen van de Landbouwhogeschool en der opzoekingsstations van den Staat te Gent*, 19: 377–408.

Orgiazzi, A., Bardgett, R.D., Barrios, E., Behan-Pelletier, V., Briones, M.J.I., Chotte, J.-L. et al. (2016). *Global Soil Biodiversity Atlas*. European Commission, Publications Office of the European Union, Luxembourg.

Ortiz-Ceballos, A.I., Hernández-García, M.E.C. and Galindo-González, J. (2009). Nest and feeding chamber construction for cocoon incubation in the tropical earthworm: *Pontoscelex corethrurus*. *Dynamic Soil, Dynamic Plant*, Special Issue 3: 115–118.

Ortiz-Ceballos, A.I., Pérez-Staples, D. and Pérez-Rodríguez, P. (2016). Nest site selection and nutritional provision through excreta: a form of parental care in a tropical endogeic earthworm. *PeerJ*, 4: e2032.

Osca, D., Templado, J. and Zardoya, R. (2015). Caenogastropod mitogenomics. *Molecular Phylogenetics and Evolution*, 93: 118–128.

Osler, G.H.R. and Beattie, A.J. (1999). Relationships between body length, number of species and species abundance in soil mites and beetles. *Pedobiologia*, 43: 401–412.

Palmer, S.C. and Norton, R.A. (1991). Taxonomic, geographic and seasonal distribution of theylotokous parthenogenesis in the Desmonomata (Acari: Oribatida). *Experimental and Applied Acarology*, 12: 67–81.

Palmiter, R.D. (1994). Regulation of metallothionein genes by heavy metals appears to be mediated by a zinc-sensitive inhibitor that interacts with constitutively active transcription factor, MTF-1. *Proceedings of the National Academy of Sciences of the United States of America*, 91: 1219–1223.

Panganiban, G., Irvine, S.M., Lowe, C., Roehl, H., Corley, L.S., Sherbon, B. et al. (1997). The origin and evolution of animal appendages. *Proceedings of the National Academy of Sciences of the United States of America*, 94: 5162–5166.

Paoletti, M.G. (1999). The role of earthworms for assessment of sustainability and as bioindicators. *Agriculture, Ecosystems and Environment*, 74: 137–155.

Paoletti, M.G. and Hassall, M. (1999). Woodlice (Isopoda: Oniscidea): their potential for assessing sustainability and use as bioindicators. *Agriculture, Ecosystems and Environment*, 74: 157–165.

Parker, G.A. and Birkhead, T.R. (2013). Polyandry: the history of a revolution. *Philosophical Transactions of the Royal Society B*, 368: 2012.0335.

Parniske, M. (2008). Arbuscular mycorrhiza: the mother of plant root endosymbioses. *Nature Reviews Microbiology*, 6: 763–775.

Pass, D.A., Morgan, A.J., Read, D.S., Field, D., Weightman, A.J. and Kille, P. (2015). The effect of anthropogenic arsenic contamination on the earthworm microbiome. *Environmental Microbiology*, 17: 1884–1896.

Paul, E.A. and Clark, F.E. (1989). *Soil Microbiology and Biochemistry*. Academic Press Inc., San Diego.

Pennington, J.T. and Chia, F.-S. (1985). Gastropod torsion: a test of Garstang's hypothesis. *Biological Bulletin*, 169: 391–396.

Perrot-Minnot, M.-J. and Norton, R.A. (1997). Obligate thelytoky in oribatid mites: no evidence for *Wolbachia* inducement. *The Canadian Entomologist*, 129: 691–698.

Pętal, J., Andrzejewska, L., Breymeyer, A. and Olechowicz, E. (1971). Productivity investigation of two types of meadows in the Vistula valley. X. The role of ants as predators in a habitat. *Ekologia Polska*, 19: 213–222.

Peters, R.H. (1983). *The Ecological Implications of Body Size*. Cambridge University Press, Cambridge.

Petersen, H. (1978). Sex-ratios and the extent of parthenogenetic reproduction in some collembolan populations. pp. 19–35. *In:* Dallai, R. (ed.). *First International Seminar on Apterygota*. Atti dell'accademia Della Scienze di Siena, Detta Dei Fisiocritici, Siena.

Petersen, H. (1980). Population dynamic and metabolic characterization of Collembola species in a beech forest ecosystem. pp. 806–833. *In:* Dindal, D.L. (ed.). *Soil Biology as Related to Land Use Practices*. US-EPA, Washington.

Petersen, H. (1981). The respiratory metabolism of Collembola species from a Danish beech wood. *Oikos*, 37: 273–286.

Petersen, H. and Luxton, M. (1982). A comparative analysis of soil fauna populations and their role in decomposition processes. *Oikos*, 39: 287–388.

Petersen, H. and Gjelstrup, P. (1987). Response of soil microarthropod populations to temporary reclamation of an old *Calluna-Deschampsia* heathland. pp. 426–430. *In:* Striganova, B.R. (ed.). *Soil Fauna and Soil Fertility*. Nauka, Moscow.

Petersen, H. (2000). Collembola populations in an organic crop rotation: population dynamics and metabolism after conversion from clover-grass ley to spring barley. *Pedobiologia*, 44: 502–515.

Petersen, L. (1991). Pedological research in Denmark. *Folia Geographica Danica*, 19: 9–48.

Petersen, S.O. and Holmstrup, M. (2000). Temperature effects on lipid composition of the earthworms *Lumbricus rubellus* and *Eisenia nordenskioeldi*. *Soil Biology and Biochemistry*, 32: 1787–1791.

Petersen, C.R., Holmstrup, M., Malmendal, A., Bayley, M. and Overgaard, J. (2008). Slow desiccation improves dehydration tolerance and accumulation of compatible osmolytes in earthworm cocoons (*Dendrobaena octaedra* Savigny). *The Journal of Experimental Biology*, 211: 1903–1910.

Petrusewicz, K. and Macfadyen, A. (1970). *Productivity of Terrestrial Animals. Principles and Methods*. IBP Handbook 13. Blackwell Scientific Publications, Oxford.

Pfander, I. and Zettel, J. (2004). Chemical communication in *Ceratophysella sigillata* (Collembola: Hypogastruridae): intraspecific reaction to alarm substances. *Pedobiologia*, 48: 575–580.

Pflug, A. and Wolters, V. (2002). Collembola communities along a European transect. *European Journal of Soil Biology*, 38: 301–304.

Philippot, L., Spor, A., Hénault, C., Bru, D., Bizouard, F., Jones, C.M. et al. (2013). Loss in microbial diversity affects nitrogen cycling in soil. *The ISME Journal*, 7: 1609–1619.

Phillips, H.R.P., Guerra, C.A., Bartz, M.L.C., Briones, M.J.L., Brown, G., Crowther, T.W. et al. (2019). Global distribution of earthworm biodiversity. *Science*, 366: 480–485.

Phillipson, J. (1971). *Methods of Study in Quantitative Soil Ecology: Population, Production and Energy Flow*. IBP Handbook 18. Blackwell, Oxford.

Phillipson, J., Abel, R., Steel, J. and Woodell, S.R.J. (1977). Nematode numbers, biomass and respiratory metabolism in a beech woodland – Wytham Woods, Oxford. *Oecologia*, 27: 141–155.

Phillipson, J. (1983). Life cycle, numbers, biomass and respiratory metabolism of *Trichoniscus pusillus* (Crustacea, Isopoda) in a beech woodland – Wytham Woods, Oxford. *Oecologia*, 57: 339–343.

Phillipson, J. and Abel, R. (1983). Snail numbers, biomass and respiratory metabolism in a beech woodland – Wytham Woods. *Oecologia*, 57: 333–338.

Phillipson, J. and Meyer, E. (1984). Diplopod numbers and distribution in a British beechwood. *Pedobiologia*, 26: 83–94.

Pianka, E.R. (1970). On r- and K-selection. *The American Naturalist*, 104: 592–597.

Pianka, E.R. (1972). r- and K-selection or b and d selection? *The American Naturalist*, 106: 581–588.

Pianka, E.R. (1976). Natural selection of optimal reproductive tactics. *American Zoologist*, 16: 775–784.

Pianka, E.R. (1978). *Evolutionary Ecology*. Harper and Row, New York.

Poinar, Jr., G., Rykken, J. and LaBonte, J. (2004). *Parachordodes tegonotus* n. sp. (Gordioidea: Nematomorpha), a hairworm parasite of ground beetles (Carabidae: Coleoptera), with a summary of gordiid parasites of carabids. *Systematic Parasitology*, 58: 139–148.

Poinsot-Balaguer, N. and Barra, J.A. (1991). L'anhydrobiose : un problème biologique nouveau chez les Collemboles (Insecta). *Revue d'Ecologie et de Biologie du Sol*, 28: 197–205.

Pokarzhevskii, A.D. (2002). Soil as an environment and its role in the evolution of insects (in commemoration of the 50th anniversary of soil ecology in Russia and the 90th anniversary of the birth of Merkurii Sergeevich Gilyarov. *Russian Journal of Ecology*, 33: 301–302.

Pokarzhevskii, A.D., Van Straalen, N.M., Zaboev, D.P. and Zaitsev, A.S. (2003). Microbial links and element flows in nested detrital food-webs. *Pedobiologia*, 47: 213–224.

Pomeroy, D.E. (1977). Biological studies on a peaty podzol III. Numbers of soil-living Collembola. *Pedobiologia*, 17: 115–134.

Ponge, J.-F., Patzel, N., Delhaye, L., Devigne, E., Levieux, C., Beros, P. et al. (1999). Interactions between earthworms, litter and trees in an old-growth beech forest. *Biology and Fertility of Soils*, 29: 360–370.

Ponge, J.-F. (2003). Humus forms in terrestrial ecosystems: a framework to biodiversity. *Soil Biology and Biochemistry*, 35: 935–945.

Ponge, J.-F. (2020). Move or change, an eco-evolutionary dilemma: the case of Collembola. *Pedobiologia*, 79: 150625.

Poole, T.B. (1961). An ecological study of the Collembola in a coniferous forest soil. *Pedobiologia*, 1: 113–137.

Poser, G. (1988). Chilopoden als Prädatoren in einem Laubwald. *Pedobiologia*, 31: 261–281.

Posthuma, L., Hogervorst, R.F., Joosse, E.N.G. and Van Straalen, N.M. (1993). Genetic variation and covariation for characteristics associated with cadmium tolerance in natural populations of the springtail *Orchesella cincta* (L.). *Evolution*, 47: 619–631.

Posthuma, L. and Van Straalen, N.M. (1993). Heavy-metal adaptation in terrestrial invertebrates: a review of occurrence, genetics, physiology and ecological consequences. *Comparative Biochemistry and Physiology*, 106C: 11–38.

Posthuma, L. and Janssen, G.M. (1995). Genetic variation for life-history characteristics in reference populations of *Orchesella cincta* (L.) in relation to adaptation to metals in soils. *Acta Zoologica Fennica*, 196: 301–306.

Potapov, A.A., Semenina, E.E., Korotkevich, A.Y., Kuznetsova, N. and Tiunov, A.V. (2016). Connecting taxonomy and ecology: Trophic niches of collembolans related to taxonomic identity and life forms. *Soil Biology and Biochemistry*, 101: 20–31.

Potapov, A., Bellini, B.C., Chown, S.L., Deharveng, L., Janssens, F., Kovác, L. et al. (2020). Towards a global synthesis of Collembola knowledge – challenges and possible solutions. *Soil Organisms*, 92: 161–188.

Price, D.W. (1973). Abundance and vertical distribution of microarthropods in the surface layers of a California pine forest soil. *Hilgardia*, 42: 121–147.

Pritchard, G., McKee, M.H., Pike, E.M., Scrimgeour, G.J. and Zloty, J. (1993). Did the first insects live in water or in air? *Biological Journal of the Linnean Society*, 49: 31–44.

Proctor, H.C. (1993). Mating biology resolves trichotomy for cheliferoid pseudoscorpions (Pseudoscorpionida, Cheliferoidea). *The Journal of Arachnology*, 21: 156–158.

Proctor, H.C. (1998). Indirect sperm transfer in arthropods: behavioral and evolutionary trends. *Annual Review of Entomology*, 43: 153–174.

Prosi, F. and Dallinger, R. (1988). Heavy metals in the terrestrial isopod *Porcellio scaber* Latr. I. Histochemical and ultrastructural characterization of metal containing lysosomes. *Cell Biology and Toxicology*, 4: 81–96.

Prpic, N.-M. (2019). A lesson in homology. *eLIFE*, 8: e43828.

Purvis, G. and Curry, J.P. (1980). Successional changes in the arthropod fauna of a new ley pasture established on a previously cultivated arable land. *Journal of Applied Ecology*, 17: 309–321.

Puurtinen, M. and Kaitala, V. (2002). Mate-searching efficiency can determine the evolution of separate sexes and the stability of hermaphroditism in animals. *The American Naturalist*, 160: 645–660.

Qu, Z., Nong, W., So, W.L., Barton-Owen, T., Li, Y., Leung, T.C.N. et al. (2020). Millipede genomes reveal unique adaptations during myriapod evolution. *PLoS Biology*, 18: e3000636.

Quaiser, A., Ochsenreiter, T., Lanz, C., Schuster, S.C., Treusch, A.H., Eck, J. et al. (2003). Acidobacteria form a coherent but highly diverse group within the bacterial domain: evidence from environmental genomics. *Molecular Microbiology*, 50: 563–575.

Rae, R.G., Robertson, J.F. and Wilson, M.J. (2008). Susceptibility and immune response of *Deroceras reticulatum*, *Milax gagates* and *Limax pseudoflavus* exposed to the slug parasitic nematode *Phasmarhabditis hermaphrodita*. *Journal of Invertebrate Pathology*, 97: 61–69.

Rajagopal, D. and Ganesha Bhat, U. (1995). Distribution and abundance of different soil microarthropod communities in different habitats. pp. 34–41. *In:* Edwards, C.A., Abe, T. and Striganova, B.R. (eds.). *Structure and Function of Soil Communities*. Kyoto University Press, Kyoto.

Reardon, W., Chakrabortee, S., Campos Pereira, T., Tyson, T., Banton, M.C., Dolan, K.M. et al. (2010). Expression profiling and cross-species RNA interference (RNAi) of desiccation-induced transcripts in the anhydrobiotic nematode *Aphelenchus avenae*. *BMC Molecular Biology*, 11: 6.

Rebecchi, L., Altiero, T., Eibye-Jacobson, J., Bertolani, R. and Kristensen, R.M. (2008). A new discovery of *Novechiniscus armadilloides* (Schster, 1975) (Tardigrada, Echiniscidae) from Utah, USA with considerations on non-marine Heterotardigrada phylogeny and biogeography. *Organisms, Diversity and Evolution*, 8: 58–65.

Rebecchi, L., Cesari, M., Altiero, T., Frigieri, A. and Guidetti, R. (2009). Survival and DNA degradation in anhydrobiotic tardigrades. *The Journal of Experimental Biology*, 212: 4033–4039.

Redmond, C.T., Kesheimer, A. and Potter, D.A. (2014). Earthworm community composition, seasonal population structure, and casting activity on Kentucky golf courses. *Applied Soil Ecology*, 75: 116–123.

Regier, J.C., Shultz, J.W., Zwick, A., Ball, B., Wetzer, R., Martin, J.W. et al. (2010). Arthropod relationships revealed by phylogenomic analysis of nuclear protein-encoding sequences. *Nature*, 463: 1079–1083.

Rehm, P., Borner, J., Meusemann, K., Von Reumont, B.J., Simon, S., Hadrys, H. et al. (2011). Dating the arthropod tree based on large-scale transcriptomic data. *Molecular Phylogenetics and Evolution*, 61: 880–887.

Reinecke, A.J. and Viljoen, S.A. (1990). The influence of worm density on growth and cocoon production of the compost worm *Eisenia fetida* (Oligochaeta). *Revue d'Ecologie et de Biologie du Sol*, 27: 221–230.

Reise, H. (2007). A review of mating behavior in slugs of the genus *Deroceras* (Pulmonata: Agriolimacidae). *American Malacological Bulletin*, 23: 137–156.

Reise, H., Visser, S. and Hutchinson, J.M.C. (2007). Mating behaviour in the terrestrial slug *Deroceras gorgonium*: is extreme morphology associated with extreme behaviour? *Animal Biology*, 57: 197–215.

Reuner, A., Hengherr, S., Mali, B., Förster, F., Arndt, D., Reinhardt, R. et al. (2010). Stress response in tardigrades: differential gene expression of molecular chaperones. *Cell Stress and Chaperones*, 15: 423–430.

Reynolds, D.R., Reynolds, A.M. and Chapman, J.W. (2014). Non-volant modes of migration in terrestrial arthropods. *Animal Migration*, 2: 8–28.

Ricker, W.E. (1946). Production and utilization of fish populations. *Ecological Monographs*, 16: 373–391.

Riddle, W.A. (1985). Hemolymph osmoregulation in several myriapods and arachnids. *Comparative Biochemistry and Physiology*, 80A: 313–323.

Riddle, D.L. (1988). The dauer larva. pp. 393–412. *In:* Wood, W.B. and the Community of *C. elegans* Researchers (ed.). *The Nematode Caenorhabditis elegans*. Cold Spring Harbor Laboratory Press, New York,

Rigaud, T. and Juchault, P. (1993). Conflict between feminizing sex ratio distorters and an autosomal masculinizing gene in the terrestrial isopod *Armadillidium vulgare* Latr. *Genetics*, 133: 247–252.

Riparbelli, M.G., Giordano, R. and Callaini, G. (2006). Centrosome inheritance in the parthenogenetic egg of the collembolan *Folsomia candida*. *Cell and Tissue Research*, 326: 861–872.

Riutort, M., Álvarez-Presas, M., Lázaro, E., Solà, E. and Paps, J. (2012). Evolutionary history of the Tricladida and the Platyhelminthes: an up-to-date phylogenetic and systematic account. *The International Journal of Developmental Biology*, 56: 5–17.

Roberts, J.M.K., Umina, P.A., Hoffmann, A.A. and Weeks, A.R. (2011). Population dynamics and diapause response of the springtail pest *Sminthurus viridis* (Collembola: Sminthuridae) in Southeastern Australia. *Journal of Economic Entomology*, 104: 465–473.

Roberts, J.M.K. and Weeks, A.R. (2011). Genetic structure and long-range dispersal in populations of the wingless pest springtail, *Sminthurus viridis* (Collembola: Sminthuridae). *Genetic Research, Cambridge*, 93: 1–12.

Robin, N., D'Haese, C.A. and Barden, P. (2019). Fossil amber reveals springtails' longstanding dispersal by social insects. *BMC Evolutionary Biology*, 19: 213.

Rodriguez, C., Prieto, G.I., Vega, I.A. and Castro-Vazquez, A. (2019). Functional and evolutionary perspectives on gill structures of an obligate air-breathing, aquatic snail. *PeerJ*, 7: e7342.

Rodriguez, C., Prieto, G.I., Vega, I.A. and Castro-Vazquez, A. (2021). Morphological grounds for the obligate aerial respiration of an aquatic snail: functional and evolutionary perspectives. *PeerJ*, 9: e10763.

Roelofs, D., Timmermans, M.J.T.N., Hensbergen, P., Van Leeuwen, H., Koopman, J., Faddeeva, A. et al. (2013). A functional isopenicillin N synthase in an animal genome. *Molecular Biology and Evolution*, 30: 541–548.

Roff, D. (1981). On being the right size. *The American Naturalist*, 118: 405–422.

Roff, D.A. (2002). *Life History Evolution*. Sinauer Associates, Inc., Sunderland.

Rohlfs, M. (2015). Fungal secondary metabolism in the light of animal–fungus interactions: From mechanism to ecological function. pp. 177–198. *In:* Zeilinger, S., Martín, J.-F. and García-Estrada, C. (eds.). *Biosynthesis and Molecular Genetics of Fungal Secondary Metabolites, Volume 2*. Springer Science+Business Media, New York.

Römbke, J. (1985). Zur Biologie eines Buchenwaldbodens. 6. Die Regenwürmer. *Carolinea*, 43: 93–104.

Romero, P.E., Weigand, A.M. and Pfenninger, M. (2016a). Positive selection on panpulmonate mitogenomes provide new clues on adaptations to terrestrial life. *BMC Evolutionary Biology*, 16: 164.

Romero, P.E., Pfenninger, M., Kano, Y. and Klussmann-Kolb, A. (2016b). Molecular phylogeny of the Ellobiidae (Gastropoda: Panpulmonata) supports independent terrestrial invasions. *Molecular Phylogenetics and Evolution*, 97: 43–54.

Rong, Y., Zeng, M., Guan, X., Qu, K., Liu, J., Zhang, J. et al. (2019). Association of HSF1 genetic variation with heat tolerance in Chinese cattle. *Animals*, 9: 1027.

Ros, V.I.D., Fleming, V.M., Feil, E.J. and Breeuwer, J.A.J. (2009). How diverse is the genus *Wolbachia*? Multiple-gene sequencing reveals a putatively new *Wolbachia* supergroup recovered from spider mites (Acari: Tetranychidae). *Applied and Environmental Microbiology*, 75: 1036–1043.

Rota-Stabelli, O., Campbell, L., Brinkmann, H., Edgecombe, G.D., Longhorn, S.J., Peterson, K.J. et al. (2011). A congruent solution to arthropod phylogeny: phylogenomics, microRNAs and morphology support monophyletic Mandibulata. *Proceedings of the Royal Society B*, 278: 298–306.

Rota-Stabelli, O., Daley, A.C. and Pisani, D. (2013). Molecular timetrees reveal a Cambrian colonization of land and a new scenario for ecdysozoan evolution. *Current Biology*, 23: 392–398.

Roughgarden, J. (1971). Density-dependent natural selection. *Ecology*, 52: 453–468.

Roure, B. and Philippe, H. (2011). Site-specific time heterogeneity of the substitution process and its impact on phylogenetic inference. *BMC Evolutionary Biology*, 11: 17.

Roy, H.E., Steinkraus, D.C., Eilenberg, J., Hajek, A.E. and Pell, J.K. (2006). Bizarre interactions and endgames: entomopathogenic fungi and their arthropod hosts. *Annual Review of Entomology*, 51: 331–357.

Rundgren, S. and Augustsson, A.K. (1998). Test on the enchytraeid *Cognettia sphagnetorum* (Vejdovsky) 1877. pp. 73–94. *In:* Løkke, H. and Van Gestel, C.A.M. (eds.). *Handbook of Soil Invertebrate Toxicity Tests*. John Wiley and Sons, Chichester.

Růžičková, J. and Veselý, M. (2016). Using radio telemetry to track ground beetles: Movement of *Carabus ullrichii*. *Biologia*, 71: 924–930.

Růžičková, J. and Veselý, M. (2018). Movement activity and habitat use of *Carabus ullrichii* (Coleoptera: Carabidae): The forest edge as a mating site? *Entomological Science*, 21: 76–83.

Růžičková, J. and Elek, Z. (2021). Recording fine-scale movement of ground beetles by two methods: Potentials and methodological pitfalls. *Ecology and Evolution*, 11: 8562–8572.

Rypstra, A.L., Wieg, C., Walker, S.E. and Persons, M.H. (2003). Mutual mate assessment in wolf spiders: differences in the cues used by males and females. *Ethology*, 109: 315–325.

Saito, F. (1969). Energetics of isopod populations in a forest of Central Japan. *Researches on Population Ecology*, 11: 229–258.

Sakai, M., Taoda, Y., Mori, K., Fujino, M. and Ohta, C. (1991). Copulation sequence and mating termination in the male cricket *Gryllus bimaculatus*. *Journal of Insect Physiology*, 37: 599–615.

Sakai, M., Kumashiro, M., Matsumoto, U., Ureshi, M. and Otsubo, T. (2017). Reproductive behaviour and physiology in the cricket *Gryllus bimaculatus*. pp. 245–269. *In:* Horch, H.W., Mito, T., Popadícm, H., Ohuchi, H. and Noji, S. (eds.). *The Cricket as a Model Organism. Development, Regeneration and Behavior*. Springer Japan KK, Tokyo.

Salmon, S. (2001). Earthworm excreta (mucus and urine) affect the distribution of springtails in forest soils. *Biology and Fertility of Soils*, 34: 304–310.

Salmon, S., Rebuffat, S., Prado, S., Sablier, M., D'Haese, C., Sun, J.-S. et al. (2019). Chemical communication in springtails: a review of facts and perspectives. *Biology and Fertility of Soils*, 55: 425–438.

Salzberg, S.L. (2017). Horizontal gene transfer is not a hallmark of the human genome. *Genome Biology*, 18: 85.

Santamaria, S., Enghoff, H. and Reboleira, A.S.P.S. (2014). Laboulbeniales in millipedes: the genera *Diplodomyces* and *Troglomyces*. *Mycologia*, 106: 1027–1038.

Santana, R.H., Catão, E.C.P., Cardoso Lopes, F.A., Constantino, R., Barreto, C.C. and Krüger, R.H. (2015). The gut microbiota of workers of the litter-feeding termites *Syntermes wheeleri* (Termitidae: Syntermitinae): archaeal, bacterial and fungal communities. *Microbial Ecology*, 70: 545–556.

Santoro, G. and Jones, H.D. (2001). Comparison of the earthworm population of a garden infested with the Australian land flatworm (*Australoplana sanguinea alba*) with that of a non-infested garden. *Pedobiologia*, 45: 313–328.

Sapkota, R., Santos, S.R., Farias, P., Krogh, P.H. and Winding, A. (2020). Insights into the earthworm gut multi-kingdom microbial communities. *Science of the Total Environment*, 727: 138301.

Sassaman, C. and Garthwaite, R. (1984). The interaction between the terrestrial isopod *Porcellio scaber* Latreille and one of its dipteran parasites, *Melanophora roralis* (L.) (Rhinophoridae). *Journal of Crustacean Biology*, 4: 595–603.

Satchell, J.E. (1977). Earthworms—the trombones of the grave. Closing presidential address. *Ecological Bulletins*, 25: 598–603.

Satchell, J.E. (1980). r Worms and K worms: a basis for classifying lumbricid earthworm strategies. pp. 848–864. *In:* Dindal, D.L. (ed.). *Soil Biology as Related to Land Use Practices. Proceedings of the VIIth International Colloquium on Soil Zoology*. Office of Pesticide and Toxic Substances EPA, Washington.

Satchell, J.E. (1983). *Earthworm Ecology. From Darwin to Vermiculture*. Chapman and Hall London.

Schaefer, M. and Schauermann, J. (1990). The soil fauna of beech forests: comparison between a mull and a moder. *Pedobiologia*, 34: 299–314.

Schaefer, I., Norton, R.A., Scheu, S. and Maraun, M. (2010). Arthropod colonization of land – linking molecules and fossils in oribatid mites (Acari, Oribatida). *Molecular Phylogenetics and Evolution*, 57: 113–121.

Schaller, F. (1971). Indirect sperm transfer by soil arthropods. *Annual Review of Entomology*, 16: 407–446.

Schaller, F. (1979). Significance of sperm transfer and formation of spermatophores in arthropod phylogeny. pp. 587–608. *In:* Gupta, A.P. (ed.). *Arthropod Phylogeny*. Van Nostrand Reinhold Company, New York.

Schat, H., Van Hoof, N.A.L.M., Tervahauta, A., Hakvoort, H.W.J., Chardonnens, A.N., Koevoets, P. et al. (2000). Evolutionary responses to zinc and copper stress in bladder campion, *Silene vulgaris* (Moench.) Garcke. pp. 343–360. *In*: Cherry, C.H., Locy, R.D. and Rychter, A. (eds.). *Plant Tolerance to Abiotic Stresses in Agriculture: Role of Genetic Engineering*. Kluwer Academic Publishers, Dordrecht.

Scheiner, S.M. and Lyman, R.F. (1989). The genetics of phenotypic plasticity. I. Heritability. *Journal of Evolutionary Biology*, 2: 95–107.

Scheiner, S.M. (1993). Plasticity as a selectable trait: reply to Via. *The American Naturalist*, 142: 371–373.

Scheller, U. (2008). A reclassification of the Pauropoda (Myriapoda). *International Journal of Myriapodology*, 1: 1–38.

Scheu, S. and Poser, G. (1996). The soil macrofauna (Diplopoda, Isopoda, Lumbricidae and Chilopoda) near tree trunks in a beechwood on limestone: indications for stemflow induced changes in community structure. *Applied Soil Ecology*, 3: 115–125.

Scheu, S. and Drossel, B. (2007). Sexual reproduction prevails in a world of structured resources in short supply. *Proceedings of the Royal Society B*, 274: 1225–1231.

Schilthuizen, M. and Rutjes, H.A. (2001). Land snail diversity in a square kilometre of tropical rainforest in Sabah, Malaysian Borneo. *Journal of Molluscan Studies*, 67: 417–423.

Schmelzle, S. and Blüthgen, N. (2019). Under pressure: force resistance measurements in box mites (Actinotrichida, Oribatida). *Frontiers in Zoology*, 16: 24.

Schmidt, N., Boll, E.S., Malmquist, L.M.V. and Christensen, J.H. (2017). PAH metabolism in the earthworm *Eisenia fetida* – identification of phase II metabolites of phenanthrene and pyrene. *International Journal of Environmental Analytical Chemistry*, 97: 1151–1162.

Schmidt-Nielsen, K. (1998). *Animal Physiology. Adaptation and Environment. Fifth Edition*. Cambridge University Press, Cambridge.

Schmidt-Rhaesa, A., Farfan, M.A. and Bernard, E.C. (2009). First record of millipeds as hosts for horsehair worms (Nematomorpha) in North America. *Northeastern Naturalist*, 16: 125–130.

Schneider, K., Migge, S., Norton, R.A., Scheu, S., Langel, R., Reineking, A. et al. (2004). Trophic niche differentiation in soil microarthropods (Oribatida, Acari): evidence from stable isotope ratios ($^{15}N/^{14}N$). *Soil Biology and Biochemistry*, 36: 1769–1774.

Scholte, E.-J., Knols, B.G.J., Samson, R.A. and Takken, W. (2004). Entomopathogenic fungi for mosquito control: A review. *Journal of Insect Science*, 4: 19.

Scholtz, G. and Kamenz, C. (2006). The book lungs of Scorpiones and Tetrapulmonata (Chelicerata, Arachnida): Evidence for homology and a single terrestrialisation event of a common arachnid ancestor. *Zoology*, 109: 2–13.

Schön, I., Martens, K. and Van Dijk, P. (2009). *Lost Sex. The Evolutionary Biology of Parthenogenesis*. Springer Science+Business Media B.V., Dordrecht.

Schoolfield, R.M., Sharpe, P.J.H. and Magnuson, C.E. (1981). Non-linear regression of biological temperature-dependent rate models based on absolute reaction-rate theory. *Journal of Theoretical Biology*, 88: 719–731.

Schratzberger, M., Holterman, M., Van Oevelen, D. and Helder, J. (2019). A worm's world: Ecological flexibility pays off for free-living nematodes in sediments and soils. *BioScience*, 69: 867–876.

Schubart, C.D., Diesel, R. and Hedges, S.B. (1998). Rapid evolution to terrestrial life in Jamaican crabs. *Nature*, 393: 363–365.

Schüßler, A., Schwarzott, D. and Walker, C. (2001). A new fungal phylum, the *Glomeromycota*: phylogeny and evolution. *Mycological Research*, 105: 1413–1421.

Schweizer, M., Triebskorn, R. and Köhler, H.-R. (2019). Snails in the sun: Strategies of terrestrial gastropods to cope with hot and dry conditions. *Ecology and Evolution*, 9: 12940–12960.

Schwander, T. and Oldroyd, B.P. (2016). Androgenesis: where males hijack eggs to clone themselves. *Philosophical Transactions of the Royal Society B*, 371: 20150534.

Schwentner, M., Combosch, D.J., Nelson, J.P. and Giribet, G. (2017). A phylogenomic solution to the origin of insects by resolving crustacean-hexapod relationships. *Current Biology*, 27: 1818–1824.

Segovia, J.M.G., Del-Claro, K. and Willemart, R.H. (2015). Delicate fangs, smart killing: the predation strategy of the recluse spider. *Animal Behaviour*, 101: 169–177.

Selden, P.A. (2005). Terrestrialization (Precambrian - Devonian). *Encyclopedia of Life Sciences*, doi: 10.1038/npg.els.0004145.

Sender, R., Fuchs, S. and Milo, R. (2016). Revised estimates for the number of human and bacteria cells in the body. *PLoS Biology*, 14: e1002533.

Senepati, B.K. and Dash, M.C. (1981). Effect of grazing on the elements of production in vegetation and Oligochaete components of a tropical pasture. *Revue d'Ecologie et de Biologie du Sol*, 18: 487–505.

Sengupta, S., Ergon, T. and Leinaas, H.P. (2016). Genotypic differences in embryonic life history traits of *Folsomia quadrioculata* (Collembola: Isotomidae) across a wide geographical range. *Ecological Entomology*, 41: 72–84.

Sfenthourakis, S. and Hornung, E. (2018). Isopod distribution and climate change. *ZooKeys*, 801: 25–61.

Sgardelis, S.P., Sarkar, S., Asikidis, M.D., Cancela da Fonseca, J.P. and Stamou, G.P. (1993). Phenological patterns of soil microarthropods from three climatic regions. *European Journal of Soil Biology*, 29: 49–57.

Shakir, S.H. and Dindal, D.L. (1997). Density and biomass of earthworms in forest and herbaceous microecosystems in Central New York, North America. *Soil Biology and Biochemistry*, 29: 275–285.

Sharma, G.D. and Kevan, D.K.M. (1963). Observation on *Pseudosinella petterseni* and *Pseudosinella alba* (Collembola: Entomobryidae) in Eastern Canada. *Pedobiologia*, 3: 62–74.

Sharma, P.P. (2017). Chelicerates and the conquest of the land: a view of arachnid origins through an evo-devo spyglass. *Integrative and Comparative Biology*, 57: 510–522.

Sharpe, P.J.H. and De Michele, D.W. (1977). Reaction kinetics of poikilotherm development. *Journal of Theoretical Biology*, 64: 649–670.

Shaw, P. (2003). Collembola of pulverised fuel ash sites in east London. *European Journal of Soil Biology*, 39: 1–8.

Shear, W.A. and Edgecombe, G.D. (2010). The geological record and phylogeny of the Myriapoda. *Arthropod Structure and Development*, 39: 174–190.

Shen, H.-P., Yu, H.-T. and Chen, J.H. (2012). Parthenogenesis in two Taiwanese mountain earthworms *Amynthas catenus* Tsai et al., 2001 and *Amynthas hohuanmontis* Tasi et al., 2002 (Oligochaeta, Megascolecidae) revealed by AFLP. *European Journal of Soil Biology*, 51: 30–36.

Shibuya, K., Chiba, S. and Kimura, K. (2022). Sexual inactivation induced by the mucus that covers land snail love darts: sexual selection and evolution of allohormones in hermaphrodites. *Journal of Experimental Biology*, 225: jeb238782.

Shresta, B., Kubátová, A., Tanaka, E., Oh, J., Yoon, D.-H., Sung, J.-M. et al. (2019). Spider-pathogenic fungi within Hypocreales (Ascomycota): their current nomenclature, diversity, and distribution. *Mycological Progress*, 18: 983–1003.

Siegfried, B.D. and Young, L.J. (1993). Activity of detoxification enzymes in aquatic and terrestrial insects. *Environmental Entomology*, 22: 958–964.

Siepel, H. (1994). Life-history tactics of Collembola; an alternative to Gisin's life forms? *Acta Zoologica Fennica*, 195: 129–131.

Silva, T.S.R. and Feitosa, R.M. (2017). Hunting for wasps in-between: the use of the Winkler extractor to sample leaf litter Hymenoptera. *Neotropical Entomology*, 46: 711–718.

Silva, T.S.R., Silva-Freitas, J.M. and Schoeninger, K. (2019). A synopsis of the Brazilian Eucharitidae (Hymenoptera: Chalcidoidea) fauna: an annotated checklist of the family in the country, with a revised key for the New World genera. *Zootaxa*, 4564: 347–366.

Silva, T.S.R. and Casadei-Ferreira, A. (2020). A species-level association in *Pheidole* Westwood (Hymenoptera: Formicidae) ants with a parasitoid wasp of the genus *Orasema* Cameron (Hymenoptera: Eucharitidae) in Brazil. *Revista Brasileira de Entomologia*, 64: e20200005.

Simkiss, K. (1979). Metal ions in cells. *Endeavour (New Ser.)*, 3: 2–6.

Simmons, L.W. (1986). Female choice in the field cricket *Gryllus bimaculatus* (de Geer). *Animal Behaviour*, 34: 1463–1470.

Simmons, L.W. (1989). Kin recognition and its influence on mating preferences of the field cricket, *Gryllus bimaculatus* (De Geer). *Animal Behaviour*, 38: 68–77.

Simmons, L.W. (1991). Female choice and the relatedness of mates in the field cricket, *Gryllus bimaculatus*. *Animal Behaviour*, 41: 493–501.

Simmons, L.W., Beesley, L., Lindhjem, P., Newbound, D., Norris, J. and Wayne, A. (1999). Nuptial feeding by male bushcrickets: an indicator of male quality? *Behavioral Ecology*, 10: 263–269.

Simmons, L.W. and Buzatto, B.A. (2014). Contrasting responses of pre- and postcopulatory traits to variation in mating competition. *Functional Ecology*, 28: 494–499.

Simon, J.-C., Delmotte, F., Rispe, C. and Crease, T. (2003). Phylogenetic relationships between parthenogens and their sexual relatives: the possible routes to parthenogenesis in animals. *Biological Journal of the Linnean Society*, 79: 151–163.

Simone, L.R.L. and Seabra, M.I.G.L. (2017). Shell and body structure of the plesiomorphic pulmonate marine limpet *Siphonaria pectinata* (LInnaeus, 1758) from Portugal (Gastropoda: Heterobranchia: Siphonariidae). *Folia Malacologia*, 25: 147–164.

Simone, L.R.L. (2021). The molluscan pallial cavity. *Malacopedia*, 4: 1–9.

Simonsen, V. and Christensen, P.G. (2001). Clonal and genetic variation in three collembolan species revealed by isozymes and randomly amplified polymorphic DNA. *Pedobiologia*, 45: 161–173.

Singh, A., Singh, D.P., Tiwari, R., Kumar, K., Singh, R.V., Singh, S. et al. (2015). Taxonomic and functional annotation of gut bacterial communities of *Eisenia foetida* and *Perionyx excavatus*. *Microbiological Research*, 175: 48–56.

Siriwut, W., Edgecombe, G.D., Sutcharit, C. and Panha, S. (2014). Brooding behaviour of the centipede *Otostigmus spinosus* Porat, 1876 (Chilopoda: Scolopendromorpha: Scolopendridae) and its morphological variability in Thailand. *Raffles Bulletin of Zoology*, 62: 339–351.

Sjögren, M., Augustsson, A. and Rundgren, S. (1995). Dispersal and fragmentation of the enchytraeid *Cognettia sphagnetorum* in metal polluted soil. *Pedobiologia*, 39: 207–218.

Sjursen, H. and Sinclair, B.J. (2002). On the cold hardiness of *Stereotydeus mollis* (Acari: Prostigmata) from Ross Island, Antarctica. *Pedobiologia*, 46: 188–195.

Slice, L.W., Freedman, J.H. and Rubin, C.S. (1990). Purification, characterization, and cDNA cloning of a novel metallothionein-like cadmium-binding protein from *Caenorhabditis elegans*. *The Journal of Biological Chemistry*, 265: 256–263.

Slotsbo, S., Maraldo, K., Malmendal, A., Nielsen, N.C. and Holmstrup, M. (2008). Freeze tolerance and accumulation of cryoprotectants in the enchytraeid *Enchytraeus albidus* (Oligochaeta) from Greenland and Europe. *Cryobiology*, 57: 286–291.

Sluys, R. (2019). The evolutionary terrestrialization of planarian flatworms (Platyhelminthes, Tricladida, Geoplanidae): a review and research programme. *Zoosystematics and Evolution*, 95: 543–556.

Smant, G., Stokkermans, J.P.W.G., Yan, Y., De Boer, J.M., Baum, T.J., Wang, X. et al. (1998). Endogenous cellulases in animals: isolation of β-1:4–endoglucanase genes from two species of plant-parasitic cyst nematodes. *Proceedings of the National Academy of Sciences of the United States of America*, 95: 4906–4911.

Smith, M.R. (2016). Evolution: velvet worm biogeography. *Current Biology*, 26: R882–R884.

Snider, R. (1973). Laboratory observations on the biology of *Folsomia candida* (Willem) (Collembola: Isotomidae). *Revue d'Écologie et de Biologie du Sol*, 10: 103–124.

Snider, R.M. and Butcher, J.W. (1973). The life history of *Folsomia candida* (Willem) (Collembola: Isotomidae) relative to temperature. *The Great Lakes Entomologist*, 6: 97–106.

Snider, R.M. (1974). The life cycle relative to temperature of *Protaphorura armatus* (Tullberg) (Collembola: Onychiuridae), a parthenogenetic species. *The Great Lakes Entomologist*, 7: 9–15.

Snider, R.J. (1983). Observations on the oviposition, egg development and fecundity of *Onychiurus (Onychiurus) folsomi*. *Pedobiologia*, 25: 241–252.

Soberón, M., Gill, S.S. and Bravo, A. (2009). Signaling versus punching hole: How do *Bacillus thuringiensis* toxins kill insect midgut cells? *Cellular and Molecular Life Sciences*, 66: 1337–1349.

Socias-Martínez, L. and Kappeler, P.M. (2019). Catalyzing transitions to sociality: ecology builds on parental care. *Frontiers in Ecology and Evolution*, 7: 160.

Soejono Sastrodihardjo, F.X. and Van Straalen, N.M. (1993). Behaviour of five isopod species in standardized tests for pH preference. *European Journal of Soil Biology*, 29: 127–131.

Sohlenius, B. (1980). Abundance, biomass and contribution to energy flow by soil nematodes in terrestrial ecosystems. *Oikos*, 34: 186–194.

Sohlenius, B. and Boström, S. (1986). Short-term dynamics of nematode communities in arable soil – Influence of nitrogen fertilization in barley crops. *Pedobiologia*, 29: 183–191.

Sohlenius, B. and Boström, S. (2001). Annual and long-term fluctuations of the nematode fauna in a Swedish Scots pine forest soil. *Pedobiologia*, 45: 408–429.

Sohlenius, B. (2002). Influence of clear-cutting and forest age on the nematode fauna in a Swedish pine forest soil. *Applied Soil Ecology*, 19: 261–277.

Solhøy, T. (1972). Quantitative invertebrate studies in mountain communities at Hardangervidda, South Norway. I. *Norsk entomologisk Tidsskrift*, 19: 99–108.

Soma, K. and Saitô, T. (1979). Ecological studies of soil organisms with references to the decomposition of pine needles. I.- Soil macrofaunal and mycofloral surveys in coastal pine plantations. *Revue d'Écologie et de Biologie du Sol*, 16: 337–354.

Sombke, A. and Edgecombe, G.D. (2014). Morphology and evolution of Myriapoda. *Arthropod Structure and Development*, 43: 3–4.

Sømme, L. (1978). Notes on the cold-hardiness of prostigmate mites from Vestfjella, Dronning Maud Land. *Norwegian Journal of Entomology*, 25: 51–55.

Sømme, L. (1996). Anhydrobiosis and cold tolerance in tardigrades. *European Journal of Entomology*, 93: 349–357.

Sømme, L. (1999). The physiology of cold hardiness in terrestrial arthropods. *European Journal of Entomology*, 96: 1–10.

Sømme, L. (2000). The history of cold hardiness research in terrestrial arthropods. *CryoLetters*, 21: 289–296.

Song, Y., Drossel, B. and Scheu, S. (2011). Tangled Bank dismissed too early. *Oikos*, 120: 1601–1607.

Southwood, T.R.E. (1967). The interpretation of population change. *Journal of Animal Ecology*, 36: 519–529.

Souty-Grosset, C., Chentoufi, A., Mocqard, J.P. and Juchault, P. (1988a). Seasonal reproduction in the terrestrial isopod *Armadillidium vulgare* (Latreille): Geographical variability and genetic control of the response to photoperiod and temperature. *Invertebrate Reproduction and Development*, 14: 131–151.

Souty-Grosset, C., Nasri, K., Mocquard, J.P. and Juchault, P. (1988b). Individual variation in the seasonal reproduction of the terrestrial isopod *Armadillidium vulgare* Latr. (Crustacea, Oniscidea). *Acta Oecologia*, 19: 367–375.

Spicer, J.I., Moore, P.G. and Taylor, A.C. (1987). The physiological ecology of land invasion by the Talitridae (Crustacea: Amphipoda). *Proceedings of The Royal Society of London*, B 332: 95–124.
Spurgeon, D., Lahive, E., Robinson, A., Short, S. and Kille, P. (2020). Species sensitivity to toxic substances: evolution, ecology and applications. *Frontiers in Environmental Sciences*, 8: 588380.
Srivastava, R., Gupta, D.K., Choudhary, A.K. and Sinha, M.P. (2013). Biomass and secondary production of earthworm *Drawida willsi* (Michaelsen) from a tropical agroecosystem in Ranchi, Jharkhand. *Nature Environment and Pollution Technology*, 12: 179–182.
Stålhandske, P. (2001). Nuptial gift in the spider *Pisaura mirabilis* maintained by sexual selection. *Behavioural Ecology*, 12: 691–697.
Stam, E.M., Van de Leemkule, M.A. and Ernsting, G. (1996). Trade-offs in the life history and energy budget of the parthenogenetic collembolan *Folsomia candida* (Willem). *Oecologia*, 107: 283–292.
Stam, E., Isaaks, A. and Ernsting, G. (1998a). Negative maternal effect revisited: a test on two populations of *Orchesella cincta* L. (Collembola: Entomobryidae). *Evolution*, 52: 1839–1843.
Stam, E.M. (1998b). *An Evolutionary Ecological Analysis of Reproduction in Springtails*. PhD thesis, Vrije Universiteit, Amsterdam.
Stam, E.M. and Hoogendoorn, G. (1999). Indirect sperm transfer and male mating strategies in soil invertebrates. *Invertebrate Reproduction and Development*, 36: 187–189.
Stam, E., Isaaks, A. and Ernsting, G. (2002). Distant lovers: spermatophore deposition and destruction behavior by male springtails. *Journal of Insect Behavior*, 15: 253–268.
Stamou, G.P. and Sgardelis, S.P. (1989). Seasonal distribution patterns of oribatid mites (Acari: Cryptostigmata) in a forest ecosystem. *Journal of Animal Ecology*, 58: 893–904.
Stearns, S.C. (1976). Life-history tactics: a review of the ideas. *Quarterly Review of Biology*, 51: 3–47.
Stearns, S.C. (1992). *The Evolution of Life Histories*. Oxford University Press, Oxford.
Steenberg, T., Eilenberg, J. and Bresciani, J. (1996). First record of a *Neozygites* species (Zygomycetes: Entomophthorales) infecting springtails (Insecta: Collembola). *Journal of Invertebrate Pathology*, 68: 97–100.
Stenersen, J. and Øien, N. (1981). Glutathione-S-transferases in earthworms (Lumbricidae). Substrate specificity, tissue and species distribution and molecular weight. *Comparative Biochemistry and Physiology*, 69C: 243–252.
Stenersen, J., Kobro, S., Bjerke, M. and Arend, U. (1987). Glutathione transferases in aquatic and terrestrial animals from nine phyla. *Comparative Biochemistry and Physiology*, 86C: 73–82.
Stenersen, J. (1992). Uptake and metabolism of xenobiotics by earthworms. pp. 129–138. *In:* Greig-Smith, P.W., Becker, H., Edwards, P.J. and Heimbach, F. (eds.). *Ecotoxicology of Earthworms*. Intercept Ltd., Andover.
Stewart, M.J., Wang, T., Koene, J.M., Storey, K.B. and Cummins, S.F. (2016). A "love" dart allohormone identified in the mucous glands of hermaphroditic land snails. *The Journal of Biological Chemistry*, 291: 7938–7950.
Stirling, G., Nicol, J. and Reay, F. (2002). *Advisory Services for Nematode Pests. Operational Guidelines*. Rural Industries Research and Development Corporation, Barton.
Stokke, K. and Stenersen, J. (1993). Non-inducibility of the glutathione transferases of the earthworm *Eisenia andrei*. *Comparative Biochemistry and Physiology*, 106C: 753–756.
Stokland, J.N., Siitonen, J. and Jonsson, B.G. (2012). *Biodiversity in Dead Wood*. Cambridge University Press, Cambridge.
Storey, K.B., Storey, J.M. and Churchill, T.A. (2007). Freezing and anoxia tolerance of slugs: a metabolic perspective. *Journal of Comparative Physiology B*, 177: 833–840.
Stoutjesdijk, P. (1977). High surface temperatures of trees and pine litter in the winter and their biological importance. *International Journal of Biometeorology*, 21: 325–331.
Stratton, G.E. and Uetz, G.W. (1983). Communication via substratum-coupled stridulation and reproductive isolation in wolf spiders (Araneae: Lycosidae). *Animal Behaviour*, 31: 164–172.
Strickberger, M.W. (2000). *Evolution. Third Edition*. Jones and Bartlett Publishers, Sudbury.
Stroomberg, G.J., De Knecht, J.A., Ariese, F., Van Gestel, C.A.M. and Velthorst, N.H. (1999). Pyrene metabolites in the hepatopancreas and gut of the isopod *Porcellio scaber*, a new biomarker for polycyclic aromatic hydrocarbon exposure in terrestrial ecosystems. *Environmental Toxicology and Chemistry*, 18: 2217–2224.

Stroomberg, G.J., Zappey, H., Steen, R.J.C.A., Van Gestel, C.A.M., Ariese, F., Velthorst, N.H. et al. (2004). PAH biotransformation in terrestrial invertebrates—a new phase II metabolite in isopods and springtails. *Comparative Biochemistry and Physiology, Part C*, 138: 129–137.

Struck, T.H., Paul, C., Hill, N., Hartmann, S., Hösel, C., Kube, M. et al. (2011). Phylogenomic analyses unravel annelid evolution. *Nature*, 471: 95–98.

Strullu-Derrien, C., Selosse, M.-A., Kenrick, P. and Martin, F.M. (2018). The origin and evolution of mycorrhizal symbioses: from palaeomycology to phylogenomics. *New Phytologist*, 220: 1012–1030.

Sturm, H. (1978). Zum Paarungsverhalten von *Petrobius maritimus* Leach (Machilidae: Archaeognatha: Insecta). *Zoologische Anzeiger*, 201: 5–20.

Sturm, H. (1992). Mating behaviour and sexual dimorphism in *Promesomachilis hispanica* Silvestri, 1923 (Machilidae, Archaeognatha, Insecta). *Zoologische Anzeiger*, 228: 60–73.

Sturm, H. (1997). The mating behaviour of *Tricholepidion gertschi* Wygod., 1961 (Lepidotrichidae, Zygentoma) and its comparison with the behaviour of other "Apterygota". *Pedobiologia*, 41: 44–49.

Stürzenbaum, S.R., Kille, P. and Morgan, A.J. (1998). The identification, cloning and characterization of earthworm metallothionein. *FEBS Letters*, 431: 437–442.

Stürzenbaum, S.R., Winters, C., Galay, M., Morgan, A.J. and Kille, P. (2001). Metal ion trafficking in earthworms. Identification of a cadmium-specific metallothionein. *The Journal of Biological Chemistry*, 276: 34013–34018.

Stürzenbaum, S., Georgiev, O., Morgan, A.J. and Kille, P. (2004). Cadmium detoxification in earthworms: from genes to cells. *Environmental Science and Technology*, 38: 6283–6289.

Sullivan, T.J., Dreyer, A.P. and Peterson, J.W. (2009). Genetic variation in a subterranean arthropod (*Folsomia candida*) as a method to identify low-permeability barriers in an aquifer. *Pedobiologia*, 53: 99–105.

Sun, L.-Y., Liu, J., Li, Q., Fu, D., Zhu, J.-Y., Guo, J.-J. et al. (2021). Cloning and differential expression of three heat shock protein genes associated with thermal stress from the wolf spider *Pardosa pseudoannulata* (Araneae: Lycosidae). *Journal of Asia-Pacific Entomology*, 24: 158–166.

Sunderland, K.D., Hassall, M. and Sutton, G. (1976). The population dynamics of *Philoscia muscorum* (Crustacea, Oniscoidea) in a dune grassland ecosystem. *Journal of Animal Ecology*, 45: 487–506.

Suring, W., Mariën, J., Broekman, R., Van Straalen, N.M. and Roelofs, D. (2016). Biochemical pathways supporting beta-lactam biosynthesis in the springtail *Folsomia candida*. *Biology Open*, 5: 1784–1789.

Suring, W., Meusemann, K., Blanke, A., Mariën, J., Schol, T., Agamennone, V. et al. (2017). Evolutionary ecology of beta-lactam gene clusters in animals. *Molecular Ecology*, 26: 3217–3229.

Šustr, V. (1996). Influence of temperature on respiration-temperature relationship in *Tetrodontophora bielanensis* (Collembola: Onychiuridae). *European Journal of Entomology*, 93: 435–442.

Šustr, V. and Pižl, V. (2009). Oxygen consumption of the earthworm species *Dendrobaena mrazeki*. *European Journal of Soil Biology*, 45: 478–482.

Šustr, V. and Pižl, V. (2010). Temperature dependence and ontogenetic changes of metabolic rate of an endemic earthworm *Dendrobaena mrazeki*. *Biologia*, 65: 289–293.

Sutton, S.L. (1968). The population dynamics of *Trichoniscus pusillus* and *Philoscia muscorum* (Crustacea, Oniscoidea) in limestone grassland. *Journal of Animal Ecology*, 37: 425–444.

Sutton, S. (1972). *Woodlice*. Pergamon Press, Oxford.

Swift, M.J., Heal, O.W. and Anderson, J.M. (1979). *Decomposition in Terrestrial Ecosystems*. Blackwell, Oxford.

Szathmáry, E. and Maynard Smith, J. (1995). The major evolutionary transitions. *Nature*, 374: 227–232.

Szeptycki, A. (1997). The present knowledge of Protura. *Fragmenta Faunistica (Warszawa)*, 40: 307–311.

Takeda, H. (1976). Ecological studies of collembolan populations in a pine forest soil. I.- The life cycle and populations dynamics of *Tetracanthella sylvatica* Yosii. *Revue d'Écologie et de Biologie du Sol*, 13: 117–132.

Takeda, H. (1983). A long term study of life cycles and population dynamics of *Tullbergia yosii* and *Onychiurus decemsetosus* (Collembola) in a pine forest soil. *Pedobiologia*, 25: 175–185.

Takeda, H. (1984). A long term study of life cycle and population dynamics of *Folsomia octoculata* Handschin (Insecta: Collembola) in a pine forest soil. *Researches on Population Ecology*, 26: 188–219.

Takeda, H. (1995a). Changes in the collembolan community during the decomposition of needle litter in a coniferous forest. *Pedobiologia*, 39: 304–317.

Takeda, H. (1995b). Templates for the organization of collembolan communities. pp. 5–20. *In:* Edwards, C.A., Abe, T. and Striganova, B.R. (eds.). *Structure and Function of Soil Communities*. Kyoto University Press, Kyoto.

Takeda, N. (1984). The aggregation phenomenon in terrestrial isopods. *Symposia of the Zoological Society of London*, 53: 381–404.

Tamura, H. (1967). Some ecological observations on Collembola in Sapporo, Northern Japan. *Journal of the faculty of Science, Hokkaido University, Series VI, Zoology*, 16: 238–252.

Tamura, H., Nakamura, Y., Yamauchi, K. and Fujikawa, T. (1969). An ecological survey of soil fauna in Hidaka-Mombetsu, Southern Hokkaido. *Journal of the Faculty of Science, Hokkaido University, Series VI, Zoology*, 17: 17–57.

Tanaka, M. (1970). The bio-economics on the populations of *Isotoma* (*Desoria*) *trispinata* Mac Gillivray (Collembola; Isotomidae) and *Onychiurus* (*Protaphorura*) sp. (Collembola; Onychiuridae) in a grassfield. *Publications from the Amakusa Marine Biological Laboratory*, 2: 51–120.

Tanaka, K. and Udagawa, T. (1993). Cold adaptation of the terrestrial isopod, *Porcellio scaber*, to subnivean environments. *Journal of Comparative Physiology B*, 163: 439–444.

Tarazona, O.A., Lopez, D.H., Slota, L.A. and Cohn, M.J. (2019). Evolution of limb development in cephalopod molluscs. *eLIFE*, 8: e43828.

Taylor, L.R. (1961). Aggregation, variance and the mean. *Nature*, 189: 732–735.

Taylor, L.R., Woiwod, I.P. and Perry, J.N. (1979). The negative binomial as a dynamic ecological model for aggregation, and the density dependence of k. *Journal of Animal Ecology*, 48: 289–304.

Taylor, M.G. and Simkiss, K. (1984). Inorganic deposits in invertebrate tissues. *Environmental Chemistry*, 3: 102–138.

Taylor, B.E. and Carefoot, T.H. (1993). Terrestrial life in isopods: evolutionary loss of gas-exchange and survival capability in water. *Canadian Journal of Zoology*, 71: 1372–1378.

Tebbe, C.C., Czarnecki, A. and Thimm, T. (2006). Collembola as a habitat for microorganisms. pp. 133–153. *In:* König, H. and Varma, A. (eds.). *Soil Biology, Volume 6*. Springer-Verlag, Berlin.

Terhivuo, J. and Saura, A. (1993). Genic and morphological variation of the parthenogenetic earthworm *Aporrectodea rosea* in southern Finland (Oligochaeta, Lumbricidae). *Annales Zoologici Fennici*, 30: 215–224.

Teuben, A. and Smidt, G.R.B. (1992). Soil arthropod numbers and biomass in two pine forests on different soils, related to functional groups. *Pedobiologia*, 36: 79–89.

Thakur, M.P., Phillips, H.R.P., Brose, U., De Vries, F.T., Lavelle, P., Loreau, M. et al. (2020). Towards an integrative understanding of soil biodiversity. *Biological Reviews*, 95: 350–364.

Thakur, S.S., Lone, A.R., Tiwari, N., Jain, S.K. and Yadav, S. (2021). Metagenomic exploration of bacterial community structure of earthworms's gut. *Journal of Pure and Applied Microbiology*, 15: 1156–1172.

Thibaud, J.-M. (1970). Biologie et écologie des Collemboles Hypogastruridae édaphiques et cavernicoles. *Memoires de Musée National d'Histoire Naturelle, Série A, Zoologie*, 61: 83–201.

Thomas, J.O.M. (1979). An energy budget for a woodland population of oribatid mites. *Pedobiologia*, 19: 346–378.

Thomas, M.L., Gray, B. and Simmons, L.W. (2011). Male crickets alter the relative expression of cuticular hydrocarbons when exposed to different acoustic environments. *Animal Behaviour*, 82: 49–53.

Thomas, C.G., Woodruff, G.C. and Haag, E.S. (2012). Causes and consequences of the evolution of reproductive mode in *Caenorhabditis* nematodes. *Trends in Genetics*, 28: 213–220.

Thompson, C.L., Vier, R., Mikaelyan, A., Wienemann, T. and Brune, A. (2012). '*Candidatus Arthromitus*' revised: segmented filamentous bacteria in arthropod guts are members of Lachnospiraceae. *Environmental Microbiology*, 14: 1454–1465.

Thorne, B.L. (1997). Evolution of eusociality in termites. *Annual Review of Ecology and Systematics*, 28: 27–54.

Timmermans, M.J.T.N., Mariën, J., Roelofs, D., Van Straalen, N.M. and Ellers, J. (2004). Evidence for multiple origins of *Wolbachia* infection in springtails. *Pedobiologia*, 48: 469–475.

Timmermans, M.J.T.N., Ellers, J., Mariën, J., Verhoef, S.C., Ferwerda, E.B. and Van Straalen, N.M. (2005). Genetic structure in *Orchesella cincta* (Collembola): strong subdivision of European populations inferred from mtDNA and AFLP markers. *Molecular Ecology*, 14: 2017–2024.

Timmermans, M.J.T.N., Roelofs, D., Mariën, J. and Van Straalen, N.M. (2008). Revealing pancrustacean relationships: Phylogenetic analysis of ribosomal protein genes places Collembola (springtails) in a monophyletic Hexapoda and reinforces the discrepancy between mitochondrial and nuclear DNA markers. *BMC Evolutionary Biology*, 8: 83.

Timmermans, M.J.T.N. and Ellers, J. (2009). *Wolbachia* endosymbiont is essential for egg hatching in a parthenogenetic arthropod. *Evolutionary Ecology*, 23: 931–942.

Timmermans, M.J.T.N., Roelofs, D., Nota, B., Ylstra, B. and Holmstrup, M. (2009). Sugar sweet springtails: on the transcriptional response of *Folsomia candida* (Collembola) to desiccation stress. *Insect Molecular Biology*, 18: 737–746.

Tiwari, S.C. and Mishra, R.R. (1995). Earthworm density, biomass and production of cast in pineapple orchard soil. *Pedobiologia*, 39: 434–441.

Todd, M.E. (1963). Osmoregulation in *Ligia oceanica* and *Idotea granulosa*. *Journal of Experimental Biology*, 40: 381–392.

Torres-Leguizamon, M., Mathieu, J., Decaëns, T. and Dupont, L. (2014). Genetic structure of earthworm populations at a regional scale: inferences from mitochondrial and microsatellite molecular markers in *Aporrectodea icterica* (Savigny 1826). *PLoS One*, 9: e101597.

Traill, L.W., Bradshaw, C.J.A. and Brook, B.W. (2007). Minimum viable population size: a meta-analysis of 30 years of published estimates. *Biological Conservation*, 139: 159–166.

Trudgill, D.L., Honek, A., Li, D. and Van Straalen, N.M. (2005). Thermal time—concepts and utility. *Annals of Applied Biology*, 146: 1–14.

Trumbo, S.T. (2012). Patterns of parental care in invertebrates. pp. 81–100. *In:* Royle, N.J., Smiseth, P.T. and Kölliker, M. (eds.). *The Evolution of Parental Care*. Oxford University Press, Oxford.

Tullgren, A. (1918). Ein sehr einfacher Ausleseapparat für terricole Tierformen. *Zeitschrift für Angewandte Entomologie*, 4: 149–150.

Tuni, C., Beveridge, M. and Simmons, L.W. (2013). Female crickets assess relatedness during mate guarding and bias storage of sperm towards unrelated males. *Journal of Evolutionary Biology*, 26: 1261–1268.

Tunnacliffe, A. and Wise, M.J. (2007). The continuing conundrum of the LEA proteins. *Naturwissenschaften*, 94: 791–812.

Ulrich, W. (2004). Soil-living parasitic Hymenoptera: comparison between a forest and an open landscape habitat. *Pedobiologia*, 48: 59–69.

Urbásek, F. and Rusek, J. (1994). Activity of digestive enzymes in seven species of Collembola (Insecta: Entognatha). *Pedobiologia*, 38: 400–406.

Usher, M.B. (1969). Some properties of the aggregations of soil arthropods: Collembola. *Journal of Animal Ecology*, 38: 607–622.

Usher, M.B. (1970). Seasonal and vertical distribution of a population of soil arthropods: Collembola. *Pedobiologia*, 10: 224–236.

Usher, M.B. (1975). Some properties of the aggregations of soil arthropods: Cryptostigmata. *Pedobiologia*, 15: 355–363.

Usher, M.B. and Hider, M. (1975). Studies on populations of *Folsomia candida* (Insecta: Collembola): causes of aggregations. *Pedobiologia*, 15: 276–283.

Uvarov, A.V. (1998). Respiration activity of *Dendrobaena octaedra* (Lumbricidae) under constant and diurnally fluctuating temperature regimes in laboratory microcosms. *European Journal of Soil Biology*, 34: 1–10.

Uvarov, A.V. and Scheu, S. (2004). Effects of density and temperature regime on respiratory activity of the epigeic earthworm species *Lumbricus rubellus*. *European Journal of Soil Biology*, 40: 163–167.

Van Breemen, N., Mulder, J. and Driscoll, C.T. (1983). Acidification and alkalinization of soils. *Plant and Soil*, 75: 283–308.

Van Dam, M.H., Trautwein, M., Spicer, G.S. and Esposito, L. (2018). Advancing mite phylogenomics: Designing ultraconserved elements for Acari phylogeny. *Molecular Ecology Resources*, 19: 465–475.

Vandekerckhove, T.T.M., Watteney, S., Willems, A., Swings, J.G., Mertens, J. and Gillis, M. (1999). Phylogenetic analysis of the 16S rDNA of the cytoplasmic bacterium *Wolbachia* from the novel host *Folsomia candida* (Hexapoda, Collembola) and its implications for wolbachial taxonomy. *FEMS Microbiology Letters*, 180: 279–286.

Vandel, A. (1928). La parthénogenèse géographique. Contribution à l'étude biologique et cytologique de la parthénogenèse naturelle. *Bulletin biologiques de la France et de la Belgique*, 62: 164–281.

Van der Have, T.M. (2002). A proximate model for thermal tolerance in ectotherms. *Oikos*, 98: 141–155.

Van der Heijden, M.G.A., Bardgett, R.D. and Van Straalen, N.M. (2008). The unseen majority: soil microbes as drivers of plant diversity and productivity in terrestrial ecosystems. *Ecology Letters*, 11: 296–310.

Van der Laan, P.A. (1963). A combination of an electric vacuum brush and an exhaustor as apparatus for catching small insects. *Entomologische Berichten*, 23: 66.

Van der Woude, H.A. (1987). Seasonal changes in cold hardiness of temperate Collembola. *Oikos*, 50: 231–238.

Van der Woude, H.A. and Verhoef, H.A. (1988). Reproductive diapause and cold hardiness in temperate Collembola *Orchesella cincta* and *Tomocerus minor*. *Journal of Insect Physiology*, 34: 387–392.

Van der Woude, H.A. and Joosse, E.N.G. (1988). The seasonality of respiration in two temperate Collembola as related to starvation, temperature and photoperiod. *Comparative Biochemistry and Physiology*, 91A: 147–151.

Van der Wurff, A.W.G., Isaaks, A., Ernsting, G. and Van Straalen, N.M. (2003). Population substructures in the soil invertebrate *Orchesella cincta*, as revealed by microsatellite and TE-AFLP markers. *Molecular Ecology*, 12: 1349–1359.

Van der Wurff, A.W.G., Gols, R., Ernsting, G. and Van Straalen, N.M. (2005). Population genetic structure of *Orchesella cincta* (Collembola; Hexapoda) in NW Europe, as revealed by microsatellite markers. *Pedobiologia*, 49: 167–174.

Van Dooremalen, C., Pel, R. and Ellers, J. (2009). Maximized PUFA measurements improve insight in changes in fatty acid composition in response to temperature. *Archives of Insect Biochemistry and Physiology*, 72: 88–104.

Van Dooremalen, C., Koekkoek, J. and Ellers, J. (2011). Temperature-induced plasticity in membrane and storage lipid composition: Thermal reaction norms across five different temperatures. *Journal of Insect Physiology*, 57: 285–291.

Van Dooremalen, C., Berg, M.P. and Ellers, J. (2013). Acclimation responses to temperature vary with vertical stratification: implications for vulnerability of soil-dwelling species to extreme temperature events. *Global Change Biology*, 19: 975–984.

Van Frankenhuyzen, K. (2009). Insecticidal activity of *Bacillus thuringiensis* crystal proteins. *Journal of Invertebrate Pathology*, 101: 1–16.

Van Gestel, C.A.M., Van Belleghem, F.G.A.J., Van den Brink, N.W., Droge, S.T.J., Hamers, T., Hermens, J.L.M. et al. (2019). *Environmental Toxicology – open online textbook*. SURF, Utrecht.

Van Meyel, S., Devers, S. and Meunier, J. (2019). Love them all: mothers provide care to foreign eggs in the European earwig *Forficula auricularia*. *Behavioral Ecology*, 30: 756–762.

Vannier, G. (1974). Calcul de la résistance cuticulaire à la diffusion de vapeur chez un insecte Collembole. *Comptes rendus hebdomadaires des séances d l'Academie des sciences. Série D, Sciences naturelles*, 278: 625–628.

Vannier, G. and Verhoef, H.A. (1978). Effect of starvation on transpiration and water content in the populations of two co-existing Collembola species. *Comparative Biochemistry and Physiology*, 60A: 483–489.

Vannier, G. and Verdier, B. (1981). Critères écophysiologiques (transpiration, respiration) permettant de séparer une espèce souterraine d'une espèce de surface chez les Insectes Collemboles. *Revue d'Écologie et de Biologie du Sol*, 18: 531–549.

Vannier, G. (1987). The porosphere as an ecological medium emphasized in Professor Ghilarov's work on soil animal adaptations. *Biology and Fertility of Soils*, 3: 39–44.

Van Straalen, N.M. (1982). Demographic analysis of arthropod populations using a continuous stage-variable. *Journal of Animal Ecology*, 51: 769–783.

Van Straalen, N.M. and Rijninks, P.C. (1982). The efficiency of Tullgren apparatus with respect to interpreting seasonal changes in age structure of soil arthropod populations. *Pedobiologia*, 24: 197–209.

Van Straalen, N.M. (1983a). Demographic analysis of soil arthropod populations. A comparison of methods. *Pedobiologia*, 25: 19–26.

Van Straalen, N.M. (1983b). Physiological time and time-invariance. *Journal of Theoretical Biology*, 104: 349–357.

Van Straalen, N.M. (1985a). Comparative demography of forest floor Collembola populations. *Oikos*, 45: 253–265.

Van Straalen, N.M. (1985b). Production and biomass turnover in stationary stage-structured populations. *Journal of Theoretical Biology*, 113: 331–352.

Van Straalen, N.M. and Joosse, E.N.G. (1985). Temperature responses of egg production and egg development in two species of Collembola. *Pedobiologia*, 28: 265–273.

Van Straalen, N.M., Burghouts, T.B.A., Doornhof, M.J., Groot, G.M., Janssen, M.P.M., Joosse, E.N.G. et al. (1987). Efficiency of lead and cadmium excretion in populations of *Orchesella cincta* (Collembola) from various contaminated forest soils. *Journal of Applied Ecology*, 24: 953–968.

Van Straalen, N.M. (1987). Turnover of accumulating substances in populations with weight-structure. *Ecological Modelling*, 36: 195–209.

Van Straalen, N.M., Kraak, M.H.S. and Denneman, C.A.J. (1988). Soil microarthropods as indicators of soil acidification and forest decline in the Veluwe area, the Netherlands. *Pedobiologia*, 32: 47–55.

Van Straalen, N.M. (1989). Production and biomass turnover in two populations of forest floor Collembola. *Netherlands Journal of Zoology*, 39: 156–168.

Van Straalen, N.M., Schobben, J.H.M. and De Goede, R.G.M. (1989). Population consequences of cadmium toxicity in soil microarthropods. *Ecotoxicology and Environmental Safety*, 17: 190–204.

Van Straalen, N.M. (1994). Adaptive significance of temperature responses in Collembola. *Acta Zoologica Fennica*, 195: 135–142.

Van Straalen, N.M. and Van Diepen, A.M.F. (1995). Evolution of the Arrhenius activation energy in soil arthropods. pp. 113–118. *In:* Sommeijer, M.J. and Francke, P.J. (eds.). *Proceedings of the Section Experimental and Applied Entomology of the Netherlands Entomological Society*. NEV, Amsterdam.

Van Straalen, N.M. and Verhoef, H.A. (1997). The development of a bioindicator system for soil acidity based on arthropod pH preferences. *Journal of Applied Ecology*, 34: 217–232.

Van Straalen, N.M. (1998). Evaluation of bioindicator systems derived from soil arthropod communities. *Applied Soil Ecology*, 9: 429–437.

Van Straalen, N.M., Timmermans, M.J.T.N., Roelofs, D. and Berg, M.P. (2008). Apterygota in the spotlights of ecology, evolution and genomics. *European Journal of Soil Biology*, 44: 452–457.

Van Straalen, N.M. and Roelofs, D. (2012). *An Introduction to Ecological Genomics*. Oxford University Press, Oxford.

Van Straalen, N.M. (2021). Evolutionary terrestrialization scenarios for soil invertebrates. *Pedobiologia*, 87–88: 150753.

Vanthournout, B., Swaegers, J. and Hendrickx, F. (2011). Spiders do not escape reproductive manipulations by *Wolbachia*. *BMC Evolutionary Biology*, 11: 15.

Van Valen, L. (1973). A new evolutionary law. *Evolutionary Theory*, 1: 1–30.

Vargas, H.C.M., Panfilio, K.A., Roelofs, D. and Rezende, G.L. (2021). Increase in egg resistance to desiccation in springtails correlates with blastoderm cuticle formation: Eco-evolutionary implications for insect terrestrialization. *Journal of Experimental Zoology B (Mol. Dev. Evol.)*, 336: 606–619.

Veerman, A. (1992). Diapause in phytoseiid mites: a review. *Experimental and Applied Acarology*, 14: 1–60.

Vegter, J.J. (1983). Food and habitat specialization in coexisting springtails (Collembola, Entomobryidae). *Pedobiologia*, 25: 253–262.

Velando, A., Eiroa, J. and Domínguez, J. (2008). Brainless, but not clueless: earthworms boost their ejaculates when they detect fecund non-virgin partners. *Proceedings of the Royal Society B*, 275: 1067–1072.

Vera, H. (1993). Demographic variation in two forest populations of oribatid mites. *Pedobiologia*, 37: 95–106.

Verhoef, H.A. and Nagelkerke, C.J. (1977). Formation and ecological significance of aggregations in Collembola. An experimental study. *Oecologia*, 31: 215–226.

Verhoef, H.A., Nagelkerke, C.J. and Joosse, E.N.G. (1977). Aggregation pheromones in Collembola. *Journal of Insect Physiology*, 23: 1009–1013.

Verhoef, H.A., Bosman, C., Bierenbroodspot, A. and Boer, H.H. (1979). Ultrastructure and function of the labial nephridia and the rectum of *Orchesella cincta* (L.) (Collembola). *Cell and Tissue Research*, 198: 237–246.

Verhoef, H.A. (1981). Water balance in Collembola and its relation to habitat selection: water content, haemolymph osmotic pressure and transpiration during an instar. *Journal of Insect Physiology*, 27: 755–760.

Verhoef, H.A., Witteveen, J., Van der Woude, H.A. and Joosse, E.N.G. (1983). Morphology and function of the ventral groove of Collembola. *Pedobiologia*, 25: 3–9.

Verhoef, H.A. (1984). Releaser and primer pheromones in Collembola. *Journal of Insect Physiology*, 30: 665–670.

Verhoef, H.A. and Prast, J.E. (1989). Effects of dehydration on osmotic and ionic regulation in *Orchesella cincta* (L.) and *Tomocerus minor* (Lubbock) (Collembola) and the role of the coelomoduct kidneys. *Comparative Biochemistry and Physiology*, 93A, 691–694.

Verhoef, H.A. (1995). Animal ecophysiology: cornerstone for soil ecosystem studies as exemplified by studies on soil arthropods. *Acta Zoologica Fennica*, 196: 176–182.

Vermeij, G.J. and Dudley, R. (2000). Why are there so few evolutionary transitions between aquatic and terrestrial ecosystems? *Biological Journal of the Linnean Society*, 70: 541–554.

Verne, S., Johnson, M., Bouchon, D. and Grandjean, F. (2012). Effects of parasitic sex-ratio distorters on host genetic structure in the *Armadillium vulgare-Wolbachia* association. *Journal of Evolutionary Biology*, 25: 264–276.

Verschoor, B.C. and De Goede, R.G.M. (2000). The nematode extraction efficiency of the Oostenbrink elutriator-cottonwool filter method with special reference to nematode body size and life strategy. *Nematology*, 2: 325–342.

Via, S. (1993). Adaptive phenotypic plasticity: target or by-product of selection in a variable environment. *The American Naturalist*, 142: 352–365.

Viana, F., Jensen, C.E., Macey, M., Schramm, A. and Lund, M.B. (2016). Earthworm ecology affects the population structure of their *Verminephrobacter* symbionts. *Systematic and Applied Microbiology*, 39: 170–172.

Viana, F., Paz, L.-C., Methling, K., Damgaard, C., Lalk, M., Schramm, A. et al. (2018). Distinct effects of the nephridial symbionts *Verminephrobacter* and *Candidatus* Nephrothrix on reproduction and maturation of its earthworm host. *FEMS Microbiology Ecology*, 94: fix178.

Viktorov, A.G. (1997). Diversity of polyploid races in the family Lumbricidae. *Soil Biology and Biochemistry*, 29: 217–221.

Villani, M.G., Allee, L.L., Díaz, A. and Robbins, P.S. (1999). Adaptive strategies of edaphic arthropods. *Annual Review of Entomology*, 44: 233–256.

Waagner, D., Holmstrup, M., Bayley, M. and Sørensen, J.G. (2013). Induced cold-tolerance mechanisms depend on duration of acclimation in the chill-sensitive *Folsomia candida* (Collembola). *The Journal of Experimental Biology*, 216: 1991–2000.

Wagner, T.L., Wu, H.I., Sharpe, P.J.H., Schoolfield, R.M. and Coulson, R.N. (1984). Modeling insect development rates: a literature review and application of a biophysical model. *Annals of the Entomological Society of America*, 77: 208–225.

Wagner, G.P. (2014). *Homology, Genes and Evolutionary Innovation*. Princeton University Press, Princeton.

Waki, T., Ohari, Y., Hayashi, K., Moribe, J., Matsuo, K. and Takashima, Y. (2021). The first detection of *Dicrocoelium chinensis* sporocysts from the land snail *Aegista vulvifaga* in Gifu prefecture, Japan. *Journal of Veterinary Medical Science*, 83: 957–961.

Waki, T., Nakao, M., Sasaki, M., Ikezawa, H., Inoue, K., Ohari, Y. et al. (2022). *Brachylaima phaedusae* n. sp. (Trematoda: Brachylaimidae) from door snails in Japan. *Parasitology International*, 86: 102469.

Waldorf, E.S. (1971). The reproductive biology of *Sinella curviseta* (Collembola: Entomobryidae) in laboratory culture. *Revue d'Écologie et de Biologie du Sol*, 8: 451–463.

Walker, C.H. (1980). Species variations in some hepatic microsomal enzymes that metabolize xenobiotics. *Progress in Drug Metabolism*, 5: 113–164.

Walker, C.H., Hopkin, S.P., Sibly, R.M. and Peakall, D. (2001). *Principles of Ecotoxicology. Second Edition*. Taylor and Francis, London.

Wallace, M.M.H. (1968). The ecology of *Sminthurus viridis* (Collembola). II. Diapause in the aestivating egg. *Australian Journal of Zoology*, 16: 871–883.

Wallwork, J.A. (1983). Oribatids in forest ecosystems. *Annual Review of Entomology*, 28: 109–130.

Wander, M. (2019). Soil organic matter fractions and their relevance to soil function. pp. 67–102. *In:* Magdoff, F. and Weill, R.R. (eds.). *Soil Organic Matter in Sustainable Agriculture*. CRC Press, Boca Raton.

Wang, Y., Slotsbo, S. and Holmstrup, M. (2022). Soil dwelling springtails are resilient to extreme drought in soil, but their reproduction is highly sensitive to small decreases in soil water potential. *Geoderma*, 421: 115913.

Wassef, M.K. (1977). Fungal lipids. *Advances in Lipid Research*, 15: 159–232.

Warburg, M.R. (1968). Behavioral adaptations of terrestrial isopods. *American Zoologist*, 8: 545–559.

Warburg, M.R. (1987). Isopods and their terrestrial environment. *Advances in Ecological Research*, 17: 187–242.

Ward, P.S. (2007). Phylogeny, classification, and species-level taxonomy of ants (Hymenoptera: Formicidae). *Zootaxa*, 1668: 549–563.

Wardle, D.A., Bardgett, R.D., Klironomos, J.N., Setälä, H., Van der Putten, W.H. and Wall, D.H. (2004). Ecological linkages between aboveground and belowground biota. *Science*, 304: 1629–1633.

Warnock, A.J., Fitter, A.H. and Usher, M.B. (1982). The influence of a springtail *Folsomia candida* (Insecta, Collembola) on the mycorrhizal association of leek *Allium porrum* and the vesicular-arbuscular mycorrhizal endophyte *Glomus fasciculatus*. *New Phytologist*, 90: 285–292.

Warren, I.A., Vera, J.C., Johns, A., Zinna, R., Marden, J.H., Emlen, D.J. et al. (2014). Insights into the development and evolution of exaggerated traits using *de novo* transcriptomes of two species of horned scarab beetles. *PLoS One*, 9: e88364.

Watson, P.J. (1991a). Multiple paternity and first mate sperm precedence in the sierra dome spider, *Linyphia litigiosa* Keyserling (Linyphiidae). *Animal Behaviour*, 41: 135–148.

Watson, P.J. (1991b). Multiple paternity and genetic bet-hedging in female sierra dome spiders, *Linyphia litigiosa* (Linyphiidae). *Animal Behaviour*, 41: 343–360.

Watson, P.J. and Lighton, J.R.B. (1994). Sexual selection and the energetics of copulatory courtship in the Sierra dome spider, *Linyphia litigiosa*. *Animal Behaviour*, 48: 615–626.

Wehner, K., Scheu, S. and Maraun, M. (2014). Resource availability as driving force of the reproductive mode in soil microarthropods (Acari, Oribatida). *PLoS One*, 9: e104243.

Wełnicz, W., Grohme, M.A., Kaczmarek, Ł., Schill, R.O. and Frohme, M. (2011). Anhydrobiosis in tardigrades – The last decade. *Journal of Insect Physiology*, 57: 577–583.

Werren, J.H., Baldo, L. and Clark, M.E. (2008). *Wolbachia*: master manipulators of invertebrate biology. *Nature Reviews Microbiology*, 6: 741–751.

West, G.B., Brown, J.H. and Enquist, B.J. (1997). A general model for the origin of allometric scaling laws in biology. *Nature*, 276: 122–126.

Weygoldt, P. (1966). Mating behavior and spermatophore morphology in the pseudoscorpion *Dinocheirus tumidus* Banks (Cheliferinea, Chernetidae). *Biological Bulletin*, 130: 462–467.

Weygoldt, P. (1969). *The Biology of Pseudoscorpions*. Harvard University Press, Cambridge.

Wharton, D.A., Young, S.R. and Barrett, J. (1984). Cold tolerance in nematodes. *Journal of Comparative Physiology B*, 154: 73–77.

Wharton, D.A. and Brown, I.M. (1991). Cold-tolerance mechanisms of the Antarctic nematode *Panagrolaimus davidi*. *Journal of Experimental Biology*, 155: 629–641.

Wharton, D.A. (1995). Cold tolerance strategies in nematodes. *Biological Reviews*, 70: 161–185.

Wharton, D.A. (2003). The environmental physiology of Antarctic terrestrial nematodes: a review. *Journal of Comparative Physiology B*, 173: 621–628.

Wharton, D.A., Goodall, G. and Marshall, C.J. (2003). Freezing survival and cryoprotective dehydration as cold tolerance mechanisms in the Antarctic nematode *Panagrolaimus davidi*. *The Journal of Experimental Biology*, 206: 215–221.

Wharton, D.A. (2015). Anhydrobiosis. *Current Biology*, 25: R1114–R1116.

White, J.J. (1968). Bioenergetics of the woodlouse *Tracheoniscus rathkei* Brandt in relation to litter decomposition in a deciduous forest. *Ecology*, 49: 694–704.

Whitehead, D.C., Dibb, H. and Hartley, R.D. (1982). Phenolic compounds in soil as influenced by the growth of different plant species. *Journal of Applied Ecology*, 19: 579–588.

Whitehouse, M.E.A. and Lubin, Y. (2005). The functions of societies and the evolution of group living: spider societies as a test case. *Biological Reviews*, 80: 1–15.

Widianarko, B., Verweij, R.A., Van Gestel, C.A.M. and Van Straalen, N.M. (2000). Spatial distribution of trace metals in sediments from urban streams of Semarang, Central Java, Indonesia. *Ecotoxicology and Environmental Safety*, 46: 95–100.

Wieser, W. (1972). O/N ratios of terrestrial isopods at two temperatures. *Comparative Biochemistry and Physiology*, 43A, 859–868.

Wignall, A.E. and Taylor, P.W. (2010). Predatory behaviour of an araneophagic assassin bug. *Journal of Ethology*, 28: 437–445.

Wijnhoven, H. and Berg, M.P. (1999). Some notes on the distribution and ecology of Iridovirus (Iridovirus, Iridoviridae) in terrestrial isopods (Isopoda, Oniscidae). *Crustaceana*, 72: 145–156.

Willows, R.I. (1987). Population and individual energetics of *Ligia oceanica* (L.) (Crustacea: Isopoda) in the rocky supralittoral. *Journal of Experimental Marine Biology and Ecology*, 105: 253–274.

Wilson, G.D.F. (2009). The phylogenetic position of the Isopoda in the Peracarida (Crustacea: Malacostraca). *Arthropod Systematics and Phylogeny*, 67: 159–198.

Winsor, L. (1998). Collection, handling, fixation, histological and storage procedures for taxonomic studies of terrestrial flatworms (Tricladida: Terricola). *Pedobiologia*, 42: 405–411.

Winsor, L., Johns, P.M. and Yeates, G.W. (1998). Introduction, and ecological and systematic background, to the Terricola (Tricladida). *Pedobiologia*, 42: 389–404.

Witte, H. and Döring, D. (1999). Canalized pathways of change and constraints in the evolution of reproductive mode of microarthropods. *Experimental and Applied Acarology*, 23: 181–216.

Witteveen, J., Verhoef, H.A. and Letschert, J.P.W. (1987). Osmotic and ionic regulation in marine littoral Collembola. *Journal of Insect Physiology*, 33: 59–66.

Witteveen, J., Verhoef, H.A. and Huipen, T.E.A.M. (1988). Life history strategy and egg diapause in the intertidal collembolan *Anurida maritima*. *Ecological Entomology*, 13: 443–451.

Wolda, H. and Dennis, B. (1993). Density dependence tests, are they? *Oecologia*, 95: 581–591.

Wolters, V. (1998). Long-term dynamics of a collembolan community. *Applied Soil Ecology*, 9: 221–227.

Wolters, V. (2000). Invertebrate control of soil organic matter stability. *Biology and Fertility of Soils*, 31: 1–19.

Womersley, C. and Smith, L. (1981). Anhydrobiosis in nematodes – I. The role of glycerol, myoinositol and trehalose during desciccation. *Comparative Biochemistry and Physiology*, 70B, 579–586.

Wong, J.W.Y. and Kölliker, M. (2013). The more the merrier? Condition-dependent brood mixing in earwigs. *Animal Behaviour*, 86: 845–850.

Wood, T.G. (1967). Acari and Collembola of moorland soils from Yorkshire, England. I. Description of the sites and their populations. *Oikos*, 18: 102–117.

Wood, P.A. and Gabbutt, P.D. (1979). Silken chambers built by adult pseudoscorpions in laboratory culture. *Bulletin of the British Arachnological Society*, 4: 285–293.

Wood, C.T., Nihei, S.S. and Araujo, P.B. (2018). Woodlice and their parasitoid flies: revision of Isopoda (Crustacea, Oniscidae) – Rhinophoridae (Insecta, Diptera) interaction and first record of a parasitized Neotropical woodlouse species. *ZooKeys*, 801: 401–414.

Wray, G.A., Hahan, M.W., Abouheif, E., Balhoff, J.P., Pizer, M., Rockman, M.V. et al. (2003). The evolution of transcriptional regulation in eukaryotes. *Molecular Biology and Evolution*, 20: 1377–1419.

Wright, J.C. and Machin, J. (1993). Atmospheric water absorption and the water budget of terrestrial isopods (Crustacea, Isopoda, Oniscidae). *Biological Bulletin*, 184: 243–253.

Wright, J.C., Westh, P. and Ramløv, H. (1992). Cryptobiosis in Tardigrada. *Biological Reviews*, 67: 1–29.

Wright, J.C. (2001). Cryptobiosis 300 years on from van Leeuwenhoek: What have we learned about tardigrades? *Zoologischer Anzeiger*, 240: 563–582.

Wu, M., Sun, L.V., Vamathevan, J., Riegler, M,. Deboy, R., Brownlie, J.C. et al. (2004). Phylogenomics of the reproductive parasite *Wolbachia pipientis* wMel: a streamlined genome overrun by mobile elements. *PLoS Biology*, 2: 0327–0341.

Xiang, Q., Zhu, D., Chen, Q.-L., Delgado-Baquerizo, M., Su, J.-Q., Qiao, M. et al. (2019). Effects of diet on gut microbiota of soil collembolans. *Science of the Total Environment*, 676: 197–205.

Xiao, R., Wang, L., Cao, Y. and Zhang, G. (2016). Transcriptome response to temperature stress in the wolf spider *Pardosa pseudoannulata* (Araneae: Lycosidae). *Ecology and Evolution*, 6: 3540–3554.

Xu, X., Liu, F., Cheng, R.-C., Chen, J., Xu, X., Zhang, Z. et al. (2015a). Extant primitively segmented spiders have recently diversified from an ancient lineage. *Proceedings of the Royal Society B*, 282: 20142486.

Xu, X., Liu, F., Chen, J., Ono, H., Li, D. and Kuntner, M. (2015b). A genus-level taxonomic review of primitively segmented spiders (Mesothelae, Liphistiidae). *ZooKeys*, 488: 121–151.

Yamamoto, T., Nakagoshi, N. and Touyama, Y. (2001). Ecological study of pseudoscorpion fauna in the soil organic layer in managed and abandoned secondary forests. *Ecological Research*, 16: 593–601.

Yang, Y., Chen, X., Cheng, L., Coa, F., Romeis, J., Li, Y. et al. (2015). Toxicological and biochemical analyses demonstrate no toxic effect of Cry1C and Cry2A to *Folsomia candida*. *Scientific Reports*, 5: 15619.

Yeates, G.W. (1981). Nematode populations in relation to soil environmental factors: a review. *Pedobiologia*, 22: 312–338.

Yeates, G.W. (1991). Nematode populations at three forest sites in New Caledonia. *Journal of Tropical Ecology*, 7: 411–413.

Yip, E.C. and Rayor, L.S. (2014). Maternal care and subsocial behaviour in spiders. *Biological Reviews*, 89: 427–449.

Yoshida, Y., Koutsovoulos, G., Laetsch, D.R., Stevens, L., Kumar, S., Horikawa, D.D. et al. (2017). Comparative genomics of the tardigrades *Hypsibius dujardini* and *Ramazottius varieornatus*. *PLoS Biology*, 15: e2002266.

Zaitsev, A.S. (1997). The communities of the oribatid mites (Acari: Oribatida) of the Zakopane environs. *Ochrona Przyrody*, 54: 131–140.

Zaitsev, A.S., Chauvat, M., Pflug, A. and Wolters, V. (2002). Oribatid mite diversity and community dynamics in a spruce chronosequence. *Soil Biology and Biochemistry*, 34: 1919–1927.

Zaitsev, A.S. and Wolters, V. (2006). Geographic determinants of oribatid mite communities structure and diversity across Europe: a longitudinal perspective. *European Journal of Soil Biology*, 42: S358–S361.

Zchori-Fein, E. and Perlman, S.J. (2004). Distribution of the bacterial symbiont *Cardinium* in arthropods. *Molecular Ecology*, 13: 2009–2016.

Zeilinger, A.R., Andow, D.A., Zwahlen, C. and Stotzky, G. (2010). Earthworm populations in a northern U.S. Cornbelt soil are not affected by cultivation of Bt maize expressing Cry1Ab and Cry3B1 proteins. *Soil Biology and Biochemistry*, 42: 1284–1292.

Zettel, U. and Zettel, J. (1994). Seasonal and reproductional polymorphism in *Ceratophysella sigillata* (Uzel)(Collembola, Hypogastruridae). *Acta Zoologica Fennica*, 195: 154–156.

Zeuthen, E. (1950). Cartesian diver respirometer. *Biological Bulletin*, 98: 139–143.

Zhang, J. and Elser, J.J. (2017). Carbon:nitrogen:phosphorus stochiometry in fungi: a meta-analysis. *Frontiers in Microbiology*, 8: 1281.

Zhang, F., Ding, Y., Zhou, Q.-S., Wu, J., Luo, A. and Zhu, C.-D. (2019). A high-quality draft genome assembly of *Sinella curviseta*: a soil model organism (Collembola). *Genome Biology and Evolution*, 11: 521–530.

Zhang, R., Chen, R., An, J. and Santamaria, C.A. (2022). Phylogeny of terrestrial isopods based on the complete mitochondrial genomes, subvert the monophyly of Oniscidea and Ligiidae up to new subfamily Ligiaidea of Isopoda. *Research Square*, doi.org/10.21203/rs.21203.rs-22549/v21201.

Zheng, H., Dietrich, C., Thompson, C.L., Meuser, K. and Brune, A. (2015). Population structure of Endomicrobia in single host cells of termite gut flagellates (*Trichonympha* spp.). *Microbes and Environments*, 30: 92–98.

Zhu, D., Delgado-Baquerizo, M., Ding, J., Gillings, M.R. and Zhu, Y.-G. (2021). Trophic level drives the host microbiome of soil invertebrates at a continental scale. *Microbiome*, 9: 189.

Zinkler, D. (1966). Vergleichende Untersuchungen zur Atmungsphysiologie von Collembolen (*Apterygota*) und andere Bodenkleinarthropoden. *Zeitschrift für vergleichende Physiologie*, 52: 99–144.

Zizzari, Z.V., Braakhuis, A., Van Straalen, N.M. and Ellers, J. (2009). Female preference and fitness benefits of mate choice in a species with dissociated sperm transfer. *Animal Behaviour*, 78: 1261–1267.

Zizzari, Z.V., Van Straalen, N.M. and Ellers, J. (2013). Male-male competition leads to less abundant but more attractive sperm. *Biology Letters*, 9: 20130762.

Zizzari, Z.V., Smolders, I. and Koene, J.M. (2014). Alternative delivery of male accessory gland products. *Frontiers in Zoology*, 11: 32.

Zizzari, Z.V., Engl, T., Lorenz, S., Van Straalen, N.M., Ellers, J. and Groot, A.T. (2017). Love at first sniff: a spermatophore-associated pheromone mediates partner attraction in a collembolan species. *Animal Behaviour*, 124: 221–227.

Zorn, M.I., Van Gestel, C.A.M. and Eijsackers, H. (2005). Species-specific earthworm population responses in relation to flooding dynamics in a Dutch floodplain soil. *Pedobiologia*, 49: 189–198.

Index

C

E

G

M

Q

R

S

T

U

V

W

X

Y

Z